SECOND EDITION

Human Aging

biological perspectives

AUGUSTINE GASPAR DIGIOVANNA

Salisbury State University

Boston   Burr Ridge, IL   Dubuque, IA   Madison, WI   New York   San Francisco   St. Louis
Bangkok   Bogotá   Caracas   Lisbon   London   Madrid
Mexico City   Milan   New Delhi   Seoul   Singapore   Sydney   Taipei   Toronto

McGraw-Hill Higher Education

A Division of The **McGraw-Hill** Companies

Vice president and editorial director: *Kevin T. Kane*
Publisher: *Colin H. Wheatley*
Sponsoring editor: *Kristine Tibbetts*
Developmental editor: *Patrick F. Anglin*
Marketing manager: *Heather K. Wagner*
Senior project manager: *Peggy J. Selle*
Senior production supervisor: *Mary E. Haas*
Coordinator of freelance design: *Michelle D. Whitaker*
Compositor: *Carlisle Communications, Ltd.*
Typeface: *10/12 Times Roman*
Printer: *Quebcor Printing Book Group/Dubuque, IA*

Cover designer: *Sean M. Sullivan*
Cover image: *©Nick White/FPG International*

Library of Congress Cataloging-in-Publication Data

DiGiovanna, Augustine Gaspar.
    Human aging : biological perspectives / Augustine Gaspar
DiGiovanna. — 2nd ed.
      p. cm.
    Includes bibliographical references and index.
    ISBN 0–07–292691–0
    1. Aging—Physiological aspects. I. Title.
QP86.D54   2000
612.6'7—dc21   99–14025
                                         CIP

CONTENTS

Preface xi

1 INTRODUCTION 1
Why Study Human Aging? 1
  Personal and Professional Reasons 1
  Population Trends 1
What is Aging? 3
  Types of Aging 4
  What Aging Is Not 10
Why Study Biological Aging? 11
How Is Biological Aging Studied? 11
  Cross-sectional Method 11
  Longitudinal Method 12
  Cross-sequential Method 13
  Nonhuman Studies 13
What We Know Thus Far 14
  What Happens during Biological Aging? 14
  Life Expectancy 18
  Quality of Life 20

2 MOLECULES, CELLS, AND THEORIES ... 22
Hierarchy of the Body 22
  Atoms, Ions, and Molecules 22
  Organelles and Cells 24
  Tissues, Organs, Systems, Organism 24
Body Chemicals 25
  Carbohydrates 25
  Nucleic Acids 26
  Proteins 28
  Lipids 28
  Molecular Complexes 28
  Free Radicals 29
  Glycation 32
Cells 33
  Cell Membrane 33
  Cytoplasm 33
  Endoplasmic Reticulum 33
  Golgi Apparatus 33
  Vacuoles 33
  Mitochondria 33
  Microtubules and Microfilaments 35
  Nucleus 35
  Genetic Control 35
  Cell Division 36
  Neoplasia 38
Apoptosis 39
Genes and Aging 39
Intercellular Materials 40
  Amorphous Materials 40
  Fibers 40
Biological Aging Theories 41
  Reasons for Theories of Aging 41

General Characteristics of the Theories 41
Evolutionary Theories 42
Physiolgical Theories 43

3 INTEGUMENTARY SYSTEM 49
Main Functions for Homeostasis 49
    Serving as a Barrier 49
    Providing Information 50
    Temperature Regulation 50
    Vitamin D Production 50
The Epidermis 50
    Keratin and Keratinocytes 50
    Melanin and Melanocytes 51
    Immune Function and Langerhans Cells 53
Age Changes in the Epidermis 53
    Keratin and Keratinocytes 53
    Melanin and Melanocytes 53
    Immune Function and Langerhans Cells 53
Epidermal Accessory Structures 54
    Hair 54
    Age Changes in Hair 54
    Nails 55
    Age Changes in Nails 55
Dermis 56
    Foundation Material 56
    Blood Vessels 58
    Sweat Glands 59
    Sebaceous Glands 59
    Nerves 60
Boundary between Epidermis
    and Dermis 60
    Age Changes in the Epidermal-Dermal
    Boundary 61
Vitamin D Production 61
    Age Changes in Vitamin D Production 61
    Subcutaneous Layer 62
    Loose Connective Tissue 62
    Fat Tissue 62
    Age Changes in the Subcutaneous Layer ... 62
Miscellaneous Cosmetic Changes 63
Abnormal Changes 63
    Effects of Sunlight 63
    Effects from Heat 65
    Bedsores 65
    Neoplasms 66

4 CIRCULATORY SYSTEM 68
Main Functions for Homeostasis 68
    Transportation 68
    Defense 70

Temperature Control 70
Acid/Base Balance 70
Heart 71
    Chambers and the Cardiac Cycle 71
    Valves 73
    Layers 73
    Coronary Blood Flow 73
    Cardiac Adaptability 74
Age Changes in the Heart 75
    Resting Conditions 75
    Cardiac Adaptability 75
    Exercise and the Aging Heart 75
Diseases of the Heart 75
    Coronary Artery Disease 76
    Congestive Heart Failure 80
    Valvular Heart Disease 80
Arteries 80
    Inner Layer 81
    Middle Layer: Large Arteries 81
    Middle Layer: Smaller Arteries 82
    Outer Layer 82
Age Changes in Arteries 82
    Middle Layer: Large Arteries 82
    Middle Layer: Smaller Arteries 84
    Number of Arteries 84
Atherosclerosis: An Arterial Disease 84
    Importance 84
    Development and Effects 85
    Mechanisms Promoting
        Atherosclerosis 85
    Prevention 87
Capillaries 87
    Age Changes in Capillaries 88
Veins 88
    Age Changes in Veins 89
    Diseases of Veins 89
Lymphatics and the Spleen 90
    Age Changes in Lymphatics and the
        Spleen 90
Blood 90
    Red Blood Cells 91
    White Blood Cells 91
    Age Changes in Blood 92

5 RESPIRATORY SYSTEM 93
Main Functions for Homeostasis 93
    Gas Exchange 93
    Sound Production 95
Ventilation 95
    Inspiration 95
    Expiration 96

Rate of Ventilation
  (Minute Volume) 96
Requirements for Ventilation 98
  Contributions by Airways 98
  Control Systems 101
Age Changes Affecting Ventilation 103
  Open Airways 103
  Defense Mechanisms 104
  Proper Pressure Changes 104
  Compliance 104
  Control Systems 104
  Consequences 105
Perfusion 106
  Age Changes in Perfusion 107
Diffusion 107
  Age Changes in Diffusion 108
Effects from Altered Gas Exchange 108
  Biological Effects 108
  Interactions 108
Diseases of the Respiratory System 109
  Lung Cancer 109
  Chronic Bronchitis 110
  Emphysema 111
  Pneumonia 111
  Pulmonary Embolism 112
  Control Errors 112
Smoking 113
Sound Production and Speech 113
  Mechanisms 113
  Age Changes 114

6 NERVOUS SYSTEM 115
Main Functions for Homeostasis 115
  Monitoring 115
  Communicating 115
  Stimulating 116
  Coordinating 116
  Remembering 117
  Thinking 117
Neurons 117
  Components 117
  Operations 117
Neuroglia 120
Schwann Cells 120
Nervous System Organization 120
  Central Nervous System 120
  Peripheral Nervous System ....
Nervous System Pathways 121
  Reflexes 121
  Conscious Sensation 123
  Voluntary Movements 123

Higher-Level Functions 125
Age Changes in Sensory Functioning 125
  Skin Receptors 125
  Sense of Smell 126
  Sense of Taste 126
  Other Sensory Neurons 127
Age Changes in Somatic Motor
  Functioning 127
  Somatic Motor Neurons 127
Age Changes in Autonomic
  Motor Functioning 128
  Autonomic Motor Neurons 128
  Sympathetic Neurotransmitters 129
Age Changes in Reflexes 129
Age Changes in Conscious Sensation
  and Voluntary Movements 130
Age Changes in the CNS 130
  Spinal Cord 131
  Brain 131
Aging of Other Brain Functions 132
  Memory 132
  Thinking 134
  Vocabulary and Conversation 135
  Supporting Memory and Intelligence 135
  Personality 135
  Sleep 135
  Biorhythms 136
Conclusion 137
Diseases of the Nervous System 137
  Strokes 137
  Dementia 140
  Multi-Infarct Dementia 140
  Alzheimer's Disease 141
  Parkinson's Disease 145
  Dementia with Lewy Bodies 147

7 EYES AND EARS 148
The Eyes 148
  Image Formation and Vision 148
  Deterrents to Clear Vision 149
Forming Images 149
  Conjunctiva 149
  Cornea 150
  Iris and Pupil 151
  Ciliary Body 151
  Lens and Suspensory Ligaments 151
  Aqueous Humor 153
  Vitreous Humor 153
Accommodation 154
  Age Changes 155
Retina 155

Layers and Regions 155
Cones.. 155
Rods.. 156
Age Changes 157
Other Eye Components 157
Choroid.. 157
Sclera 157
External Muscles 158
Adipose Tissue 159
Eyelids 159
Lacrimal Gland 159
Two Eyes 160
Depth Perception (Binocular Vision) 160
Age Changes in Vision 160
Light Intensity 160
Quality of Light 161
Visual Acuity 161
Depth Perception (Binocular Vision) 162
Field of View 162
Optimizing Vision 162
Diseases of the Eyes 162
Minor Diseases 162
Cataracts.. 163
Age-Related Macular Degeneration 163
Glaucoma 164
Diabetic Retinopathy 164
The Ears 165
Hearing 165
External Ear 165
Middle Ear 166
Inner Ear 167
Localizing Sound 167
Central Nervous System 169
Disorders in Hearing 169
Presbycusis 169
Tinnitus.. 170
Detecting Other Stimuli................................ 170
Gravity and Changes in Speed 170
Rotation 171
Age Changes (Gravity, Changes
in Speed, and Rotation) 171
Dizziness and Vertigo 172

8 MUSCLE SYSTEM.................................... 173
Main Functions for Homeostasis................. 173
Movement 173
Support 173
Heat Production 174
Age Changes Versus Other Changes 174
Muscle Cells 175
Structure and Functioning 175
Age Changes in Muscle Cells....................... 177

Nerve-Muscle Interaction 178
Motor Units 178
Changes in Motor Units 179
Other Nerve-Muscle Interactions............ 179
Blood Flow in Muscles 179
Changes in Muscle Mass 180
Effects of Mass on Strength 180
Other Effects 180
Muscle System Performance 181
Reaction Time and Speed of Movement.... 181
Skill 181
Stamina 181
Staying Physically Active.............................. 183
Specific Effects 183
Overview 183
Starting or Increasing Exercise 184
Effects 184
Exercise Recommendations 186
Set Goals.. 186
Evaluate and Individualize 187
Plan a Program 187
Minimize Problems...................................... 187
Consider Alternatives 187
Driving Motor Vehicles.................................. 188

9 SKELETAL SYSTEM ....
Main Functions for Homeostasis................. 189
Support 189
Protection from Trauma 190
Movement 191
Mineral Storage 192
Blood Cell Production................................. 192
Bones 192
Bone Tissue Components 192
Bone Tissue Types 194
Other Tissues 195
Age Changes in Bones 195
Bone Matrix 195
Osteoporosis: A Bone Disease 198
Type I, or Postmenopausal,
Osteoporosis... 198
Type II, or Senile Osteoporosis 198
Incidence 198
Effects... 198
Causes 200
Diagnosis 200
Modifiable Risk Factors 200
Intrinsic Risk Factors.................................. 200
Treatments 201
Joints 202
Functions 202
Immovable Joints... 202

Slightly Movable Joints 202
  Freely Movable Joints 203
Diseases of Joints 204
Osteoarthritis 205
  Rheumatoid Arthritis 205

10 DIGESTIVE SYSTEM 207
  Main Functions for Homeostasis 207
    Supplying Nutrients 207
    Eliminating Toxins and Wastes 209
    Other Functions 209
  The Chemistry of Digestion 209
    Using Water and Enzymes 209
  Age Changes versus Other Changes 209
  Oral Region 210
    Oral Mucosa 210
    Teeth 210
    Salivary Glands 211
    Muscles 212
    Bones and Joints 212
    Abnormal Changes 212
  Esophagus 214
    Age Changes 214
    Abnormal Changes 214
  Stomach 215
    Secretion and Absorption 215
    Movements 215
    Age Changes 215
    Abnormal Changes 215
  Small Intestine 216
    Secretion 217
    Absorption 217
    Movements 217
    Age Changes 217
    Abnormal Changes 218
  Large Intestine 218
    Secretion and Absorption 218
    Movements 220
    Defecation 220
    Appendix 221
    Age Changes 221
    Abnormal Changes 221
  Liver 225
    Blood Flow 225
    Bile Flow 226
    Functions 227
    Age Changes 229
    Abnormal Changes 229
  Gallbladder 230
    Age Changes 231
    Abnormal Changes: Gallstones 231

Pancreas 231
  Age Changes 232
  Abnormal Changes 232

11 DIET AND NUTRITION 234
  Need for Nutritional Homeostasis 234
  Relationships between Diet and
    Nutrition 234
  Problems from Malnutrition 235
    Diversity of Problems 235
    Onset of Problems 235
    Nature of Problems 235
    Consequences 235
  A Proper Diet 236
    A Word of Caution 236
    Diet Based on Food Selection 236
    Diet Based on Chemical Composition 236
    Comparing Proper Diets
      for Younger and Older Adults 238
  Malnutrition among the Elderly 239
    Reducing and Preventing Malnutrition 239
  Energy and Body Weight 240
    Energy Uses and Storage 240
    Dietary Sources of Energy 240
    Energy Balance 241
    Overweight and Obesity 243
    Underweight 244
  Carbohydrates 244
    Digestible Carbohydrates 244
    Indigestible Carbohydrates: Fiber 245
    Uses 245
    Recommended Dietary Intakes 245
    Carbohydrate Deficiencies 245
    Carbohydrate Excesses 245
  Lipids 246
    Tri-, Di-, and Monoglycerides
      and Fatty Acids 246
    Cholesterol 246
    Uses 246
    Recommended Dietary Intakes 246
    Lipid Deficiencies 247
    Lipid Excesses 247
  Proteins 247
    Uses 248
    Recommended Dietary Intakes 248
    Protein Deficiencies 248
    Protein Excesses 248
  Water 248
    Uses 248
    Recommended Dietary Intakes 249
    Water Deficiencies 249

Water Excesses 249
Vitamins 249
   Characteristics 249
   Sources 249
   Deficiencies and Excesses 250
Minerals 250
   Characteristics 250
   Sources 250
   Deficiencies and Excesses 250
Nutrition and Alcohol 251
Nutrition and Medications 251
Nutrition and Disease 251
Nutrition and Maximum Longevity ... 251

12 URINARY SYSTEM 253
Main Functions for Homeostasis 253
   Removing Wastes and Toxins 253
   Regulating Osmotic Pressure 253
   Maintaining Individual
      Concentrations 254
   Maintaining Acid/Base Balance 256
   Regulating Blood Pressure 256
   Activating Vitamin D 257
   Regulating Oxygen Levels 257
Kidneys 257
   Blood Vessels 257
   Tubules and Collecting Ducts 257
   Nephrons 257
   Overall Functions 257
Age Changes in Kidneys 258
   Blood Vessels 258
   Renal Blood Flow 258
   Glomerular Filtration Rate 259
   Tubules and Collecting Ducts 259
   Other Changes 259
   Consequences 259
Abnormal and Disease
   Changes in Kidneys 260
Ureters 260
Urinary Bladder 260
   Age Changes 261
Urethra 261
   Age Changes 262
Urination 262
   Age Changes 262
   Urinary Incontinence 263

13 REPRODUCTIVE SYSTEM 265
Main Functions 265
   Gonads 265

Accessory Structures 265
Male System 266
   Testes 266
   Ducts 268
   Glands 269
   Penis 269
   Pubic Hair 270
Age Changes in the Male System 270
   Testes 270
   Ducts 271
   Glands 271
   Penis 271
Male Sexual Activity 271
   Excitement Phase 272
   Plateau Phase 272
   Orgasmic Phase 272
   Resolution Phase 272
   Refractory Period 272
Age Changes in Male Sexual Activity ... 273
   Excitement Phase 274
   Plateau Phase 274
   Orgasmic Phase 274
   Resolution Phase 274
   Refractory Period 274
Female System 275
   Ovaries 275
   Oviducts 277
   Uterus 277
   Vagina 278
   External Structures (Genitalia) 279
   Breasts 280
Age Changes in the Female System ... 280
   Ovaries 281
   Oviducts 281
   Uterus 281
   Vagina 282
   External Structures (Genitalia) 282
   Breasts 282
Female Sexual Activity 282
   Excitement Phase 282
   Plateau Phase 283
   Orgasmic Phase 283
   Resolution Phase 284
Age Changes in Female Sexual Activity ... 284
   Excitement Phase 284
   Plateau Phase 284
   Orgasmic Phase 284
   Resolution Phase 284
Frequency and Enjoyment of Sexual
   Activity 284
   Trends 284
   Contributing Factors 285

Making Adjustments 285
Abnormal and Disease Conditions 285
  Benign Prostatic Hypertrophy 285
  Impotence 286
  Prostate Cancer 287
  Vaginal Infections 288
  Breast Cancer 288
  Endometrial Cancer 289
  Ovarian Cancer 289
  Cervical Cancer 289
  Uterine Fibroids 290
  Sexually Transmitted Diseases 290

14 ENDOCRINE SYSTEM 291
Main Functions for Homeostasis.................. 291
Comparing Endocrine and
    Nervous Systems 293
  Coordinated Operation 293
Hormones.. 293
  Control of Secretion................................. 293
  Hormone Elimination 294
  Receptors and Responses 294
  Hormone Effectiveness 294
Specific Hormones 294
Growth Hormone 295
  Source and Control of Secretion 295
  Effects... 295
  Age Changes 295
  Growth Hormone Supplementation........ 296
Antidiuretic Hormone 296
  Source and Control of Secretion 296
  Effects... 296
  Age Changes 296
Melatonin 297
  Source and Control of Secretion 297
  Effects... 297
  Age Change 297
  Melatonin Supplementation 297
Thyroid Hormones (T_3 and T_4)................. 297
  Source and Control of Secretion 297
  Effects... 297
  Age Changes 297
Calcitonin (Thyrocalcitonin) 298
  Source and Control of Secretion 298
  Effects... 298
  Age Changes 298
Parathormone 298
  Source and Control of Secretion 298
  Effects... 298
  Age Changes 298
Thymosin 299

Source and Control of Secretion 299
  Effects.. 299
  Age Changes 299
Cholecystokinin 299
  Source and Control of Secretion 299
  Effects.. 299
  Age Changes 299
Glucocorticoids 299
  Source and Control of Secretion 299
  Effects.. 300
  Age Changes 300
Mineralocorticoids (Aldosterone) 300
  Source and Control of Secretion 300
  Effects.. 301
  Age Changes 301
Sex Hormones in Men 301
  Sources and Control of Secretion 301
  Forms of Testosterone 302
  Effects.. 302
  Age Changes 302
Sex Hormones in Women 303
  Sources and Control of Secretion 303
  Effects.. 303
  Age Changes 304
  Effects of Age Changes 304
  Estrogen Replacement Therapy 305
Insulin and Glucagon 306
  Sources and Control of Secretion 306
  Effects.. 306
  Blood Glucose Homeostasis 306
  Age Changes 307
  Abnormal Changes 307
Diabetes Mellitus 308
  Definition and Types 308
  Incidence 308
  Causes 309
  Main Effects and Complications 309
  Prevention.. 310
  Treatment 310

15 IMMUNE SYSTEM 312
Main Functions for Homeostasis.................. 312
Unique Characteristics 312
Development of the Immune System 313
  Macrophages and Langerhans Cells........ 313
  Thymus and T Cells 313
  B Cells 315
Immune Responses 315
  Processing and Presentation 316
  T-Cell Participation 317
  B-Cell Participation 320

Memory 321

Age Changes 322

Trends 322

Developmental Changes 322

Immune Responses 323

Consequences 324

Minimizing Consequences 325

Abnormal and Disease Conditions 325

16 INTO THE FUTURE 327

GLOSSARY 331

BIBLIOGRAPHY, Second Edition 347

BIBLIOGRAPHY, First Edition 361

INDEX 369

PREFACE

Aging in humans is becoming an increasingly important topic. In recognition of this expanding importance, research on all aspects of aging including the biological, psychological, sociological, and economic aspects, continues to increase. Since these areas are interrelated, individuals seeking to understand any one of them must have at least some understanding of the others.

In response to this growing need, courses and curricula on aging in humans are being developed. Though many books have been published that provide some support for these courses and curricula, few are satisfactory for use as textbooks by nonscience majors in introductory courses that emphasize biological aspects of human aging. This book has been written because of the dearth of such texts.

The intended audience for this book consists of students who have little or no background in science, particularly in biology. These students may want to learn about the biology of aging in humans for personal reasons or as part of their preparation for careers that involve working with older people, such as health care, sociology, social work, recreation, education, and psychology. Such students may also need to know about diseases that commonly affect the elderly. This book will serve as a text for these students in courses that emphasize or are devoted entirely to the biology of aging in humans. This book can enable students to do the following:

1. Demonstrate knowledge of demographic information pertaining to the elderly in the United States and an understanding of the significance of this information.
2. Name and describe factors that are believed to cause or influence the process of aging.
3. List and describe theories of aging.
4. Define and describe the concept of homeostasis and explain its importance and how it is maintained.
5. Describe the normal structure, functioning, and contributions to healthy survival for each body system in young adults.
6. Describe age changes in each body system and the interactions among these systems in older adults.
7. Describe certain abnormal changes in body systems and the interactions among these systems in older adults.
8. Describe interactions among biological, psy-

chological, social, and economic factors in older adults.

9. Relate and use this knowledge in their personal and professional lives.

The organization of this book represents an effective and successful strategy for achieving these goals. The first four objectives are addressed in the opening chapters. The remaining chapters are devoted to achieving the fifth through seventh objectives. The eighth and ninth objectives are addressed periodically throughout the book to provide a holistic perspective. The concept of homeostasis and the effects of aging on the ability to maintain homeostasis are the unifying themes which tie together all the sections of the book. Descriptions of changes in reserve capacity and methods of preventing abnormal changes and disease also permeate the text.

Chapter 1 includes background material about why aging should be studied, homeostasis, types of aging, methods of study, differences between age changes and other age-related changes, heterogeneity, and longevity.

Chapter 2 describes structure and functioning of cells and intercellular materials. This information provides the foundation for a discussion of theories of aging and for the later chapters. Most of the information about age changes and age-related abnormal changes at the cellular and molecular levels is incorporated into the discussions of body systems in Chapters 3 through 15, where such information is related to structural and functional alterations that affect homeostasis and the quality of life.

Except for Chapter 11, which deals with nutrition, each of the remaining chapters discusses one of the systems of the body. The organization within these chapters is consistent. Each begins with a discussion of the ways in which the system or aspect of the body being discussed contributes to homeostasis and therefore to the healthy survival of the individual. This is followed by a section on anatomy and physiology which contains information pertinent to the subsequent sections of the chapter. Students who are well versed in these topics may be able to move quickly to the second section of each chapter.

The second section describes changes in structure and function that occur with aging and alter the ability to maintain homeostasis or a high qual-

ity of life. The third part of each chapter presents information about abnormal changes and diseases that commonly occur in the elderly and are usually considered important. This section includes information about causes, development, signs and symptoms, and consequences. Much information about preventive measures is included, and some treatment strategies are mentioned. In chapters where many different parts of a system and lengthy discussions about one part of a system are presented, the information about age changes, abnormal changes, and diseases in each part follows the material on the anatomy and physiology of that part. Interspersed within each chapter are references to interrelationships between age changes, abnormal changes, diseases, and psychological, sociological, and economic factors involving the elderly.

Additional features of this book that will enable students to achieve the objectives enumerated above include the following:

1. The table of contents provides an outline for each chapter.
2. The level and style of writing are friendly and readable.
3. The use of technical terms and scientific jargon are minimized, and terms that are included are clearly defined in the text and in the glossary.
4. The discussions have enough depth and breadth to be complete without being overwhelming.
5. The organization of topics allows instructors to delete selected sections when course time is limited.
6. The text includes illustrations and tables designed to clarify the narrative.
7. Age changes are distinguished from abnormal and disease changes wherever such distinctions have been discovered. Places where these distinctions have not yet been elucidated are clearly indicated.

The second edition was written for two reasons. One was to provide an updated version of my first edition, which was greeted by overwhelmingly positive responses by reviewers, colleagues, and students. The second reason was to address as best I could the shortcomings mentioned by these same audiences. Still, I wanted to remain true to the overall purposes of the book, which are stated in the first edition's preface and encapsulated in the title. Of course, these goals had to be reached within the limits of space and economics imposed upon any author. Here is how I pursued these goals.

First, I compared the entire book's content with current literature. Content that was reaffirmed remained. Outdated content was updated or eliminated. Beyond this, the second edition contains all the content from the first edition because I received very few suggestions to shorten or omit content.

Then I expanded the sections that seemed most wanting. You will see meaningful expansion in: demographics; methods of studying aging; reserve capacity; differences between aging, disease, and environmental insults; genetics of human aging; basic chemistry; cell division; theories of aging; cognitive functions (i.e., memory, intelligence); Alzheimer's disease; photoaging of the skin and treatments; antioxidant supplements; dietary supplements; hormone supplements; caloric restriction; immune system imbalance; use of theories of aging within chapters; and incidences of diseases. I used theories of aging in explaining age changes in body systems (e.g., atherosclerosis, dementia, muscle, eyes, glaucoma, nutrition, diabetes, immune dysregulation).

I also added new sections on: quality of life; free radicals and antioxidants; evolutionary theories; telomeres and telomerase; Hayflick limit; replicative senescence; Werner's syndrome; progeroid syndromes; chronic heat exposure; heat shock protein; elastin peptides; homocysteine; smoking; dementia with Lewy bodies; biorhythms and circadian rhythms; types of muscle cells; driving motor vehicles; Dietary Reference Intakes (DRIs); Viagra; melatonin supplements; DHEA; Grave's disease; insulin-like growth factors; future problems, challenges, and opportunities; and world population growth.

Finally, I added illustrations on: demographics; body reserve capacity; molecules; cell membrane; chromosome and telomere; cytoskeleton; cell cycle; lung capacities; primary and secondary immune responses; effects from altering longevities; and world population growth.

Adding content and illustrations required adding space. To keep the size and cost of this edition manageable, page margins were made smaller. To conserve more space, content that might have been placed in tables was placed within the text. Listed items are separated by semicolons (see above.)

It was impossible to add color, more illustrations, more content, or valuable pedagogical aids (e.g., outlines, class activities, statistical data, lecture notes, chapter bibliographies, chapter summaries, practice questions). However, I hope to serve as a resource clearing house for faculty and students interested in the biology of aging, especially human aging. Therefore, I am developing pedagogical aids, and I plan to establish a web site where they will be available. I also intend to make aids available by E-mail.

If you want access to these items, contact me at AGDIGIOVANNA@SSU.EDU. You can also check the WCB/McGraw-Hill web site for more information at http://www.mhhe.com/biosci/abio/ which is the applied biology webpage. Click on "SEARCH" and search for DIGIOVANNA.

ACKNOWLEDGMENTS

I received wonderful assistance in developing this second edition. I extend my sincere gratitude to the following colleagues who provided excellent constructive criticisms. Cindy Beck - The Evergreen State College: Annette Benedict - Ramapo College of New Jersey: Susan H. Franzblau - Fayetteville State University: Madhu N. Mahadeva - Sanford Center for Aging; University of Nevada, Reno: Dale S. Mazzoni - Colorado State University: Donald J. Nash - Colorado State University: John P. Walsh - Andrus Gerontology Center; University of Southern California. I wish I could have included all the additions they requested.

I am also grateful to Lois Bromley - SSU interlibrary loan department, for cheerfully obtaining countless articles: to Dr. Jeffrey A. Chesky - University of Illinois, Springfield and Dr. Ronald Lucchino - Utica College of Syracuse University, for inspiring some of the new illustrations: to Pat Anglin and Kris Tibbetts - WCB/McGraw-Hill, for patience, guidance, and encouragement: to the SSU publications department, for advice and for use of publishing equipment: to my wife Linda L. DiGiovanna - SSU publications department, for turning the manuscript into the camera-ready copy.

Augustine Gaspar DiGiovanna
Salisbury State University

ABOUT THE AUTHOR

Dr. Augustine G. DiGiovanna is Professor of Biology in the Richard A. Henson School of Science and Technology at Salisbury Sate University. He received his B.S. degree in biology from St. John's University in New York and his M.S. and Ph.D. degrees from the University of Maryland College Park.

Since becoming a faculty member at SSU in 1972, Dr. DiGiovanna's main teaching responsibilities have been in the areas of human biology including anatomy, physiology, pathology, embryology, and aging. In the 1980s, he was a Visiting Professor at the University of Maryland College Park during the summers. He was awarded SSU's Distinguished Faculty Award in 1995 and the School of Science Excellence in Academic Advising award in 1997.

Early in the 1980s, Dr. DiGiovanna developed an interest in the biology of aging, and he created a course in the Biology of Human Aging. Teaching this course prompted him to write the first edition of this book. He is a member of the Gerontological Society of America (GSA) and the Human Anatomy and Physiology Society (HAPS). He is also active in the Maryland Consortium for Gerontology in Higher Education and the Association for Gerontology in Higher Education (AGHE).

1

INTRODUCTION

WHY STUDY HUMAN AGING?

Personal and Professional Reasons

Why have you chosen to study human aging? Why have others done so? For some people the answer is based on personal reasons. Younger individuals may expect to live long enough to reach old age and may wonder what will happen to them as they get older. They also may want to learn about aging so that they can improve their chances of aging happily and with good health. Curiosity or interest in bolstering one's well-being may be a prime reason older individuals study aging. Still other individuals may have family members, friends, colleagues, or acquaintances who are experiencing aging. Their interest may spring from curiosity about what is happening to those other people. Beyond being curious, individuals may want to be better able to interact with and care for older people.

On a professional level, some individuals study human aging because their careers involve working with or caring for older people. The careers of others may entail carrying out research on or educating people about human aging.

Whatever your reasons for studying human aging, you should be aware that people have many reasons for doing so and that those who are studying human aging are being joined by a growing number of people.

Population Trends

Why has the study of aging become so important during the last few decades? One main reason is the rapid increase in the number of elderly people. According to current projections, this will continue until about A.D. 2030, after which the number of elderly people will rise more slowly (Fig. 1.1). The proportion of elderly persons in the total population is also rising and will probably continue to grow for several decades. For example, in 1990, 21.2 percent of the population was over age 54. This number should rise slowly to about 21.4 percent by the year 2000, but it will probably increase to about 25.1 percent by the year 2010 and grow to about 30.5 percent by 2030.

There will also be an increase in both the number and proportion of people of higher ages (Figs. 1.1 and 1.2). For example, consider all people over age 65. In 1990, they numbered 31.6 million, made

FIGURE 1.1 Past and projected changes in the population 55 years old and over by age, 1900-2050.

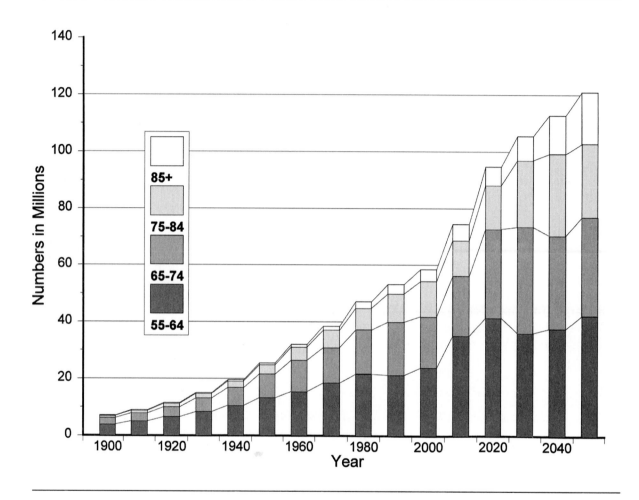

FIGURE 1.2 Elders as percentages of the U.S. population; past and projected 1900-2050.

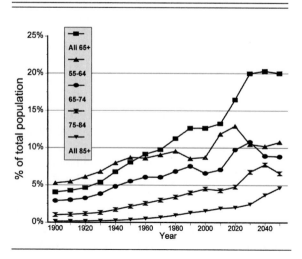

up 12.7 percent of the total population, and represented 59.6 percent of the population over age 55. By the year 2010 the corresponding statistics will probably change to 39.4 million, 13.2 percent, and 52.6 percent, respectively. By the year 2030 these values should reach 69.4 million, 20 percent, and 6 percent, respectively. During this period and extending to the year 2050, people over age 85 will make up the fastest growing group both in numbers and as a proportion of the population. As a percent of the total population, they are increasing 2.7 times as fast as the population over age 64. In 1994, those over age 84 comprised 1 percent of people over age 64. By the year 2040, those over age 84 may make up 15 percent of the total population and 24 percent of people over age 64.

Four factors explain these population changes. One factor is the high birthrates before 1920 and between 1946 and 1964, followed by a decrease after 1964 (Fig. 1.3). A second is the high number of births between 1946 and 1964 (Fig. 1.4). A third factor is the decline in childhood death rates, especially during the first year of life (Fig. 1.5). Since the childhood death rate in 1940 was already low compared with 1900 and since the childhood death rate dropped substantially between 1940 and 1955, a much higher percentage of those born during these latter years survived into adulthood. The last factor is the increase in life expectancy at all ages, including middle age and old age. (Fig. 1.6). Between 1900 and 1940, life expectancy for those over age 64 increased by less than one year, while it increased almost two years between 1940 and 1954. Life expectancy for those over age 64 increased more than 2.6 years since then. Because of these circumstances, a large group is now entering old age while a smaller group is replacing them as the younger segment of the population. This large group has become known as the "baby boomers" (Fig. 1.7). Life expectancy for all adults including those over age 64 is expected to continue increasing for decades. Therefore, a larger percentage of those reaching old age will remain alive longer. In terms of populations and elders, the "baby boomer bump" is the wave of the future.

The significance of these increases in the number and proportion of older people is that the elderly will have an ever greater influence on many aspects of society. As a group they will spend larger amounts of money, use more services, and have more political power. Therefore, an understanding of aging processes and other age-related changes is vital if society is to adapt to the changes that will accompany this phenomenon. Such an understanding may be especially important for those who, by virtue of their leadership positions, make decisions that have a broad impact, such as corporate and political decision makers.

WHAT IS AGING?

Exactly what is *aging*? How would you define it? Do all people use the same definition? Is aging different from other changes that occur as people get older? The term *aging* is difficult to define because it has diverse meanings for different people. However, one definition will be selected here to help in the study of aging. To understand this definition, we must first understand developmental changes.

Developmental changes are irreversible normal changes in a living organism that occur as time passes. The same changes can be expected to occur in all members of a particular type of organism. Developmental changes are neither accidental nor a result of abuse, misuse, disuse, or disease. They occur in humans from the moment of conception to the moment of death. Familiar examples include growth in height, sexual maturation, and graying of the hair. The field of biology in which developmental changes are studied is called *developmental biology*.

Note that developmental changes are irreversible or at least rarely reversible. Conversely, bodily changes that occur in one direction for a while and then reverse direction are called *physiological changes*. Some physiological changes, such as increases and decreases in the rate of breathing, are rapidly reversible. Others, such as fluctuations in weight and physical fitness, are reversed more slowly.

Developmental changes can be divided into three categories. The first consists of changes that occur before birth or during childhood. Examples include the formation of specialized organs from a single-celled fertilized egg and increases in muscle coordination. Collectively, these early changes are usually called *development*. Studies of development before birth constitute the segment of biology called *embryology*.

The second category includes changes that result in the transformation of a child into an adult. These changes make up what is frequently called *maturation*. Puberty is an example of maturation. Both development and maturation consist of changes that usually improve the ability of a person to survive. Examples include the strengthening of muscles and bones and increases in intellectual ability.

The third category-*aging*-refers to the group of developmental changes that become most evident in the later years; these are also called *age changes*. Examples described in later chapters include stiffening of the lungs, thinning of the bones, and a decline in the sense of smell. For practical purposes, the later years of life are considered in this book to begin about age 50. However, no one knows when many age changes that become evident in the later years actually start. Many age

FIGURE 1.3 U.S. birthrates, 1900-1998.

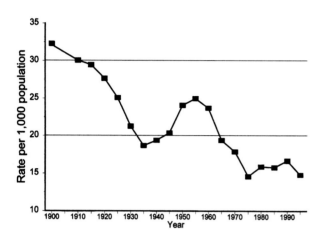

FIGURE 1.5 U.S. death rates for different age groups, 1900-1998.

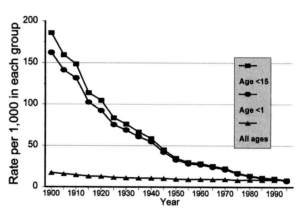

FIGURE 1.4 U.S. births, 1900-1998.

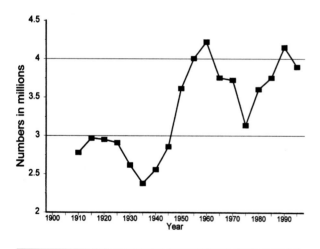

FIGURE 1.6 Life expectancies, 1900-2050.

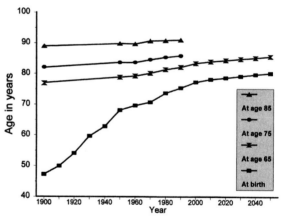

changes, such as reductions in kidney function, begin as early as age 20. Unlike development and maturation, almost all age changes reduce a person's ability to maintain healthy survival and a high quality of life. The term *senescence* includes only those age changes that have such detrimental effects. However, there are beneficial age changes, such as certain changes in the sweat glands, the heart, and the brain. These and other examples of positive age changes are described in Chaps. 3, 4, and 6.

Types of Aging

Biological Aging   Aging includes several different kinds of changes. One group of changes—*biological aging*—involves aging in the physical structures and functioning of the body that affects a person's ability to survive or a person's appearance. Biological aging is the main topic of this book. To understand its significance, one must first understand what is required for the survival and well-being of the body.

The human body, like most living things, is made up of small units called *cells* and materials that the cells produce. For example, muscles are made up mostly of muscle cells (Fig. 1.8*a*). By contrast, bones contain some bone cells but consist mostly of materials that those cells secrete (Fig. 1.8*b*).

The cells do more than furnish the substance of the body: They also perform all of its functions. Every thought and movement a person has actually results from nerve cells producing and carrying impulses and muscle cells moving. If the cells stopped working, there would be no bodily activity.

The cells of the body must have just the right set of conditions virtually all the time to build and maintain the structure of the body and carry out its functions. The state of having proper and fairly steady conditions is called *homeostasis*. It involves many conditions, such as temperature, nutrient levels, and water content. Each condition may change slightly from time to time; such small changes occur because being alive means doing things such as growing and moving, and doing things causes changes in body conditions (Fig. 1.9). For example, an ordinary activity such as walking raises body temperature, burns nutrients for energy, and results in water loss by evaporation from breathing and perspiring. Even the environment surrounding the body tends to cause changes within the body. An example is the tendency of body temperature to drop when a person is in a cool room because warm objects lose heat to a cool environment.

For a person to stay alive and well, each condition must not be allowed to stray above or below an acceptable range. If one of them, such as temperature, deviates too far, the cells will be injured and begin to malfunction. This means that the body is malfunctioning. Its well-being, and perhaps its very survival, is then jeopardized. The greater the number of injured cells and the more severe the injury, the greater the decline in bodily functioning and well-being and the greater the danger to the body.

If the errant condition is out of the acceptable range for only a brief period or to only a small degree, the cells can often recover once conditions are again favorable. However, if the deviation is present for an extended period or is extreme, some cells may be permanently altered or killed. The body has then lost the contributions which those cells should be making (Fig. 1.9). Again, depend-

FIGURE 1.7 U.S. "baby boomers" at different years, 1980-1995.

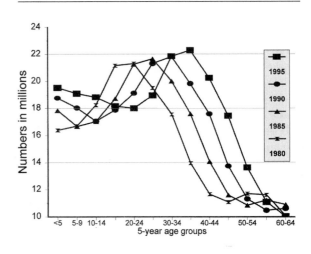

ing on the amount of injury, the result can range from barely noticeable discomfort to death.

Consider what happens to a person whose body temperature is dropping. As body temperature falls, the heart cells and brain cells slow down. If the chill is not severe or long-lasting, the person will recover completely once the body is warmed again. However, if the temperature drops too far or if the person stays chilled for a long time, as can happen when a person falls into icy water, cell functioning becomes so slow that the person dies of hypothermia.

Since many activities are occurring inside the body and many changes in its surroundings, one might ask how conditions for the cells are kept proper and fairly stable. Part of the answer involves the ability of the body to provide materials and structures that tend to prevent changes in these conditions. For example, fat under the skin helps prevent cooling by slowing heat loss from the body. The other part of the answer is the process of *negative feedback*, which involves three steps. The first step is detecting the presence of deviations from homeostasis. The next is informing the parts of the body that some condition is unacceptable and telling them how to correct the problem. The nervous system contributes to these steps by continuously monitoring conditions such as body temperature. For example, if the nervous system detects a drop in temperature, it sends impulses to several parts of the body (Fig. 1.10).

FIGURE 1.8 The structural basis of the body: (*a*) Cross section of skeletal muscle. (*b*) Cross section of bone.

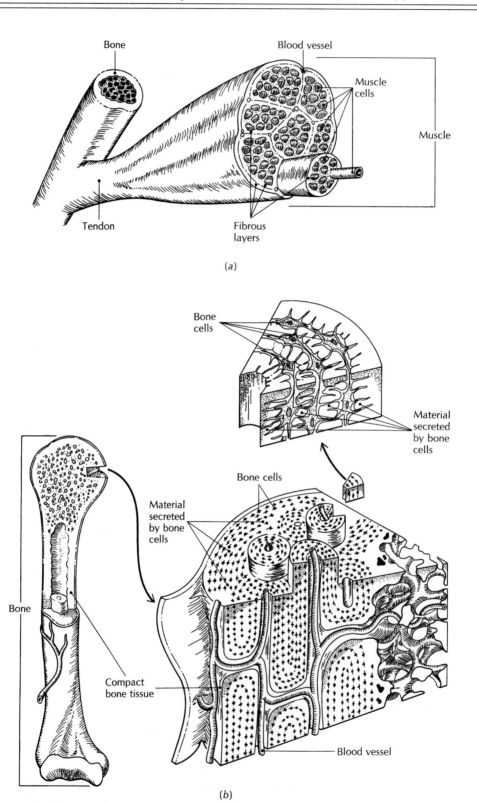

FIGURE 1.9 Homeostasis and unacceptable alterations in body conditions.

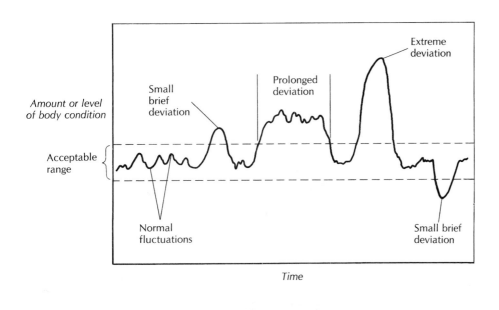

The brain is informed about the problem, and the person is warned of the danger by the feeling of being cold. The skin and the muscles are directed to correct the deviation from homeostasis.

The third step in negative feedback is making the necessary adjustment to restore the condition to a normal level. Many body systems contribute to this process. For example, when the body becomes chilled, the blood vessels in the skin become narrow to reduce the loss of heat. The muscles may then cause shivering as they contract and relax quickly and repeatedly to produce more heat to warm the body. Using the muscles and bones, the person may move to a warmer location or turn on a heater. These and other activities can restore normal body temperature before any cells are significantly affected. Homeostasis is maintained, and the person stays alive and well.

Having developed an appreciation for how the body keeps itself alive and well, one can understand the importance of biological aging. With few exceptions, biological aging reduces the ability of the body to maintain homeostasis and therefore to survive. This happens in two main ways.

First, some biological age changes allow more rapid or extreme alterations in body conditions to occur. For example, thinning of the insulating layer of fat under the skin allows the body to chill faster.

Second, other biological age changes reduce the functioning of negative feedback systems. There is a decline in the ability of certain parts of the body to detect alterations in body conditions and notify other parts that the body is threatened. Age changes in the nervous system are among the most important in this category. With aging, there is a decrease in the number of nerve cells that monitor conditions, and the nerve cells that remain often function weakly. Thus, the detection of deviations from homeostasis, such as a lowering of body temperature, is reduced. The ability to notify and activate parts of the body that can correct the problem also declines. This is especially pronounced when several parts of the body must act in a coordinated fashion. For example, there is a decline in coordinating the many complex muscle contractions needed to maintain balance while one is standing on a moving surface such as a boat deck. Finally, the structures that should restore conditions to an acceptable level are less able to do so. For example, as aging causes a decrease in the amount of muscle, there is a reduced ability to produce heat to raise body temperature back to normal.

FIGURE 1.10 Negative feedback for thermoregulation.

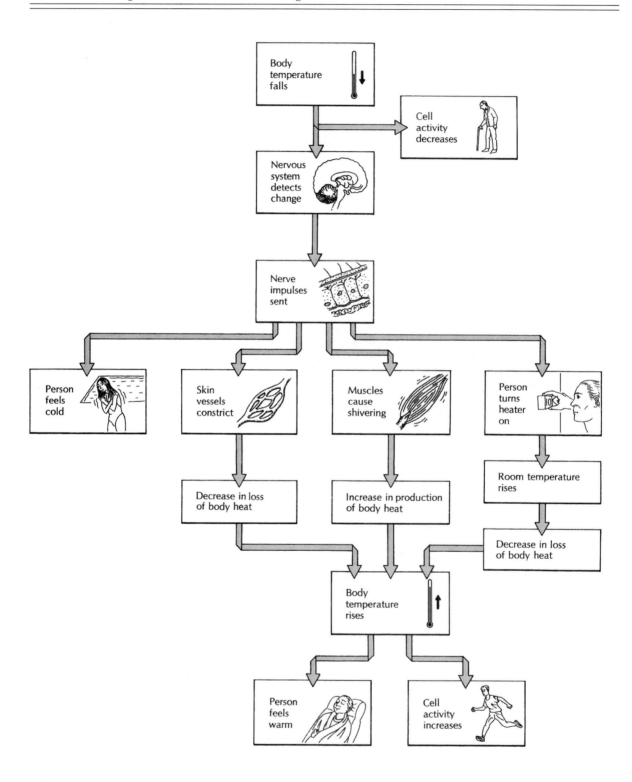

In summary, most biological aging allows more of the conditions in the body to stray farther from the acceptable range and to stay beyond the normal range longer. This causes more cells to be injured and fail in their functions. When many cells are affected to a large degree, the person feels less well and does not function as well. When too many cells are no longer able to perform adequately the person becomes ill and dies.

Chronological Aging   The simplest type of aging is *chronological aging*, which refers to the passage of time since birth. It is usually measured in years, though sometimes decades are used. While chronological age can be useful in estimating the average status of a large group of people, it is a poor indicator of an individual person's status because there is tremendous variation among individuals in the rate biological age changes occur. For example, on the average aging results in people losing much of their ability to perform strenuous activities, yet some elderly individuals are excellent marathon runners.

Cosmetic Aging   *Cosmetic aging* consists of changes in outward appearance with advancing age. This includes changes in the body and changes in other aspects of a person's appearance, such as the style of hair and clothing, the type of eyeglasses worn, and the use of a hearing aid. Like chronological aging, it is frequently used to estimate the degree to which other types of aging have occurred. It is even used to guess a person's chronological age. However, it is an inaccurate indicator for either purpose because of variation among individuals and because a person's appearance is affected by many factors that are not part of aging, including illness, poor nutrition, and exposure to sunlight.

Although cosmetic aging provides little evidence about other forms of aging, it can have profound effects on many aspects of life. For example, people who notice that their hair is turning gray may begin to think of themselves as old, and this may result in withdrawal from physically demanding activities, loss of appetite, depression, and subsequent declining health. Since people in this situation may lose interest in their appearance and may look worse because of ill health, they may be entering a vicious cycle of decline. The time, effort, and money that people spend trying to look young provides further evidence for the importance of appearance.

Social Aging   Another type of aging is *social aging*, which consists of age changes in the interactions people have with others. The birth of grandchildren, for example, can alter the ways in which the new parents interact with the new grandparents and even the ways in which the maternal and paternal grandparents relate to each other.

As with chronological and cosmetic aging, social aging has an impact on other age changes. The death of a spouse, for example, may decrease a person's interest in his or her own appearance, leading to cosmetic changes. The loneliness that often follows the loss of a spouse may cause stress, which in turn may result in a more rapid decline in the ability to fight off infection.

Psychological Aging   *Psychological aging* consists of age changes that affect the way people think and behave. It often results from other types of aging. For example, biological aging of the brain directly affects the speed of learning and the ability to remember some types of information. Examples involving other types of aging were mentioned above.

Psychological aging also contributes to other types of aging. Memory loss can result in forgetting to keep an appointment with a friend or a physician. Slowed thinking can prevent a person from retaining certain types of employment, such as jobs requiring rapid decision-making involving many variables (e.g., flight controller, fighter pilot, emergency room staff, crisis situation manager), especially in unfamiliar situations (e.g., newly employed or promoted).

Economic Aging   *Economic aging* consists of age changes in a person's financial status. Like psychological aging, economic aging can result from other types of aging. For example, in spite of laws against discrimination based on chronological age, some older people find it difficult to retain a job or obtain a new one simply because of their age. The resulting loss of income can cause difficulty in obtaining proper medical treatment or purchasing adequate food. Loss of contact with business colleagues and lowered self-esteem can have social and psychological effects.

FIGURE 1.11 Interactions among types of aging.

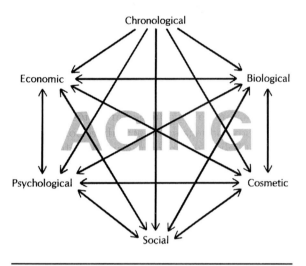

Social

Interactions among Types of Aging As is seen from the above examples, most types of aging can result from any of the other types. Also, each type can influence the others, and complex series of interactions can develop. The one exception is that chronological aging cannot be altered by the other types of aging (Fig. 1.11).

What Aging Is Not

It is important to note that many changes in the elderly are thought to be age changes but are abnormal rather than true age changes. Some abnormal changes result from abuse or misuse of parts of the body. Examples include skin wrinkling caused by sunlight (farmers), hearing loss caused by loud noise (factory workers), and joint stiffness caused by repeated traumatic injury (athletes). Other abnormal changes that are often thought to be age changes result from disuse. Examples include reductions in the pumping capacity of the heart, muscle power, and bone strength caused by inadequate exercise. Changes from extrinsic factors such as abuse, misuse, and disuse frequently accompany or amplify true age changes. For example, aging does cause some skin wrinkling, hearing loss, joint stiffness, muscle weakening, and bone weakening. Since many extrinsically caused abnormal changes in aging individuals are not severe enough to be considered disease, they are called "usual" age changes by some authors. People who avoid these abnormal changes while undergoing normal age changes are said to have achieved "successful" aging.

In addition to abnormal changes mentioned above, many abnormal changes that accompany aging are not part of true aging but are aspects of a disease. Examples include disabilities caused by heart attacks and strokes. However, aging is not a disease, does not mean disease, and does not automatically include disease. The elderly are more susceptible than the young to certain diseases, but no diseases occur only in the elderly or occur in every elderly person.

Why then is aging often equated with disease? This probably stems from the much higher incidence of diseases among the elderly. One reason for this increase in disease is that most age changes reduce the ability of the body to keep conditions within the normal range. As examples, timing mechanisms may only delay diseases under genetic control, the sensory function of the nervous system declines, reflexes become slower and weaker, and immune responses against infection dwindle. However, there are compensating mechanisms that make up for many of these detrimental changes. Something as simple as wearing warmer clothing can compensate for the reduced ability to maintain an adequate body temperature. The use of eyeglasses and brighter lighting can restore much of the decline in vision. Allowing more time for tasks can make up for slower reactions and slower learning or remembering. Avoiding exposure to infectious agents places less demand on defense mechanisms. If one creatively develops and uses compensating strategies, many undesirable consequences of aging that increase the likelihood of disease can be reduced or eliminated.

A second reason for the increase in disease with advancing age is that years must pass before some diseases become serious enough to be noticed. Sometimes this is because of the reserve capacity found in many parts of the body. Having reserve capacity means that under normal resting conditions, only a fraction of the full functional capacity of certain organs is needed to maintain homeostasis. For example, up to 50 percent of the functional capacity of the kidneys may be lost before a person notices that something is wrong. When body structures have little reserve capacity, a disease is not noticed because it progresses very slowly. For example, osteoarthritis, the most common type of arthritis, seems to require the cumulative effect of years of abuse of the joints before it becomes a problem. Also, atherosclerosis, which is a type of hardening of the arteries (arteriosclerosis), frequently begins before age 25. However, because deterioration of the arteries occurs slowly, the heart attacks and strokes it causes usually do not occur until several decades later.

A third reason for the age-related increase in disease is that as time passes, there is a greater chance that a person will be subjected to factors that promote disease and that these exposures will occur many times and for longer periods. Examples include physical trauma, infectious organisms, air pollution, and harmful radiation.

These facts indicate a very important point: Many abnormal changes associated with aging can be prevented, and the progress of many other diseases can be slowed enough so that their detrimental effects may be delayed for many years. It is even possible that their effects will not become apparent before death from other causes occurs. Of course, not all cases of every disease are preventable. Diseases such as Alzheimer's disease and rheumatoid arthritis cannot be prevented at all. However, for many age-related diseases, the use of disease prevention strategies before the disease begins often reduces the seriousness of its effects. For example, avoiding cigarette smoking reduces the effects of emphysema caused by other types of air pollution or by genetic factors.

Usually, all a person needs to do is avoid the factors that increase the risk of developing abnormal changes and the diseases that cause many of them. For risk factors, such as air pollution, that cannot be completely avoided, reducing their intensity or the frequency of exposure can help. This can reduce or nearly eliminate the chances of developing certain abnormal changes.

To be most effective, the avoidance of risk factors must begin early in life, but changing bad habits will probably help at any age. Even when a person begins to develop an abnormal change or disease, reducing risk factors can slow its progress so much that the change or disease may never become a significant problem. Some of the most important risk factors are smoking, stress, poor nutrition, inadequate exercise, and excessive exposure to harmful chemicals and sunlight. Others can be identified only by a medical checkup, including high blood pressure and high levels of cholesterol in the blood.

Finally, there is good news for those who develop a disease. Many diseases, including serious ones such as certain types of cancer and dementia, can be cured. Many others, such as arthritis, can be treated so that they have a minimal impact on a person's lifestyle. Early detection is important because it greatly increases the success achieved by treatment.

WHY STUDY BIOLOGICAL AGING?

We have discussed several broad reasons for studying aging in general and biological aging in particular. We will now examine more specific reasons for studying biological aging. One of the most important is being able to distinguish true age changes from changes that occur by chance or are caused by abuse, misuse, disuse, or disease. Individuals with this ability will be able to recognize changes in the body that represent the beginning of an abnormal condition. Then effective steps can be taken to prevent or combat undesirable changes that are not inevitable results of aging. Effort will not be wasted worrying about or attempting to alter conditions resulting from aging. Having knowledge about biological aging also makes it easier to select appropriate preventive or corrective measures. Furthermore, if one knows the course of age changes, the effects and effectiveness of new treatments can be better evaluated. There will be less chance of confusing the effects of a treatment with the effects of aging. Knowing the timing and nature of age changes also provides some predictability. Better estimates of a person's future biological or medical status can be made, and it is easier to predict the life expectancy of an individual or a group of people.

HOW IS BIOLOGICAL AGING STUDIED?

Two methods are commonly used to study biological aging, and each has advantages and disadvantages. The most reliable conclusions regarding aging are those supported by both types of study or a combination of the two.

Cross-Sectional Method

A *cross-sectional study* starts with a group of people of different ages who are placed into age categories. In some cases each category may contain all individuals who have reached the same age in years; in other cases each category may contain all individuals whose age in years falls within a selected range. For example, each range may span five years. Thus, one category may include all those between the ages of 45 and 49; the next category may include all those between the ages of 50 and 54; and so on. Alternatively, the age ranges may be of different sizes, such as all those age 50 through 59 and all those age 60 and above.

Once the categories have been established, the researchers measure characteristics such as intelligence, muscle strength, or heart rate for each individual. The data for individuals in each category are then compared with the data from individuals in the other categories. In this way, correlations between differences in characteristics and increases in age can be identified. If a trend is observed, the researchers conclude that it is caused by increasing age.

Cross-sectional studies are very popular for several reasons. First, they can be done quickly, so there is no need to wait for years while the subjects in the study age. Second, since each subject needs to be evaluated only once, many subjects can be tested and then released from the study. Therefore, this procedure is relatively inexpensive. Third, since many subjects are included in the study, the results are statistically reliable. Finally, these studies largely eliminate the problem of a period effect. A *period effect* is the influence of events or conditions during the study on the people being studied. For example, changes in the employment status of the subjects during an economic depression or a war cause period effects.

There are several drawbacks to cross-sectional studies. A very important one is that such studies do not really measure changes that occur as time passes. It is only inferred that the differences among the age groups result from the passage of time. These differences could be caused by other factors that affected the subjects before the study. This flaw in the basic design of cross-sectional studies is called a *birth-cohort effect*. For example, the individuals in certain age categories might have been different from those in the other age categories at birth. This could have resulted from immigration or from relocation of large segments of the population. Thus, one age category may have an overly large representation of individuals of one nationality. These individuals could be genetically different from individuals in another age category composed largely of people with a different nationality and genetic makeup. As another example, some cross-sectional studies show that there is a decrease in intelligence with aging. This difference may be due not to aging but to less opportunity and encouragement for those in the old-age categories to have attended school in their youth. This last problem can be identified by performing a *time-lag study*. It carries out the same cross-sectional study procedure after many years and makes comparisons between two groups of the same age category. For example, measurements of people who are 65 years old in 1990 could be compared with measurements of people who are 65 years old in 2010. Differences between these groups would reveal effects from differences in historical conditions.

Another design flaw in these studies is called *differential mortality*. It means that because of inborn differences in susceptibility to certain causes of death (e.g., certain infectious diseases), specific groups of individuals who would have been included in certain age categories have been inadvertently selected out of their categories because they died before the study began. Thus, there is a built-in bias among the age categories that has nothing to do with aging. Another problem with cross-sectional studies is that they measure only average changes. They cannot detect change in a single individual.

Overall, though cross-sectional studies sometimes detect true age changes, investigators using this technique may believe that they have found an age change where none exists. They also may conclude that an age change occurs faster or slower than it truly does.

Longitudinal Method

Another method for studying biological aging is the *longitudinal study*, in which a group of individuals of similar or identical chronological age is selected. Each individual is evaluated for the characteristics that are to be studied. Then, at specified intervals, the same individuals are evaluated in the same ways for the same characteristics. The intervals may be short, such as 1 year, or longer.

Longitudinal studies have several advantages over cross-sectional studies. First, they actually measure changes that occur as time passes; the relationship of the changes to aging is not simply inferred. Second, though they establish averages for a group as cross-sectional studies do, longitudinal studies can also detect age changes within the individual and can even establish the rate of change for each person. As a result, longitudinal studies reveal that different individuals age at different rates. As we will see later in this chapter, this is a very important finding.

By evaluating people periodically, longitudinal studies can also identify and measure the influence on aging of sudden events such as an accident or of long-term treatments or diseases. Alternatively, these studies can investigate the effects of aging on the course of a disease. Through careful analysis, longitudinal studies can establish the complicated interactive effects of several variables, such as the effects of changes in body weight on the way in which exercise affects the regulation of blood glucose. Finally, these studies can discover the predictive value of conditions present at one period of life on parameters such as future health and time of death.

Despite their many advantages, longitudinal studies on humans are not done as frequently as are cross-sectional studies because longitudinal studies have several negative characteristics. Of prime importance is the length of time needed to carry out such a study. It may be necessary to evaluate subjects over a period of many years. For example, if the study attempts to measure certain age changes from age 50 to age 80, the study must be conducted continuously for 30 years. During this period, many subjects may lose interest in the study, move away from the area where it is conducted, or die. The investigators themselves also face these problems. In addition, to achieve scientific reliability, the techniques for performing measurements of the characteristics of interest must remain basically the same despite technological advances. These factors cause a second drawback: Longitudinal studies usually cost a great deal more than do cross-sectional studies. Because of the expense, longitudinal studies usually include fewer subjects. Thus, after all the work, the results are not as statistically valid as those garnered from cross-sectional studies.

Longitudinal studies also contain certain design flaws. One is the period effect. For example, the results from a longitudinal study during a time of economic prosperity may be quite differ-ent from those obtained during a period of economic hardship. There is even a birth-cohort effect. This effect can be substantially reduced in longitudinal studies, but only by extending the studies over much longer periods. Finally, test results may be affected as people being tested become more familiar with the procedures.

While longitudinal studies can provide more and better information about biological aging than can cross-sectional studies, they do so only at great human and financial expense. Therefore, few long-term longitudinal studies have been conducted.

Baltimore Longitudinal Study of Aging   One of the largest and most successful continuing longitudinal studies is the Baltimore Longitudinal Study of Aging (BLSA). The BLSA is conducted by the National Institute on Aging, which is part of the National Institutes of Health, an agency of the federal government.

The BLSA started in 1958. At first it had only a few hundred male subjects, most of whom were at or beyond middle age. It now includes more than 700 volunteer subjects, both female and male, ranging in age from the twenties through age 90. Subjects receive a thorough evaluation, including numerous biological and psychological characteristics, every two to three years.

Cross-sequential Method

A third study method combines cross-sectional and longitudinal studies, forming a *cross-sequential study*. In this method, a cross-sectional study is performed and is then repeated after some years have passed. For example, people in five-year age categories extending from ages 40 to 70 could be evaluated in 1990, in 2000, and in 2010. Separate and combined comparisons could then be made among the groups at each time and among the three times. Using this complex method helps reduce problems from period effects, birth-cohort effects, and differential mortality.

Nonhuman Studies

Besides being studied in humans, biological aging is studied in many animals, including mice, rats, flies, and worms. In addition, individual cells, both human and animal, are grown in nutrient materials to study biological aging. Though many of these studies have little or no immediate application to biological aging in humans, many others are used either as preliminary studies for future human studies or as experiments to support the results from human studies.

Animal and cell studies are very useful and important in the investigation of biological aging in humans since studying humans presents several problems. One is the genetic heterogeneity among people. This high degree of intrinsic variability results in considerable difficulty when one is interpreting data. Obviously, one cannot selectively breed people to achieve more genetic similarity among them, but selective breeding can be done with animals. Furthermore, environmental factors such as diet, temperature, and exercise can be controlled in animals to a degree that would be impossible in people. Even more control and freedom for study and experimentation are possible when one is dealing with individual cells.

Other factors make studying animals and cells desirable. Laboratory animals and cells take up less space, are less expensive to maintain, and have much shorter life spans than do humans. Unlike humans, certain lower animals (e.g., flies, worms) and cells apparently are not affected by psychological and emotional factors. Animals and cells can be humanely sacrificed for detailed anatomic and chemical analysis. Finally, unlike animals and cells, many elderly people have diseases and receive treatments that would affect the outcome of studies in which they might be subjects.

Still, humans must be studied directly if we are to increase our knowledge and understanding of the biology of human aging. It is known that aging in animals and individual cells differs from aging in humans in a variety of ways. For example, as some commonly used laboratory animals get older, they develop specific types of cancer and kidney disease that are not found in humans. Also, as some animals age, they have changes in hormone levels that are not seen in humans. The hormones involved may alter the aging processes in many parts of the body. Finally, certain chemicals and dietary substances are known to affect aging in animals differently from the way they affect aging in humans.

WHAT WE KNOW THUS FAR

The study of biological aging became a topic of great interest only recently compared with areas of biology such as embryology, anatomy, and genetics. Although many fundamental questions about aging have not been addressed and others have been only partially answered, much has been learned in a short time. Research is continuing, expanding, and creating synergistic interdisciplinary bridges as the study of aging - *gerontology* - is undertaken by experts in more disciplines. Aging studies are also becoming more cross-cultural and international. These studies are revealing similarities and differences in aging among people of different cultures, races, regions, and national origins. Results from these comparisons are beyond the scope of this book. It deals with human aging in advanced Western civilizations, especially in the U.S.

What Happens during Biological Aging?

When Aging Begins   The ability of the body to maintain homeostasis seems to reach peak capacity during the third decade of life, after which age changes begin and bodily functioning starts to decline. (Fig. 1.12). There seems to be no plateau period during which the body retains its maximum level of performance. The effects of aging are not immediately apparent, however.

Why Aging Appears Well after It Begins   One reason for delay in the appearance of age changes is the large reserve capacity present in many parts of the body. The heart, which can increase the amount of blood it normally pumps fivefold, and the respiratory system, which can move six times the amount of air normally breathed, provide two examples. Detrimental age changes draw first on such reserve capacities. The effects become apparent only when the body is called upon to function near peak capacity but much of the reserve is gone. Since biological aging occurs slowly and the body is rarely called upon to function at peak capacity, it takes many years for the reserves to become noticeably low. For example, some individuals reach age 40 or beyond before significant age changes are noticed. Additional normal aging must occur before body capacities become so low that a person seems impaired most of the time. As aging continues after that, impairments become so severe that they are classified as diseases. Eventually, body capacities are so low that homeostasis cannot be maintained even with

FIGURE 1.12 Body capacity and health status from birth to death under optimum conditions and with detrimental factors.

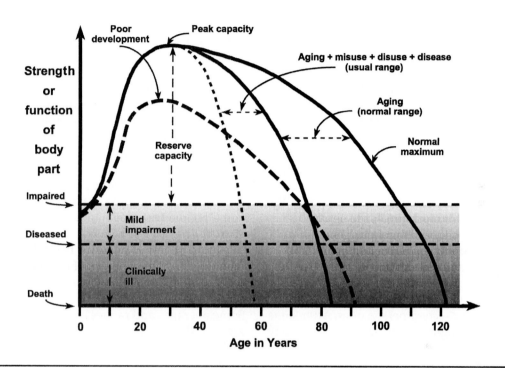

medical assistance, and the person dies. Death is inevitable, but since the person has made the most of their biological life, they have aged normally.

The body behaves like a person who develops a large savings account when his or her income is high. That person can continue to live well by drawing on the savings when his or her income goes down. When savings become low, the person may not have enough funds to afford ordinary recreational activities, but essentials are still affordable. As funds dwindle, essentials become unaffordable, and the person may depend upon loans or other financial support. Eventually, funds become so low that the person is bankrupt.

Aging is like finances in two other ways (Fig. 1.12). First, if a person does not develop to his or her peak potential during youth, they enter the declining years of aging with less reserve capacity. Malnutrition, poor health care, or other adverse circumstances during youth may produce this effect. With less reserve capacity when aging

begins, less time is needed for the body to reach the impaired or diseased levels. Second, an adult with a high peak capacity may have their bodily reserves ravaged by aging plus by other adverse factors (i.e., misuse, disuse, disease). Again, impairment and disease develop more quickly. This latter scenario is common, so it has been called "usual" aging.

A second reason for the delay in the appearance of age changes in normal aging and in usual aging is the use of compensatory mechanisms. Some adjustments bolster diminishing functions. For example, greater amounts of chemicals (e.g., norepinephrine) that stimulate the heart are produced. This helps maintain pumping as the intrinsic strength of the heart declines. Some adjustments involve using more efficient ways to accomplish goals. People can learn how to pace themselves or use tools more effectively and thus continue to perform very difficult tasks. Finally, through changes in lifestyles and goals, many people tend to adjust their activities by participating in activities where they are comfortable and capable while shying away from activities that become too difficult and burdensome.

Variability in Aging   There is considerable variability among people in both the age at which age changes are noticed and the rate at which these changes progress. This variability derives from several differences in aging (Fig. 1.13). First, aging of a particular part of the body starts at different times for different people. Second, once a part has begun to age, it does so at different rates in different people. For example, bone strength declines faster in some individuals than in others. Third, the parts that age fastest in one person may not be the ones that age fastest in another person. Thus, one person's heart may have the most age changes, while in another person the lungs may be aging faster than the heart. Fourth, the rate of aging of most of one person's body parts is faster than the rate in another person. In other words, throughout the body, some people age faster than others. Because of all four differences, one person may begin aging or show signs of aging before a second person does. After some years, however, aging in the second person may surpass that in the first.

Two other types of variability in aging make this matter even more complicated. First, certain body parts usually seem to age faster than others do; for example, the lungs age faster than the blood. Second, though aging generally progresses steadily, the aging of some body parts in some individuals may speed up for a while, become quite slow for a while, and stop or show a reversal for a while.

Many factors combine in each person to affect the specific time of onset and rate of aging for each body part. Each individual's sex, genetically determined condition, and intrinsic compensatory powers when aging begins are unalterable factors. Occasional occurrences such as accidental injury and short-term diseases, along with long-term aspects of a person's life such as education, diet, exercise, occupation, air quality, and protracted diseases, also play a role both before and after aging begins. The rate and the degree of effects from the progressive changes caused by aging are altered as these factors change.

Therefore, though the rate of aging is determined in large part by conditions over which a person has no control, it is also heavily influenced by modifiable factors. As we identify and learn more about the factors that can be altered, we can gain more control over the progress of biological aging. We will also be better able to ward off the abnormal changes and diseases that become more likely as aging progresses.

Heterogeneity among the Elderly   Every individual is subjected to a unique combination of the factors that affect aging. Each factor acts at various ages to different degrees and for different lengths of time. The complex interactions among these factors add even more diversity. As a result, the older people get, the more different from one another they become. For many body parts, differences among those who have reached age 50 are already great. As more years pass, the heterogeneity among people expands more quickly. Some people become impaired or seriously ill at an early age, while other people remain hardy beyond age 100 (Fig. 1.12). The elderly are the most diverse age group. Therefore, this book avoids using numerical values, which may be erroneously interpreted as ideals, norms, or goals. Some averages and ranges of values for the body are included, but only to provide approximations and trends within an expanding and increasingly diverse group.

An important consequence of heterogeneity among the elderly is the need to provide individualized treatment for them. As the age of a group of people increases, generalities apply less and less well to the individuals in that group. Any planning for elderly persons must consider this individuality. This would include, for example, evaluating eligibility for employment or educational opportunities, designing housing, developing nutritional programs, planning physical fitness programs, and providing health care. More attention to the increased differences among aging people would assure not only that more individuals will receive proper consideration but also that fewer will be subjected to detrimental care and therapy.

Another significant conclusion derived from increased heterogeneity with aging is that there is no set age at which a person becomes "elderly." Although this book describes biological age changes observed frequently in people above age 50, that age was selected because most research on human aging has been done on people above

FIGURE 1.13 Aging variables: (solid line = person A, dashed line = person B (*a*) Age at onset. (*b*) Aging rate of the same structure. (*c*) Aging rate of different structures. (*d*) Average aging rate. (*e*) Aging rate in the same person. (*f*) Aging rate at different ages.

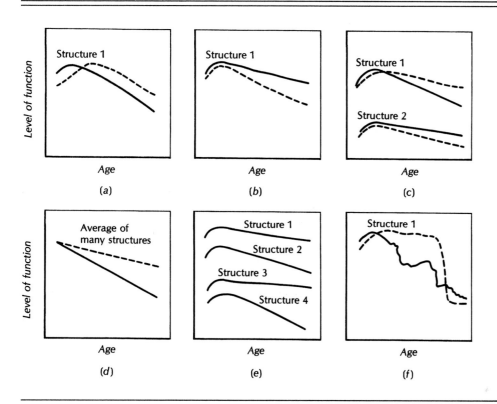

age 50 and because many age changes do not become significant until after that age. Many individuals consider "old age" to begin at age 60, age 70, or even age 80 and beyond. Age 65, a figure commonly used to denote the onset of old age, was first used when the Social Security system was established. It was based on estimates of how long people should be fully employed so that there would be enough revenue to pay benefits to those who retired. Choosing age 65 really had nothing to do with aging. With changes in populations and government policies, the standard retirement age under Social Security has been changed to 67. This change occurred for demographic, economic and political reasons, not because it takes two additional years for people born after 1941 to become "old."

The Concept of Biological Age Although there is no specific chronological age at which a person becomes biologically old, some researchers believe that determining a person's biological age is possible. While there are several ways to do this, all of them start by attempting to determine average values for normal people at each chronological age. In one method, the levels of functioning of organs or systems are measured under resting conditions. In another technique, the levels of functioning are measured under stressful or maximum operating conditions such as during vigorous exercise. A third procedure measures the ability of the body to maintain normal conditions under adverse conditions, for example, the ability to maintain temperature while in a cold environment. Still another approach is to find the rate at which the body returns to resting conditions after being exposed to an adverse situation such as an excessive intake of salt.

Once the average normal values have been obtained, the measurements for the individual whose biological age is being determined are compared with those values. Scores for different functions may be considered individually, or a figure calculated from a combination of scores may be employed.

A simple procedure for carrying out the comparison would be to find the normal group whose average score equals that of the individual being considered. The individual's biological age could then be said to equal the chronological age of that group. Other types of comparisons of biological status among people of the same chronological age can establish the percentile rank of an individual within the group, as is done in comparisons of intelligence test scores.

The value of determining an individual's biological age can be greatly increased by repeating the procedure periodically, such as annually. This will provide information about the individual's rate of aging.

While any of these techniques can produce seemingly meaningful results, there is a lack of consensus about the validity of the procedures. Disagreements arise over which approach should be used. There is also the question of whether all the functions tested are equally important. If they are not, attempts to select the useful ones or to rank those that are used result in more discord. For example, should the ability to feel vibrations be included? What about clarity of eyesight? If these are included, is either of them more important than the resting heart rate? Or is maximum heart rate a better indicator of biological age? Perhaps there is no such thing as an overall biological age but only separate biological ages for the various parts of the body.

Although this problem is far from resolved, attempts to find solutions are worthwhile for reasons similar to those that justify the study of biological aging. Once a biological age is determined for a person or a group of individuals, the factors that modify the aging processes can be discovered. This can lead to the formulation of improved care plans and can even lead to predictions of a person's life expectancy.

Life Expectancy

Maximum Longevity   How long can a person expect to live? The answer depends on many factors. The first factor to address is the one that establishes the longest life possible for humans. The longest life achieved by the members of a species is called the *maximum longevity (XL)* of that spe-

cies. According to scientific records, the maximum longevity for humans is 122 years, the age attained by Jeanne Calment of France. In 1999, at age 118, the oldest person was Sarah Knauss of Pennsylvania. Analysis of census and mortality data for the U.S. suggests that the human XL is probably 130 years. Human maximum longevity and the XL of other animals seem to be determined by genes. As described in Chap. 2, these genes may control activities such as the timing of life events, the time of death, the correcting of errors in other genes, and the repairing of molecules that carry out genetic instructions.

While the maximum longevity in some animal species can be changed by selective breeding and genetic manipulations, some scientists believe that maximum longevity for humans probably cannot be altered. This is due largely to several limitations to altering human genes. First, ethical considerations make selective breeding of humans impossible. Second, the genes that determine maximum longevity have not been identified. Even if they were identified, the ways by which they control life span are not known and therefore are not subject to manipulation. Third, if the information and techniques needed to perform the required genetic engineering are discovered, the question of whether such interference should be carried out remains. Ethical, social, political, and economic problems will need to be addressed.

Another reason militating against extending human XL derives from the techniques that might be required. The major discomfort or alteration in lifestyle caused by some procedures, such as the severe diet restrictions that lengthen the life span of some animals, might not be worth the possible gain in human life span. Some scientists believe that if these problems were solved, others, such as late life diseases not yet recognized, would become limiting factors. Finally, one would have to consider if having a longer life, with its many inevitable and unwanted age changes and increased likelihood of disease, is desirable.

Mean Longevity   Though humans can live to an age of 122 years, this rarely happens. Even reaching the age of 100 is considered remarkable. One reason people live to different ages is the variation in genes controlling life span. Additionally, people are subjected to many other causes of death, such as accidents and disease. These other causes act before the genes determining life span have an opportunity to do so. Therefore, a statistic that is more useful for most people than maximum longevity is *mean longevity (ML)*, the average age at which death occurs for the members of a population; this is also called the *life expectancy of the population*. The conditions that determine mean longevity provide the second part of the answer to the question of how long a person can expect to live. These conditions reduce life expectancy to a value less than maximum longevity.

Statistically speaking, all people in a population have the same mean longevity at the time of birth. However, different populations have different MLs. One reason for this is the historical period in which birth occurred. For example, the mean longevity in America in 1776 was 35 years of age. By 1900 it had increased to about 47 years. It reached slightly over 68 by 1950 and climbed to almost 74 years of age by 1980.

Between 1900 and 1970 the increase in mean longevity was due mostly to a decrease in the death rates of infants and children. Early in this period, poor provisions for public health (e.g., sanitation) and weak control of infectious diseases (e.g., vaccinations, antibiotics) were the main causes of high infant and child mortality (Fig. 1.5). Harsh working conditions and limited education further shortened the lives not only of children but also of adults. The result was that few people lived long lives. As environmental and other external conditions improved, many more people survived the first few decades of life, and this led to a dramatic increase in mean longevity (Fig. 1.6).

Mean longevity in the United States has continued to rise since 1980. It reached 75.4 years by 1990 and is expected to reach 77.3 by A.D. 2000. It will probably rise slowly but steadily well beyond the year 2030, reaching as high as age 80 by the year 2050. Most of this increase is due more to decreased death rates for those above age 35 than to changes in death rates among younger people. The reason is that so much progress has been made in improving the extrinsic conditions that affect younger people that few advances in this area can be expected. Intrinsic factors and chronic diseases, which come into play in the later years of life, now have a more predominant influence on ML because they have become the main causes of death. This situation is expected to continue as long as human activity does not cause additional deterioration of the environment or become more self-destructive.

Reasons other than historical periods cause differences among populations in mean longevity at birth. For example, gender affects mean longevity (ML). The population consisting of all women has a higher ML than does the population of all men. One factor contributing to this higher mean longevity seems to be that higher levels of certain hormones (estrogen and progesterone) help protect women from specific serious diseases (e.g., heart attacks). Another possible factor among women may be that female cells can use more of the genetic material (i.e., sex chromosomes) they contain. A third possible factor is that women have less iron before menopause due to periodic menstruation. With lower iron, women may sustain less damage to their molecules from free radicals (see Chapter 2). A fourth factor may be that over the past decades, lifestyles and careers traditionally involving primarily women provided less danger and stress than did those involving primarily men. Finally, men may be more willing to take serious physical risks.

Another important factor affecting mean longevity is race. The white population has a higher mean longevity at birth than does the black population. Like the differences in mean longevity between women and men, these differences are probably due to differences in both genetics and lifestyle factors (e.g., nutrition, education, employment).

The differences in mean longevities between sexes and among races and cultures have always existed in the United States, but the degrees of difference have not always been the same. Most recently the differences have been decreasing. It is uncertain whether these differences are more likely to decrease or increase in the next several decades.

While all members of a population have the same mean longevity at birth, the mean longevities of individuals of different ages in that population are different (Fig. 1.6). This is the case because as time passes, the death of some members of each birth cohort selects out those who do not survive well. This selection process spares those in the population who have better intrinsic characteristics for survival, better living conditions, or better mechanisms to adapt to life-shortening situations. These survivors thus have higher life expectancies. Thus, the life expectancy of those who were born in 1990 was age 75.4, while the life expectancy of those who were 65 years of age was 82.2, and was 91.1 for those age 85.

Thus far we have looked at life expectancy in broad terms. We will now point out a few examples of additional factors that help determine how long a person can expect to live.

Some factors that influence life expectancy are fixed at birth. Here we must again mention genes. People who have parents who lived long lives tend also to live long lives. In fact, the more blood relatives with long lives a person has, the greater the chances that person has of living a long life. This is due only in part to the genes passed from one generation to the next, however. It is also due to the nurturing and culture that members of a family share.

Two other influential characteristics that are not easily changed are intelligence and personality. Overall, people with higher intelligence and people with personalities that result in lower stress levels tend to live longer. By contrast, those whose personalities provide more stress, especially highly competitive perfectionists with a persistent sense of lacking sufficient time to accomplish their goals, tend to have lower life expectancies.

While intelligence and personality are partly established by the time of birth, they can be modified by the environment to which an individual is exposed. Good education and positive social relationships can significantly shape and strengthen intelligence and personality, resulting in an increase in life expectancy.

Though we have virtually no control over which genes people inherit, we can exert much influence over the environment in which people develop and live by providing conditions and opportunities known to increase life expectancy. People tend to live longer if they have proper nutrition, housing, and health care. Being employed, being married, having an adequate income, and receiving more education also increase life expectancy. Avoiding or reducing exposure to environmental insults (e.g., air pollution, smoking, excessive alcohol, toxic chemicals, radiation) are also positive influences. Living in areas where accidents are minimized also improves the chances for a longer life. Of course, preventing diseases makes a substantial contribution in this regard. Note that disease prevention is only one of many factors that increase life expectancy. It has been estimated that even if the diseases that are the top 10 causes of death were eliminated, mean longevity would increase only 11 years.

There are, then, many conditions affecting life expectancy over which we have considerable control. Interestingly, these factors not only increase life expectancy, they also greatly affect the quality of life, including life in the later years. Modifying conditions to improve the quality of life is perhaps an even more important goal than modifying them simply to extend the length of life. Furthermore, just as with planning for our financial future and retirement, the sooner we get started and the more regular our contributions, the greater the chances for happy and successful aging.

Status of an Individual   Thus far, we have dealt with mean longevity, the average life expectancy for a group of people. Attempting to estimate the life expectancy of one individual would require considering all factors affecting the group. However, more information about the current biological status of the individual would also be very helpful. This is where a medical checkup or a determination of the person's biological age becomes quite useful. An even better estimate of life expectancy can be formed if the individual is evaluated regularly to detect changes in the ability to maintain homeostasis. This can identify problems early in their development. Then steps can be taken to ward off or to compensate for the oncoming difficulty. An increase in life expectancy and in the quality of life in the years remaining could result.

Just how long can a person expect to live? There are certain limits within which the answer lies. Although rough estimates can be made, finding the answer with accuracy is difficult. The answer depends on the unique combination of several factors that are present in a person's life. Also, as the types and intensities of these factors change, the answer also changes. Perhaps we should be satisfied with the rough estimates and devote more time and energy to improving the quality of life we have left as we age.

Quality of Life

Quality of life can be evaluated in several ways. When determining the quality of life of others, evaluators usually use quantitative observable parameters and use tests and interviews. A person's status in several areas may be evaluated including physical health, ability to perform activities of daily living (e.g., dressing, bathing, eating, mobility), psychological status, emotional status, economic status, social functioning, and involvement with life activities. When determining one's own quality of life, many elders use parameters different from those used by others who evaluate them. Elders often consider factors related to self-identity, sense of independence, sense of self-efficacy, sense of control of one's environment and life, and life satisfaction.

Evaluating quality of life and determining how to evaluate it for individuals is important in developing public policies (e.g., health care, retirement plans) and individual courses of action (e.g., purchases, finances, health care, family matters). Also, determining quality of life is needed when assessing outcomes. Through the interactions between the biology of the body and perceptions, quality of life affects health status, mean longevity, and how much contribution elders can make to society.

2

MOLECULES, CELLS, AND THEORIES

HIERARCHY OF THE BODY

The importance of understanding body materials and cells becomes evident when one realizes that the human body has a hierarchy of structure and functioning, with these materials and cells at its foundation. This chapter will take a broad view of this hierarchy and then examine major body substances and cells in greater detail.

Atoms, Ions, and Molecules

Like all physical objects, the body is made of matter composed of units called *atoms*. Each atom has at its center a nucleus that contains one or more small particles called *protons* and *neutrons* and one or more smaller particles called *electrons*, which move about in regions some distance away from the nucleus (Fig. 2.1).

An atom is the smallest unit of an element that has the properties of that element. For example, all the carbon atoms that make up a lump of coal can burn and produce carbon dioxide. Most types of atoms lack complete sets of electrons in their outer regions. Therefore, these atoms may react chemically with one another in ways that produce complete electron sets. Atoms engaged in one type of chemical reaction may lose or gain electrons and become *ions* (Fig. 2.2); atoms engaged in other chemical reactions end up sharing electrons with neighboring atoms (Fig. 2.3). Ions that attract each other because they have opposite charges and atoms that share electrons may be bonded together in specific ratios, forming groups called *molecules*. Furthermore, fragments of some molecules may lose or gain electrons. Such molecular fragments are also called ions (Fig. 2.4).

Each type of molecule has its own properties, which may be very different from those of the atoms composing it. For example, hydrogen atoms form a gas that can burn explosively and oxygen atoms support combustion, but when hydrogen atoms are chemically bonded to oxygen atoms in a 2:1 ratio, they form water molecules, which can extinguish fires. Molecules also undergo chemical reactions, which may produce still other molecules. For example, during digestion water molecules may react with starch molecules and produce sugar molecules.

Besides producing new substances, each chemical reaction either releases or absorbs *energy*. The

FIGURE 2.1 The structure of atoms.

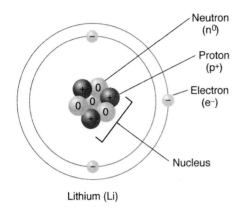

Lithium (Li)

FIGURE 2.2 A chemical reaction with the giving and taking of electrons, forming molecules with ions.

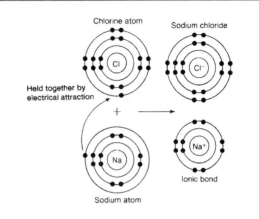

FIGURE 2.3 A chemical reaction with the sharing of electrons, forming molecules.

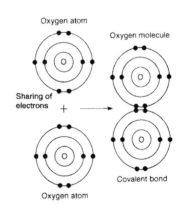

FIGURE 2.4 Forming ions from molecules. The H^+ is a form of acid, and can affect hydrogen bonds in other molecules.

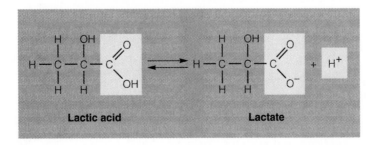

energy given off by chemical reactions can be used to power other activities, including other chemical reactions that absorb energy. For example, the energy given off by burning fuel can be used to power an automobile. The energy in foods, measured as calories, is given off when the foods are metabolized. This energy powers all body functions and keeps us warm.

Organelles and Cells

Many atoms, ions, and molecules in the body are arranged in combinations that form structures of various shapes, such as sheets, granules, and tubes. These structures make up components called *organelles* (e.g., cell membrane, ribosomes, microtubules) (Fig. 2.5). Organelles in turn are in highly complicated and organized structures called *cells*. The organelles are like the parts of an automobile in that each organelle can carry out only a few specialized functions. However, like the parts of an automobile that is being driven correctly, a complete and properly assembled set of organelles that has proper guidance can operate on its own.

Cells are the smallest units of the body that can survive on their own under favorable conditions (i.e., homeostasis) and have all the characteristics of life. These characteristics include organization, constant chemical activity, external or internal movement, and an active response to stimulation. By possessing and carrying out the characteristics of life, cells and the substances they produce constitute all the larger components in the body and perform all body functions. Though cells in the body have many features in common, they are also specialized in a variety of ways and therefore can perform different functions. For example, muscle cells can contract to provide gross movements, and bone cells can secrete materials that make hard and rigid bones, which provide support and protection.

Tissues, Organs, Systems, Organism

Cells of the same type plus some materials they secrete are found together in organized groups called *tissues* (Fig. 2.5). Different tissues are organized in groups called *organs*, organs are organized in groups called *systems*, and systems are found in an organized group called the *organism*, a human body. Just as each cell performs certain

FIGURE 2.5 The hierarchy of body structure.

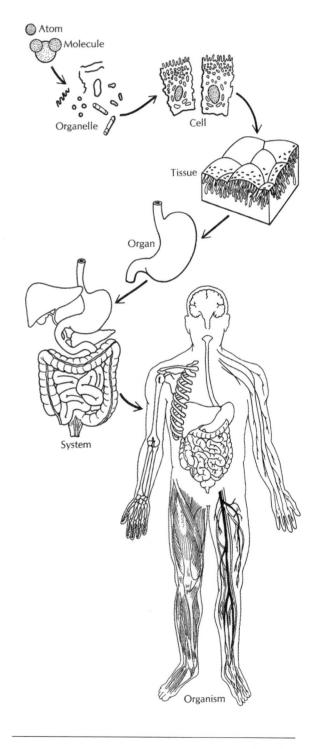

functions, so also does each tissue, organ, and system. When these functions are combined and coordinated, homeostasis can be maintained and the individual can survive. Combining and coordinating male and female functions can result in reproduction, which maintains the survival of the human species. There are additional hierarchical levels beyond the body (e.g., family, community, nation) which are studied in disciplines other than biology (e.g., sociology and political science).

BODY CHEMICALS

The atoms, ions, and molecules contained in the body or making up its parts are derived from atoms, ions, and molecules in the diet. However, many dietary substances are modified before entering the body and becoming integral parts of it. The following discussion focuses on body chemicals, which are chemicals that are already internal or compose parts of the body. Dietary chemicals and their relationships to body chemicals are discussed in Chap. 11.

Approximately 70 percent of the body consists of *water* molecules (Fig. 2.6). Water dissolves and transports materials, lubricates and cushions structures, regulates temperature, and modulates osmotic pressure and acid/base balance. The other body chemicals are of myriad variety and complexity. Some of them are minerals that are present as ions (e.g., sodium ions). The bulk of the remaining body chemicals are molecules, many of which share common features and therefore can be grouped into broad categories. These categories are *carbohydrates*, *lipids*, *proteins*, and *nucleic acids*.

Carbohydrates

The smallest and simplest carbohydrate molecules are simple sugar molecules called *monosaccharides*. One of the most abundant monosaccharides is *glucose* (Fig. 2.7). Other carbohydrate molecules consist of monosaccharide molecules linked together. A carbohydrate containing two monosaccharides is called a disaccharide, and carbohydrates consisting of many monosaccharides are called *polysaccharides*. Polysaccharides may contain from dozens to thousands of monosaccharide molecules, which are arranged to form straight or branched chains (Fig. 2.8). A common polysaccharide is *glycogen*, which is made of glucose.

FIGURE 2.6 Water molecules. In water and other molecules, oxygen keeps the electrons most of the time. This produces positive (+) and negative (-) regions and hydrogen bonds. Hydrogen bonds form within and between many body molecules (e.g., DNA, proteins), affecting shapes and functions.

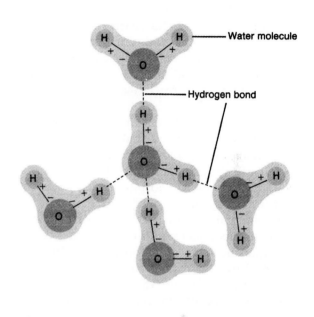

FIGURE 2.7 Glucose molecule.

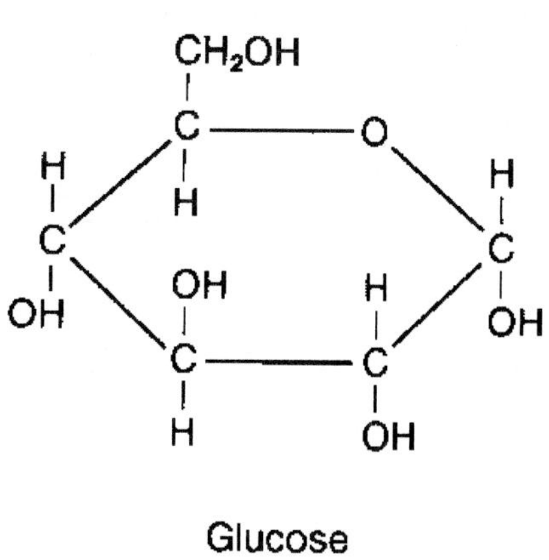

Glucose

Carbohydrates are used for storing and supplying energy and for building materials and receptor molecules for communication.

Nucleic Acids

Nucleic acid molecules contain dozens to thousands of small molecules linked to form chains (Fig. 2.9). The small molecular units in nucleic acids are called *nucleotides*. Each nucleotide contains a sugar molecule with five carbon atoms, a molecular fragment containing phosphorus (a *phosphate group*), and a molecular fragment containing nitrogen (a *nitrogen base*). There are five types of nitrogen bases, which can be designated by the initials A, G, C, U, and T (adenine, guanine, cytosine, uracil, and thymine). Though many nucleotides are linked to form nucleic acids, others remain separate. These individual nucleotides often have additional phosphate groups or other modifications and are used to transfer energy from chemical reactions to other activities. Two of the most abundant and widely used nucleotides are *adenosine diphosphate* (*ADP*) and *adenosine triphosphate* (*ATP*) (Fig. 2.10).

The nucleic acids are divided into two main classes based on the type of sugar used in their nucleotides. The sugars in nucleic acids called *ribonucleic acid* (*RNA*) have one more oxygen atom than do the sugars in nucleic acids called *deoxyribonucleic acid* (*DNA*). A second difference between RNA and DNA is that nucleotides in RNA contain A, G, C, and U, while nucleotides in DNA contain A, G, C, and T. Most DNA molecules contain two chains of DNA linked side by side by matching nitrogen bases in complementary pairs. The double strand of DNA is twisted in the form of a double helix, which resembles a twisted ladder (Fig. 2.9). Other combinations of DNA, RNA, and individual nucleotides are also made by matching complementary nitrogen bases.

Though there are only four different nucleotides in RNA and DNA, nucleic acid molecules in both classes have enormous variety. The reasons are the same as those permitting the writing of an essentially limitless variety of sentences using the limited number of letters in the alphabet. That is, nucleic acid molecules may be of different lengths, and more important, the nucleotides may be arranged in a virtually infinite number of ratios and sequences. As with the letters in a sentence, the nucleotide sequence in each nucleic acid molecule contains a message. Also like letters in sentences, changes in either the nitrogen bases employed or the sequence of the bases can change the message or make it meaningless (e.g., dog - dig - god - dgo). When complementary nitrogen bases are matched, the encoded message can be deciphered in cells and the information it contains can be used as instructions. Many of these instructions direct the formation of protein molecules by a process called *translation*.

FIGURE 2.8 Carbohydrate molecules of different sizes.

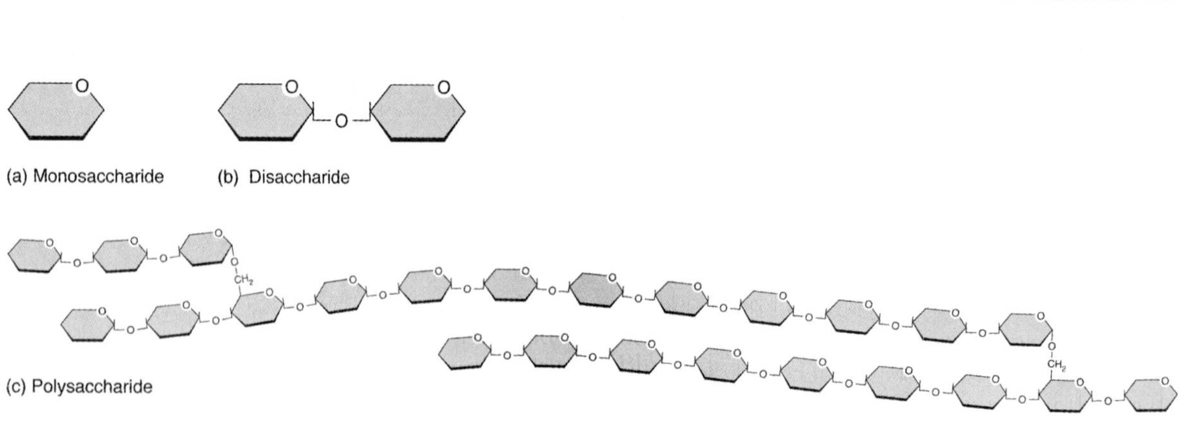

(a) Monosaccharide   (b) Disaccharide

(c) Polysaccharide

FIGURE 2.9 Deoxyribonucleic acid (DNA).

FIGURE 2.11 Chain structure of a protein molecule.

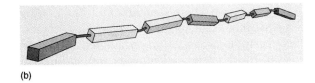

(a)

(b)

FIGURE 2.12 Complex structure of a protein molecule. Hydrogen bonds, cross-links, and other bonds form this complex structure from the protein chain.

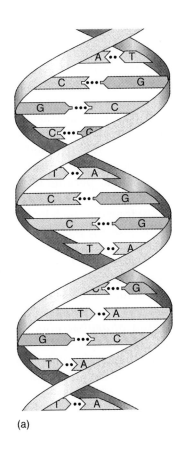

(a)

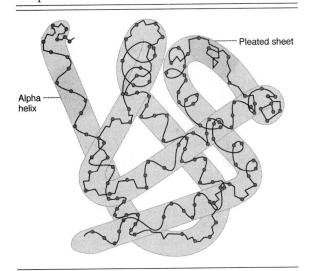

Pleated sheet

Alpha helix

FIGURE 2.10 Adenosine triphosphate (ATP).

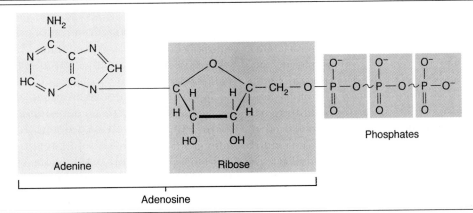

Adenine

Ribose

Phosphates

Adenosine

Proteins

Like carbohydrate and nucleic acid molecules, protein molecules consist of many small molecules linked to form chains (Fig. 2.11). The small molecular units in proteins are called *amino acids*, of which there are 20 types.

Compared with many polysaccharides and nucleic acid molecules, protein molecules show much more variety in structure and functioning. This variety exists for five reasons. First, the different amino acids may be combined in a virtually infinite variety of lengths, ratios, and sequences. Second, polysaccharides and nucleic acids contain few different types of units, while each protein molecule may contain all 20 types of amino acids. Third, the shapes of different amino acids cause protein chains to form various twists and bends. Fourth, some amino acids link to others in a protein, causing it to take on and maintain other twists and bends (Fig. 2.12). Fifth, some amino acids link to amino acids in adjacent protein molecules, causing additional changes in configurations and positions.

Many twists, bends, and links, and therefore the shape and position of each protein, are determined by the conditions surrounding it (e.g., temperature). When these conditions change, the shape and position of a protein may shift, and slight changes in these conditions cause dramatic shifts in some proteins. Furthermore, as with any tool or device, the shape and position of each protein determine what functions it can perform and how well it can perform them (Fig. 2.13). This is a major reason body structures retain their normal shapes and proper functioning only under homeostatic conditions. Proteins serve as building materials, receptor molecules, and hormones for communication; enzymes for regulating reactions; and antibodies for defense.

Lipids

Lipid molecules are placed into the same category because at least a large portion of each molecule does not dissolve well in water. We will mention only a few types of body lipid molecules.

Among the most water-repellent lipid molecules are the *triglycerides*, also called *fat*. Triglycerides contain a backbone made of a *glycerol* molecule with three fatty acid arms protruding from it (Fig. 2.14a). Fatty acids contain up to 20 carbon atoms in a row (Fig. 2.15). The body also contains glycerol combined with only one or two fatty acid molecules, forming *monoglycerides* and *diglycerides*, respectively. Glycerides store and supply energy.

The carbon atoms in some fatty acids are linked to the maximum number of hydrogen atoms; such fatty acids are called *saturated fatty acids* (Fig. 2.15). In other fatty acids, additional hydrogen atoms can be linked to the carbon atoms, and these fatty acids are called *unsaturated fatty acids*. *Monounsaturated fatty acids* have only one location that permits the addition of hydrogen atoms, while *polyunsaturated fatty acids* have more than one such location. Similar terms are applied to triglycerides with fatty acids able to hold zero, one, or more than one additional hydrogen atom (i.e., *saturated fat*, *monounsaturated fat*, and *polyunsaturated fat*).

Often, glycerol linked to two fatty acids is also linked to a molecular fragment containing phosphorus, forming *phospholipids* (Fig. 2.14b). While the regions of phospholipids containing fatty acids repel water, the region containing phosphorus attracts water. These properties cause phospholipids to align and form double-layered membranes in the watery internal environment of the body.

A third group of lipid molecules are called *steroids*. Their carbon atoms are linked to form rings (Fig. 2.16). Well-known examples of steroids include cholesterol, which is used as a building material, and sex steroids such as testosterone, estrogen, and progesterone, which serve as hormones for communication.

Molecular Complexes

Though many carbohydrate, nucleic acid, protein, and lipid molecules are not joined to any others in these categories, many molecules link together and form molecular complexes. Combinations of carbohydrate and protein are called *glycoproteins* or *mucopolysaccharides*, depending on whether carbohydrate or protein predominates. Combinations of nucleic acids and proteins are called *nucleoproteins*, and combinations of lipids and proteins are called *lipoproteins*. The formation of molecular complexes can modify the physical properties (e.g., flexibility) and activities (e.g., accessibility) of the molecules involved.

Free Radicals

A *free radical* (*FR*) is an atom or molecule with an unpaired electron (* = an unpaired electron). For example, an ordinary oxygen molecule is made of two oxygen atoms, and it contains 16 electrons. (Fig. 2.3). If another electron is added to the molecule, one electron would be unpaired. The resulting molecule would be a free radical called a *superoxide radical*. Some free radicals are made from highly reactive substances that contain oxygen, and some free radicals produce such highly reactive substances. These substances, which are not free radicals themselves, are called *reactive oxygen species* (*ROS*).

Small *FRs in the body include the *superoxide radical* ($*O_2^-$), the *hydroxyl radical* (*OH), and the *nitric oxide radical* (*NO). Larger free radicals contain an organic molecule, such as a fatty acid, combined with extra oxygen. Examples include the alkoxyl radical (*RO) and the peroxyl radical (*ROO), where the R represents the original organic molecule. Peroxyl radicals containing a fatty acid are also called *lipid peroxides* (*LPs*).

Reactive oxygen species in the body include *hydrogen peroxide* (H_2O_2), *peroxynitrite anion* ($ONOO^-$), organic hydroperoxide (ROOH), plus certain amino acids (e.g., tryptophan) and other substances produced during cell metabolism. An organic hydroperoxide containing a fatty acid is also called a *lipid hydroperoxide*.

These *FRs and ROS are not equally important. Superoxide radicals and H_2O_2 result in damage only when they fuel reactions that produce hydroxyl radicals (*OH) or $ONOO^-$, because these latter two substances are among the fastest acting and most toxic to the body. Free radicals containing fatty acids react much slower than these.

Importance in Aging Free radicals seem to be implicated in aging. Reasons include the apparent negative correlations between the following; mean longevities (MLs) and maximum longevities (XLs) of species and their rates of forming *FRs and ROS; MLs and XLs and their rates of developing damage from *FRs; age and the level of *FR defenses in some species; and age and the rate of repairing damage from *FRs.

Other reasons include the apparent positive correlations between the following; MLs and XLs of some species and their levels of *FR defenses; MLs and anti-*FR supplements (i.e., antioxidants); age and the rate of *FR formation; age and the amount of damage from *FRs; age-related diseases and *FRs (e.g., atherosclerosis, heart attacks, strokes, Alzheimer's diseases, parkinsonism, cataracts, kidney failure, cancers).

Formation of *FRs and ROS Some *FRs and ROS produced by the body are useful. Examples include; *NO for signals among neurons; *NO to cause blood vessel dilation for blood pressure regulation; H_2O_2 to destroy bacteria; and other defense *FRs and ROS produced during defense processes such as inflammation and immune responses. Many *FRs and ROS are produced as by-products from other useful reactions. Examples include cells producing $*O_2^-$, H_2O_2 and *OH and other *FRs and ROS when cells obtain energy from nutrients; detoxifying certain plant materials; breakdown of dopamine (DOPA) and fatty acids;

FIGURE 2.13 Importance of proper protein structure for function. Correct structure of the protein enzyme is necessary to help molecules (i.e., substrates) react. Cofactors (e.g., minerals, vitamins) help proteins work effectively.

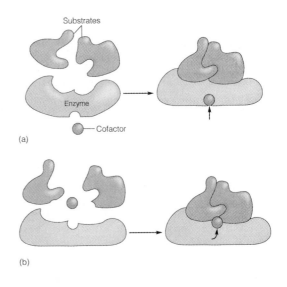

FIGURE 2.14 Lipids: *(a)* Triglyceride. *(b)* Phospholipid.

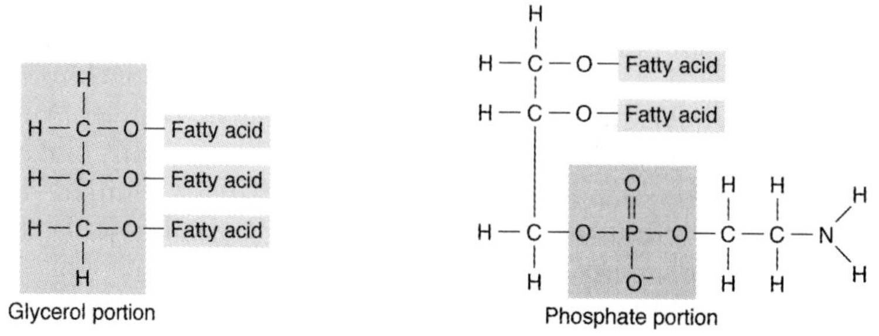

Glycerol portion

(a) A fat molecule

Phosphate portion

(b) Cephalin (a phospholipid molecule)

FIGURE 2.15 Fatty acids: *(a)* Saturated. *(b)* Unsaturated. The molecule's length and the locations of unsaturated spots affect its shape and function.

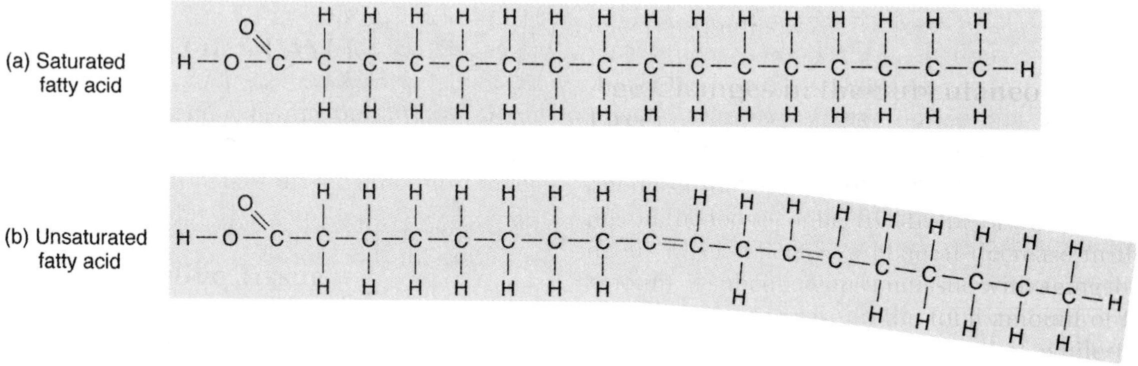

(a) Saturated fatty acid

(b) Unsaturated fatty acid

FIGURE 2.16 Steroids: *(a)* General structure. *(b)* Cholesterol.

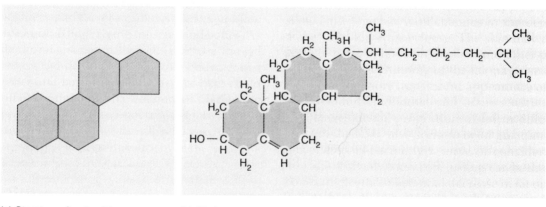

(a) Structure of a steroid

(b) Cholesterol

using iron or copper in reactions; and performing reactions peculiar to their special functions. Finally, *FRs and ROS result from unwanted conditions including exposure to ultraviolet light, internal bleeding, reversing unduly restricted blood flow, and reactions between glucose and proteins (see Glycation, below).

Conditions that increase *FR production include elevated amounts of O_2 in the body; high blood LDLs; high blood glucose levels; excess vitamin C; very high level of exercise; skin photosensitizers including certain cosmetics, medications, and air pollutants; menopause; and smoking. Conditions that decrease *FR production include reduced intake of polyunsaturated lipids; moderate exercise; increased intake of cruciferous vegetables (e.g., broccoli, cauliflower, cabbage); reduced intake of copper, iron, or magnesium; reduced intake of certain amino acids (e.g., histidine, lysine); and increasing age.

Common chemical reactions by which *FRs and ROS are formed include the following, where e- represents an electron, H^+ represents a hydrogen ion, and H- represents a hydrogen atom sharing electrons with another atom or molecule.

$$O_2 + e\text{-} \rightarrow \text{*}O2^-$$
$$\text{*}O2^- + e\text{-} + 2H^+ \rightarrow H_2O_2$$
$$H_2O_2 + e\text{-} \rightarrow OH^- + \text{*}OH$$

These reactions are common when cells use oxygen to derive energy from nutrients (see Mitochondria, below); where iron or copper ions exist; in skin struck by ultraviolet light; and when enzymes in the brain use monoamine oxidases (MAOs).

Common reactions with *FRs involving fatty acids are shown next, where PUFA represents a polyunsaturated fatty acid. These reactions occur in three processes called *initiation*, *propagation*, and *termination*. The reactions are *chain reactions* because propagation or reinitiation may occur repeatedly before termination occurs. Each time propagation or reinitiation occurs, another damaged fatty acid is produced and a new *FR is formed.

initiation

$$H\text{-PUFA}^1 + \text{*FR*} \rightarrow \text{PUFA}^1 + H\text{-R}$$

*PUFA^1 molecule rearranges itself

rearranged *PUFA^1 + $O_2 \rightarrow$ *PUFA1-O-O (peroxyl radical = *LP)

propagation

*PUFA^1-O-O + H-PUFA$^2 \rightarrow$ H-PUFA-O-O (damaged fatty acid) + *PUFA^2 (new *FR)

repeated propagations

*PUFA^2 + $O_2 \rightarrow$ *PUFA^2-O-O

*PUFA^2-O-O + H-PUFA$^3 \rightarrow$ H-PUFA^2-O-O (damaged fatty acid) + *PUFA^3 (new *FR)

etc., etc. $\rightarrow$ many damaged PUFAs + *PUFA^n

termination (either of two methods)

*PUFA^n + H-R $\rightarrow$ H-PUFA^n + R

*PUFA^n + *PUFA$^n \rightarrow$ R-O-O-O-O-R (toxic substances) + O_2

reinitiation

R-O-O-O-O-R + iron or copper $\rightarrow$ *PUFA-O or *PUFA-O-O

Effects from *FRs   Free radicals damage body molecules by taking electrons from molecules, a process called *oxidation*. This process alters the shapes and functions of the molecules, causing the body to sustain structural damage and malfunctions.

The main types of molecules affected are nucleic acids, proteins, and lipids. The consequences from even a small alteration in DNA can be devastating because the effect is multiplied during protein production. Also, damaged DNA may be unable to be replicated, preventing cells from reproducing by mitosis. Alternatively, some types of DNA damage promote cancer.

Damage to proteins and lipids causes abnormal cell and body structures and operations. Protein molecules such as those in tendons and ligaments become excessively joined together, and the functions of enzymes and cell membranes become abnormal. Damaged mitochondria are often unable to produce adequate energy for maximum cell activity. *FR damage can initiate inflammation, cause excess blood clotting, and promote several diseases, especially cataracts and atherosclerosis. Some *FR effects on proteins and lipids compound problems by increasing the rate of *FR production, reducing the body's ability to eliminate *FRs, and decreasing the ability to repair or remove damaged molecules.

Clearly, free radicals damage a variety of essential bodily components and alter body functions. To defend against *FR damage, the body has mechanisms to eliminate free radicals and to remove and repair molecules damaged by them. Substances called *antioxidants* destroy them by helping to create pairs of electrons. Examples of dietary antioxidants include vitamin E, vitamin C, beta-carotene (a vitamin A precursor), and substances in fruits and vegetables. The body also makes many antioxidants (e.g., melatonin, glutathione, albumin, uric acid), and can even recycle some of these antioxidants after they have neutralized *FRs.

The body also makes enzymes to divert *FR production and to speed up *FR elimination. Extremely important examples remove superoxide radicals (superoxide dismutase) and H_2O_2 (glutathione peroxidase, catalase). These enzymes are especially important because they prevent the formation of *OH, the most reactive and harmful free radical. Selenium is a vital dietary constituent because it helps an enzyme (glutathione peroxidase) remove H_2O_2.

Glycation

Glucose joins chemically with certain amino acids in proteins. No enzymes are needed for this reaction, and they usually occur with the side groups of arginine and lysine (Fig. 2.11, see R-). The products are altered amino acids attached to glucose (i.e., Amadori products). The amino acid/glucose portion may break down to form a distorted protein plus an ROS (e.g., H_2O_2), and the ROS may form *FRs.

Alternatively, the amino acid/glucose portion may join with others on the same or different protein molecules. This is called a *Maillard reaction*. The results are cross-links among the protein strands, and they then resemble strips of tape that have become stuck and tangled together. The cross-links are extremely stable and long-lasting, and common ones are called *pentosidine*.

The reactions forming glucose cross-links between proteins are called *nonenzymatic glycation* because glucose reacts without the use of enzymes. They are also called *glycation*, or *glycoxydation* because *FRs help as oxidizing substances during the process. The glycated proteins are damaged and distorted, turn a darker color, and are called *advanced glycation end-products* (*AGEs*). Other sugars in the body perform similar reactions that produce altered proteins, and AGEs may be formed by other chemical pathways. Most types of protein in the body are subject to glycation.

Effects Studies on AGEs began in the 1970s, and studies on their relationship to aging began in the 1980s. To date, little is known about the exact identity and characteristics of AGEs formed by glycation. Though they do not seem to cause aging, they accumulate with aging and seem to contribute to aging and age-related diseases. For example, the rate of glycation is inversely correlated with the XL of animals, though the amount of AGEs formed is not related to XLs. Also, *FRs and glycation are related in at least four ways; *FRs increase glycation; glycation increases *FRs; both processes increase the effects from *FRs damaging defense and repair enzymes.

There is no known benefit from glycation or from AGEs. However, glycation adversely affects all body parts and functions directly by distorting the proteins involved. Other adverse effects from glycation include indirect damage to DNA and lipids through the *FRs produced from glycation and by the abnormal operations of damaged proteins; body stiffening, poor movement of materials between cells, and distorted signaling of cells (see Intercellular Materials, below); reduced ability to control blood vessels and blood pressure; damage to blood vessel linings and atherosclerosis; increased blood clotting; increased development of Alzheimer's disease; kidney injury and eye damage from AGEs in blood vessel walls; amplification of most effects from diabetes mellitus; and tissue damage and inflammation by activating defense cells. For example, macrophages and monocytes are activated when AGEs attach to their cell membrane receptors for AGEs. These receptors are called *receptors for AGES* (*RAGEs*).

Keeping blood glucose levels within normal levels seems to be a main way to minimize glycation and its adverse effects.

CELLS

We can now use our understanding of body chemicals to examine cells in greater detail. In doing so, we will focus on the structures and corresponding functions in one cell that has the general features found in most cells (Fig. 2.17). Specializations in cell structures and age changes in cells are described in Chaps. 3 through 15, along with the systems in which they are found.

Cell Membrane

The outer boundary of the cell is called the *cell membrane*. It consists of a double layer of phospholipid molecules that contains other lipid molecules and protein molecules. Some of these proteins have carbohydrate molecules extending outward from them (Fig. 2.18).

The lipids and some proteins give the membrane strength to hold the cell contents together and regulate the continuous passage of substances into and out of the cell through the membrane. Other proteins and carbohydrates serve as identification markers for the cell, attach the cell to other cells or neighboring structures, or serve as cell membrane *receptors*. Receptors are like antennae that receive messages by binding messenger molecules. The cell membrane may also engulf particles and take them into the cell, a process called *phagocytosis*.

Cytoplasm

A very soft gel called *cytoplasm* lies within the cell membrane. Cytoplasm is mostly water, and it contains a large quantity and variety of dissolved ions and molecules. Its gel-like consistency supports organelles, allowing them to function and interact properly. Cytoplasm also stores dissolved materials and granular substances. Finally, many chemical reactions occur in the cytoplasm. Some of them release energy, which may be transferred by ATP or other modified nucleotides to energy-consuming reactions and activities in the cell.

Endoplasmic Reticulum

Membranes similar to the cell membrane extend throughout much of the cytoplasm. These membranes, called *endoplasmic reticulum (ER)*, parti-

tion the cytoplasm much as walls, floors, and ceilings divide the inside of a building into rooms and corridors. The ER compartmentalizes the cytoplasm and regulates the movement of materials within it. *Smooth ER* also manufactures lipids (e.g., steroids). The other type of ER is called *rough ER* because its coating of granular structures makes it appear like rough sandpaper. The granules are called *attached ribosomes*, and they manufacture proteins that will be secreted from the cell. Other ribosomes, called *free ribosomes*, are suspended in the cytoplasm and manufacture proteins for use within the cell. Proteins destined for secretion are transported between layers of ER to a packaging area.

Golgi Apparatus

The protein-packaging area is an organelle called the *Golgi apparatus*, which consists of stacks of containers made of membranes arranged like stacks of flattened bags. Like grocery bags being packed at a checkout counter, Golgi apparatus containers are filled with proteins destined for secretion. The Golgi apparatus also manufactures carbohydrates, some of which combine with proteins as they are packaged. Filled Golgi containers are transported to the cell membrane, where, like bubbles, they burst open and release their contents from the cell.

Vacuoles

Some proteins in the cell are stored in droplets of fluid surrounded by membranes. Other manufactured materials, as well as dissolved substances or particles taken in by the cell membrane, may be stored in a similar way. Such storage containers are called *vacuoles*. Vacuoles with special functions may have other names. For example, vacuoles containing proteins that help digest materials (i.e., digestive enzymes) are called *lysosomes*, and vacuoles containing substances that destroy certain toxins are called *peroxisomes*.

Mitochondria

Each *mitochondrion* consists of a double layer of membrane enclosing a small amount of liquid (Fig. 2.19). *Mitochondria* are of various shapes and sizes.

Many of the numerous chemical reactions in

FIGURE 2.17 Cell structure.

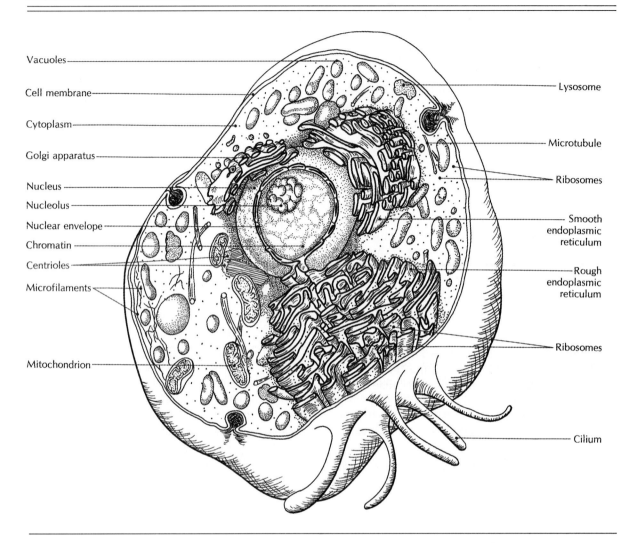

mitochondria convert one type of molecule to another. This activity helps provide a balance of molecules in the cell. Other related chemical re-actions release energy. Mitochondria that receive oxygen release much more energy than is released by the cytoplasm. However, as in the cytoplasm, most of the energy released in mitochondria is placed into ATP molecules for transfer to energy-consuming activities throughout the cell.

The energy in the ATP molecules originates in nutrient molecules in food. As the cell breaks down the molecules, carbon dioxide and other wastes are released. During these processes, elec-trons and hydrogen ions removed from the nutri-ents carry energy from the nutrients into the mi-tochondria. Then the electrons and ions are moved

by regulated mechanisms along and through the inner mitochondrial membrane. The mechanisms are called *electron transport* and *oxidative phos-phorylation*. At the end of these mechanisms, the electrons and hydrogen ions are combined with oxygen to form water while the energy they car-ried is used to make ATP.

A small percentage of the electrons and ions escape the ATP-producing mechanisms and com-bine with oxygen to form free radicals and ROS (e.g., $*O_2^-$, H_2O_2, $*OH$). From less than 1 percent to 5 percent of the oxygen used by mitochondria ends up in *FRs and ROS. The amount is deter-mined by several factors including the type of cell, chemical conditions in the cell, and the age and condition of the mitochondria. Damaged and old

FIGURE 2.18 Molecules composing the cell membrane. Some membrane regions also contain cholesterol, which affects membrane properties.

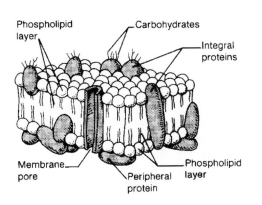

FIGURE 2.19 Mitochondrial structure. Mitochondrial DNA is in the matrix.

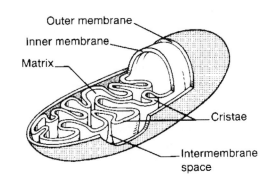

mitochondria produce more *FRs and ROS, though mitochondria contain antioxidants and enzymes to eliminate them. Some *FRs and ROS escape from the mitochondria and cause damage in other parts of the cell or the body. The *FRs and ROS also damage the mitochondria, especially their inner membrane and their DNA. Mitochondrial DNA damage is greatest in active non-dividing cells (e.g., heart, muscle, brain). Damaged mitochondria also produce less ATP. All these changes increase with age and may be a main cause of aging. (see Mitochondrial Theory and Mitochondrial DNA Theory, below).

Microtubules and Microfilaments

Besides membranous organelles, the cytoplasm has long thin organelles that consist mainly of protein. Those shaped like tubes are called *microtubules*, and those shaped like fibers are called *microfilaments*. Like tent poles and ropes, both types of organelles provide internal support for the cell, forming the *cytoskeleton* (Fig. 2.20). They also help the cell change shape and move, and help transport materials from place to place within the cell.

Nucleus

The cytoplasm and its organelles are separated from an inner region of the cell by a double layer of membrane called the *nuclear membrane*. This membrane and the materials it surrounds constitute the *nucleus*. The soft gel in the nucleus (i.e., *nucleoplasm*) resembles cytoplasm. The highly convoluted DNA molecules it contains have several names, including *chromosomes*, *chromatin*, *hereditary material*, and *genetic material*. The information encoded in the DNA directs the construction and activities of the cell. The portion of the DNA directing the production of ribosomes is called a *nucleolus*. The portion of the DNA at one end of each chromosome is called a *telomere*. The telomeres are of different lengths on different chromosomes.

Genetic Control

Human cells have 46 chromosomes. Like the sentences in each chapter of a lengthy instruction manual, each chromosome has thousands of instructions. When an instruction is to be carried out, the nucleus makes an RNA copy of the instruction contained in the DNA. Accurate transcription is achieved by complementary base pairing. The RNA copy, called *messenger RNA* (*mRNA*), may be edited in the nucleus before being transported to the cytoplasm. Using other RNA molecules in ribosomes and in the cytoplasm, the instruction in the mRNA directs the assembly of amino acids to form a chain with a specific length, ratio, and sequence. These characteristics help determine the final shape and functions of the amino acid chain. After twisting, bending, and possibly combining with other amino acid chains, the amino acid chain is a finished protein molecule.

The length of DNA used to direct the formation of an amino acid chain is called a *gene*. Only some genes are used at any given time. One way a cell can prevent a gene from operating is by winding the DNA for that gene tightly. Masses of tightly wound DNA are called *heterochromatin*. Other gene activity is controlled by other genes, by messenger substances, and by conditions in and around the cell.

Some protein molecules are called *structural proteins* because they become structural components of the cell. Other protein molecules (i.e., *enzymes*) control the production of non-protein substances and regulate cell activities. Therefore, by directing the manufacture of structural proteins and enzymes, DNA controls all the structures and functions of the cell.

Cell Division

If a cell continues enlarging, it must eventually divide. Otherwise, the amount of cytoplasm and organelles it contains will become too large to be adequately served by its cell membrane and nucleus.

DNA Duplication   In preparation for division, a growing cell makes a copy of its DNA (Fig. 2.21). Occasional errors occur during this process, but certain enzymes, acting like proofreaders, identify and correct nearly all these errors before duplication of the DNA is completed. It is noteworthy that similar enzymes can maintain the genes in an error-free condition by repairing the DNA if it is damaged afterward.

One error that is usually not corrected is the omission of part of each telomere. As a cell divides repeatedly during life, its telomeres become ever shorter in an age-related manner. Shortening of telomeres occurs at different rates among the chromosomes. Once the telomeres reach the minimum critical length, the cell is unable to divide again because it cannot make a complete copy of the remaining DNA, which contains essential genes. The reasons are not clear, but they may involve deleterious effects on chromatin structure or signals that inactivate genes needed for cell division.

Human telomeres consist of repeated segments made of six nucleotides (i.e., TTAGGG). Other animals and other organisms have different sequences and lengths in their telomere repeat units. Telomeres have diverse functions besides permitting complete replication of the essential DNA in a chromosome. Examples include attaching chromosomes to the nuclear membrane; preventing chromosomes from attaching to each other; protecting DNA from enzymatic attack; and influencing genetic activities.

Some cells can prevent shortening of their telomeres with each cell division. Examples include embryonic cells, sperm-producing cells, and cancer cells. Usually, such cells use the enzyme *telomerase* to rebuild the telomere during DNA replication. Some cells use mechanisms not requiring telomerase.

The effects of telomere shortening and telomerase on human aging are not yet known. Rapid telomere shortening is found in Werner's syndrome and in Down's syndrome, which are two syndromes that mimic rapid aging. Persons born with low birth weights who experience growth bursts (i.e., bursts of cell division) after birth, or people who experience growth spurts for any reason, may have especially short telomeres in their rapidly

FIGURE 2.20 The cytoskeleton.

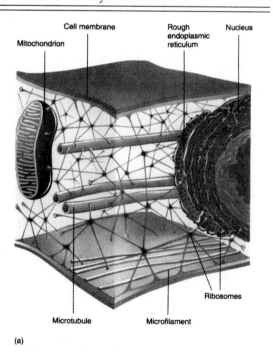

Cell membrane · Rough endoplasmic reticulum · Nucleus · Mitochondrion · Ribosomes · Microtubule · Microfilament

(a)

growing body parts because of the decline in telomerase after birth. Results for such individuals may include increased risk of high blood pressure because their kidneys may grow faster than normal or increased risk of atherosclerosis due to rapid cell proliferation in arteries injured by high blood pressure.

Maintaining or reinitiating telomerase production may be a main cause of cancer. If telomerase activity could be stopped in cancer cells, the cancer might be curable. Alternatively, if telomerase could be reactivated in specific cells where injury has occurred, better healing might result.

Mitosis   Once the cell has copied its DNA and has enough cytoplasm and organelles to divide, it partitions the DNA into two identical sets of genes. The cell then pinches itself into two cells, each of which contains one set of genes plus some of all the other cell components. This process, which results in cell reproduction, is called *mitosis* (Fig. 2.21). In the early phase of mitosis, the cell winds its duplicated DNA strands into tightly coiled chromosomes. (Fig. 2.22) At the end of mitosis, portions of each chromosome are unwound so many genes become active again.

Besides maintaining efficient cells, the growth and reproduction of cells allow for growth, replacement, and repair of parts of the body. It is through these processes that a single microscopic cell (a fertilized egg) develops into a full-grown person composed of trillions of cells. Cell growth and reproduction also allow for the replacement of skin cells and red blood cells, which are dying continuously, and for the healing of skin that has been cut or scraped.

Hayflick Limit   In 1961, Hayflick and Moorhead reported that cells removed from the dermis of human skin and grown in laboratory vessels divide a certain number of times, stop dividing, and gradually die. Since then, this characteristic has been observed in many cell types from humans and other animals. It is now commonly called the *Hayflick limit*. Cells in laboratory vessels that stop dividing and eventually die are said to undergo *replicative senescence* (*RS*).

The Hayflick limit shows three properties that seem to be related to aging. First, number of divisions is negatively correlated with the XL of the species from which the cells originate (e.g., mouse 10-15 divisions, humans 60 divisions, Galápagos

turtles 100 divisions or more). Second, the number of divisions is inversely proportional to the age of the person from whom the cells were taken. Third, the number of divisions is lower for cells from people who undergo changes like aging but at an abnormally young age (i.e., progeroid syndromes). Because of these and other factors, many scientists believed that the Hayflick limit was the key to age changes.

It has since become evident that aging does not result from a loss of ability of body cells to divide. Cells from even the oldest humans can divide 20 or more times when grown in laboratories. Also, many body cells lose their ability to divide in the process of differentiation (e.g., neurons, muscle), so loss of ability to divide does not equal old age or the approach of death of the cells. However, age-related changes occur in cells as they approach their Hayflick limit. Examples include enlargement, less motion, altered chromatin and nucleoli, and very short or absent telomeres. Therefore, aging occurs in cultured cells, and reaching the Hayflick limit is one indication that the cells are aging.

FIGURE 2.21 Cell life cycle leading to cell reproduction.

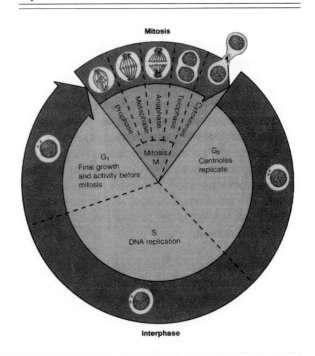

Scientists continue to debate the importance of the Hayflick limit to human aging. Some say it is an artificial phenomenon that occurs only in laboratory vessels. Others state that cells undergoing RS are different from cells that remain in the body. Also, the proportion of cells in the skin that have some features like RS cells is not related to the age of the person. Certain researchers believe that RS-like cells in the skin exist only in pathological conditions (e.g., arthritis, atherosclerosis, Werner's syndrome), not in normal aging.

Despite this controversy, the Hayflick limit and the process of replicative senescence has been studied intensively. Much research has focused on discovering how RS is controlled. Telomeres and telomerase may be an important key. Recent research showed that cells do not reach a Hayflick limit and RS when active genes for telomerase are placed into the cells. Many other factors, regulators, and possible mechanisms have been scrutinized and seem to be involved. As a result, this research has been helped by theories of aging and has contributed to further development and modification of these theories (see Biological Aging Theories, below). These pathways seem to be related to age changes in the body and to regulating the growth of cancer.

Neoplasia

A third age-related cell process involving cell division is continuous uncontrolled cell reproduction, called *neoplasia. Benign neoplasia* occurs when neoplastic cells remain in one mass. When neoplastic cells spread to other areas of the body, the disease is called *cancer*.

There are age-related increases in incidence rates and mortality rates for cancer. People over age 65 have 60 percent of all cases of cancer; have an incidence 11 times greater and have a mortality rate from cancer 15 times greater than those under age 65. The twelve leading types of cancer have a mean age at diagnosis of age 63 or greater.

For those over age 64, major cancers, in decreasing order of occurrence, are cancers of the prostate; colon; pancreas; urinary bladder; stomach; rectum; lungs and bronchi; leukemia; uterus; non-Hodgkin's lymphomas; breast; and ovary. Except for lung cancer, which has a peak mortality rate in the 75-79 age group, there is an age-related increase in mortality rates from most of these cancers. Rates of colon cancer are expected at least to double by the year 2030 due to the increase in the number and percentage of elders in the population.

Elders have greater problems from cancer because of having higher frequency and severity of coexisting diseases plus other age-related problems. Care and treatment for elderly cancer patients needs to be more individualized and modified based on age-related changes.

The rate of cancers among the elderly is increasing in numbers and as a percent of the elder population. This may be due in part to the decreasing rates of death from heart attacks, resulting in more people living long enough to develop cancer. Barring significant discoveries about cancer, or other

FIGURE 2.22 Chromosome structure during mitosis. Each side has a complete copy of the chromosome's genes. The two sides separate at the centromere during mitosis, and each new cell receives a side. The DNA in one telomere region is shown uncoiled, as it might exist outside of mitosis.

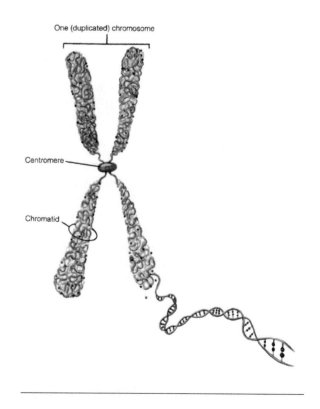

One (duplicated) chromosome

Centromere

Chromatid

major changes in society, these trends will continue and will become worse for the elderly and for the society as whole. Therefore, the need for research on cancer in the elderly continues to grow. Aging and cancer should be studied together because many factors believed to underlie aging also correlate positively with the onset and spread of cancer (e.g., genetic regulation and damage, *FRs, telomeres, and immune responses).

APOPTOSIS

Another cellular process whose study may increase the understanding of aging is deliberate programmed death of cells, called *apoptosis*. The name comes from the Greek meaning "to fall off." Apoptosis is helpful in removing unwanted cells or extra cells during development (e.g., unused neurons, webbing between fingers and toes); removing damaged cells; and balancing cell reproduction with cell removal to maintain homeostasis. It is a deliberate energy-requiring process. Apoptosis may be regulated by genes or damaged organelles within the cells, by signals from other cells, by environmental factors, or by a combination of such factors. Not all types of cells undergo apoptosis, and cells reach apoptosis at different rates and at different times (e.g., prenatal, postmenopause). The significance of apoptosis to aging is not known.

GENES AND AGING

Genes play several significant roles in aging. Many genes seem to influence the very different life spans among different species (e.g., flies, mice, humans). Of the estimated 100,000 genes in humans, scientists estimate that 70 to 7,000 of them may influence aging itself.

Genes related to aging help determine an organism's ML and XL. Evidence for this includes differences (e.g., weeks to decades) in ML and XL among different species; effects from selective breeding; effects from placing new genes into animals (i.e., transgenic animals); and effects from specific gene mutations. In humans, genetic impact on ML and XL is also shown by the similarity in life spans within families and the even greater similarity in life spans among twins. Genes also influence the onset and nature of age-related diseases plus the effects of lifestyle and environmental factors in aging processes. In ad-

dition, genes undergo significant accumulated alteration and damage during life, and portions of genes move from place to place within cells (i.e., transposable elements). These two alterations also seem to influence aging and diseases.

Though genes have a major role in aging and age-related diseases, scientists estimate that only 35 percent of the variance among human life spans can be accounted for by variation among genes. For identical twins, the figure is 40-70 percent. The same seems true for many other animals. The remaining variation between ML and age-related diseases is due to lifestyle, environmental factors, and occasional incidents (e.g., accidents). For example, life spans and age-related diseases are not identical for identical twins, who have identical genes. For identical twins, the more differences between their lifestyles, the greater the differences in aging and genetically caused age-related diseases (e.g., Alzheimer's disease). Finally, the impact of genes on ML and age-related diseases decreases with advancing age, especially at very advanced ages.

Nature and Nurture   Genes influence how the environment affects aging, and the environment influences how genes affect aging. Environmental factors may have different effects on genes at different times because age changes make genes more or less sensitive as time passes. Environmental factors include external influences and internal ones (e.g., hormones). Genes influencing age-related diseases may also be affected by environmental factors, and these effects may be different at different ages (e.g., pre- or postmenopause). Human ML, health in later years, and perhaps even XL may be increased by avoiding harmful lifestyle and environment factors while increasing beneficial ones.

Some disease-promoting genes become active or important only at older ages (e.g., Alzheimer's disease, certain cancers). These genes may be time-dependent or may be triggered by lifestyle or environmental factors. When human ML was low, few people got these diseases. Since human ML is increasing, these genes will have a growing impact. Some unknown or unnoticed genetically-induced diseases may become significant. The effect may be to put additional limitations on the increase in ML. Alternatively, since mortality rates decrease after age 100, genes that promote high ML may start to overpower such late-acting detrimental genes.

Methods of study One way to identify genes affecting human aging involves finding similar genes in animals. Popular animals for such studies include a small worm found in soil (*C. elegans*), fruit flies (*D. melanogaster*), and laboratory mice. Diverse techniques are used to study the genetics of aging in these and other animals. Examples include selective breeding; placing genes from one animal into another (i.e., transgenic research); and mutating genes.

The genes that control normal aging and human XL are still unknown, though genes affecting the immune system (e.g., MHC genes), blood pressure regulators (e.g., ACE), and brain neurons (e.g., APOE) are good candidates. Genes that influence aging, ML and XL in some nonhuman species (e.g., *C. elegans*, *D. melanogaster*) have been identified.

Age-related Abnormalities Genes that influence or cause age-related human disease have been studied in detail (e.g., cancer, Alzheimer's disease). Controversy exists about whether such genes should be viewed as normal variations of age-related genes or as abnormal disease conditions. Among these are genetic conditions that promote *progeroid syndromes*. People with these syndromes show changes that resemble aging but that occur at much earlier ages than normal. Examples include Down's syndrome and Werner's syndrome.

Down's syndrome occurs in people having an extra chromosome 23. The extra chromosome is present because of abnormal formation of an egg cell in the ovary. Progeroid symptoms include more rapid shortening of telomeres; shorter Hayflick limits in skin cells; increased risk of brain changes resembling Alzheimer's disease; and a shorter ML.

Werner's syndrome (WS) is caused by loss of a region of chromosome 8. The condition develops only if a person has the mutation on both copies of chromosome 8 (i.e., autosomal recessive mutation). People with only one mutation will not show the disease, but they can transmit the disease to their children.

Indications of WS appear during adolescence. Manifestations include slow DNA synthesis; rapid telomere shortening; abnormal chromosome structure; increased DNA damage from *FRs; lower Hayflick limits in skin cells; premature graying (age 20); hair thinning (age 25); thinning skin; premature skin aging; atrophy of the subcutaneous tissues of the limbs; loss of fat from arms and legs; calcification of soft tissues and blood vessels; heart disease; muscle thinning; osteoporosis; cataracts (age 30); diabetes (age 34); skin ulcers (age 33); increased cancers; aging voice (age 27); death (age 30-50).

INTERCELLULAR MATERIALS

Since higher levels in the hierarchy of body structure are composed both of cells and of the materials they secrete, we will now examine the more abundant of these materials. These substances are found between cells and are therefore called *intercellular materials*. Some intercellular materials are amorphous (i.e., lack organized structure) and contain proteins or carbohydrate/protein complexes dissolved in water; others are fibers. In many body structures the substances between cells contain a mixture of amorphous materials and fibers.

Amorphous Materials

Amorphous materials vary in consistency depending on the amount and types of materials present in the water. For example, the intercellular material in blood (plasma) is a liquid because it contains approximately 90 percent water and its other molecules are not tightly bound together. Other amorphous intercellular materials are soft slippery gels because they contain more protein. Much of this type of material is under the skin. Other amorphous materials, such as that in cartilage, are firm gels containing much mucopolysaccharide. Finally, amorphous materials that contain much mineral, such as in bone, may be hard and rigid.

Fibers

Collagen Fibers Many fibers in intercellular materials are made of a protein called *collagen*, the most abundant protein in the body. The molecules in a collagen fiber are aligned parallel to each other and are twisted and bound together. The resulting fiber is thick, flexible, and strong and stretches little when pulled. Therefore, a collagen fiber is like a string, rope, or cable. Woven mats of collagen fibers can form tough sheets (e.g., flat tendons), while bundles of collagen fibers can form strong cable-like connectors (e.g., ligaments).

Elastin Fibers Another common type of fiber is made of a protein called *elastin*. As in a collagen fiber, the molecules in an elastin fiber are aligned and twisted together. Though an elastin fiber is flexible, the nature and arrangement of the elastin molecules produce a fiber that stretches easily when pulled. The fiber also snaps back to its original length when the tension is released. Therefore, an elastin fiber resembles a rubber band. The ability to be stretched and snap back is called *elasticity*. Structures containing many elastin fibers are often resilient (e.g., outer ear, dermis of skin).

BIOLOGICAL AGING THEORIES

Reasons for Theories of Aging

The theories of aging are general statements proposed to summarize and explain some observations about aging. The theories are tested by additional research, after which they are modified to include the new information. While each theory may be valid for some observations about aging, none of them explain completely all aspects of aging. The theories are actually hypotheses in the scientific method of inquiry. As with all hypotheses, they are valuable in giving broad logical perspectives to diverse bits of information; giving direction to additional research; and helping to develop practical applications of knowledge.

General Characteristics of the Theories

Developing good theories of aging is difficult for diverse reasons. Examples include the lack of agreement on a definition of aging; the multitude of possible causes of aging; the complexity of aging within one organism or species of organisms; the diverse ways aging reveals itself in different species; the interactions among aging processes, environmental influences and diseases (i.e., nature versus nurture); the limited information about aging processes; the diverse expertise or perspectives among people who develop the theories; and societal, cultural, and economic conditions. As a result, dozens of theories of aging have been proposed over the passed 100 years. Their diversity, complexity, and interconnections can be intimidating and confusing. While realizing this, we will look at some of them. First we will examine broad groups of theories, and then look at more specialized ones.

One group of theories, the *evolutionary theories*, attempts to explain evolutionary aspects of how and why aging exists in living things. Another group, the *physiological theories*, focuses on how and why aging occurs within present day animals. They explain structural and functional age changes in animal bodies. Physiological theories may concentrate on one aspect or one structural level of an animal. Examples include genes and genetic mechanisms (e.g., senescence genes); molecules and their chemical reactions (e.g., glycation); activities of cell organelles or entire cells (e.g., mitochondria, cell division); signaling among cells (e.g., interleukins); whole body regulatory and control systems (e.g., immune system, nervous system, endocrine system); or behavioral and psychological characteristics. Some physiological theories attempt to explain all age changes based on only one or a very few phenomena (e.g., free radical theory, cross-linkage theory). Others use a combination of factors (e.g., immuno-neuro-endocrine theory).

A dichotomy between the physiological theories separates *programmed theories* from *stochastic theories*. All programmed theories maintain that aging occurs because of intrinsic timing mechanisms and signals. Stochastic theories (i.e., probability theories) maintain that aging occurs by accidental chance events. Some theories include both aspects. For example, one theory proposes that aging occurs because timed programmed genetic signals make an organism more susceptible to accidental events. Another dichotomy between physiological theories separates universal theories of aging, which apply to all organisms, from theories that apply to only one type of organism (e.g., mammals, humans).

Theories of aging are difficult to categorize because many of them overlap. As examples, evolutionary theories may incorporate aspects of genetics and behavior; mitochondrial theories may incorporate aspects of free radicals, energy metabolism, membranes, and mitochondrial genetics. Scientists who focus on the interrelatedness of body structures and functions have proposed *network theories*, which combine physiological theories from the molecular to the system level of the body.

Here are some popular broad categories of aging theories. Some of them are discussed next. Genetic theories may emphasize nuclear gene mutations or damage; mtDNA mutations or damage; programmed aging genes; senescence genes; decreased DNA repair; error catastrophe; or dysdifferentiation. Altered molecule theories may emphasize damaged and abnormal proteins; cross-linkage; glycation; waste accumulation; or general molecular wear and tear. Free radical theories may emphasize free radical formation or free radical defense mechanisms. Cell and cell organelle theories may emphasize the cell membrane; mitochondria; telomeres; or heterochromatin. Signaling theories may emphasize intercellular signals or individual hormones. System theories may emphasize the nervous system, the endocrine system, the immune system, or combinations such as the immuno-neuro-endocrine theory.

Evolutionary Theories

Aging must have evolved because not all species and not all cells show aging. To understand evolutionary theories of aging, one must be familiar with the theory of evolution of living things. The theory of evolution states that early in the Earth's history there were no living things. Over millions of years and through chance chemical reactions, larger and more complex molecules were formed. Through continued chance events, some of these molecules grouped together into organized clusters that could reproduce. These were the first cells. Some of these cells developed cooperative interactions as they formed colonies. Over many generations, these multicellular colonies evolved into today's plants and animals.

As time passed, some molecules, cells, and organisms had characteristics making them better able to survive environmental problems and competition from others. They were able to produce more offspring with similar characteristics. The instructions to produce these characteristics were contained in their genes. These successful well-adapted offspring were able to produce the next generation, and so forth. Molecules, cells, and organisms with characteristics that made them less able to survive and reproduce became less common and, finally, extinct because their genes were not sustained through continuing generations. This is the process of natural selection.

During evolution by natural selection, environmental conditions and interactions among living things changed. At the same time, chance events caused alterations in genes such as mutations and recombinations. These alterations led to organisms with different characteristics. Natural selection allowed organisms with beneficial genetic changes to reproduce more successfully, so their genetic characteristics became more common. Genes in organisms that allowed less reproduction became less common but remained among the organisms because as generations passed, there were fewer of these organisms. Genes that allowed little or no reproduction gradually decreased and disappeared. These genes were not passed to future generations. Since these processes occurred in different environments (e.g., hot, cold; dry, wet; light, dark), different genes and types of organisms survived in different places.

Thus, chance events, genetic changes, and natural selection produced the great variety of living things present today. These processes are still happening. For example, selective breeding produces new varieties of plants and animals, and altering the environment causes extinction of species. Natural events and people are part of evolution by natural selection.

One evolutionary theory of aging is based on the ***disposable body theory***. Once an organism has reproduced successful offspring that can eventually reproduce, its body is no longer needed, and it ages and dies. Here is why. Resources are limited. An organism must partition its resources between survival of its body and reproduction. Resources for survival are used for defense and repair activities, which would also slow or prevent detrimental aging. If an organism's genes allocate most of its resources toward its own survival, and thus limit or prevent aging, the organism could not allocate enough resources to reproduce. Its genes would not survive by natural selection. If an organism's genes allocate inadequate resources to survive, it would not live long enough to produce ample successful offspring. Its genes would not survive by natural selection. An organism whose genes allocate ample resources to survive long enough to produce many successful offspring would have its genes survive by natural selection. However, because genes limited the resources allocated to defense and repair mechanisms, these mechanisms begin to fail, aging occurs, and the organism dies. In this theory, genes do not cause aging, they just do not prevent it after successful reproduction.

A second evolutionary theory, the ***antagonistic pleiotropy theory***, states that effects from certain genes may be beneficial early in life but detrimental later in life. These detrimental effects result in aging. For example, certain genes may promote rapid metabolism leading to rapid successful reproduction early in life. However, rapid metabolism may also cause damage to body molecules. The damaged molecules may accumulate. The continued accumulation of damaged molecules causes aging. According to this theory, genes actively cause aging.

A third evolutionary theory of aging, the *accumulation of late-acting error theory*, states that natural selection has permitted the evolution of genes that cause aging. During evolution, any genetic changes that caused detrimental changes prior to successful reproduction would be eliminated by natural selection. However, new detrimental genetic changes that do not cause harm until after adequate successful reproduction would not be eliminated by natural selection. Over evolutionary time, these late-acting genes have accumulated. The detrimental effects from these genes are what we know as aging. This theory also concludes that genes actively cause aging.

Other evolutionary theories of aging try to explain why not all species seem to have a maximum longevity, why there are such diverse maximum longevities among species within the same group (e.g., among mammals), and if aging is advantageous in any way.

Physiological Theories

Genetic Theories Genetic theories suggest that aging is heavily influenced by or actually caused by genes (see Genes and Aging, pg. 39). Finally, though certain genes that influence longevities and aging in research animals have been identified, no one has identified aging genes in humans (see Genes and Aging, pg. 39).

Genetic Timers One genetic theory of aging states that the genes are used in a specific sequence, much as a book would be read from the first sentence of the first chapter through the last sentence of the final chapter. Each section of this *genetic biography* would direct the body's activities during a specific stage of life including fetal development and maturation. The sections would also contain instructions on how to progress to the next stage. The sections for later stages in the individual's life would provide instructions on how to carry the changes called biological aging. The person's life ends when aging limits adaptation enough that homeostasis cannot be maintained and death occurs. Another genetic biography theory, the *antagonistic pleiotropy theory*, was described with evolutionary theories of aging

A modified version of the genetic biography theory, the *genetic clock theory*, suggests that some genes keep track of the body's progress and perhaps the passage of time or number of cell divisions. In this way, genes can control the age at which certain events occur. For example, when grown experimentally, some types of cells can reproduce only a certain number of times, after which they die. Furthermore, the number of times the cells can divide decreases as the age of the person from whom the cells were extracted increases.

Another modified version of the genetic biography theory, the *death gene theory*, states that the last chapter in the genetic instruction manual contains genes called *death genes* that tell the body to deteriorate and die. Some scientists consider genes that cause fatal age-related diseases to be death genes.

One way cells seem to keep track of their age is through shortening of their telomeres as they divide. The *telomere theory* states that shortening of the telomeres alters the expression of other genes, perhaps those closest to the telomeres. This might happen if heterochromatin near the telomere unwinds, allowing detrimental genes to become active. Different rates of aging could occur in different cells or parts of the body because the telomeres in some cells shorten faster than those in other cells. The *heterochromatin loss theory* suggests that unwinding of chromosomes happens at many areas in a cell. This activates detrimental genes that cause aging.

Limited Gene Usage Other genetic theories of aging suggest that genes are used over and over during adult life rather than being used in a specific sequence. One of these theories, the *limited gene usage theory*, suggests that there is a limited number of times that the instructions in genes can be read. The reading somehow alters or damages the genes. After many years of being read and reread, some genes may become unreadable so that instructions are lost. Other genes may be read poorly, resulting in mistakes by the body. In either case, the results are the detrimental changes that are biological aging. A sufficient number of these changes weaken the body so much that it can no longer maintain itself, and death occurs.

There are two suggested explanations why genes can be read only a limited number of times. One explanation, the **somatic mutation theory**, proposes that harmful factors injure the genes. Possible environmental factors include radiation, toxic chemicals, and free radicals. Within the cell, genetic disruption can occur when movable parts of the genetic material, called transposable elements, shift positions. In humans and other organisms, transposable elements move between the mitochondrial DNA and the nuclear DNA. The other explanation for limited gene usage, the **faulty DNA repair theory**, states that though genes are being damaged throughout life, cells also have mechanisms to repair the damage as quickly as it occurs. At first, then, damage to the genes has little effect. After many years, though, the repair mechanisms begin to fail. With either of these explanations, the result is the same. The damage that accumulates over the years reduces the genes' ability to properly direct the body's activities, and biological age changes begin to occur.

Error Catastrophe Theory   A related theory, the **error catastrophe theory**, states that the damage is not to the genes themselves but to the RNA and protein molecules that read the genes and carry out their instructions. These damaged molecules spread increasing numbers of mistakes throughout the cell and the body, causing biological age changes.

Rate of Living Theory   The **rate of living theory** of aging states that aging is determined by the rate of metabolism because metabolism causes damage. The higher the rate of metabolism, the faster the rate of aging and the shorter the mean and maximum longevity. The rate of metabolism in animals can be measured by the rate of oxygen use. This theory proposes that an animal can use only a certain amount of oxygen per unit of body mass in a lifetime. The animal can use the oxygen quickly and have rapid aging and a short life, or use it slowly and have slow aging and a long life.

Most animals follow this rule. Two major exceptions are mammals and birds, which have life spans longer than their rates of metabolism would predict. Proposed explanations for these discrepancies suggest that birds and mammals have more

efficient metabolism resulting in less damage or that these animals have better repair mechanisms.

Free Radical Theory   In 1956, Harman proposed the *free radical theory* following research on how radiation causes damage to organisms. He used the research on free radicals from radiation to include other sources and effects of free radicals, including aging. The theory states that free radical damage is a main reason or the main reason for true aging and for age-related diseases. No one knows which, if any, of the many types of free radicals are more important in promoting aging.

This theory is now based on several observations. Main examples include positive correlations between the following; metabolic rate and free radical production; age and rate of free radical formation; age and amount of free radical damage; free radicals and many age-related diseases (e.g., atherosclerosis, heart attacks, strokes, Alzheimer's disease, parkinsonism, cataracts, renal failure, and certain cancers); and, sometimes, mean longevity and antioxidant supplements. Other examples supporting the free radical theory include negative correlations between longevity and free radical production; and between age and free radical defenses. Thus, the free radical theory of aging became the most likely explanation for the former rate of living theory.

Free radicals seem to contribute to aging and age-related disease primarily by damaging DNA, proteins and lipids. The exact effects and the relative importance of effects from free radicals on DNA, proteins, and lipids are not known, though the general effects seem to be aging and an increase in certain age-related diseases. For example, damage to DNA slows DNA production for cell reproduction; adversely affects cell processes; and promotes cancer. Damage to proteins disturbs and distorts much bodily structure; reduces enzyme activity; makes proteins more susceptible to enzymatic destruction; promotes inflammation; and upsets signaling and control mechanisms for homeostasis. Damage to lipids reduces the effectiveness of cellular membranes to regulate the movement of substances; reduces energy production by mitochondria; promotes atherosclerosis and blood clotting; and promotes additional free radical production.

In spite of all the body's efforts, free radical production and damage from *FRs occurs. Further, their rate of production, rate of causing damage, and amount of damage all increase with age. This free radical damage adversely affects a variety of essential bodily components and alters body functions. Many scientists believe that years of such damage are a main cause, if not the most important cause, of what we know as biological aging.

Mitochondrial Theory The *mitochondrial theory* was proposed by Ozawa and colleagues in 1989. This theory states that mitochondrial activities and damage to mitochondria cause aging. The theory developed from the free radical theory when scientists combined numerous discoveries. As examples, mitochondria are main sources of free radicals; mitochondria are severely damaged by free radicals; mitochondria are easily affected by harmful environmental agents (e.g., radiation, pollutants, medications); damaged mitochondria accumulate with aging; cells that do not divide (e.g., muscle, neurons) or that divide slowly (e.g., bone, liver) accumulate many damaged mitochondria; damaged mitochondria produce greater amounts of free radicals; mitochondria release substances that produce free radicals in other parts of the cell, other cells, and the blood; damaged mitochondria have less ability to regulate signaling substances (e.g., calcium more susceptible to free radical damage than is nuclear DNA; mitochondria cannot repair their DNA; unlike nuclear DNA, cells use virtually all their mtDNA, so any mtDNA damage causes problems; mitochondrial transposable elements, including damaged mtDNA, move to the nucleus; and certain mitochondrial diseases promote specific age-related diseases (e.g., atherosclerosis, types of neuron degeneration).

Mitochondrial DNA Theory Scientists who focus more on genetic mechanisms of aging have also focused on the age-related changes in mtDNA to develop the *mitochondrial DNA theory*. According to this theory, mtDNA damage occurs much faster than does damage to nuclear DNA. mtDNA sustains damage 10 to 20 times faster than does nuclear DNA because mtDNA is not protected by proteins; it is attached to the inner mitochondrial membrane, where most free radicals are produced; and it cannot repair itself. Furthermore, damaged mtDNA accumulates in cells because damaged mitochondria replicate faster than undamaged mitochondria; mitochondria that rep-

licate retain damaged mtDNA; mitochondria with damaged mtDNA are eliminated slower than normal mitochondria; and non-dividing cells or slowly dividing cells accumulate high percentages of damaged mitochondria. Some scientists believe that the XL for humans is approximately 130 years because by that age, all mtDNA in the body would have some type of damage from *FRs. Death would result soon after that from mitochondrial failure.

The damaged mtDNA leads to more age changes than does damage to nuclear DNA because each cell uses almost all its mtDNA genes while using only approximately 7 percent of its nuclear DNA genes. Thus, nearly any adverse change in mtDNA will have adverse effects on the mitochondria. These effects include less energy production; more free radical formation; reduced control of other cell processes; and accumulation of damaged harmful molecules. These changes lead to aging and certain age-related diseases.

Clinker Theories Potentially harmful substances are known to accumulate in the body over a period of years. Because these materials interfere with the body passively, theories that claim that they cause aging are called *clinker theories*. A material first proposed to cause aging is *lipofuscin*. It is a mixture of chemical waste products from normal cell activities, including those in mitochondria. As time passes, lipofuscin becomes more concentrated inside cells, such as those of the heart and the brain, because the cells cannot effectively eliminate it. When cells have accumulated a great deal of lipofuscin, they appear darker in color. Because of the gradual darkening, lipofuscin has been called *age pigment*. Though lipofuscin now seems unimportant in aging, some people believe that it contributes to aging by interfering with cell activities.

Other materials that accumulate and seem to get in the way include a protein called *amyloid*. It is found between cells in the heart, the brain, and other organs. In some disease conditions, called amyloidosis, it becomes excessive and severely interferes with the operation of these organs. For example, amyloid is found in great abundance between the nerve cells in the brains of people with Alzheimer's disease. Accumulation of glucose has also been implicated as causing age changes. It binds to molecules (e.g., collagen, hemoglobin), causing them to stick closer together, restricting their movements, and altering their functions. Collagen with many glucose cross-links also becomes darker and contributes to age pigments.

Cross-linkage Theories   The fact that collagen molecules and other chemicals in the body become linked together as time passes has led to another group of theories of aging, the *cross-linkage theories*. Free radicals, glucose, and even light seem to promote the formation of bonds between molecules. The cross-linkage theories maintain that such bonding reduces the movement of molecules for chemical reactions, the movement of materials through the body, and the movement of body parts. The result is malfunctioning and aging. In addition, when glucose cross-links proteins by glycation, free radicals are produced. These free radicals may also contribute to aging. The *glycation theory* proposes that most aging is caused by glycation and the resulting free radicals.

Hormone Theories   Some research provides evidence that aging is caused by *hormones*. Hormones are chemical messengers produced in the body by structures called endocrine glands. Hormones are carried by the blood and give instructions to almost all body cells.

Some hormone theories attribute human aging to only one hormone. An example is the *insulin theory*. It proposes that when cells are subjected to high levels of insulin, they become less sensitive to insulin. Also, elevated levels of insulin reduce the production of growth hormone (GH) from the pituitary gland. Combining the proposed effects of high insulin and low GH resulted in the *insulin/growth factor imbalance theory*. It proposes that aging results from excess stimulation of growth by insulin and other growth-promoting substances including GH and glucocorticoids. The results are faster cell reproduction, larger cell size, larger body size, and the decline in physiological reserve that marks aging. This theory is reflected in the disposable body theory.

The *glucocorticoid theory* focuses on glucocorticoid hormones, which are secreted by the adrenal glands. It states that if glucocorticoid levels are slightly elevated frequently or steadily, aging is inhibited because such levels keep the body's adaptive mechanisms at peak performance. The slightly elevated levels of glucocorticoids also prevent damage from excess inflammatory responses or immune responses. Aging results from improper levels of glucocorticoids. When levels are low, adaptive mechanisms become inadequate. When levels are high, damage results from excess suppression of defense mechanisms (e.g., inflammation, immune function) and stimulation of high blood glucose levels. In addition, high levels of glucocorticoids cause damage to certain areas in the brain. Since mild stress promotes slightly elevated levels of glucocorticoids, proponents of this theory suggest that life expectancy and perhaps quality of life are maximized in people with mildly stressful lives. In other words, having realistic challenges throughout life is beneficial.

Another hormone theory, the *reproductive hormone theory*, states that aging results when reproductive hormone levels decline after reproductive years. With lower sex hormones, genes receive inadequate or detrimental signals. The result is declining production of desirable proteins and excess production of deleterious proteins, leading to aging.

More complex hormone theories combine interactions and effects from insulin, GH, glucocorticoids, melatonin, and other hormones.

Calcium Theory   Some scientists have used portions of several theories to develop the *calcium theory*. This theory states that abnormal concentrations and movements of calcium occur from several factors including free radical damage to membranes in cells; inadequate energy supply from damaged mitochondria; accumulations of amyloid protein; and elevated levels of glucocorticoids. Since many of the body's regulatory mechanisms rely on calcium as a signaling substance, abnormal levels of calcium lead to cell malfunctions and inadequate regulation of adaptive mechanisms. Examples include the functions of many enzymes, muscle cells, nerve cells, and blood vessels. The result is aging.

Immune System Theories   Other theories of aging focus on the immune system. The *immune system* consists of cells found in many parts of the body. Some of the cells are grouped in structures like the lymph nodes. The lymph nodes may become noticeable when an ill person has "swollen glands." Other immune system cells are concentrated in the outer layer (i.e., epidermis) of the skin. Many immune system cells are carried from place to place in the body by the blood and the lymphatic fluid. Most of these mobilized cells are lymphocytes.

The immune system is one of the major defense systems in the body. This system operates by identifying many large molecules and all cells in the body. Those identified as normal parts of the body are left undisturbed. However, anything identified as not belonging in the body is attacked and destroyed by the immune cells.

One *immune theory* of aging is like the error catastrophe theory in that it focuses on making mistakes. This immune theory states that as a person gets older, the ability of the immune system to distinguish normal from foreign materials weakens. The immune cells begin to attack and destroy important bodily components, thereby causing changes associated with aging. An example of such changes would be inflammation of the joints (i.e., arthritis). Because the body's immune system is attacking the body itself, this theory is called the *autoimmune theory*.

Other types of immune theories are the *immune deficiency theories*. One version resembles the disposable body theory. It states that an immune system with indefinite capabilities never evolved because such ability is not needed for reproductive success. The other version resembles the limited gene usage theory. It states that the immune system becomes weaker as it is used. With either theory, after many years the immune system is not able to defend the body against foreign molecules and microbes. These noxious agents are allowed to injure the cells of the body and to disrupt their functioning. Detrimental age changes are the result.

Both types of immune theories have been combined into a more unified *immune dysregulation theory*. It states that both changes in the immune system occur and cause aging because regulating signals among immune system functions become disproportionate. Additional age changes occur because of imbalanced signals sent by the immune system to other cells.

Wear and Tear Theory The *wear and tear theory* suggests that aging is nothing more than the accumulation of injuries and damage to parts of the body. Use, accidents, disease, radiation, toxins, and other detrimental factors adversely affect parts of the body randomly. The result of years of such abuse is aging. This theory was once quite popular, but it has fallen into disrepute. A major reason for its demise is it cannot account for the rather regular and universal nature of biological age changes in humans. However, as shown above, more focused forms of this theory have appeared in stochastic theories, such as the somatic mutation theory, the free radical theory, and the cross-linkage theory.

Network Theories Many scientists believe that aging results from combinations of phenomena like those in the above theories. Scientists who believe that there are interactions among these phenomena have developed *network theories*. Sometimes, these interactions seem to interact in a positive feedback fashion, producing an expanding spiral of damage and leading to aging. For example, free radicals from mitochondria damage the mitochondria, cause somatic mutations, cause leakage of calcium from the mitochondria, and promote glycation. These changes reduce the production and effectiveness of enzymes that remove free radicals and that repair molecules damaged by free radicals. Also, glycation increases free radical production. With more free radicals and less free radical defenses, the rate of damage to mitochondria increases, leading to faster free radical production, and so forth. A network theory for human aging may include all these phenomena plus their effects on the immune, nervous, and endocrine systems. Damage to these systems leads to disruption in regulatory and defense systems needed for homeostasis. Some theories of human aging may also include social and cultural factors. Aging is a result of degeneration or breakdown in proper integration and regulation among all these levels.

In conclusion, the number and diversity of theories attempting to explain biological aging can be confusing. Each theory is supported by some evidence, and each can explain certain aspects of biological aging. However, none tells the complete story. This may be because aging results from a combination of causes. Some causes may be more important than others, and some may act at different times. Others may affect different parts of the body or may be effects rather than causes. It is also possible that none of the theories are correct.

However, it is still important to formulate, test, and revise theories if the cause or causes of aging are to be discovered. Once they are known, influencing the processes in biological aging might also be possible. It might also be possible to identify undesirable but not inevitable changes that frequently occur along with aging and to direct more attention to them. Much progress has already been made in this direction. Many diseases that are not part of aging but that are associated with aging provide excellent examples. Some of the more common ones include heart attack, stroke, osteoporosis, emphysema, and cancer. Numerous oth-

ers will be mentioned in the following chapters. Through good nutrition, exercise, timely health care, and avoidance of the risk factors for these diseases, many cases can be prevented, improved, or at least have their progress slowed.

We now turn to a detailed examination of biological aging in humans. We will start on the outside of the body and examine the integumentary system.

3

INTEGUMENTARY SYSTEM

The *integumentary system* is made up of the *skin* and the *subcutaneous layer* that underlies it. This system makes up most of the external surface of the body. Because of its position, the integumentary system is always in direct contact with the external environment and lies between it and the internal environment of the body. Therefore, this system plays a major biological role in maintaining a person's homeostasis and thus that person's happy and healthy survival. The major functions of the integumentary system are serving as a barrier between the body and its surroundings, providing information about the external environment, regulating body temperature, and starting the process of vitamin D production. In addition, because it is a highly visible system, it frequently affects the social, psychological, and economic aspects of a person's life.

MAIN FUNCTIONS FOR HOMEOSTASIS

Serving as a Barrier

The integumentary system must provide a barrier against several types of factors because unrestricted entry or exit of these factors can cause a deviation from homeostasis.

Microorganisms One of these factors consists of *microorganisms* (bacteria and viruses) that cause infections.

Chemicals A second factor is chemicals. Many chemicals, such as vitamins, are within the body because they are needed by the cells. They must not be allowed to dissipate into the external environment. By contrast, noxious and toxic substances in the external environment must not be allowed to come into contact with the body's delicate living cells.

Water Though also a chemical, *water* is considered a third factor because it is an especially important substance and is present in such large quantities. Its movement between the interior of the body and the external environment must be restricted. The cells of the body must have enough water to maintain their shape, size, and activities but not so much that they swell or burst.

Light   *Light* is essential to healthy survival because it provides energy needed by the integumentary system to produce vitamin D. Still, only a small amount of light should be allowed to penetrate the body because the energy introduced by excess light damages several of the most important large molecules, including proteins, RNA, and DNA. The outcome is obvious to anyone who has had a sunburn.

Trauma   The fifth factor to which the integumentary system is a barrier is *trauma*. Direct contact of the cells with rough or sharp materials in the external environment immediately cuts the cells open. The result is the death of those cells and the portion of the body they constitute. Pressure or sharp blows can also injure or kill cells.

Providing Information

Gathering information about conditions in and around the body is essential for survival. It is the first step in negative feedback systems, which help maintain homeostasis.

Since the integumentary system is located between most cells of the body and the external environment, it is able to supply information about factors that might alter the internal conditions of the body even before those factors have an opportunity to do so. Many of the abundant nerve cells in the integument continuously monitor the external environment and send messages to other parts of the nervous system. As a result, the person knows much about the area surrounding the body, including the location, size, shape, texture, movement, and temperature of objects and materials (e.g., clothes, furniture, water, air). Then the person can take steps to avoid or correct any threatening features, perhaps before harm is done.

Temperature Regulation

The maintenance of a proper and relatively stable temperature within the body is essential not only for comfort and satisfactory performance but also for sustaining life. The body must be sufficiently warm that its chemical reactions proceed quickly enough. Cooling an individual's body can slow chemical reactions so much that that person will die of hypothermia. By contrast, if the temperature rises too high, certain types of molecules (e.g., proteins) in and around the cells will be damaged

or die. If it is extensive enough, this damage can result in the death of the individual.

Vitamin D Production

Vitamin D is necessary for healthy survival because it allows the intestines to absorb calcium from food. The calcium is then used in numerous vital ways.

Clearly, the integumentary system performs a wide variety of necessary jobs. To carry them out, it has a multitude of component structures and materials organized to operate efficiently and effectively. This chapter will examine the contributions of each of those components, paying particular attention to those known to change with the passage of time.

Where possible, changes occurring because of aging will be distinguished from those caused by environmental factors or disease. However, these distinctions often cannot be made. For one thing, the skin is subjected to many environmental influences over long periods of time, including sunlight and physical wear. Also, different parts of the body receive vastly different doses of these influences; perhaps the best example is differences in exposure to sunlight. Finally, there is tremendous disparity among the lifestyles of individuals. For example, the skin of a farmer or a fisher receives very different treatment from that of a person who works in an office.

THE EPIDERMIS

Keratin and Keratinocytes

The *epidermis* is the outermost layer of the skin (Figs. 3.1 and 3.2). It is composed mostly of a thin sheet built up from many layers of flat cells produced continuously in the deepest region of the epidermis. As they accumulate, many of these cells are pushed upward toward the outside of the body. As they move farther away from the favorable internal environment of the body, they begin to produce a protein called *keratin*. Therefore, these cells are called *keratinocytes*. Finally, these cells are moved so close to the air that they die, leaving behind their keratin as the outermost layer of the epidermis, the *stratum corneum*. Starting with the formation of a new cell in the deep region of the epidermis, this process takes about a month.

FIGURE 3.1 Components of the integumentary system.

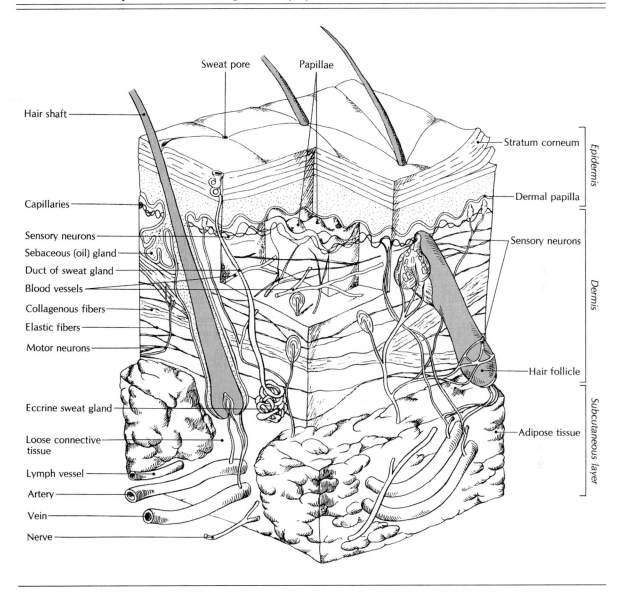

Sweat pore   Papillae

Hair shaft

Stratum corneum

Epidermis

Dermal papilla

Capillaries

Sensory neurons
Sebaceous (oil) gland
Duct of sweat gland
Blood vessels
Collagenous fibers
Elastic fibers
Motor neurons

Sensory neurons

Dermis

Hair follicle

Eccrine sweat gland

Loose connective tissue

Lymph vessel

Artery

Vein

Nerve

Adipose tissue

Subcutaneous layer

Keratinocytes produce several substances that help regulate inflammation and immune responses. The stratum corneum provides a barrier against microbes; many chemicals, including water; and abrasion. Though it is always being gradually worn away at the outer surface, it is maintained by having new keratin produced at the same rate by the next generation of keratinocytes. The stratum corneum serves well as long as it remains thick enough and is not broken by cuts, tears, scrapes, or burns.

If the keratin is injured, the keratinocytes re-produce at a faster rate and repair the damage, as occurs when a cut heals. The epidermis can even form a thick pad of keratin—a callus—in areas that are regularly subjected to physical abuse. Thus, the keratinocytes provide adaptation to maintain homeostasis.

Melanin and Melanocytes

The keratinocytes are aided by other types of cells. One type is located in the innermost layer of the

FIGURE 3.2 Structure of the epidermis and underlying region.

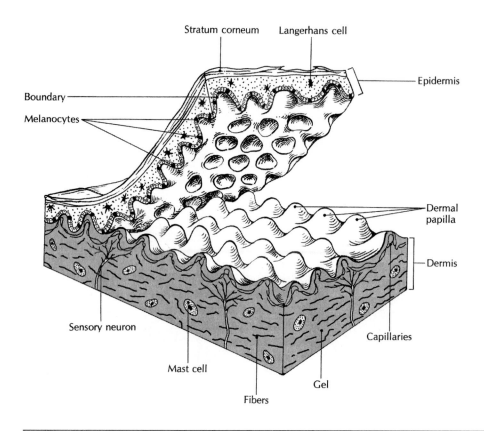

epidermis and produces the brown skin pigment *melanin*. Cells of this type are called *melanocytes*. They make up 2 to 3 percent of the cells in the epidermis.

Melanocytes send melanin into keratinocytes that are about to begin migrating toward the outer surface of the skin. The melanin helps maintain homeostasis by absorbing excess light. Recall that excess light can cause great harm to the body. This is especially true of ultraviolet (UV) light, which contains a great deal of energy and is particularly well absorbed by proteins, RNA, and DNA. UV light is in high concentration in sunlight and in the light from certain bulbs, such as those used for getting a tan.

Excess light is hazardous to the body because it changes the structure of molecules. There are three main undesirable outcomes. One is damage to the protein molecules making up much of the inner layer on the skin (i.e., dermis), which are discussed later in this chapter. This damage may contribute to wrinkling of the skin. The second outcome is injury to the protein molecules inside the cells. The result can be painful and severe sunburn with blistering and peeling off of the epidermis. In addition to causing pain, such injury detracts from the barrier functions provided by the keratin. The third outcome is alteration of DNA, which can lead to skin cancer.

Fortunately, enough melanin can usually be produced to significantly reduce the incidence of these problems. Like the keratinocytes, the melanocytes are adaptable, and they increase their production of melanin when the skin is exposed to excess light. This is evident as the development of a tan. Melanocytes also decrease melanin production when less light is encountered. Then, as the melanin-rich keratinocytes die and are sloughed off at the surface of the skin, melanin-poor keratinocytes replace them and the tan fades.

Immune Function and Langerhans Cells

The third type of important cell in the epidermis is one that starts a body defense strategy called an *immune response* (Chap. 15). These cells are called *Langerhans cells* and make up less than 1 percent of the epidermal cells. Their specific function is to monitor substances throughout the epidermis to determine whether they are native to the body or foreign. These substances may be free molecules between the cells, molecules on the surfaces of microbes, or viruses. Of course, being near the surface of the body, Langerhans cells are positioned to guard against the entry of harmful materials and organisms.

When a Langerhans cell determines that a substance is foreign and therefore does not belong in the body, it alters that substance so that other cells in the immune system can attack and eliminate it. This helps maintain proper chemical conditions within the body and greatly reduces the risk of infection and cancer.

AGE CHANGES IN THE EPIDERMIS

Keratin and Keratinocytes

Thickness   The overall thickness of the epidermis changes little with advancing age, though the epidermis becomes slightly uneven in different parts of the body. Of greatest importance is the fact that the stratum corneum in protected areas, such as those usually covered by clothing, becomes only slightly thinner. Therefore, the keratin retains most of its ability to serve as a barrier against microbes and viruses, water, and abrasion. However, the slight thinning allows certain substances to penetrate the epidermis more easily. Therefore, the elderly should be careful about skin contact with chemicals and topically applied medications.

Structure   Two other microscopic changes in the epidermis have been observed. One is the increase in variability of the size, shape, and internal structure of keratinocytes in the deeper regions. These irregularities may indicate initial abnormalities in cells that are precursors to skin cancer. Skin cancers are among the most common types of cancer in the elderly.

The second structural change is a decrease in the strength of attachment and an increase in the spacing between cells and between keratin materials. These changes may be an additional reason for the increased chemical permeability of the epidermis. The separation of the scaly bits of keratin, together with their flattening and broadening, also contributes to the age-related increase in the scaliness of the skin.

Replacement   There is an age-related decrease in the secretion of some signaling substances by keratinocytes. The rate of new keratinocyte production also decreases. The amount of decline is different in different individuals, and the rate of decline becomes much faster after age 50. By age 75, the rate of cell production may drop to 50 percent of the rate in youth. These changes cause a decrease in the speed of wound healing, leading to an increase in the risk of infection.

Melanin and Melanocytes

Aging affects the melanocytes also. The somewhat uneven distribution found in youth becomes much more pronounced because while the total number of melanocytes decreases, certain areas of the skin develop clumps of melanocytes. These clumps form dark "age spots" or "liver spots" that are noticeable against the gradually fading coloration of the rest of the skin. By contrast, the number of dark moles decreases because of the overall reduction in melanocytes. Still, the final result is that the elderly have a paler but increasingly mottled skin coloration.

The widespread reduction in the number of melanocytes due to their shorter life spans and slower production, combined with a cessation of melanin production in increasing proportions of the remaining melanocytes, causes a decline in protection from excess light. Therefore, elderly people who are exposed to sunlight cannot develop as dark a tan as they did when they were younger. It also takes them longer to develop a tan. This places the elderly at much higher risk of suffering sunburn and skin cancer from exposure to sunlight.

Immune Function and Langerhans Cells

Langerhans cells decrease dramatically with aging. By the time of very old age, the number of these cells declines to less than half the number present in youth. The reduction is greatest in areas of the skin chronically exposed to sunlight. Because of declining numbers of Langerhans cells, one of the body's first lines of defense is largely crippled. This leaves the elderly much more susceptible to infections and skin cancers that would otherwise be eliminated by the immune system. It also decreases allergic reactions by the skin. This change may seem beneficial since such reactions can be uncomfortable, but allergic reactions serve as a warning sign that the body has come into contact with an injurious agent (e.g., noxious chemicals). Without this warning sign, steps to avoid or correct problems are not taken, and this may place a person in jeopardy from these agents.

EPIDERMAL ACCESSORY STRUCTURES

There are numerous places in the skin where groups of epidermal cells have sunk into the underlying dermis so that they may form additional helpful structures. Two of these structures are hair and nails, which will be discussed here because they are visible on the surface of the skin. The others (i.e., sweat glands, sebaceous glands) will be discussed as part of the dermis because they are located under the epidermis proper.

Hair

Recall that the upper layer of the epidermis consists of a thin layer of keratin. Hair is also made of keratin. Each hair is formed at the bottom of a deep pit of epidermal cells called a *hair follicle*, which extends down into the dermis (Fig. 3.1). At the base of the follicle, the same processes that produce the stratum corneum occur, leaving the keratin behind as the shaft of the hair.

Each follicle is not always making hair. On a fairly regular basis the production of cells in a follicle slows and may even stop. When this occurs, the hair falls out and the follicle enters a resting period. After a while the follicle will begin producing cells and melanin again, and a new hair will emerge.

Hair is found on almost all parts of the skin. In areas such as the forehead, it is sparse, thin, and light in color and has only a slight value. However, the scalp, eyebrows, and eyelids have dense,

thick, and long hairs that contribute substantially to a person's well-being in a variety of ways.

Perhaps the most obvious characteristic of hair is its cosmetic value. To appreciate this, one need only notice how much time, energy, attention, and money people spend on their hair. A person's appearance has great social, psychological, and economic impact on the quality of his or her life; this holds true for the elderly as well as the young.

Hair also has several biological functions. For example, scalp hair shades the head from sunlight, provides a thermal insulating layer, and cushions the head against bumps. The hair around the eyes shades them and filters out dust and other small particles. The hair in the openings to the nose and ears also serves as a filter.

In addition, hair helps increase the skin's sensitivity to touch. Since hairs jut out from the surface of the body, any object or material that is about to touch the skin or is moving along its surface collides with these hairs. When such collisions move a hair, its motion travels down the shaft to nerve endings around the follicle. These nerve endings detect the motion and send impulses to the brain, informing the person of the presence and motion of the object or material. Recall that such monitoring is the first step in the negative feedback processes necessary for healthy survival. The person can then take the necessary steps to avoid or remove the object or material. Alternatively, if the object or material causing the motion is not harmful, the person may derive pleasure from the sensations, such as those from a caress or a gentle breeze.

Age Changes in Hair

Aging results in four changes that decrease the amount of visible hair. First, there is a decrease in the number of follicles which decreases the total number of hairs present. Second, increasing proportions of the remaining follicles spend longer periods in the resting stage. This further reduces the amount of hair, since follicles have no hair present during the resting stage. Third, when follicles reenter the active stage, they produce hair more slowly, and so it takes more time for a new hair to emerge. Fourth, almost all the hairs that are produced are thinner.

Age-related decreases in the levels of sex hormones are the main reason for the decline in armpit and pubic hair. However, relatively high levels of male sex hormones (e.g., testosterone) cause more rapid loss of hair from the scalp. Since aging men retain relatively high levels of sex hormones, they lose much scalp hair. Furthermore, men who have inherited the genes for pattern baldness lose increasing amounts of hair from the crown of the head. Women also have some male sex hormone, which is produced by the adrenal gland. Since the level of male sex hormone in women is low before menopause, loss of scalp hair in women is low at first. After menopause, this loss increases dramatically because menopause is accompanied by a rise in male sex hormone.

While a decrease in both the amount and thickness of hair occurs in most areas of the body, some exceptions occur. In aging women these include an increase in facial hair and thickening and lengthening of some hairs on the chin and upper lip. In aging men, thicker and longer hairs are produced in the eyebrows, on the external ears, and within the ear canals and nostrils. All these alterations can be cosmetically troublesome.

Other cosmetically important changes include the development of air pockets within hairs and decreases in the amount of oil secreted onto the hair, resulting in a loss of softness and luster. In addition, the number of melanocytes in each follicle declines, resulting in a decline in the intensity of hair color. As more follicles lose all their melanocytes, increasing numbers of hairs become white. With declining pigment in each hair and fewer hairs containing any pigment, the hue of a person's hair becomes gray and finally white.

The time and rate of graying are determined mostly by genes. Both the time of onset and the rate of graying of scalp hair are not well correlated with chronological age, and graying of scalp hair shows wide variation among individuals. Therefore, graying of the scalp hair is a very poor indicator of chronological age. By contrast, the initiation and progress of graying of axillary hair is a very good indicator.

The consequences of age changes in hair vary. The amount of cushioning provided for the head remains high. Shading of the eyes, as well as the filtering action and the contributions to touch sensation hair provides, may improve when hairs thicken and lengthen. By contrast, the decline in the abundance of hair results in decrements in shading and thermal insulation for the scalp. However, wearing a hat can provide the same type of protection for the scalp.

Because of decreases in both the number and length of hairs in most areas, there are widespread reductions in the contributions hair makes to touch perception. This reduction is exacerbated by the decline in both the number and functioning of sensory nerve cells.

All these biological effects may seem slight compared to the variety and degree of social, psychological, and economic effects caused by the appearance of becoming old. While there may be some positive effects from appearing to be older or more mature, most of the effects are negative.

Nails

Like hair, fingernails and toenails are made of keratin that is produced by essentially the same process as is the keratin in the stratum corneum and in hair. However, no melanin is incorporated into nails.

Nails serve primarily to protect the fingers and toes from traumatic injuries such as crushing, cuts, and scrapes. They can also be used like tools to pick up small objects or scratch irritants off the skin. In addition, though the toenails are usually hidden from view, the fingernails are usually very visible and therefore can have a significant impact on a person's appearance.

Age Changes in Nails

As a person ages, the rate of growth of nails decreases by as much as 50 percent and the thickness and strength of the nails also decrease. The keratin plate becomes less clear, longitudinal grooves develop on its upper surface, and the growth zone at the base of the nail decreases in size. Though these changes are partly due to aging, they can also be caused by trauma and reduced blood flow to the extremities. Changes are greater in the toenails than in the fingernails because there is a greater age-related decline in the blood supply to the feet compared with the hands.

Age changes in nails have several undesirable consequences. The structural weakening of the nails makes them susceptible to injury and disfigurement. The declining growth rate means that the damage is present for a longer period before the injured part grows out and is worn off or cut away. Therefore, nails are less able to perform their functions. Furthermore, because of the higher incidence, severity, and duration of nail injuries, fungal infections of the nails become more common.

Such infections cause the nails to thicken, become opaque, and become misshapen. Infected nails may become unsightly, causing cosmetic problems. In addition, curing fungal nail infections takes longer because a declining blood supply to the nails causes medications to be delivered more slowly. Eventually fungal infections of the toenails may become impossible to cure.

Aging and disease changes in the toenails can have a more serious biological impact than can those in the fingernails because the toenails are out of sight most of the time and therefore often do not get proper care. Furthermore, because of age changes in the eyes and the skeletal system, it is increasingly difficult for aging individuals to see and reach their toenails, resulting in further decrements in toenail care. For example, toenails may become so large that they interfere with the proper fit of shoes, making walking difficult or painful. Toenails are also a common site of infection in diabetics.

DERMIS

The skin layer under the epidermis is called the dermis (Fig. 3.1). It is considerably thicker than the epidermis and contains many different types of structures, including glands, blood vessels, nerve cells, and small muscles. These structures are embedded in a foundation material consisting mostly of fibers and some cells suspended in a small quantity of soft gel. The gel consists mostly of water with some complex proteins and carbohydrates. Because of the variety of its structures, the dermis makes a greater number of contributions to three of the four main functions of the skin discussed earlier in this chapter.

Foundation Material

Fibers Fibers made of protein are the most abundant material in the dermis. The protein fibers are mainly of two types. Collagen fibers constitute approximately 80 percent of the fiber materials, and the others consist of elastin fibers.

The collagen fibers are tangled with each other to form a dense mat. This ensures that the skin will not split open or tear when it is subjected to pulling or twisting forces or is cut. Therefore, it is like a rip-stop fabric. Still, the mat is very flexible so that parts of the body can bend freely.

Elastin fibers are mixed in among the collagen fibers. Because of their elasticity, these fibers cause the skin to automatically return to its original position after it has been pulled, bent, or twisted out of shape.

Cells Scattered among the dermal fibers are cells of various types. The most abundant type, accounting for about 60 percent of these cells, are the *fibroblasts*. Fibroblasts produce and secrete the proteins that form collagen and elastin fibers. They are regulated by chemical and physical signals. Sunlight inhibits collagen production.

Macrophages ("large eaters") constitute 20 to 40 percent of the dermal cells. They wander about among the fibers, engulfing and digesting unwanted materials, including cellular debris, foreign substances, and bacteria. Macro-phages are also important as defense cells because they function like the Langerhans cells of the epidermis.

Additional defense cells in the fibers include white blood cells and mast cells. White blood cells attack and remove harmful materials in a variety of ways (see Chaps. 4 and 15). Mast cells release a substance called *histamine* whenever there is injury to the dermis. Histamine starts the process of *inflammation* (Fig. 3.3).

Inflammation in any part of the body involves an increase in the diameter of blood vessels and in the porosity of the smallest vessels (capillaries). These changes deliver more white blood cells to the injured area to protect the body from infection and remove damaged body cells. More oxygen and nutrients are also brought to the area to supply body cells with all the materials they need to repair the injury. The extra fluids that arrive cause swelling and usually help flush away toxins and debris. In more serious vessel damage, movement of the fluid is inhibited by clotting materials that leak out of the vessels. The redness, swelling, and pain that accompany inflammation serve as warning signs that an injury has occurred. In addition, the pain encourages the person to avoid the circumstances that caused the injury and limit the use of the damaged area until healing has occurred.

Beyond their role in defense, mast cells release a substance called *heparin*, which stimulates the migration of cells that form new blood vessels in

FIGURE 3.3 Inflammation

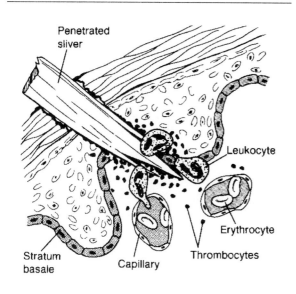

Penetrated
sliver

Leukocyte

Erythrocyte

Stratum
basale

Capillary

Thrombocytes

the dermis. These new vessels are important when new dermal components are formed during the healing of a wound and when the skin grows.

Gel As was mentioned above, the fibers and cells of the dermis are surrounded by a small amount of soft gel material. This gel is made up mostly of water but also contains a variety of large and small molecules dissolved in the water. The water provides a favorable environment for the cells, allows materials to get to and from the cells, and maintains the firmness of the skin. The large molecules provide firmness by keeping enough water in the dermis. They also bind together the other structures in the dermis as a soft glue would. Most of the small dissolved molecules are either nutrients moving to the cells to supply their needs or waste products moving to the blood vessels for removal from the skin. Some inactive vitamin D is also located in the water. The vitamin D moves out of the dermis after being acted on by light (see Vitamin D, below).

Age Changes in Foundation Material The collagen fibers of the dermis undergo substantial changes with increasing age, and these changes have a profound effect on the properties of the

dermis and the ability of the skin to perform normally. One change is a gradual decrease in the amount of collagen. The remaining collagen fibers become thicker and less well organized and form larger bundles of fibers. These changes may be due to the increase in the number of cross-links between the fibers. The increasing cross-links make the fibers stiffer and less able to move, leaving the skin stiffer and less able to stretch. Pulling forces are then more likely to cause injury to the skin because the skin yields less when pulled.

The progressive cross-linking of collagen does not continue throughout life, however. In very old age enzymes in the skin break down the cross-links faster than they can form. Then the strength of the collagen mat decreases, and the skin can be torn more easily.

Age changes in elastin fibers are not as well documented, partly because of the difficulty of distinguishing age changes from changes caused by exposure to sunlight. In any event, elastin fibers become thicker, stiffer, more tightly bound by cross-links, less regular in their arrangement, and, in some cases, impregnated with calcium. The changes in dermal elastin fibers are virtually identical to the changes that occur in the elastic fibers in arteries altered by atherosclerosis (Chap. 4).

While changes in elastin fibers do not alter the ability of the skin to be stretched, they reduce its tendency to return to its original shape and size after being pulled. The skin also does not regain its normal thickness as well after being compressed. Overall, then, the skin seems to become a looser covering that hangs from the body.

With advancing age, the number of fibroblasts increases. This increase may result from an accumulation of old cells and a decrease in the ability to produce new ones. The old fibroblasts may also have less ability to produce new fibers to replace older fibers. Therefore, age changes in the fibroblasts may permit the accumulation of age changes in the fibers and the resulting alterations in the properties of the skin. Furthermore, the deterioration of fibroblasts seems to contribute to the gradual reduction in the speed and strength of skin healing.

Unlike the fibroblasts, the numbers of dermal macrophages and white blood cells seem to decrease with age. The result is a reduction in the defense functions performed by the skin, including a lowered ability to prevent infection, remove harmful chemicals and debris, and initiate immune responses. Therefore, the healthy survival of the entire body is at greater risk.

The age-related decline in the number of dermal mast cells, which may reach 50 percent, also reduces the defense function of the skin. With fewer mast cells, there is a lowering of both the speed and the intensity of inflammatory responses. Therefore, there is both less warning that injury is occurring and a reduced defense against further damage. Furthermore, the declining population of mast cells cannot produce as much heparin. This results in a declining ability to produce new blood vessels in areas of healing and may be a main contributing factor to the normal decrease in the number of dermal blood vessels.

Another age change in the dermal foundation material involves large molecules called mucopolysaccharides, which hold much water. The amount of mucopolysaccharides in the dermis decreases slightly. Therefore, the amount of bound water also declines, and so the firm consistency of the skin diminishes. The skin becomes more easily compressed and returns to its original thickness more slowly. The decrease in bound water may also reduce the movement of small molecules through the dermis. This means that skin cells are not as well serviced. Finally, the reduction in the amount of water held in the dermis probably contributes to the general thinning of the dermis and the thinner appearance of the skin in many elderly people.

Blood Vessels

The blood vessels in the dermis are numerous, although some areas of the body (e.g., scalp) have more of them. Like all vessels that carry blood, the dermal vessels deliver useful materials to the cells and carry away manufactured substances that can be used elsewhere in the body. These vessels also remove wastes produced by the cells. Furthermore, blood flow in these vessels delivers white blood cells and antibodies for defense of the area they serve.

Dermal vessels also help regulate body temperature. They widen when the temperature in the body rises above the desirable level. This widening is called *dilation* (or *vasodilation*), and it allows more warm blood to flow close to the surface of the body. Much of the heat in the blood passes out of the body to the cooler surrounding environment. The result is a lowering of the body temperature so that it is again in the desirable range.

Conversely, if body temperature begins to drop below the normal level, the dermal vessels become narrower. This is called *constriction* (or *vasoconstriction*), and it reduces the amount of blood flow through the vessels. With less warm blood flowing near the surface of the body, the rate of heat loss is reduced. The body can then become warmer as its muscles and other active cells produce more heat.

Age Changes in Dermal Vessels With aging, the number of dermal blood vessels decreases substantially, particularly in the layer just below the epidermis and in skin chronically exposed to sunlight. The remaining vessels often show irregularities in structure. These changes cause a decrease in blood flow to the dermis.

This reduction in blood flow leads to a reduction in the delivery of nutrients to all dermal structures. This may be a main reason for the age-related shrinkage and decline in function of many skin structures. There is also slower removal of material (e.g., wastes, vitamin D), and the delivery of white blood cells and antibodies declines. Even the epidermis, which has no vessels of its own and therefore depends on blood flow in the dermis, is serviced less well. Reduced blood flow can also cause paleness of the skin. Furthermore, reduced dermal blood flow can be of great importance for elderly individuals who use topically applied medications, which can reach dangerously high concentrations in the skin. Meanwhile, the rest of the body, which may need the medication, receives less because the medication remains in the skin.

Adding to the problems caused by reduced dermal blood flow is the increase in thickness of a layer of material that surrounds the capillaries. This layer—*the basement membrane*—is normally quite thin. It allows certain white blood cells and many substances to enter and leave the capillaries freely so that the areas near the capillaries are well serviced.

The age-related thickening of the basement membrane inhibits the movement of white blood cells and chemicals, further reducing the ability of the circulatory system to provide for the needs of the skin.

The aging of dermal vessels also adversely affects the thermoregulatory function of the skin.

As blood flow declines, there is a reduction in the ability to release excess heat from the body. The vessels also constrict and dilate more slowly and to a lesser degree. The result is a reduced ability not only to release heat but to slow heat loss when the body begins to get chilled. These changes constitute a major reason why the elderly have difficulty maintaining normal body temperature when the external temperature deviates from a moderate level or when activities such as vigorous exercise cause an alteration in body temperature.

Sweat Glands

The dermis contains two types of sweat glands. *Eccrine sweat glands* secrete a watery material which is the visible perspiration (i.e., sweat) seen when a person becomes uncomfortably warm. The other type of sweat gland is the *apocrine sweat gland*.

Each eccrine sweat gland is a tubular gland with a highly coiled portion, located in the dermis, which produces most of the perspiration (Fig. 3.1). The coiled portion leads into a fairly straight portion that extends upward through the dermis and epidermis and finally opens onto the surface of the skin.

The purpose of the perspiration produced by these glands is to cool the body. When the brain detects an abnormal rise in body temperature, it sends nerve impulses to the glands, stimulating them to secrete perspiration. The water in the perspiration evaporates when it reaches the surface of the skin. As it evaporates, it carries heat away from the skin, resulting in a lowering of body temperature.

Perspiration also contains a number of substances, including useful ones such as salt, which are dissolved in the water. However, the secretion of these useful substances is not beneficial to the body. Serious problems such as muscle cramps and dizziness can develop if a person perspires abundantly without replacing the useful substances by drinking beverages or eating foods that contain more of these substances.

Unlike the eccrine sweat glands, which are widespread, most apocrine sweat glands are located in the skin of the armpits and in the genital area. These glands secrete a small amount of thick materials that do little to promote healthy survival. This secretion is a main source of unpleasant body odor.

Apocrine gland activity is controlled largely by the level of sex hormones in the body, though the nervous system may increase the amount of secretion during periods of stress or intense emotions.

Age Changes in Sweat Glands   With aging, the number of eccrine sweat glands decreases dramatically in all parts of the body except the scalp. The remaining glands are reduced in size and produce perspiration at a decreasing rate.

The result of these changes is a reduction in the ability of the glands to cool the body. This puts the elderly at an ever increasing risk of becoming overheated in particularly warm environments or during vigorous exercise.

Though changes in the number of apocrine sweat glands have not been well studied, it is known that they shrink and that their rate of secretion diminishes significantly. This is probably due to age-related reductions in sex hormone levels. This is one of the few alterations with advancing age that most people agree is desirable since diminishing apocrine gland secretion leads to a substantial decrease in unpleasant body odor. While apparently having no biological importance, this change can have positive effects on other parameters (e.g., social).

Sebaceous Glands

In addition to sweat glands, the dermis contains glands that produce an oily substance called *sebum*; these glands are called *sebaceous glands* (Fig. 3.1). Sebaceous glands are usually located beside hair follicles and secrete sebum into the follicles. The sebum coats the hair, and as it leaves the follicles, it spreads out to form a thin coating on the epidermis.

Sebum contributes to the maintenance of homeostasis mainly by improving the ability of the skin to act as a barrier. Because sebum is an oily material, it helps make the keratin of the epidermis more impermeable to water. It also helps keep the keratin pliable so that the stratum corneum does not crack when it is bent. Cracks in the keratin could allow water and other chemicals to leak into and out of the body and permit the entrance of harmful microbes. In addition, certain materials in the sebum inhibit the growth of fungi that could break down keratin.

Sebum is also cosmetically important because it gives skin keratin a smoother appearance and adds luster to the hair. Finally, by keeping keratin pliable, sebum reduces the breaking and splitting of hair.

Age Changes in Sebaceous Glands   Though there seems to be no age change in the number of sebaceous glands and though they increase in size, there is a decrease in the production of sebum. This decline seems to result from declining levels of the sex hormones that normally stimulate sebum production.

The reduction in sebum production lowers its contributions somewhat, though the amount produced is usually sufficient to prevent serious biological problems. However, the cosmetic contributions made by sebum decline substantially. The results are the clearly visible signs of aging of the skin and hair. The widespread use of skin and hair lotions that augment the diminished contributions of sebum attests to the importance of these cosmetic changes.

Nerves

The two types of nerve cells (neurons) are **sensory neurons** and **motor neurons** (Fig. 3.1). Sensory neurons monitor conditions in and around the dermis, including conditions in the epidermis and the external environment. Sensory neurons send information about these conditions to the brain and spinal cord. Motor neurons control the functioning of blood vessels and eccrine sweat glands by relaying instructions from the brain and spinal cord to the skin.

Sensory Neurons   There are several types of sensory neurons in the dermis, each of which is specialized to monitor a single kind of stimulus (e.g., light touch, heat, pressure). An additional type is activated when conditions vary greatly from normal or there is injury to the skin. This type of sensory neuron warns of the danger by providing the sensation of pain.

Beyond determining if there has been a change in conditions near the surface of the body, sensory neurons provide information about what type of change has occurred and the location of that change. The nervous system can then initiate the proper type of response to preserve the well-being of the body.

Motor Neurons   Information about the structure and function of the motor neurons of the skin

and the effects of aging on these neurons is presented in Chap. 6.

Age Changes in Sensory Neurons   As a person ages, there is little change in the number or structure of the sensory neurons for pain and the touch receptors connected to hair follicles. Conversely, the numbers of touch receptors not connected to hair follicles and of pressure receptors decrease dramatically. In addition, there are alterations and distortions in the structure of both types of receptors. Little is known about the effects of aging on the other types of sensory neurons.

As a result of age changes in sensory neurons, there is decreased sensitivity to touch, pressure, and vibration. This is especially evident in the fingers, the palms of the hands, and other areas of the body lacking hair (e.g., the penis). In addition, there is a decreased ability to detect the exact location of touch and pressure stimuli and therefore to determine the shapes of objects by touching them. One practical consequence is a reduction in manual dexterity. Interestingly, the thinning of the skin with age compensates to some degree for the changes in sensory neurons by making it easier for stimuli to reach these neurons.

These sensory decrements reduce the ability of the skin to inform the body about conditions on and just outside its surface. The person is then less able to respond negatively to dangerous or harmful stimuli and positively to helpful or pleasurable stimuli.

The ability of the skin to perform its monitoring function is adversely affected by many factors beyond changes in the number and shape of its sensory neurons. Some of these factors include the consistency of the skin and the subcutaneous layer, the ability of the neurons to conduct impulses to the brain and spinal cord, and the ability of the brain and spinal cord to process and interpret those impulses (Chap. 6).

BOUNDARY BETWEEN EPIDERMIS AND DERMIS

The boundary between the epidermis and the dermis is important for the maintenance of the structure and functioning of the skin. For example, it is the region through which nutrients pass upward to the epidermis and wastes pass downward to the dermis. This exchange is essential for the epidermis, which has no blood vessels to service its cells. Keratinocytes in the epidermis have special projections that attach to the dermis to help in this exchange, which is also assisted by many blunt projections from the dermis that extend up into the epidermis. These projections, called *dermal papillae*, increase the rate of exchange of materials by increasing the amount of contact (i.e., surface area) between the two layers. The dermal papillae also contain special tufts of capillaries that further increase the ability of the dermis to service the epidermis (Fig. 3.1).

In addition to improving the exchange of materials, the boundary between the epidermis and the dermis provides a strong attachment between the layers. The keratinocyte projections help by gripping the dermis, and the dermal papillae also help in this regard. As a result, the boundary is able to prevent separation of the epidermis from the dermis when sliding or pulling forces are applied to the surface of the skin.

The dermal papillae are usually scattered about in most areas of the skin. However, they are in very regular rows in the skin on the front of the hands and the bottom of the feet. These rows produce the ridges known as fingerprints, which make the skin less slippery and improve a person's ability to grip objects.

Age Changes in the Epidermal-Dermal Boundary

As a person ages, the projections from the keratinocytes decrease in number and both the number and length of the dermal papillae decrease. The distribution of small vessels in the papillae becomes uneven. The result is a reduction in the functions of the boundary. First, there is less exchange of materials between the epidermis and dermis. Therefore, the epidermis is weakened, is injured more easily, and heals more slowly. Second, the weakening of the connection between the epidermis and dermis leads to easier blister formation in the elderly when the skin is subjected to physical forces. Such forces are encountered when one performs ordinary activities such as sweeping and gardening. They are also present when the skin is pulled, as occurs during the removal of adhesive bandages. The resulting injuries not only are painful but also increase the risk of skin infection. Finally, as the fingerprints

become less prominent, it is more difficult to keep a firm grip on objects.

VITAMIN D PRODUCTION

Though the production of vitamin D begins in both the epidermis and the dermis, most of it occurs in the epidermis. Skin cells start the process by modifying cholesterol molecules. When the modified molecules are struck by light, they are altered again to form an inactive form of vitamin D. Ultraviolet light seems to be the best type of light for this process, and sunlight is the natural source of UV light for the body. Exposure of the hands and face to only 10 to 15 minutes of summertime sunlight provides enough light for the skin to produce all the vitamin D needed by the body. Inactive vitamin D is carried away by the blood in dermal blood vessels.

The inactive form of vitamin D can also be obtained from foods such as fish and vitamin D-enriched milk. Whether from the skin or from the diet, inactive vitamin D is sent to other parts of the body (i.e., liver and kidneys) for additional modification and final activation. It is then transported throughout the body. Vitamin D influences movement of calcium into and out of bones directly and indirectly. Vitamin D reaching the intestines helps absorb calcium from food.

Calcium performs many essential functions in the body. It is a main building material in bones and teeth and is essential for the contraction of muscles, the passage of impulses in the nervous system, and the clotting of blood. Calcium also controls many chemical reactions in cells.

Age Changes in Vitamin D Production

The ability of the skin to produce inactive vitamin D decreases with age. This seems to result from several factors. For example, there may be a decrease in the delivery of cholesterol-like molecules to the skin because of the decrease in blood flow in the skin. Also, the skin cells seem to be slower at converting this material. Furthermore, the process powered by light becomes less efficient. Therefore, an older person must get more exposure to sunlight to produce the same amount of vitamin D. Finally, the slower movement of materials through the skin and the decrease in dermal blood flow may slow the removal of inactive vitamin D from the skin.

The overall result of reduced vitamin D production by the skin is an increased risk of vitamin D deficiency with age. The risk is further increased because the elderly often have less exposure to sunlight as a result of reduced mobility, social customs, and the higher risk of developing sunburn. There is also a decline in the ability of the kidney to complete vitamin D activation. If a vitamin D deficiency develops, the absorption of calcium will become inadequate. All body functions that depend on calcium will then become abnormal.

Fortunately, many ordinary foods (e.g., bread and milk) have vitamin D added to them. Incorporating such foods into the diet can largely eliminate the risk of vitamin D deficiency. In situations where the diet cannot provide the necessary vitamin D, vitamin supplements such as vitamin pills can do so.

SUBCUTANEOUS LAYER

The *subcutaneous layer* lies under the dermis (Fig. 3.1). While this layer usually is not considered part of the skin, it makes up the innermost layer of the integumentary system.

Loose Connective Tissue

The foundation material of the subcutaneous layer is made of loose connective tissue that contains a soft gel consisting of a large amount of water with some protein and other substances dissolved in it. Within the gel are various types of cells and widely scattered collagen and elastin fibers.

The soft gel serves as a cushion under the skin. It is slippery and therefore allows the skin to slide easily over the underlying muscles and bones. As in the dermis, some of the cells produce the gel and fibers, while others defend the body against microbes and harmful chemicals. The fibers hold the other components in place and attach the skin to the body. Since these fibers are relatively low in number, they allow the skin to move easily, though only a limited distance over the underlying structures.

Fat Tissue

Fat tissue is also found within the subcutaneous layer. Some areas of the body (e.g., buttocks) have a thick layer of fat, while other areas (e.g., hands) have a thinner layer. There is also a wide degree of difference among individuals in terms of the amount of fat in the subcutaneous layer.

Fat tissue is a very important component of the subcutaneous layer. First, the soft but somewhat firm consistency of fat allows it to cushion the inner body parts, protecting them from injury by pressure and forceful blows. Second, because of its firmness, fat helps maintain the contour of the skin. Thus, in moderate amounts and when well distributed, fat contributes to a pleasing appearance. Third, since fat is a thermal insulator, it helps maintain proper body temperature by reducing the rate of heat loss through the skin. Finally, fat is a nutrient storage material. If the diet does not provide enough energy or building materials for the body, the cells break down fat molecules. Body cells can obtain a great deal of energy and raw materials in this way.

Age Changes in the Subcutaneous Layer

There is little information about the effects of aging on the loose connective tissue of the subcutaneous layer. There is a general decrease in the amount of subcutaneous fat tissue with aging, but there is little decrease in the total amount of fat in the body. The explanation is that while the amount of subcutaneous fat is decreasing, the amount of fat increases in the inner regions of the body, such as around the organs inside the abdomen. One effect of the generalized thinning of subcutaneous fat is a decrease in the ability of an older person to stay warm in a cold environment. Another effect is a reduction in the support of the skin. This, together with changes in the dermis, makes the skin appear loose. The skin may even seem to hang in folds on the face and other parts of the body. The thinning of the fat may also contribute to the more translucent appearance of the skin of elderly individuals.

Changes in the proportion of fat in different areas of the body also seem to be of great importance. This is currently a topic of considerable research, but only a few conclusions can be drawn. For example, the substantial decrease in fat on the bottom of the feet reduces their ability to cushion the body. It also seems that the distribution of body fat is related to the incidence of certain diseases (e.g., coronary artery disease, diabetes).

Though the immediate biological effects of an altered distribution of subcutaneous fat on healthy survival are uncertain, the cosmetic effects are very apparent. The subcutaneous fat of the arms and legs thins, as do the muscles in those areas. At the same time, there is a thickening of the fat in the trunk of the body. The result is a dramatic change in body proportions, with the waist seeming to get much larger with age. This change occurs to a greater degree in women than in men. Weakening of the abdominal muscles and other muscles important in maintaining good posture can exaggerate the result.

These changes in subcutaneous fat are believed to occur in most people because of aging. However, there is an enormous degree of variability among individuals because numerous factors besides aging can affect fat tissue. For example, alterations in diet, exercise, and hormone levels can profoundly change the amount and distribution of fat. Therefore, as individuals get older, a considerable change in their appearance may alter their social interactions, psychological health, and economic status.

MISCELLANEOUS COSMETIC CHANGES

Three additional cosmetic changes in the integumentary system will be mentioned here because they are often associated with aging. One is the gradual increase in the width of the nose and size of the ears. Most of the perceived lengthening of the nose is due to age-related changes in the skin and shrinkage of muscle and bone near the nose. A second cosmetic change is wrinkling of the skin. Wrinkling has been attributed to changes in the fibers in the dermis, changes in the foundation material in the dermis, fat loss from the subcutaneous layer, and the pull of muscles on the skin. All these factors may contribute somewhat to wrinkling, but none has been shown to be the fundamental cause. The third change is drying of the skin. Changes in keratin and reductions in the production of sweat and sebum may enhance the dry appearance of the skin, but the actual cause of skin dryness has not been identified.

ABNORMAL CHANGES

The ability of the integumentary system to serve the body is reduced by factors other than aging.

While all these factors can affect the integument in the young as well as the elderly, they are more relevant for the elderly. One reason is that older individuals have had more opportunities to be exposed to harmful environmental factors. In some cases a cumulative effect develops; an excellent example is the effect of sunlight.

Another reason is that the elderly more often have a decline in the functioning of body systems on which the integumentary system depends. For example, the ability of the nervous system to control the size of dermal blood vessels diminishes with age, further reducing the ability of the skin to regulate body temperature. A third reason is the increasing incidence of diseases in body systems on which the skin relies. A common example is circulatory system diseases such as atherosclerosis, which reduces blood supply to the skin. The skin then becomes thinner, weaker, and more susceptible to injury and infection.

Another effect of such factors is that the elderly have a higher incidence of abnormal changes in the integumentary system. Studies have shown that up to 40 percent of otherwise healthy individuals between ages 65 and 74 have at least one skin disorder serious enough to require treatment by a physician. Many of these individuals have more than one skin disorder at the same time. It is noteworthy that all these disorders can also be found at least occasionally in the young.

Though almost none of these abnormalities are fatal and almost all are preventable or treatable, they are important in several ways. First, some integumentary system problems alter the structure and functioning of the integument so that it is less able to perform its usual functions. For example, bedsores increase the risk of infection. Second, some problems produce a considerable degree of discomfort. For example, excessively dry skin causes intense itching, which can be so distracting that it disrupts normal daily activities. Finally, some problems, such as excessive wrinkling from prolonged exposure to sunlight, adversely affect the appearance of the skin.

Effects of Sunlight

Sunlight can cause numerous skin abnormalities. For example, exposure to very strong sunlight for even a few hours can cause sunburn. However, of more concern here are problems that take years to develop because each exposure advances the problem only slightly. The results are apparent only after they have accrued for decades. They are so subtle and widespread that until recently they were widely thought to be age changes. Many researchers believe that the ultraviolet light in sunlight causes these long-term effects, but other components of sunlight may be more to blame. Energy from UV light damages DNA directly, and UV light promotes free radical (*FR) production in the skin while reducing its *FR defenses. Even short doses (e.g., minutes) of low intensity UV light, less than enough to cause skin reddening, causes damage to fibroblasts and increases elastin synthesis. Smoking increases the adverse effects from sunlight, probably by reducing blood flow to the skin and by increasing *FR production. Certain cosmetics, medications, and chemical air pollutants also increase *FR formation by UV light.

Chronic exposure to sunlight affects the epidermis in several ways. The keratinocytes reproduce irregularly, and the new cells produced are uneven in shape. This makes the epidermis appear to be uneven in thickness and rough in texture. The irregularity of the cells also seems to contribute to the higher incidence of epidermal skin cancer in the elderly. In addition, the melanocytes become more unevenly distributed, increasing the number of age spots and intensifying the blotchy appearance of the skin. Langerhans cells decrease in number, leading to a reduction in their defense capability. Finally, sweat gland function declines.

The dermis is also changed by years of exposure to sunlight. There is a net loss in collagen, and the remaining collagen becomes weaker. Elastin fibers become more numerous but also become very irregular in shape and arrangement from excess cross-links, and many develop unusual thickenings. Production of abnormal molecular complexes in the gel reduces its ability to hold water. These changes may be a main reason for the excessive wrinkling of sun-exposed skin.

Unlike elastin fibers, dermal blood vessels in sun-exposed skin decrease in number, leading to a reduced blood supply to the skin. The capillaries that remain have thicker walls, and this may further reduce the vital movement of material between the blood and skin cells. In addition, certain materials, such as topically applied chemicals and antibodies produced by the immune cells, tend to accumulate within the skin. These mate-

rials can injure and irritate the skin, leading to discomfort and blistering.

Sunlight also affects the sebaceous glands, causing them to enlarge considerably. Some become so large that they become visible as unattractive comedones (blackheads).

Obviously, all the effects of long-term exposure to sunlight are detrimental. All can be prevented by shading the skin from repeated and prolonged exposure to sunlight. This can be done easily by wearing appropriate clothing, hats, and sunscreen lotions that block most of the harmful rays. Protection while in water is also important because water blocks only some UV light. Sunscreen lotion with an SPF15 is adequate to absorb almost all harmful UV light. Lotions with higher SPF provide very little additional protection. The benefits from protecting skin from excess sunlight include more attractive and healthier skin and a reduced risk of cancer.

It may be possible to prevent UV and other types of oxidative damage to the skin by using topical or oral supplements to increase the skin's *FR defenses. Research suggests that the best method may be a combination of oral supplements of selenium, vitamin C, vitamin E, and β-carotene. Supplements must be used carefully to avoid some toxic effects and to prevent additional *FR formation by imbalancing *FR defenses.

Treatments for cosmetic effects from photoaging of skin include alpha-hydroxy acid peels (e.g., glycolic acid), carbon dioxide laser treatment, and cryotherapy for epidermal color problems. *Glycolic acid* treatment requires months of regular applications and visits to a dermatologist's office. Benefits may include smoothing and thickening of the epidermis; reduction in comedones, small wrinkles, and age spots; and thickening of the dermis. Undesirable side effects can include redness, itching, burning, scabbing, pain, and tightness, which may take from a few days to a week to subside after each treatment. Other possible problems include scarring and reactivation or spreading of Herpes I sores and warts.

Topical application of a vitamin A derivative called *tretinoin* (i.e., all-trans tretinoic acid) can help reverse the effects of photoaging, and it also reverses normal age changes. Benefits in the epidermis include thicker, smoother, and more dense epidermis; reduction in abnormal keratinocytes and uneven skin color (e.g., age spots); and faster healing when used for weeks or months before surgery or injury. Thickening of the epidermis is temporary. Benefits in the dermis include increases in normal collagen; in capillaries; in dermal vessels dilation; in number and length of dermal papillae; and in attachment of the dermis to the epidermis. Tretinoin treatment also reverses normal age changes and adverse effects from reduced blood flow.

Tretinoin seems to act by stimulating DNA synthesis and tissue growth factors. There seems to be no risk of abnormal cells, precancerous cells, or cancer. The new cells seem to be even more "normal" than the normal but somewhat altered keratinocytes and melanocytes in photoaged skin.

Tretinoin may be applied topically or by injections. A combination of injections plus topical treatment may be best in some situations. Treatment can cause some temporary redness and discomfort, and treatments may require months to complete.

Effects from heat

Chronic exposure to heat produces the same effects as photoaging except that chronic heat does not cause formation of excess and abnormal elastic fibers. Chronic exposure to heat can occur in workplaces, in unevenly heated living spaces, and when using localized heaters (e.g., heating pads).

Bedsores

Another largely preventable skin problem is *bedsores* (*decubitus ulcers*), patches of skin that have died because they received insufficient blood flow. The main cause is pressure, which compresses the blood vessels in the skin so that little or no blood flows through them. If blood flow is reduced for more than 2 hours at a time, the skin cells die and peel away, leaving an open wound (Fig. 3.4).

Many factors contribute to the formation of bedsores. Because the elderly are more likely than the young to encounter a greater number of these factors at more intense levels, there is an increased incidence of bedsores among the elderly. The most important of these factors is immobility, because the weight of the body puts enough pressure on the skin to cut off blood flow through skin ves-

sels. The most susceptible parts of the body are the buttocks area and the heels because sitting or lying puts pressure on these areas. The elderly are more likely to find themselves in these positions for long periods because of disabling diseases such as strokes.

Other factors that increase the possibility of developing bedsores include (1) normal weakening of the skin, (2) thinning of the subcutaneous fat, (3) diseases of the circulatory system that reduce blood flow, (4) poor nutrition, and (5) poor skin hygiene. All these factors are more prevalent among the elderly. Physical forces on the skin, such as friction and uneven distribution of weight, also contribute to the formation of bedsores.

Once a bedsore has formed, it may become deeper and penetrate through the dermis and the subcutaneous layer. Bedsores heal very slowly if at all. Those which heal are likely to recur.

Bedsores often become infected because the barrier against microbes has been broken. The reduced blood flow also leaves the skin with weak defense mechanisms. Finally, bedsores can be quite painful and can be repugnant for care givers and others.

With proper preventive measures, bedsores can be largely avoided. Frequent changes in position, the use of soft supporting materials that distribute body weight evenly, and good hygiene can greatly reduce the occurrence of these undesirable skin afflictions.

FIGURE 3.4 Bedsores.

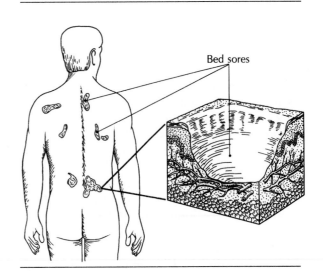

Neoplasms

Sometimes the production of new cells in the skin gets out of control. Instead of producing the number of cells needed and then stopping, cell production continues unabated. This condition is called a *neoplasm*. If the extra cells stay tightly together in one place, the mass is called a benign neoplasm (Fig. 3.5a). This type of neoplasm usually is not very harmful. However, if the cells begin to spread out or move to other parts of the body, they constitute a *malignant neoplasm* or *cancer* (Fig. 3.5b). Cancer is much more likely to cause serious problems because it disrupts the structure and functioning of any body part it invades.

In the skin, both types of neoplasm occur considerably more frequently as people get older. Furthermore, the elderly have more cases of cancer of the skin than cases of all other forms of cancer combined.

There are numerous reasons for the high incidence of skin neoplasms among the elderly, and they correlate with other age-related changes in the skin. These changes include age changes such as increased irregularity in the cells produced; reduced number of Langerhans cells; decreased amounts of melanin; a decreased inflammatory response, which can warn of the presence of noxious carcinogens; and slower removal of materials such as carcinogens. Note that all these changes are amplified by exposure to sunlight and that sunlight itself causes neoplasms. Therefore, protecting the skin from long-term exposure to sunlight can significantly reduce the risk of developing skin neoplasms.

Benign Skin Neoplasms   Common benign neoplasms of the elderly include *basal cell papilloma*, also called actinic keratosis, keratoses, senile warts, and seborrheic warts, which appears as round somewhat elevated flattened darker spots; *squamous papilloma* and *clear cell acanthoma*, which appear as small round elevations. Other benign neoplasms of the skin usually appear as small protrusions of the epidermis. These neoplasms are usually of only cosmetic importance and can be easily removed with simple surgical procedures. Removal may be desirable for cosmetic reasons and to avoid possible injury to the protruding skin, which could lead to discomfort and infection. Finally, removal of benign neo-plasms is often recommended because they may become malignant.

Malignant Skin Neoplasms   Common malignant skin neoplasms include *basal cell carcinoma* and *squamous cell carcinoma.* The first type is the most common. It appears as a slow growing light-colored spot, which develops into a sore that will not heal. Squamous cell carcinoma appears as thickened areas with irregular surfaces. Both types develop from keratinocytes. Because their cells are not well attached to each other, these malignant neoplasms can weaken the skin, greatly increasing the risk of injury and infection. As they spread, they affect larger areas of the skin. Fortunately, they are easily detected while still in the early stages of growth and can then be removed by means of simple surgery.

Malignant melanoma is a third type of skin cancer. It is usually caused by exposure to sun-

FIGURE 3.5 Skin neoplasms: (*a*) Benign neoplasm. (*b*) Malignant neoplasm.

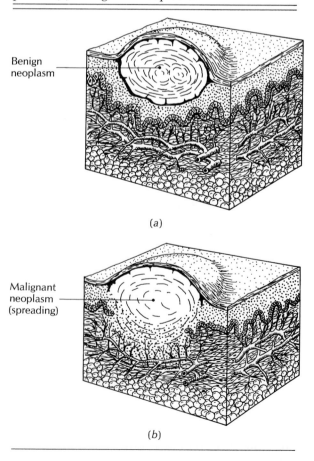

Benign neoplasm

(a)

Malignant neoplasm (spreading)

(b)

light. Malignant melanoma derives from the melanocytes, and often appears as dark irregular mottled spots that enlarge. Though less common than the other two skin cancers, it is a very serious and often life-threatening cancer. It grows very rapidly and spreads quickly to many other organs. Wherever it is found, it displaces the normal cells in the area, causing that part of the body to stop functioning normally. It also weakens body parts so that there is an increased risk of infection. Malignant melanoma can be cured if it is removed before it enlarges and spreads.

Melanoma causes more deaths from skin cancer than all other types of skin cancer combined. The number and rates of death from melanoma have increased severalfold over the passed 50 years, including among the elderly. Still, the elderly have a greater age-related increase in deaths from non-melanoma skin cancers and a greater total mortality from non-melanoma skin cancers than from malignant melanoma. This trend may result from earlier deaths of those most susceptible to melanoma.

Since skin cancers can become dangerous quickly, early detection and treatment are essential. Knowing the warning signs of skin cancer and noticing them when they appear can help. The signs include any unusual lump or thickening, any sore or wound that does not heal quickly, the appearance of dark spots, and any change in the shape or size of a wart, mole, or other dark spot. Any dark spot that develops a rough texture or an irregular outline is especially noteworthy. All suspicious areas should be reported immediately to a physician for further diagnosis and appropriate treatment.

4

CIRCULATORY SYSTEM

The circulatory system contains several very different components, including the *heart*, a hollow muscular pump that stands at the operational center of the system that pumps liquid *blood* throughout the body through three types of flexible tubes, the *blood vessels* (Fig. 4.1). The *arteries* channel blood from the heart to all parts of the body needing service. Once there, the blood passes through narrow arteries and enters the *capillaries*, which are the narrowest blood vessels. Many substances and some blood cells pass into and out of the blood by moving through the thin porous capillary walls. The blood is then carried through the *veins*, which return the blood to the heart. The passage of blood through the vessels in a part of the body is called *perfusion* of that part.

Some materials that are carried away from a region of the body do not pass into the blood but are collected by vessels called *lymph capillaries* (Fig. 4.2). The materials in these vessels make up a liquid called *lymph*, which is carried through the *lymph vessels* toward the heart. Along the way, the lymph passes through *lymph nodes* where harmful chemicals and microbes that might have entered it are removed. The lymph is finally added to the blood in the veins shortly before the blood enters the heart.

MAIN FUNCTIONS FOR HOMEOSTASIS

Transportation

One main function of the circulatory system is transportation of materials within the body. Transportation helps maintain homeostasis by ensuring that the concentrations of substances surrounding body cells are kept at proper and fairly steady levels. Materials consumed by the cells are immediately replenished, and materials produced by the cells are swept away before their concentrations become too high.

The flowing blood also transports useful materials from their point of entry into the body to the organs that need them. For example, oxygen from the lungs and nutrients from the digestive system are delivered to the muscles. Furthermore, some cells manufacture substances (e.g., hormones) needed by cells in other organs, and the circulatory system provides the delivery service for them.

FIGURE 4.1 The circulatory system, showing the pathway for blood flow.

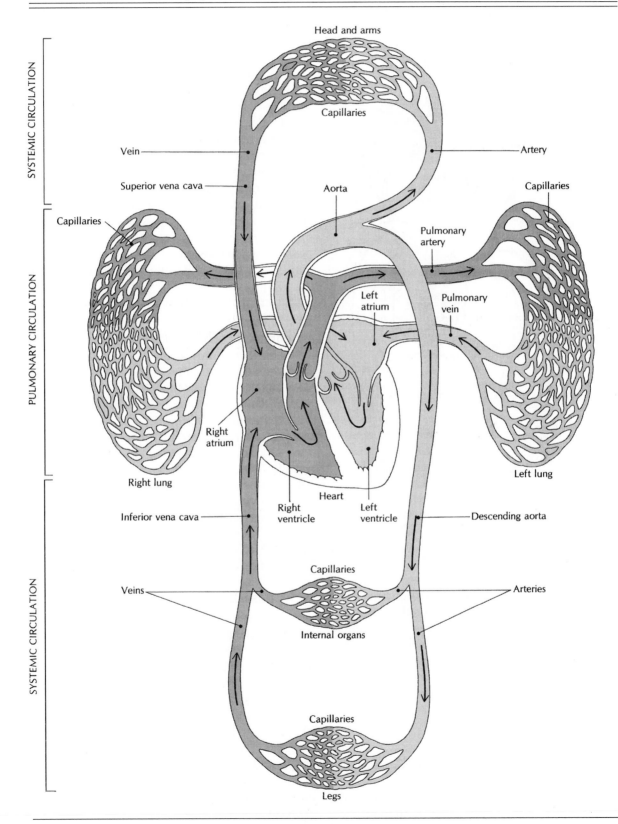

FIGURE 4.2 The circulatory system showing pathways for blood flow and lymph flow.

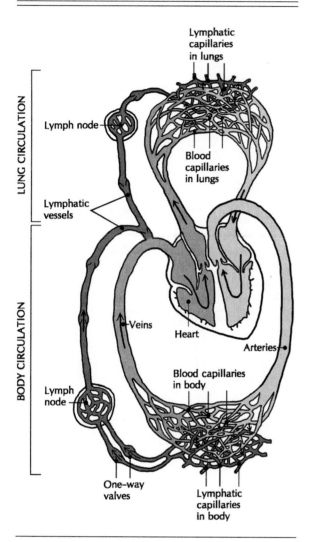

body. For example, toxins and bacteria that enter the lymph from an infected wound are inactivated as they pass through the lymph vessels. Blood and lymph contain several types of *white blood cells* (*WBCs*). Some WBCs eliminate dangerous materials contained in blood and lymph, while others leave the blood in the capillaries and travel among the cells of the body to seek out and destroy noxious materials and microbes. Other defense cells, located on the inner walls of blood vessels, monitor the contents of the flowing blood and remove undesirable materials.

Like transportation, defense is increased or decreased to meet changing needs. The number and speed of movement of WBCs and their rates of producing defensive chemicals increase temporarily whenever harmful microbes or foreign materials are detected within the body.

Temperature Control

Another homeostatic function of the circulatory system is temperature control. Temperature regulation and the role of dermal blood vessels in this process were discussed in Chap. 3.

Another way in which the circulatory system contributes to thermal regulation is by distributing heat from heat-producing sites to areas that cannot keep themselves warm. For example, muscles produce much heat, and blood carries some of it to smaller structures such as the spinal cord, slower-acting organs such as bones, and cooler areas such as the skin. In this way heat distribution helps prevent overheating in any single area of the body while sustaining activities in all its regions.

Acid/Base Balance

Besides being sensitive to heat, many substances in the body are altered by the balance between *acidic* and *basic* (*alkaline*) materials in their surroundings. The relative amounts of acids and bases are usually indicated by a numerical value called *pH.* The normal range for the body is about pH 7.35 to pH 7.45. An *acid/base balance* resulting in a pH within this range preserves the proper shape and activities of molecules in the body. Deviations from this range adversely alter these molecules, leading to malfunction, damage, and even the death of cells.

As a person's rate of activity changes, the rate of activity of that person's body cells also changes. This causes the rate of consumption of nutrients and the production of wastes and hormones to fluctuate. The circulatory system helps the body adapt to these changes by altering the rate of blood flow through each part of the body.

Defense

The circulatory system also makes an important contribution to the defense of the body. Lymph nodes trap and destroy dangerous chemicals and microbes before they can spread throughout the

FIGURE 4.3 The internal structure of the heart and adjoining blood vessels.

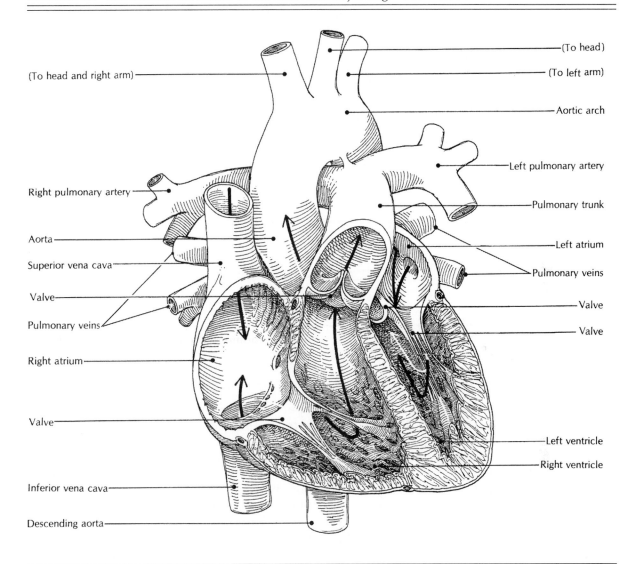

Maintaining proper acid/base balance requires the constant action of negative feedback systems because the ongoing activities of most cells result in the formation of acids. Foods and beverages can also add acids or bases to the body. Excess amounts of acid or base must be neutralized or eliminated to maintain a proper pH. This is where the circulatory system makes a major contribution. Certain minerals and protein molecules in blood plasma and red blood cells—*buffers*—act as reservoirs for acids. These buffers absorb and store excess acids. When bases become too abundant, some stored acid is released to balance them, preserving the acid/base balance. These buffers have

a limited capacity for balancing pH, and acid/base balance also depends on the activities of the respiratory and urinary systems.

We will now examine components of the circulatory system in greater detail, beginning with the heart.

HEART

Chambers and the Cardiac Cycle

The heart consists of four chambers (Fig. 4.3). Blood from the veins enters the two upper chambers, called *atria*. Blood from the lungs returns

FIGURE 4.4 (a) Cardiac cycle and (b) blood pressure.

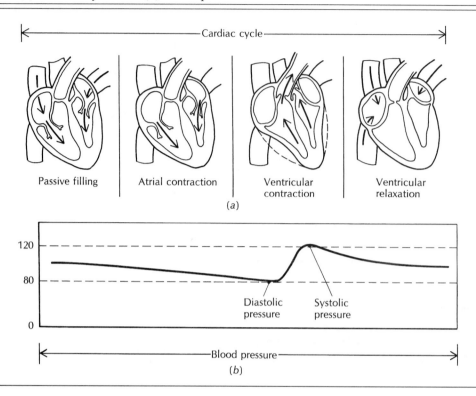

Cardiac cycle

Passive filling   Atrial contraction   Ventricular contraction   Ventricular relaxation

(a)

120

80

0

Diastolic pressure   Systolic pressure

Blood pressure

(b)

to the heart through several *pulmonary veins,* which deliver it to the *left atrium.* This blood has a high concentration of oxygen, which was added as the blood passed through the lungs. The oxygen is needed by all the cells in the body.

While blood from the lungs is entering the left atrium, blood from the body is flowing into the *right atrium* via two large veins. This blood has had most of its oxygen removed by body cells and contains a high concentration of a waste product called *carbon dioxide,* which was produced by body cells. It also carries many useful substances (e.g., nutrients and hormones) added by various organs.

The blood flows easily from each atrium into the *ventricle* just below it because the ventricles relax and tend to widen at this time (Fig. 4.4a). The flow is aided by a relatively weak contraction of the atria. Once the ventricles have been filled, they contract powerfully, squeezing the blood and pumping it into the arteries. The ventricles contract for a fraction of a second and then relax again.

The blood from the *left ventricle* is pushed very forcefully into a large artery, the *aorta.* Branches from the aorta deliver this oxygen-rich and nutrient-rich blood to all parts of the body except the lungs.

Special branches from the aorta—*coronary arteries*—transport some blood to the walls of the heart.

The right *ventricle* pumps blood through the *pulmonary arteries* to the lungs. Most of the carbon dioxide in this blood is removed while the blood is in the lungs. At the same time, oxygen is added to the blood for delivery to the rest of the body.

When the ventricles contract and force blood into the arteries, the blood pressure rises quickly to a peak value called *systolic pressure* (Fig. 4.4b). When the ventricles relax and blood in the arteries flows into the capillaries, the arterial blood pressure drops to a low value called *diastolic pressure.* Diastolic pressure does not reach zero because the ventricles remain relaxed for only a fraction of a second before contracting again. Also, as will be described later, the elasticity of large arteries helps prevent it from falling too low.

While the ventricles are contracting, the atria relax and then begin to fill with the next volume of blood that will enter the ventricles and be pumped to the body.

This completes one heartbeat, or *cardiac cycle.* By repeating this process over and over, the heart

FIGURE 4.5 Layers of the heart.

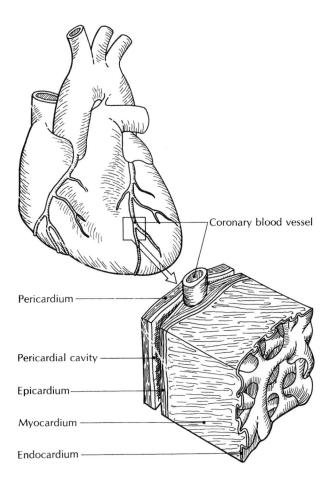

Pericardium

Pericardial cavity

Epicardium

Myocardium

Endocardium

Coronary blood vessel

sometimes used to restore the proper heart rate to a diseased heart. Other cells send the signal through the atria and then the ventricles. As the signal spreads, causing other muscle cells to contract, the contracting cells produce electrical impulses which can be detected and recorded. The recording is called an *electrocardiogram* (*ECG* or *EKG*).

Valves

Valves are located within the openings leading from the atria to the ventricles and from the ventricles to the arteries. The movement of blood from the atria into the ventricles and from the ventricles into the arteries pushes the valves open. When the ventricles begin to contract, some of the blood within them begins to move backward toward the atria. Similarly, when the ventricles relax, blood in the arteries starts to flow back into them. This causes the valves to swing shut, stopping the backward flow of blood. Thus, the valves ensure that the blood moves only in the correct direction (Fig. 4.4*a*).

Layers

The heart wall is composed of three layers: the endocardium, myocardium, and epicardium.

Endocardium   The inner lining of the heart is called the *endocardium* (Fig. 4.5). This layer must be very smooth and must have no gaps that allow blood to contact the underlying collagen. Blood that contacts rough spots or collagen will clot, and clots formed in the heart can move into arteries and block blood flow.

Myocardium   The middle layer of the heart—the *myocardium*—is a thick layer that constitutes most of the wall of the heart. The myocardium consists mostly of heart muscle (*cardiac muscle*), though it may also contain fat tissue and collagen fibers. Contraction of the cardiac muscle provides the force that pumps the blood.

The myocardium in the atria is thin because the atria pump blood only into the neighboring ventricles. The myocardium of the right ventricle is of moderate thickness because it must pump blood somewhat farther through the lungs. The myocardium of the left ventricle is much thicker because it pumps blood farther and through many

keeps the blood circulating. The blood must pass through the heart twice to make one complete circuit around the body (Fig. 4.1). The rate of flow depends on the amount pumped per minute: the *cardiac output* (*CO*). Cardiac output equals the amount pumped by each beat of either the left or the right ventricle [*stroke volume* (*SV*)] times the number of beats per minute [*heart rate* (*HR*)]. Therefore, CO = SV × HR.

The highly coordinated and well-timed operation of the heart chambers is controlled by special muscle cells. A patch of these cells in the right atrium signals when each beat is to begin. For this reason, the patch of cells is called the *pacemaker*, a name shared by the artificial electronic devices

FIGURE 4.6 Coronary arteries.

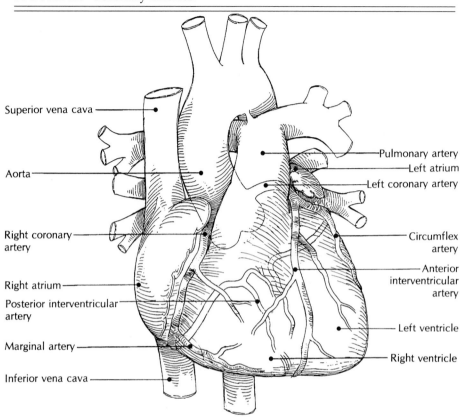

Superior vena cava

Aorta

Right coronary artery

Right atrium

Posterior interventricular artery

Marginal artery

Inferior vena cava

Pulmonary artery

Left atrium

Left coronary artery

Circumflex artery

Anterior interventricular artery

Left ventricle

Right ventricle

vessels in all other regions of the body.

Epicardium   The outer layer of the heart—the *epicardium*—contains some connective tissue coated with a smooth, slippery layer of epithelial cells. This coating allows the beating heart to move easily within the pericardial cavity. At the top of the heart, the epicardium tethers the heart to other structures in the chest so that it does not shift out of position.

Coronary Blood Flow

The heart muscle must have a steady supply of energy to pump blood continuously. It gets this energy through a complicated series of chemical reactions that combine oxygen with nutrients such as blood sugar. These materials must be delivered to the myocardial cells by the blood flowing through the *coronary arteries* (Fig. 4.6). The heart muscle cannot get materials directly from the blood inside the heart chambers because mol-

ecules do not pass easily through the thick wall of the heart.

In addition to producing useful energy, the reactions in heart cells produce wastes such as water and carbon dioxide, which are removed from the heart by blood in the coronary capillaries and veins. These wastes are finally eliminated by the lungs and kidneys.

If myocardial cells do not get enough oxygen for their energy requirements, they malfunction and the heart cannot pump blood adequately. People in this condition get out of breath easily. They feel weak and lethargic, tire quickly, may become dizzy and faint, and can suffer heart attacks. Therefore, the coronary arteries must deliver plenty of oxygen-rich blood to the myocardium.

Cardiac Adaptability

Recall that as the rate of activity of body cells changes, the amount of blood flow around the cells must also change to provide for their varying needs. This is especially important when levels of physical activity change because active muscles use materials and produce wastes much faster than resting muscles do. One way in which blood flow is adjusted is an alteration in cardiac output caused by changes in stroke volume or heart rate. Since alterations in CO must be made to maintain homeostasis, the heart is controlled by negative feedback systems. The nervous system detects changes in internal body conditions when exercise begins or ends. Cardiac output is then adjusted through changes in the nerve impulses sent to control the heart. Levels of hormones that influence the heart are also adjusted. Finally, the heart has intrinsic mechanisms to increase or decrease its own stroke volume as needed. As a result, the parts of the body receive the right amount of blood.

AGE CHANGES IN THE HEART

Resting Conditions

Though aging causes several changes in the heart, these age changes do not result in an alteration in cardiac output when a person is at rest. This is the case because the changes are slight and because adjustments that compensate for detrimental changes occur. These adjustments include changes in the atria and the myocardium that increase heart strength and increases in blood levels of *norepinephrine*, which stimulates the heart.

Cardiac Adaptability

As aging occurs, changes occur in the way the heart adjusts CO to meet the varying demands of the body. However, as with resting conditions, the changes are not very great and most detrimental changes in the heart are overshadowed by compensatory changes. For example, the heart compensates for an age-related decrease in maximum heart rate by increasing the amount it pumps per beat. Therefore, the maximum cardiac output which can be achieved when a person is exercising as vigorously as possible (cardiac reserve capacity) remains essentially unchanged. The compensatory change that seems most important involves norepinephrine: As age increases, its blood level rises faster and reaches a higher peak value after vigorous activity begins.

An adverse age change in the heart for which

there is no compensatory adjustment is an increase in the amount of blood remaining in the left ventricle after contraction. This residual blood causes a slight inhibition of blood flow from the lungs, resulting in an accumulation of blood in the lungs—*pulmonary congestion*—which raises the blood pressure in lung capillaries and forces extra fluid out through the capillary walls. This fluid accumulation (*pulmonary edema*) reduces respiratory functioning and causes people to feel out of breath sooner and more intensely when they exercise strenuously.

Another important age change involves the declining efficiency of the heart. A stiffer, dilated and thickened older heart consumes more oxygen to pump the same amount of blood pumped by a younger heart. This is not important as long as the coronary arteries remain completely normal because these arteries widen and allow blood to flow adequately when the heart needs more oxygen. However, most people do not have completely normal coronary arteries. In these cases, the decreased efficiency of the heart and the resultant increased demand for oxygen can become serious. In fact, individuals who show even slightly low coronary blood flow when exercising are very likely to have a heart attack.

In summary, because there are both positive and negative age changes in the normal heart, its ability to adjust the pumping of blood to supply the varying needs of the body remains essentially unchanged. However, the maximum rate of exercise people can perform normally declines with advancing age. This chapter and Chaps. 5, 6, 8, and 9 will show how this is due to a variety of factors outside the heart, including age changes in other parts of the circulatory system and in the respiratory, nervous, muscle, and skeletal systems. The maximum rate of physical activity decreases even more when diseases of the heart, blood vessels, or other body systems are present.

Exercise and the Aging Heart

Though the effects of exercise programs on the hearts of younger people have been well studied, only a few of the effects of training on the older heart have been elucidated. For the older heart, these effects include lack of change in the resting heart rate and the maximum heart rate attainable, decreases in the maximum heart rate required for maximum activity, and increases in stroke volume and cardiac efficiency. However, it is expected that other beneficial effects of regular exercise on the older heart will be discovered. We will see later in this chapter that regular exercise has beneficial effects on other parts of the circulatory system.

To affect an older heart, an exercise program must involve fairly vigorous exercise performed for an extended period during each session. The sessions must occur frequently, at least once every few days. The degree of improvement is proportional to the intensity, duration, and frequency of activity. Exercise at a low level, performed for a short time, or conducted infrequently has no effect on the aged heart. Beneficial changes in cardiac output can be observed within days to a few weeks after beginning an exercise program, and regressive detrimental changes in CO occur within the same time frame when a person ends the program.

DISEASES OF THE HEART

Thus far we have seen that aging of the heart does not significantly alter its ability to meet the needs of the body. However, few adults have a completely normal heart, and in most individuals some degree of disease adversely affects the heart. Both the incidence and seriousness of such disease increase with age.

The reasons for these increases are the same as the reasons for those which lead to an age-related rise in other diseases: higher chances for exposure to disease-producing factors, increasing occasions and durations of exposure to such factors, more time for the development of slowly progressing diseases.

Heart disease is the fourth leading chronic disease among people between the ages of 45 and 64 and ranks second among those over age 64. It is the leading cause for seeking medical care among those over age 64 and is a major cause of disability and altered lifestyle. Though the incidence of death from heart disease among the elderly has been declining for decades, this is still by far the leading cause of death for people 65 and over.

Coronary Artery Disease

Though several different heart diseases become more common and more serious with age, disease of the coronary arteries stands out as the most common of these disorders.

FIGURE 4.7 Atherosclerosis.

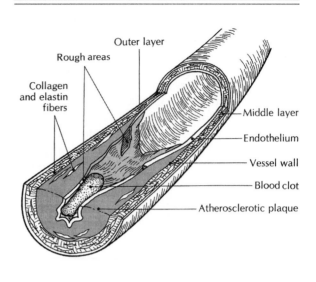

Functions of Coronary Arteries When a person is at rest, the coronary arteries are normally wide enough to allow ample blood to pass through to the cardiac muscle cells. However, the demand of cardiac muscle for oxygen goes up and down as the amount of work performed by the heart rises and falls. Conditions requiring more work and higher amounts of oxygen include increases in heart rate, stroke volume, width, and thickness and in blood pressure. Such increases occur when a person becomes physically active and as part of aging. The heart normally accommodates these increases by dilating its arteries to allow more blood to flow through them.

Effects of Atherosclerosis The coronary arteries are prevented from supplying adequate blood flow to the heart by a disease called *atherosclerosis*. Atherosclerosis, which is described in greater detail later in this chapter, involves the formation and enlargement of a weak scar-like material called *plaque* in the walls of arteries. Plaque causes coronary arteries to become narrower and thus reduces blood flow (Fig. 4.7). It also stiffens the arteries, reducing their ability to dilate when more oxygen is needed by the heart muscle. Finally, plaque causes roughening of the inner lining of the arteries and exposure of the underlying collagen. Roughness and collagen cause the blood in arteries to form clots; clots clog arteries and can stop blood flow quickly and completely.

Whenever the amount of oxygen needed by the heart is lower than the amount supplied, the cardiac muscle cells become weak and cannot pump enough blood to body organs. In addition, the muscle cells begin to produce a waste product called *lactic acid*, which upsets the normal acid/base balance. This imbalance injures the muscle cells, which become even weaker, and blood flow to the body drops further. All the organs begin to perform less well. The brain, kidneys, lungs, and heart are especially in danger because these organs require high levels of blood flow. A person in this condition often feels weak and out of breath and frequently experiences chest pain as injury to the heart cells develops. In mild cases the pain will subside if the person rests because the oxygen demand of the heart drops back to the level being supplied. Such temporary pain is called *angina*.

If oxygen demand is brought back into balance with oxygen supply soon enough, the heart begins to function normally again. Of course, the arteries are still diseased and the problem will most likely recur. The incidents may become more severe as the degree of coronary artery disease increases.

If the oxygen supply is very low for just a few minutes, the cardiac muscle cells begin to die. This condition is a true *heart attack*, also called a *myocardial infarction* (*MI*) (Fig. 4.8). The heart becomes much weaker, and the pumping of blood drops precipitously. Cells in the brain and other organs deteriorate, and the person is in danger of dying. In fact, first-time heart attacks are fatal 65 percent of the time. Individuals who survive the initial effects of a heart attack still face many problems. The heart attack can cause damage to the heart valves or may produce a hole between the left and right ventricles, in which case the blood will flow in the wrong direction within the heart. Incorrect blood flow tends to overwork the heart, causing more heart disease and often preventing the lungs from functioning properly. An MI can also cause blood clots to form inside the heart chambers and then be pumped to other organs. If clots travel to the brain, they can cause a stroke. Finally, the heart can become so weak that the person may require lengthy medical treatment and face long-term disability. The person's social contacts, sense of well-being, normal daily rou-

FIGURE 4.8 Coronary artery disease.

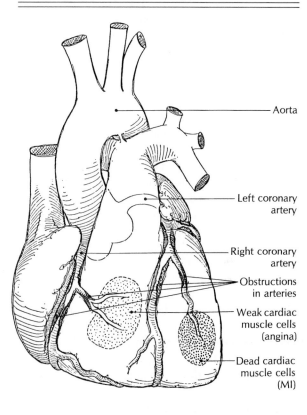

Aorta

Left coronary artery

Right coronary artery

Obstructions in arteries

Weak cardiac muscle cells (angina)

Dead cardiac muscle cells (MI)

tines, and employment often undergo radical undesirable changes.

Risk Factors All that has been said thus far about coronary atherosclerosis may seem like bad news, but there is also good news about this disease. Most of the factors contributing to the development of atherosclerosis have been identified, and many of them can be avoided or greatly reduced. This is the main reason for the dramatic decline since 1950 in the incidence of deaths from heart disease. Furthermore, reducing or eliminating one or more of the risk factors reduces the chances of being affected by this disease regardless of the age at which the decrease in risk factors occurs. Of course, the earlier the risk-reducing steps are taken, the greater is the benefit.

Another important fact regarding risk factors must be highlighted here. Having two or more risk factors drastically boosts one's chances of developing coronary atherosclerosis because the detrimental effects of each risk factor are multiplied by the effects of the others. For example, smoking almost doubles the risk. Having high blood pressure multiplies the risk four times. A person who smokes and has high blood pressure has an eightfold greater risk. Adding high blood lipoproteins, which increases the risk threefold, increases the total risk to 24-fold. Therefore, reducing or eliminating even one risk factor can cause a manyfold decrease in the risk of developing coronary artery disease.

Some risk factors create more problems than others do. The following six factors provide the highest levels of risk. The actual amount of increase in risk from each one depends on when the risk factor first existed; its intensity, frequency, and duration; and its interaction with other risk factors.

Smoking   Inhaling tobacco smoke increases blood pressure and adds substances to the blood that seem to promote the formation of plaque. The effect of smoking on arteries is greatly magnified in women who take birth control pills. The combination of smoking and taking birth control pills increases the risk of having a heart attack almost 18-fold. The solution is to not smoke.

High Blood Pressure   High blood pressure seems to cause repeated minor injuries to the arteries. As the arteries try to repair the damage, they form scar tissue and plaque. High blood pressure also makes the heart work harder, increasing the amount of oxygen it needs, and eventually weakens the heart.

Having blood pressure checked regularly and, if it is high, seeking professional advice on how to reduce it are especially important as people get older. Blood pressure tends to rise with age and an abnormal increase has more of an effect on the arteries as a person ages.

High Blood LDLs   Blood contains a variety of lipoprotein molecules. The lipids in these lipoproteins are obtained from the diet and made by the body. Most of the lipid in blood lipoproteins is cholesterol and triglycerides (fats), and lipoproteins containing predominately cholesterol are called *low-density lipoproteins* (**LDLs**). When LDLs are in high concentrations, the cholesterol can accumulate in the walls of arteries and contribute to the formation of plaque. The accumulation of cholesterol is reduced by other lipoproteins called *high-density lipoproteins* (**HDLs**). As

age increases, many individuals have an increase in the concentrations of LDLs with a simultaneous decrease in HDLs.

To reduce the risk of developing high blood LDLs, the amounts of cholesterol and saturated fats in the diet should be kept low. Foods containing high amounts of these lipids include egg yolks, dairy products containing milk fat or cream, red meats such as beef and pork, solid shortening, and oils such as palm oil and coconut oil. High alcohol consumption should be avoided since it promotes the formation of LDLs. However, consuming low or moderate levels of alcohol, eating foods containing certain dietary oils (e.g., safflower oil), and exercise can reduce blood LDLs while increasing HDLs. Blood lipoprotein levels should be checked, and professional guidance should be followed if the ratio of LDLs to HDLs is found to be too high.

Diabetes Mellitus   *Diabetes mellitus* is a disease that alters many aspects of the body, including blood glucose levels and the maintenance and repair of arterial walls. In so doing, it promotes the formation of plaque.

Individuals should be aware of the warning signs of diabetes mellitus, which include excessive hunger and thirst, fatigue, unusual weight gain or loss, excessive formation and elimination of urine, and slow healing of wounds. Suspected cases require diagnosis and treatment by a qualified professional.

Family History   Several unidentified genes increase the chances of developing coronary atherosclerosis. The mechanism by which these genes act is not known.

Individuals from families with a history of atherosclerosis may have inherited the genes that predispose them to this disease. Though these individuals cannot alter their genes, they should try to reduce or eliminate as many other risk factors as possible. They should also inform their health care providers of their family history so that problems can be detected and necessary treatments can be initiated early.

Advancing Age   Advancing age increases the risk of problems from coronary artery disease in several ways. First, aging causes arterial stiffening. Second, there is an increase in the heart's oxygen demand because the heart becomes less efficient.

Third, aging is associated with higher blood pressure, elevated blood cholesterol and LDLs, lowered blood HDLs, an increased incidence of diabetes, and decreased physical activity. Fourth, increasing age provides more time for other risk factors to take effect and for the slow process of plaque formation to progress significantly.

Though nothing can be done to alter the passage of time, people of advanced age should reduce other risk factors as much as possible.

Other risk factors are of moderate importance compared with the six just discussed. They include the following.

High Blood Homocysteine *Homocysteine (Hcy)* is produced and released into the blood when the body breaks down an amino acid called **methionine**. Having high blood levels of Hcy increases the risk of developing atherosclerosis. Blood levels of Hcy rise with increasing age and when women pass through menopause. High blood levels of Hcy also develop in people with deficiencies in *vitamin B6* or the *vitamin B12 (cobalamin)*. These vitamins are essential for adequate disposal of Hcy. Finally, some people are born with a metabolic abnormality that causes them to produce excess Hcy.

Usually blood levels of Hcy can be kept low by eating a diet with adequate vitamin B6 and vitamin B12. Since there is little vitamin B12 in plants, vegetarians are at risk for vitamin B12 deficiency. People with abnormalities of the stomach may be unable to absorb adequate vitamin B12. Vitamin supplements can help people who do not get adequate vitamin B6 or vitamin B12 from foods.

Physical Inactivity The cells of people who are physically inactive require less blood flow, and the heart therefore gets less exercise because it does not have to work hard. Like every other muscle, heart muscle that gets little exercise becomes weaker and less efficient and loses some of its blood vessels. Such a heart is unprepared to increase the pumping of blood when a person suddenly begins strenuous activity. When such activity begins, an imbalance in the oxygen demand and supply of the heart occurs. In addition, lack of exercise promotes increases in blood pressure and in the ratio of LDLs to HDLs; both changes promote the development of atherosclerosis.

Engaging in a regular program of vigorous physical activity or in an occupation or hobby that includes such activity greatly reduces or eliminates this risk factor. This occurs because the heart is strengthened and develops more blood vessels, blood pressure is kept low, blood levels of HDLs are increased, body weight is less likely to become excessive, and psychological stress is minimized. All these effects reduce the risk of coronary artery disease. Planning involvement in physical activity is especially important for persons of advanced age because of the tendency toward age-related reductions in physical activity.

Obesity Being very overweight weakens the heart and makes it less efficient because the heart is being overworked and tends to become invaded with fat. Obesity also promotes high blood pressure, high levels of blood cholesterol and LDLs, diabetes mellitus, and low levels of physical activity. Obesity can be prevented or reduced by participation in a planned program of diet modification and regular exercise.

Stress A sustained high level of emotional tension or stress promotes atherosclerosis by causing prolonged periods of high blood pressure.

Emotional stress can be reduced in many ways. One way is to avoid stress-inducing situations. When this is not possible, taking breaks or vacations from such situations helps. Exercise, hobbies, and other diversions can also provide relief. Talking with a trusted confident can help, and some individuals can benefit from professional counseling.

Menopause The decline in estrogen and progesterone that occurs at menopause ends the protective effect those hormones have on the arteries. Surgical removal of the ovaries has the same effect.

Some women can benefit from hormone replacement therapy. Such therapy should be conducted only on the advice and under the continued direction of a physician knowledgeable in this area.

Male Gender Coronary artery disease occurs more frequently in men than in women, probably because the lifestyle of men generally includes more and higher levels of the risk factors associated with this disease. The incidence rate for women has been approaching that for men as women have become more involved in the same activities. Furthermore, the incidence among women approaches that of men as the age of a population increases because of losing the protective effects of female hormones after menopause.

Personality   Individuals with certain personality characteristics seem to be at higher risk for developing coronary atherosclerosis. These characteristics include being highly competitive, striving for perfection, and feeling that there is never enough time to accomplish one's goals. These characteristics indicate a high level of stress. Recent evidence suggests that stress is the key feature and that personality contributes little if any risk.

High Blood Iron Levels   Another possible risk factor, which some scientists think provides a risk exceeded only by that from smoking, is having higher than average levels of iron in the blood. The iron may promote atherosclerosis by increasing the accumulation of LDLs in arteries and causing cell damage by promoting the formation of free radicals. If having relatively high blood levels of iron is shown conclusively to be a significant risk factor, steps to lower these levels might include reducing the consumption of foods high in iron (e.g., red meat, liver, spinach, iron-fortified foods); not drinking water with a high iron content; taking iron supplements only when absolutely necessary; and giving blood regularly.

Periodontal Disease   Periodontal disease is associated with increased risk of atherosclerosis, heart attack, and stroke. The mechanism by which periodontal disease contributes to atherosclerosis is not known. They may involve a genetic predisposition to both periodontal disease and atherosclerosis; toxins and chemical signals produced at the teeth; and effects from bacteria spreading from the teeth throughout the body. Some of these mechanisms may affect endothelial function.

Congestive Heart Failure

Congestive heart failure (CHF) is another disease that becomes more common and serious with age. Approximately three million people in the U.S. have CHF, and there are approximately 400,000 new cases each year. More than 75 percent of cases are in people age 65 and over. Incidence rates double for each decade over age 45, and approximately 10 percent of elders over age 80 have CHF. Congestive heart failure is the leading cause of hospital admissions for people 65 and over, and

it is a major cause of disability, reduced independence, and death. The number of cases is expected to double by the year 2040.

Main causes of CHF are factors that weaken the heart. The most frequent causes are coronary artery disease, high blood pressure, disease of the heart valves, obesity, and kidney disease. The underlying problem is years of overworking the heart. An overworked heart tends to strengthen itself by dilating and thickening. At first these changes increase heart strength, but if the heart continues to be overworked, it continues to dilate and thicken. Excessive amounts of these changes weaken the heart. Then the heart chambers contain a great deal of blood but cannot pump it effectively. The flow of blood diminishes, and organs begin to malfunction.

In addition, fluid accumulates in the lungs (pulmonary edema). Affected individuals have difficulty breathing and may feel out of breath after the slightest exertion or even when resting. Poor circulation in other areas, especially the legs, causes swelling and discomfort and promotes the formation of varicose veins. Many ordinary activities become difficult or impossible.

The heart tends to solve these problems by dilating and thickening even more, but this exacerbates the situation. Unless steps are taken to strengthen the heart and reduce its workload, the heart gradually becomes so weak that it fails completely and the individual dies.

Valvular Heart Disease

Untreated serious *valvular heart disease* causes detrimental changes similar to those resulting from congestive heart failure. This disease usually develops after coronary artery disease or rheumatic fever, which can prevent the valves from closing properly. Then some of the blood in the heart flows backward during each beat. Rheumatic fever can also prevent the valves from opening properly, and so blood does not flow forward as easily as it should. In either case the heart is overworked.

ARTERIES

Arteries are flexible tubes that carry blood from the heart to every region of the body. Arteries have special properties that ensure that they perform this task effectively. These properties derive from the three layers composing the arterial wall.

FIGURE 4.9 Structure of (a) arteries and (b) veins.

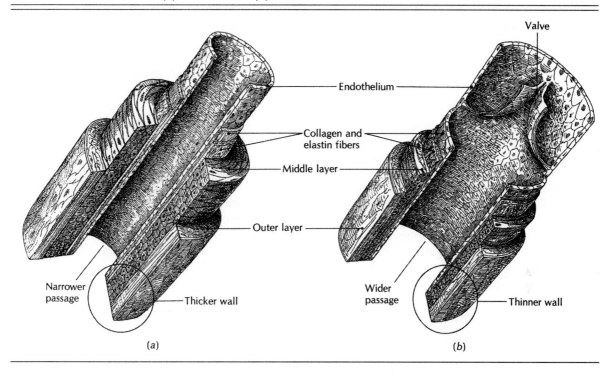

Inner Layer

The innermost layer of an artery is called the *endothelium*. It is supported by a thin underlying layer that contains collagen and a mat of elastin fibers (Fig. 4.9a). Like the endocardium in the heart, the endothelium provides smoothness by forming a continuous glistening layer that coats the collagen and other materials in the arterial wall. In so doing, it permits blood to flow easily without clotting.

The endothelium also secretes several signaling materials including *nitric oxide* (*NO). *endothelin*, *prostacyclin,* and *angiotensin converting enzyme* (*ACE*). Nitric oxide and prostacyclin promote vasodilation in many arteries. Nitric oxide also limits vessel thickening by inhibiting the growth of smooth muscle, and it inhibits clot formation and plaque formation. Endothelin and ACE stimulate vasoconstriction. The effects of *NO usually dominate, keeping vessels adequately dilated.

Middle Layer: Large Arteries

The middle layer in the largest arteries consists mostly of a thick layer of elastic fibers that make the arteries strong. Most of the elastic fibers in the aorta are produced before birth or during childhood, but some new elastin is produced throughout life. Since these arteries are closest to the heart, strength is necessary to withstand the high blood pressure produced by each heartbeat.

This layer also provides elasticity, which allows the arteries to be stretched outward somewhat each time the ventricles pump blood into them. The extra space provided by the stretching prevents the systolic pressure from rising too high, and the work the heart must perform is kept reasonably low, just as it is easier to blow up an easily stretched balloon than a stiff one. Preventing excessive pressure also keeps the arteries from being injured by the accompanying extreme forces.

The elasticity of large arteries helps to prevent blood pressure from rising too high in yet another way. Unusually high blood pressure immediately causes normal arteries to be stretched outward excessively. Nerve cells in the walls of arteries detect this abnormal stretching and send signaling impulses to the blood pressure control center in the brain. Other factors from the artery that influence the sensory neurons include prostacyclin, which increases the signals, and reactive oxygen species, which reduce them. The brain then sends impulses to the heart telling it to pump less blood. It also tells blood vessels in various areas of the body to dilate to provide more space for the blood coming from the heart. As a result, blood pressure decreases and the large arteries return to their normal size. The nerve cells are then no longer activated, and blood pressure stabilizes at the normal level. Note that this is a negative feedback system that maintains proper and fairly stable conditions in the body.

The nerves in blood vessels send different impulses to the brain when blood pressure is too low and arteries are not stretched enough. The result is the sending of norepinephrine and related substances to the heart and vessels. These substances raise blood pressure by several means, including stimulating the heart and causing the smaller arteries to constrict.

Elasticity also causes the arteries to snap back to their original diameters when the ventricles are relaxing. This elastic recoil helps maintain diastolic pressure between beats by squeezing the blood. Diastolic pressure keeps the blood moving forward steadily while the heart rests briefly after each beat. Thus, elastic recoil serves the same purpose as the spring that keeps a watch ticking between windings.

Middle Layer: Smaller Arteries

The middle layer of smaller arteries contains some elastic fibers but is composed mostly of smooth muscle. When the muscle contracts, it causes the arteries to constrict, reducing the flow of blood.

As a rule, the smooth muscle in most smaller arteries contracts weakly, providing some resistance to flow while allowing ample blood flow through the arteries. This resistance is important because without it, blood would flow from the arteries into the capillaries so quickly that blood pressure would drop too low, especially between heartbeats. The same effect is observed with tires. A tire with a tiny hole loses pressure so slowly that it can be reinflated before any harm is done, while a tire with a large leak loses all its pressure and goes flat very quickly.

In normal situations the constriction of small arteries increases whenever blood pressure begins to drop too far. This additional constriction increases resistance and raises blood pressure back to the proper level. Conversely, if blood pressure

rises too high, the smooth muscle relaxes, the arteries dilate, resistance drops, and blood pressure decreases back to normal levels.

In addition to regulating blood pressure by constricting or dilating as a group, individual arteries can constrict to reduce blood flow to organs that need little flow while others dilate to increase flow to more active organs. Thus, the smaller arteries act like a set of valves or traffic signals to make sure that each part of the body receives only as much blood flow as it needs.

The constriction and dilation of smaller arteries are controlled by negative feedback systems. The arteries respond to several factors, including nerve impulses, hormones, temperature, and chemical conditions in their vicinity. These mechanisms help maintain normal blood pressure and blood flow to each body structure.

Outer Layer

The outer layer of arteries consists largely of loose connective tissue containing soft gel and scattered fibers. This layer loosely attaches arteries to other structures, enabling arteries to be shifted as parts of the body move while preventing the arteries from moving too far out of position.

AGE CHANGES IN ARTERIES

There are no important age changes in the endothelial structure, its supporting layer, or the outer layer of arteries. These layers function well regardless of age with one major exception. There is an age-related decline in the ability of the endothelium to regulate blood vessels and blood pressure. The cause is not clear. It may be due to aging of endothelial cells, to free radical damage, or to age-related increases in blood pressure. Nitric oxide reacts with superoxide radicals ($*O_2$), producing toxic $ONOO^-$ (peroxynitrite). This reaction reduces the amount of *NO available to regulate vessels and blood clotting, and the $ONOO^-$ can slow *NO production further by injuring endothelial cells. These changes may contribute to high blood pressure and to atherosclerosis.

Middle Layer: Large Arteries

Numerous age changes occur in the middle layer

of large arteries. Age changes in elastic fibers include breakage, glycation, accumulation of calcium and lipid deposits, and faster breakdown by enzymes. Old damaged elastic fibers accumulate. The increase in many substances, including smooth muscle, collagen, calcium deposits, and cholesterol and other fatty materials, causes thickening and stiffening of the arteries. These changes amplify the decline in elasticity caused by the altered elastic fibers. Since the arteries are less able to be stretched by each pulse of blood, systolic pressure tends to rise.

Since much elastin in the aorta is produced before birth, children with low birth weights may have less aortic elastin, resulting in less aortic strength and elasticity. This can speed up aortic thickening and stiffening during childhood and adulthood, resulting in a greater risk of high blood pressure and related diseases (e.g., atherosclerosis, congestive heart failure). These effects highlight the importance of events in youth or even before birth to age-related changes and disease later in life.

At the same time the arteries are stiffening, years of containing blood under high pressure causes them to gradually widen and lengthen. This is especially evident in the aorta. These changes provide more space for blood. At first this compensates for the declining ability of large arteries to be stretched, and consequently it keeps the tendency toward increases in systolic pressure in check. Eventually, however, the elastic fibers are stretched so much that they can yield no further. Then each contraction of the heart produces a rapid and dramatic rise in systolic blood pressure. This can increase cardiac oxygen demand by almost 30 percent. At the same time, the high blood pressure and thickening of the heart reduce the amount of *NO in coronary vessels. This limits the vessel dilation required to increase oxygen supply to the heart muscle.

Once the arteries no longer stretch much with each heartbeat, sensory nerve cells that detect vessel stretching are not activated as much. Age changes in the endothelium that reduce prostacyclin and increase reactive oxygen species (ROS) also reduce the nerve cell activation. The reflex to prevent abnormal increases in blood pressure is suppressed, and the pressure remains high. In most cases sensory nerve cells are fooled and respond as though blood pressure were too low. Age changes in the brain's blood pressure control center amplify this effect. The final result is the release of norepinephrine, which augments the high pressure but also stimulates the heart. In this way, the extra norepinephrine seems to be compensatory because it helps the aging heart maintain cardiac output.

As with all age changes, there is much variation among individuals with respect to the rise in systolic pressure. While the elevations are modest in most people, about 40 percent of the elderly have systolic pressures above the safe maximum for those of advanced age (160 mmHg). Recall that elevated blood pressure increases the heart's workload and oxygen needs and the risk of developing atherosclerosis. Therefore, it is important for the elderly to have their blood pressure checked and, when necessary, receive therapy to keep it within safe limits. However, elevated blood pressure in older individuals must not be lowered too quickly or too far, since the result can be weakness, fainting, or more serious damage to the heart, the brain, and other parts of the body.

In addition to restricting stretching, stiffening of the arteries diminishes their elastic recoil. The slow decline in recoil does not cause a substantial change until about age 60, after which diastolic pressure declines slightly. The result is a slowing of blood flow through coronary arteries and other small arteries between heart beats. Normally, this decline is not large enough to cause significant effects, though it brings a person closer to having inadequate blood flow during each diastole.

In a large longitudinal study of people with no diseases of the circulatory system, systolic BP does not change until approximately age 40 in women and age 50 in men. Then BP increases approximately 5-8 mmHg per decade. In women, the systolic BP may stop rising and may even begin to decrease after age 70, while in men the systolic BP continues to rise throughout life. The overall increase in systolic pressure averages 21 mmHg in women and 15 mmHg in men. Diastolic pressure in women increases from ages 40 to 60, but then levels off or declines. Diastolic pressure in men increases 1 mmHg per decade. Overall, diastolic pressures increase 5 mmHg in women and 3.5 mmHg in men. Studies that include people with diseases or who take medications show greater changes in BP, and the women may not have the leveling and decline in BP after age 70.

Middle Layer: Smaller Arteries

Aging causes little if any change in the overall resistance provided by the smaller arteries. Their thickening seems to help prevent overstretching as systolic blood pressure increases with age. Older arteries do not respond quite as well when conditions such as chemical levels begin to change. This seems to be due in part to decreased functioning of the nervous system and altered levels of the hormones that control the vessels. The vessels also seem to have reduced sensitivity or a reduced ability to respond to control signals. Therefore, the arteries do not dilate as well when the areas they supply need more oxygen. This decrease in supply tends to reduce the maximum rate of work that certain organs (e.g., muscles) can perform.

The decreased ability of the arteries to respond to rising or falling body temperature is an even greater problem, leaving older people less able to prevent themselves from overheating or becoming chilled. For example, inadequate dilation of dermal vessels prevents the extra heat produced during exercise from leaving the body quickly. This can lead to excessively high body temperature, damage to body molecules and cell parts, malfunctioning of organs such as the brain, illness, or even death. Poor constriction by dermal vessels when a person is in a cold environment can cause excessive loss of body heat and a drop in body temperature. Not only will such an individual feel uncomfortably cold, but because of slowing cell activities and malfunctioning of organs such as the muscles and the heart, he or she may also become ill.

The declining ability to maintain normal body temperature as age increases is due not only to age changes in the middle layer of smaller arteries but also to age changes in the integumentary system (e.g., sweat glands, fat tissue), the nervous system (e.g., sensory neurons), and the muscle system (e.g., muscle mass).

Because of reduced thermal adaptability, older individuals should avoid environments and activities that tend to cause significant elevation or depression of body temperature. Hot weather or very warm indoor areas, hot baths or showers, the use of numerous blankets or electric blankets, and strenuous physical activity tend to cause overheating. Cold weather or cool rooms, cool water for swimming or bathing, exposure of the skin, inadequate clothing, and restricted physical activity increase the risk of developing hypothermia.

Number of Arteries

The number of larger arteries remains the same throughout life. The number of smaller arteries remains about the same or increases slightly in some areas of the body (e.g., heart and brain). This slight increase helps sustain normal blood flow by compensating for the development of somewhat irregular arteries. Other areas (e.g., skin, kidneys) have decreasing numbers of smaller arteries with age.

Fortunately, the adverse effects on blood flow caused by the aging of arteries can be largely overcome through steps such as receiving proper medical care, pacing activities, and avoiding situations that place a person in danger of overheating or chilling. Unfortunately, for most individuals, aging arteries are affected not only by age changes but also by arterial diseases.

ATHEROSCLEROSIS: AN ARTERIAL DISEASE

By far the most common arterial disease is *atherosclerosis*, which is one of a group of arterial diseases called *arteriosclerosis*. Because atherosclerosis is very common, some people mistakenly use these two terms interchangeably. The incidence of atherosclerosis and the serious difficulties it causes rise with age for the same reasons that cause the age-related increase in heart disease.

Importance

Some statistics on the importance of atherosclerosis were presented earlier in this chapter. In addition to causing most heart attacks, atherosclerosis causes most strokes. A *stroke* is injury to or death of brain cells caused by low blood flow or bleeding in the brain (Chap. 6). For those over age 65, strokes are now the third leading cause of death, days in the hospital, and days in bed. Strokes also cause many cases of dementia and other forms of disability.

Atherosclerosis is also a major contributor to kidney disease, problems in the legs (e.g., weakening of muscles and skin, pain during exertion), and male impotence. Such outcomes not only affect an individual's health and survival but also have an impact on all other aspects of life. For example, dietary restrictions may become necessary, demanding occupational or recreational activities may have to be curtailed, and interpersonal relations between affected men and their spouses can suffer dramatically.

Development and Effects

Atherosclerosis begins as small streaks of fatty tissue within the inner layer of arteries. Gradually, the streaks widen and thicken as they accumulate a variety of other materials, including smooth muscle cells, collagen fibers, cholesterol, and calcium deposits. The resulting masses—*plaques*—protrude inward and narrow the passageway in the artery (Fig. 4.7). The plaques often grow completely through the endothelium and replace regions of it. Both the roughness and the collagen fibers of the plaques cause the blood to form clots. As a result, the narrowing of the artery lead to reductions in or complete blockage of blood flow. In addition, pieces of the plaque sometimes break off, move down the artery, and block the artery where it branches to form smaller arteries.

These plaques usually grow outward and infiltrate the middle layer of the artery, causing it to stiffen. When this occurs in larger arteries, they are less able to be stretched outward to accommodate pulses of blood from the heart and systolic pressure can skyrocket. Since the arteries are also less able to spring back when the heart relaxes, diastolic pressure drops and the flow of blood becomes less regular. When plaque grows outward in smaller arteries, the stiffening and replacement of the smooth muscle prevent them from adjusting blood pressure and blood flow to suit body needs. The cells do not receive adequate oxygen and nutrients, and waste materials accumulate. The resulting loss of homeostasis injures or kills cells, and the organs they compose malfunction.

The outward growth of plaque also causes weakening of the middle layer, and affected arteries begin to bulge outward from blood pressure. The outpocketings, called *aneurysms*, can disturb nearby structures by pressing on them (Fig. 4.10). Additionally, blood flowing past aneurysms tends to swirl and form clots. Some arteries become so weak that they rupture, causing severe internal bleeding that can lead to the most serious strokes.

Mechanisms Promoting Atherosclerosis

Several factors seem to cause atherosclerosis or to promote its development. These include endothelial dysfunction, free radicals, blood lipoproteins, elastase, glycation, heat shock proteins, and insulin-like growth factors (IGFs). Some of these may interact synergistically.

Endothelial Dysfunction   Endothelial dysfunction may cause or result from endothelial aging, high blood pressure, or atherosclerosis. Endothelial dysfunction increases the adverse effects from high BP and from atherosclerosis. Part of the effect may be from an age-related increase in $*O_2$, which reduces *NO by reacting with it to form ONOO⁻. With less *NO, vessel dilation is reduced and vessel smooth muscle growth and clot formation increase. At the same time, ONOO⁻ may initiate or promote plaque formation by injuring the vessel wall.

Free Radicals   Free radicals may also contribute to atherosclerosis by increasing the formation of lipid peroxides (LPs) from blood lipoproteins. Blood LPs increase with age and after menopause, and also with increases in blood LDLs, blood pressure, stress, diabetes mellitus, and smoking. Lipid peroxides may promote atherosclerosis in several ways. Examples include increasing the absorption of LDLs by vessel macrophages, converting them to cholesterol-filled *foam cells*; injuring vessel cells directly; attracting monocytes and macrophages into vessel walls, which promote inflammation and cell damage; promoting vessel constriction; and promoting blood clot formation.

Blood LDLs   Blood LDLs may promote atherosclerosis by increasing LPs and by increasing elastase.

Elastase   *Elastase* is an enzyme that breaks down elastic fibers into elastin peptides. Elastin peptides are also formed during elastin synthesis. Elastase increases with age and with higher LDL levels. Elevated levels of elastin peptides seem to promote more elastase production by promoting the binding of calcium and lipids to elastin fibers.

Elastase may increase atherosclerosis by

FIGURE 4.10 Aneurysms.

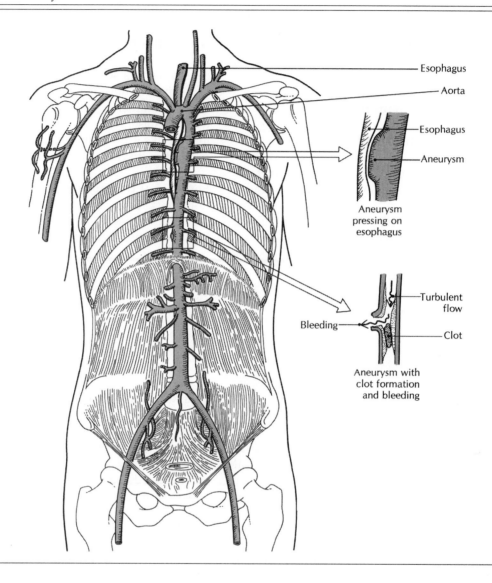

- Esophagus
- Aorta
- Esophagus
- Aneurysm

Aneurysm pressing on esophagus

- Turbulent flow
- Bleeding
- Clot

Aneurysm with clot formation and bleeding

reducing elastic fibers in arteries, making vessels more susceptible to damage by blood pressure. The effects from the elastin peptides produced seem to increase *NO. Results include benefits such as vasodilation, and drawbacks such as vessel damage by stimulated monocytes. Research has provided contradictory results regarding the effects from elevated elastin peptides on promoting or reducing atherosclerosis.

Glycation   Glycation of proteins in arteries produces age-related glycation end-products (AGEs) and *FRs. The *FRs may promote atherosclerosis directly. The AGEs bind in fatty streaks and stimu-

late inflammation and *FR formation by macrophages. Glycated collagen in arteries is distorted and stiffer causing adverse effects. These include reduced effectiveness of nitric oxide as a vasodilator; detachments of endothelium from the vessel wall; and increased clot formation.

Heat Shock Proteins   *Heat shock proteins* are produced when cells are stressed or injured. These proteins received their name because they were first discovered in cells subjected to abnormally high temperatures. Heat shock proteins seem to protect cells from a variety of harmful environmental factors. An immune response to heat shock

FIGURE 4.11 Capillary exchange and lymph formation.

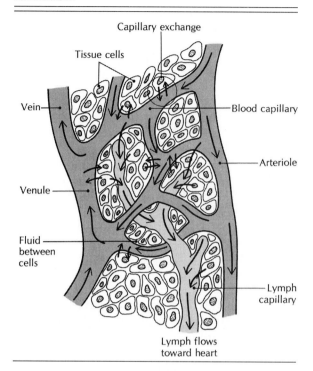

FIGURE 4.12 Capillary structure and capillary exchange.

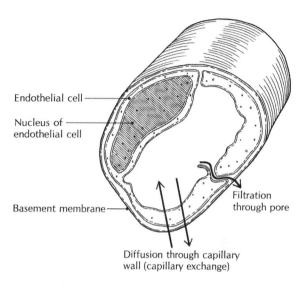

protein in damaged arteries may be the initial event in atherosclerosis.

Insulin-like Growth Factors *Insulin-like growth factors (IGFs)* from cells stimulate growth and regulate other cell activities. The distribution and effectiveness of IGFs are altered when they bind to *insulin-like growth factor binding proteins (IGFBPs)*. Research suggests that different ratios of IGFs and IGFBPs influence the development of atherosclerosis, possibly by altering blood levels of lipoproteins and *NO and by affecting the growth of vessel smooth muscle.

Prevention

The incidence and seriousness of atherosclerosis can be greatly lowered by avoiding or reducing the known risk factors, which were discussed in the section on coronary artery disease.

CAPILLARIES

The *capillaries*, which receive blood from the smallest arteries, are often no more than 1 mm in length. They are so narrow that blood cells can pass through only in single file and often must

actually fold to pass through. The capillaries run among the body cells and are so numerous and so close together that no cell is very far from a capillary (Fig. 4.11).

Capillaries are porous vessels through which materials in the blood move out to the surrounding body cells and many of the materials produced by body cells (e.g., wastes, hormones) move back into the blood. Since materials are moving in both directions, this process is called *capillary exchange*. However, a portion of the material that moves out of the capillaries and some of the material produced by the cells do not travel back into the capillaries. This material passes instead into *lymph capillaries*, where it is known as *lymph*. The lymph then passes through lymph vessels, which deliver it into large veins near the heart.

The structure of capillaries is well suited for capillary exchange (Fig. 4.12). The wall of each capillary is composed of a single layer of thin cells that are supported by a thin layer of material they secrete (*basement membrane*). Many small atoms and molecules pass through the capillary wall quickly and easily by the process of *diffusion*, which involves the movement of materials from an area of higher concentration to an area of lower concentration. Therefore, substances that are abundant in the blood diffuse outward to the cells, while other substances diffuse from the cells into the blood.

Capillary walls have pores between the cells that constitute them, making it even easier for substances to diffuse between the blood and body cells. In addition, blood pressure pushes many atoms, ions, and small molecules out of the capillaries through the pores, leaving large molecules and cells within the blood. This separation of small substances from large ones by fluid pressure is called *filtration*.

Age Changes in Capillaries

With increasing age, many capillaries become narrower and irregular in shape, and this retards the flow of blood. Some capillaries become so narrow that blood cells get stuck in them, further inhibiting blood flow. In some organs (e.g., heart, muscles) blood supply is further reduced because of a decrease in the number of capillaries. Finally, capillary walls become thicker and have a decrease in the number of pores; both changes inhibit capillary exchange.

Age changes reduce the ability of capillaries to meet the needs of body cells quickly. Therefore, while the cells may be able to function well at low levels of activity, both the ability to sustain vigorous activity and the maximum rate of physical activity may be lowered. This becomes evident when people tire more quickly while performing vigorous work or experience a gradual drop in the maximum speed of activity they can attain while performing vigorous activities such as running or riding a bicycle.

VEINS

Blood passing through capillaries flows into very small *veins*. The small veins join to form larger veins as they transport the blood back to the heart (Fig. 4.1).

Veins are made up of the same three layers found in smaller arteries (Fig. 4.9b). The layers in veins are thinner and weaker, however, since venous blood pressure is much lower than arterial pressure; therefore, there is no need for thick strong walls in veins. Veins also tend to be somewhat larger in diameter than arteries in the same area of the body. This extra internal space, along with the greater ability of veins to expand outward, allows the veins to serve as a reservoir for storing blood.

The inner layer provides smoothness to prevent blood clots, and the middle layer contains smooth muscle that regulates the diameter. When the muscle relaxes and the veins dilate, they can hold a considerable amount of blood. When the muscle contracts and constricts the veins, a great deal of blood is squeezed out and sent to the heart. These changes in diameter are useful in regulating blood pressure. For example, if blood pressure rises excessively, dilation allows the veins to store much of the extra blood from the arteries. The blood pressure will then return to normal. Conversely, if higher blood pressure is needed, the muscle layer contracts, squeezing more blood back to the heart. The heart immediately pumps this extra blood into the arteries, filling them further and increasing blood pressure and blood flow to the desired levels.

Since the blood pressure in veins is so low, blood flow tends to be sluggish. Gravity increases this tendency by pulling blood in veins below the heart downward, away from the heart. To prevent such backward flow, veins below the heart and in the arms contain *valves*. These valves consist of flaps of tissue extending from the walls of the veins into the blood (Fig. 4.9b). The valves operate in the same way as do those in the heart.

The movement of blood in veins is greatly aided by the alternating contraction and relaxation of nearby muscles, such as occurs during exercise involving body movement. During contraction, muscles widen and press on neighboring veins, forcing blood to move along the veins. During relaxation, the muscles become thinner, allowing the veins to expand and fill with blood from below. Therefore, exercise promotes blood flow in veins.

Age Changes in Veins

Several age changes occur in veins, including accumulations of patchy thickenings in the inner layer and fibers in the middle layer and valves. However, these changes do not alter the functioning of veins because veins have such a large diameter to begin with that slight narrowing is unimportant. Veins have thin walls and are able to expand easily and compensate for narrowing, and there are often several veins draining blood from each area of the body, which can provide ample alternative routes for blood.

Diseases of Veins

Some disease changes in veins occur with increasing frequency and severity as age increases. One of the most common is varicose veins, which now ranks as the tenth leading chronic condition among people above age 64.

Varicose Veins A *varicose vein* is a vein that has developed a much larger diameter than normal because blood has accumulated in the vein, stretching it outward. If the vein is stretched frequently and for prolonged periods, it loses its elasticity and remains permanently distended.

Varicose veins frequently develop in the legs. Conditions promoting their development in this area include standing still for long periods, sitting in a posture that reduces circulation, wearing tight clothing, and having certain diseases (e.g., congestive heart failure). Varicose veins are also found inside the abdomen; for example, cirrhosis of the liver is a common cause of varicose veins in the digestive system.

Varicose veins cause several problems. Affected veins close to the skin can be cosmetically undesirable because they appear as irregular bluish vessels. When veins remain engorged with blood for long periods or become inflamed, they can be very painful. They can even become sites of infection and, in extremely serious cases, sites of bleeding. Bleeding is the main problem when a person has varicose veins from cirrhosis. Very wide varicose veins also prevent the valves from stopping backward blood flow, since the valve flaps are too far apart to meet and blood slips back through the opening that remains between them. The blood backs up into the capillaries, slowing flow there. When this happens in the legs, swelling in the area below the varicose vein develops. Slow flow also prevents the capillaries from serving the needs of body cells, and the cells become weak and injured and may even die. Infection often adds to the resulting skin, nerve, and muscle problems.

Another undesirable result from varicose veins occurs because blood flow through these veins is fairly sluggish and the blood tends to clot. A stationary blood clot inside a vessel is called a *thrombus*. As in arteries, a thrombus in a vein can block blood flow. Frequently, blood flow propels the thrombus within the vein, in which case it is called a *thromboembolus* or simply an *embolus*. An embolus can cause serious problems when it moves to the heart and is pumped into the arteries, because as the arteries branch into narrower ones, the embolus will finally reach an artery through which it cannot pass and will block blood flow through that artery.

Almost all varicose veins develop in systemic veins such as those in the legs and the digestive system. Therefore, most emboli from varicose veins enter the right atrium and are pumped by the right ventricle into the pulmonary arteries. Such emboli are called *pulmonary emboli*.

A small pulmonary embolus causes death of the area of the lung normally serviced by the artery that has become blocked. If only a very small artery is blocked, the area that dies may be so small as to go unnoticed. However, repetition of this type of event or blockage of a larger pulmonary artery by a more substantial embolus may kill a considerable portion of the lung, significantly reducing the ability of the lung to serve the needs of the body. Dead lung tissue can become infected and form a pocket of pus called a *pulmonary abscess*. These infections and abscesses can make a person ill and can even be fatal.

A large pulmonary embolus can obstruct blood flow from the right ventricle to the lungs to such an extent that the right ventricle can no longer empty adequately and becomes overfull. This overfilling, coupled with the high pressure developed as the right ventricle attempts to pump blood through the blocked arteries, causes the heart to fail completely. The result can be sudden death.

Since varicose veins cause such a variety of undesirable and serious consequences, slowing or preventing their formation can help maintain the quality and length of life. When possible, people who stand or sit for long periods of time should move about or change position frequently. When one is standing, alternately tensing and relaxing the leg muscles periodically can help pump blood out of the veins. Support stockings or tights that apply an even pressure over the legs also help prevent the expansion of veins. Elevating the legs for short periods allows accumulated blood to drain out of the veins. In addition, certain situations should be avoided. For example, sitting for long periods with the legs crossed or in a tightly bent position inhibits blood flow out of the veins. Clothing that is tight in the upper regions of the legs should be avoided for the same reason. Excessive habitual consumption of alcoholic beverages should be avoided because this is the most common cause of liver cirrhosis. Individuals who have a weak heart or are developing congestive heart failure should pay particular attention to these suggestions.

Hemorrhoids   One type of varicose vein is singled out here because of its location; it is found in the area of the anus and is called a **hemorrhoid** (Fig. 4.13). Hemorrhoids may remain small for long periods, may enlarge slowly, or may become large in a short time. Some may reach the size of Ping-Pong balls.

Like other types of varicose veins, hemorrhoids can be painful and may bleed, become infected, develop thrombi, and require surgery. These consequences can cause substantial disability.

Factors that promote the formation of hemorrhoids include chronic constipation, forced bowel movements, chronic cough, and cirrhosis of the liver. The first two factors are often found among disabled individuals and people whose occupations limit the availability of toilet facilities. Chronic cough is associated with smoking and other forms of air pollution, chronic bronchitis, and emphysema.

Several strategies can be used to decrease the chance of developing hemorrhoids. Adequate amounts of fiber and water in the diet help because these substances promote regular and relatively easy bowel movements, as does exercise. Adequate access to toilet facilities and timely use of those facilities are important.

FIGURE 4.13 Hemorrhoids.

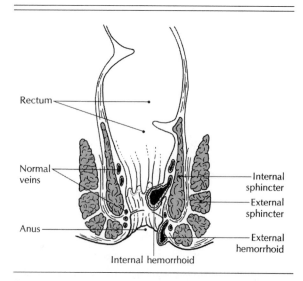

Smoking, breathing polluted air, and consuming alcohol habitually should be avoided.

LYMPHATICS AND THE SPLEEN

The structure and valves of lymph vessels are very much like those of veins. Lymph flowing through lymph vessels passes through *lymph nodes*, which are spongy structures ranging up to the size of a large bean (Fig. 4.2). Lymph nodes contain defense cells that neutralize or remove harmful chemicals and microorganisms. By acting like purifying filters, lymph nodes reduce the risk of spreading dangerous materials from a site of injury or infection in one area of the body to another region.

Much lymph passes through the *spleen*, which is toward the back of the abdominal cavity near the lowest rib on the left side of the body. The spleen serves as a very large lymph organ, stores blood, and removes old and damaged red blood cells from the circulation.

Some defense cells in the lymph nodes and spleen are called *macrophages* because they ingest substances. Others are called *lymphocytes*. Macrophages and lymphocytes are parts of the immune system.

Age Changes in Lymphatics and the Spleen

Aging seems to cause little significant change in the structure and functioning of lymph vessels, lymph nodes, and the spleen. However, there are important age changes in the *immune system*.

BLOOD

Blood is a complex fluid containing many different types of substances and cellular components. Approximately 55 percent of the blood consists of a pale yellow liquid called *plasma*; the other 45 percent is made up of the blood cells and platelets, which are suspended in the plasma (Fig. 4.14). A person maintains normal numbers of RBCs and platelets by balancing their rapid destruction with rapid production in the red bone marrow.

Plasma

About 90 percent of blood plasma consists of water. The properties of water allow it to dissolve most substances and flow easily through the circulatory system, transporting materials and blood cells. Water also provides an excellent medium for distributing heat from warmer to cooler areas. The water and buffers in the plasma help maintain proper acid/base balance in body cells. Finally, plasma contains defense substances (e.g., antibodies). Therefore, plasma contributes to all four functions of the circulatory system.

Red Blood Cells

Red blood cells (*RBCs*) are the most numerous blood cells. They contain a great deal of a red material called *hemoglobin*. Because hemoglobin can bind oxygen and carbon dioxide and can act as a buffer, RBCs can transport much oxygen from the lungs to body cells, transport some carbon dioxide from body cells to the lungs, and help regulate acid/base balance.

Platelets

Platelets are cell fragments that form when small pieces break off from large parent cells in the red bone marrow. Platelets start the formation of blood clots.

Because platelets are fragments of cells, they are fragile and burst open when they come into contact with rough spots, collagen, or unusual chemicals. This occurs, for example, when blood leaks out of a damaged vessel (Fig. 4.15). The bursting platelets release substances that help form a sticky fibrous material called *fibrin*, which begins to plug the hole in the vessel. As blood cells become trapped in the fibrin mesh, the spaces among the fibrin threads are filled. The resulting blood clot forms a leakproof seal, stopping the bleeding.

Unfortunately, platelets burst open and start the clotting mechanism whenever roughness, collagen, or unusual chemicals are encountered. When this happens on an atherosclerotic plaque, the resulting clot may block the artery and lead to a heart attack or stroke. In addition, since some platelets are bursting at all times, slow blood flow, as occurs in varicose veins, permits the substances released by platelets to accumulate at that location. When enough platelet material builds up, a thrombus forms.

White Blood Cells

The *white blood cells* (*WBCs*) are also called *leu-*

FIGURE 4.14 Components of blood.

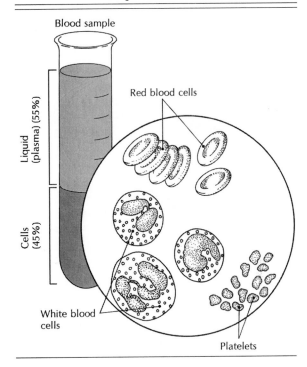

kocytes, which means "white cells." Blood contains only 1 WBC for every 500 RBCs.

WBCs can be divided into two main groups. The cells in one group are called *polymorphonuclear leukocytes* ("many structured nucleus, white cell") (*PMNs*) or *granulocytes* ("granular cells"). The other main group of WBCs consists of the *agranulocytes* ("cells without granules").

There are three types of PMNs. The most numerous type by far are *neutrophils*, which are important defense cells because they *phagocytize* (ingest) undesirable materials. Because they can move about like amoebas and can travel through capillary walls, they are found not only in the circulating blood but also among body cells. These and other phagocytic WBCs resemble vacuum cleaners as they move into every area of the body, sucking up debris.

Many neutrophils are stored within blood vessels, especially in the bone marrow. Stored neutrophils are mobilized into the circulating blood when there is a need for additional defense activity, as occurs when an infection develops.

The other two types of PMNs are *basophils* and *eosinophils*. Basophils produce histamine, which initiates inflammation whenever body cells are injured or killed. Eosinophils seem to help mod-

FIGURE 4.15 Formation of blood clots. (See text for explanation.)

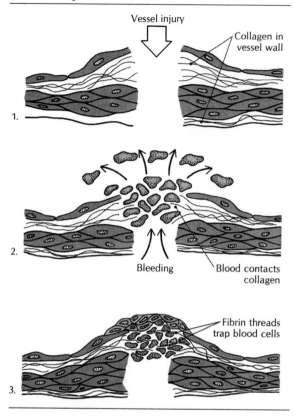

erate inflammation. Eosinophils seem to be defense cells too, because their number increases in certain situations, such as during an allergic reaction or when small parasites invade the body. Eosinophils and basophils are thought to have several additional functions.

There are only two types of agranulocytes in the blood. *Lymphocytes* are the more numerous and function as part of the immune system. *Monocytes* function like neutrophils and participate in immune responses. Lymphocytes and monocytes are discussed in Chap. 15.

Age Changes in Blood

The total amount of blood per unit of body mass and the relative amounts of plasma and cellular components remain constant regardless of age. Though some of the reserve capacity of the bone marrow to produce blood cells and platelets declines with aging, the marrow always retains enough power to supply as many blood cells and platelets as needed. Information about aging and blood components follows.

Plasma Negligible changes occur in the chemical composition of plasma, though there is an increase in certain waste products (e.g., urea, creatinine). Increases in these wastes are probably due to the decline in the ability of the kidney to remove them from the blood.

There is an age-related increase in the viscosity or "thickness" of blood. Reasons for the increased viscosity include an increase in clotting factors and broken fibrin strands; elevated norepinephrine, which promotes clot formation; and stiffening of RBCs, especially in blood with high blood cholesterol. Certain clotting factors increase dramatically at menopause, causing a rapid rise in blood viscosity. Other factors that increase blood viscosity include reduced blood oxygen; inadequate exercise; stress; and smoking, including secondhand smoke. These factors can be reduced or eliminated. Effects from higher blood viscosity include slower blood flow, increased risk of clot formation, and more rapid development of atherosclerosis.

Red Blood Cells No substantial changes occur in RBCs, though there are some indications of a decrease in the concentration of hemoglobin in men over age 65. Overall, however, aging causes no changes in the ability of the RBCs to function.

Platelets The number of platelets circulating in the blood remains essentially unchanged, and the platelets retain the ability to initiate clot formation. An age-related increase in the tendency of platelets to clump together may cause a slight increase in the risk of thrombus formation.

White Blood Cells Age changes in PMNs include decreases in the number and rate of release of stored cells, rate of movement, ability to be chemically attracted to areas, and proportion of cells capable of performing phagocytosis. The number of PMNs capable of performing phagocytosis seems to decline especially rapidly after age 60. The net effect of age changes in PMNs is a decrease in their ability to defend the body against infection, which helps explain the age-related increase in susceptibility to infections in areas such as the respiratory, urinary, and integumentary systems.

Little has been reported on age changes in monocytes that circulate in the blood. Age changes in lymphocytes are discussed in Chap. 15.

5

RESPIRATORY SYSTEM

The principal organs of the *respiratory system* are the two *lungs*, which are in the right and left sides of the chest (*thoracic cavity*) and are separated from each other by the heart. Air passes into and out of the lungs through a series of passages and tubes called the *upper airways*. The flow of air depends on other organs, including the muscular *diaphragm* and the muscles and bones that make up the wall of the thoracic cavity (Fig. 5.1).

Part of each lung consists of tubes called the *lower airways*; they end in the microscopic sac-like *alveoli*, which make up most of the lungs. The lower airways transport air to and from the alveoli. Many *pulmonary vessels* transport blood throughout the lungs.

MAIN FUNCTIONS FOR HOMEOSTASIS

Working in a coordinated fashion controlled mostly by the nervous system, these structures perform the two functions of the respiratory system: *gas exchange* and *sound production*. Gas exchange involves two processes: obtaining oxygen and eliminating carbon dioxide.

Gas Exchange

The respiratory system obtains oxygen by providing conditions that allow the oxygen contained in air to pass into the blood flowing through the lungs. The circulatory system then transports the oxygen throughout the body. Oxygen (O_2) must be supplied to body cells because it is a raw material used by mitochondria to obtain energy from nutrients. This energy provides the power needed to perform all essential bodily activities.

The respiratory system eliminates carbon dioxide (CO_2) by providing conditions that allow it to move out of the blood in pulmonary vessels and into the atmosphere. Carbon dioxide is transported from body cells to the lungs by the circulatory system.

Carbon dioxide must be eliminated because it is a waste product from the series of chemical reactions in mitochondria that release energy from nutrients. When CO_2 accumulates within the body, it can interfere with body functions because CO_2 combines with water to produce *carbonic acid*. The excess carbonic acid upsets the acid/base balance of the body. This disturbance can alter body proteins. Body structures and functions

FIGURE 5.1 The respiratory system and associated structures.

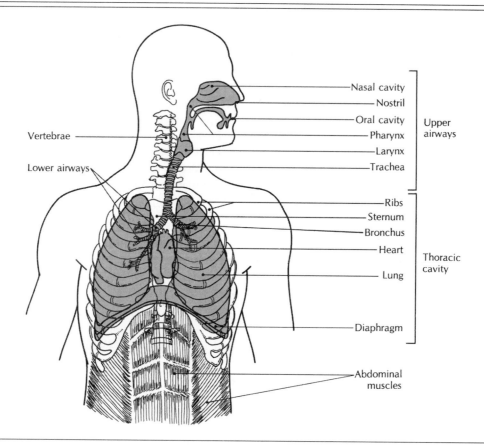

can be adversely affected, and serious illness or death can follow. Still, some acidic materials must be present for the body to achieve a normal acid/base balance, and a deficiency in acids can be as disastrous as an excess. Furthermore, some CO_2 in the blood is used to make a buffer. Therefore, the respiratory system must eliminate some but not too much CO_2.

Many other acidic materials in the body contribute to the acidic side of the acid/base balance. If there is an increase in acidic substances other than CO_2, the respiratory system can help maintain acid/base balance by eliminating more CO_2. This occurs in individuals whose kidneys do not eliminate acids adequately.

The rate of bodily activities changes from time to time. These changes cause fluctuations in the rates of O_2 use and CO_2 production and the amount of other acids. To maintain homeostasis, negative feedback systems employing the nervous system normally ensure that the rate of gas ex-

change by the respiratory system increases or decreases to meet these fluctuations. This adaptive mechanism occurs when a person begins to breathe more heavily soon after beginning vigorous physical activity.

The maximum amount that gas exchange can be increased to compensate for increases in bodily activity constitutes the reserve capacity of the respiratory system. The limited nature of respiratory capacity seems to contribute to setting a maximum limit on how vigorously a person can exercise. This limit is experienced as the sensation of feeling completely out of breath while exercising. Limitations in the maximum functional capacities of the circulatory, nervous, and muscle systems may also play a role in establishing the maximum rate of physical activity attainable.

Three operations are involved in carrying out gas exchange. *Ventilation* (breathing) involves moving air through the airways into and out of the lungs. *Perfusion* of the lungs involves the movement of blood through the pulmonary vessels. *Diffusion* causes the O_2 in inhaled air to move into the blood while CO_2 exits into the air in the lungs.

Sound Production

Sound production, which is the second main function of the respiratory system, is important because it helps people communicate. A short section on sound production, including the effects caused by aging, is presented at the end of this chapter.

VENTILATION

Ventilation involves two phases: inhaling (*inspiration*) and exhaling (*expiration*). Inspiration moves air into the nostrils and down the airways to the deepest parts of the lungs, where the O_2 it contains can diffuse into the blood. Expiration moves air containing CO_2 from the innermost parts of the lungs up and out of the body.

To understand how ventilation occurs, one must realize that air around the body is under *atmospheric pressure*. Materials normally move from areas of higher concentration or pressure to areas of lower concentration or pressure. For example, air moves into a balloon and inflates it when more air pressure is applied to its opening than is already in the balloon. Conversely, releasing the opening of an inflated balloon results in air rushing out because the pressure within the balloon is higher than atmospheric pressure.

Inspiration

Inspiration occurs for the same reason that a balloon becomes inflated. Air moves into the body when the air pressure outside the body is greater than that inside the respiratory system. A person creates this difference in pressure by contracting muscles to move the floor or walls of the thoracic cavity.

The floor of the thoracic cavity consists of the dome-shaped *diaphragm*, a thick sheet whose edge is muscle and whose center is fibrous material. The muscular edge slants downward sharply and is attached around its perimeter to the body wall. When the muscle contracts, the rounded central region is pulled downward within the body wall, moving like a piston downward in a

FIGURE 5.2 Inspiration.

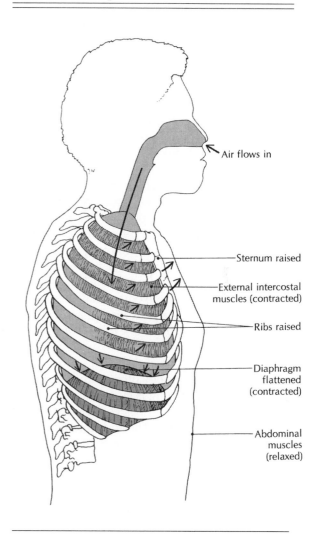

Air flows in

Sternum raised

External intercostal muscles (contracted)

Ribs raised

Diaphragm flattened (contracted)

Abdominal muscles (relaxed)

cylinder. When the central region is pulled downward and the diaphragm flattens somewhat, the pressure within the thoracic cavity decreases. Because moisture on the outer surface of the lungs causes them, in effect, to adhere to the diaphragm, a similar decrease in pressure occurs within the lungs. Since the pressure in the lungs is then lower than atmospheric pressure, air flows through the airways and into the lungs, resulting in inspiration (Fig. 5.2).

The walls of the thoracic cavity contain many bones, including the ribs, the sternum, and some vertebrae. The cartilage and joints that connect these bones allow them to move somewhat when the muscles attached to the bones contract. When muscles move the ribs and sternum upward and outward, pressure in the thoracic cavity decreases. The lungs, whose surfaces are stuck to the thoracic walls by fluid, also have a decrease in internal pressure, and inspiration occurs.

Inspiration usually involves simultaneous movement of both the diaphragm and the bones of the thorax. Some individuals rely mostly on movement of the diaphragm (*diaphragmatic breathing*), while in others movement of the ribs (*costal breathing*) makes the major contribution. Inspiration ends when parts of the body stop moving and enough air has come in to raise the pressure in the lungs to atmospheric pressure. If the muscles are held in this position, no further air movement occurs and the lungs remain inflated. People in this state are truly holding their breath.

Inspiration is called an *active* process because it requires the use of energy. Obtaining this energy uses some of the oxygen that inspiration helps obtain. The energy expended and the oxygen used are called the *work of breathing*. Usually not more than 5 percent of the oxygen brought in by inspiration is consumed in this process; the rest is available for use by other body cells. Since diaphragmatic breathing is more efficient and requires less energy than does costal breathing, it consumes less oxygen. These differences leave more of the oxygen obtained from inspiration for use by other body cells.

Expiration

Expiration for a person who is resting and breathing quietly normally requires no muscle contraction because the movements of inspiration set up conditions that allow it to occur automatically. For example, when the diaphragm moves downward, it pushes on the organs below it in the abdominal cavity, and this increases the pressure in the abdominal cavity. Also, the movements of the ribs and sternum stretch and bend elastic and springy structures in the thoracic wall such as ligaments, cartilage, and the ribs themselves. Finally, the lungs, which are elastic, are stretched outward.

As a result, as soon as the muscles of inspiration relax, the abdominal organs, structures in the thoracic wall, and the lungs start to spring back to their original positions. This *elastic recoil* increases the pressure in the lungs. The pressure quickly rises above atmospheric pressure, and

expiration occurs. The process is similar to what happens when the opening of an inflated balloon is released. Each expiration is followed shortly by the next inspiration (Fig. 5.3).

Since normal quiet expiration requires no muscle contraction or energy, it is called a *passive* process. However, when a person becomes very active, passive expiration occurs too slowly to meet the needs of the body. Then respiratory muscles and energy can be used to perform active *forced expiration*. For example, abdominal muscles can squeeze on the abdominal organs, causing them to push upward on the diaphragm more forcefully, and chest muscles can pull the ribs downward. The resulting increase in pressure in the lungs pushes air out of the respiratory system quickly (Fig. 5.4).

Rate of Ventilation (Minute Volume)

Ventilation usually occurs continuously to provide ongoing replacement of the O_2 being consumed and elimination of the CO_2 being produced. The rate of ventilation must be high enough to maintain homeostatic levels of these gases in the body. The rate of ventilation is called the respiratory *minute volume*, the volume of air inspired per breath times the number of breaths per minute. The number of breaths per minute is called the *respiratory rate*. Minute volume can be expressed mathematically:

Minute volume =
volume per breath × breaths per minute

Lung Volumes   The volume of air inspired equals the amount of air expired. When a person is at rest and breathing quietly, this volume is called the *tidal volume* (*TV*).

When a person is active and has to exchange gases more quickly, inspiratory and expiratory volumes can be increased considerably by increasing the distance the respiratory muscles contract. The extra amount a person can inspire is called the *inspiratory reserve volume* (*IRV*); the extra amount a person can expire is called the *expiratory reserve volume* (*ERV*). The combination of tidal volume, inspiratory reserve volume, and expiratory reserve volume is called the *vital capacity* (*VC*). Vital capacity is the most air a person can expire after taking the deepest possible inspi-

FIGURE 5.3 Passive expiration: (a) Elastic recoil collapses chest partially, causing expiration. (b) Elastic recoil collapses balloon, forcing air out.

FIGURE 5.4 Forced expiration: (a) Muscle contractions and elastic recoil collapse chest partially, causing rapid expiration. (b) Pressure from hands and elastic recoil collapse balloon, forcing air out rapidly.

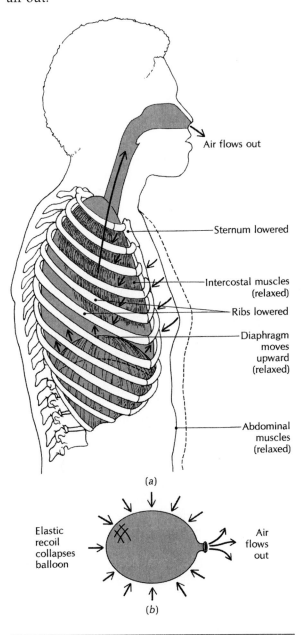

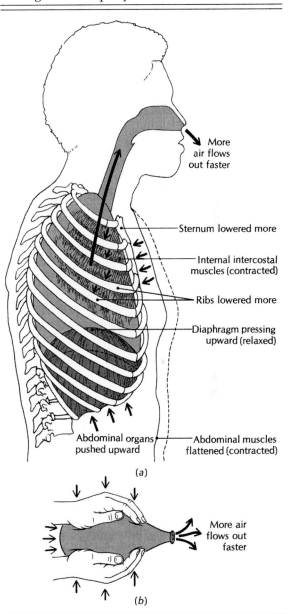

ration. This can be expressed mathematically:

$$TV + IRV + ERV = VC$$

Besides increasing the volume of air respired with each breath, a person can increase the speed at which the air flows. This is accomplished by increasing the speed and force of respiratory muscle contractions, which can magnify pressure changes in the lungs more than 25-fold.

A person who expires as much as possible still has some air left in the lungs. This volume is called the *residual volume* (*RV*). Thus, the total amount of air the lungs can hold equals TV + IRV + ERV +

RV and is called *total lung capacity* (*TLC*). A small amount of the TLC does not reach the alveoli but remains in the lower airways. This volume of air—the *dead space*—cannot be used for gas exchange because only the alveoli are thin enough for this process to occur.

Respiratory Rate   A normal person may have a respiratory rate of 15 to 20 breaths per minute, but this rate can change as needed. If the volume per breath remains high, an increase in the rate of respiration increases the minute volume and therefore the rate of gas exchange. Decreases in the respiratory rate have the opposite effect on the minute volume and the rate of gas exchange. Such adjustments in the respiratory rate occur as changes in bodily activity alter the need for gas exchange.

When the rate of respiration increases, there is less time for each inspiration and expiration. If a person does not increase the rate of airflow, breathing becomes rapid but shallow. Such breathing delivers little fresh air to the lungs for gas exchange.

REQUIREMENTS FOR VENTILATION

The major features necessary for proper ventilation include (1) *open airways* for easy air movement, (2) *defense mechanisms* that assure that only clean, moist, warm air reaches the lungs, (3) *proper pressure changes* in the thoracic cavity and lungs to make the air move, (4) *compliance* in thoracic and lung components so that pressure changes cause them to expand easily to accept incoming air, and (5) *control systems* that ensure that the process occurs successfully and at the correct rate.

Contributions by Airways

The contributions made by muscles and skeletal components to the first four of these requirements were described previously. We will now consider the ways by which the airways contribute to these requirements.

Nasal Cavities   Inspired air entering the nostrils passes through the *nasal cavities* above the hard palate. These cavities are held open by the bones of the skull. As air passes through the nasal cavi-

ties, it is cleaned, moistened, warmed, and monitored so that it does not harm the delicate structures deep within the lungs (Fig. 5.5).

The air is cleaned because dust and other particles are trapped by hairs inside the nostrils and by the sticky mucus that coats the inside of the cavities. Microscopic hairlike *cilia* on the cells lining the cavities wave back and forth, causing the mucus to glide back toward the throat. Then the mucus and its trapped debris can be harmlessly swallowed.

The air is moistened by the mucus to prevent drying of the lungs. Heat from the blood in the walls of airways warms the air so that the lungs are not chilled. Finally, sensory nerve cells monitor the chemical contents of the air and send impulses to the brain. The presence of such chemicals is perceived as aromas. The nervous system may cause inspiration to slow or stop if harmful chemicals or particles are detected. Forced expiration (e.g., sneezing) may be initiated in an attempt to blow the noxious materials out of the respiratory system.

A person may inspire some or all of the air he or she breathes through the mouth rather than through the nasal passages. This can increase the rate of airflow, but it reduces the amount of cleaning, moistening, and warming of inspired air. Injury to the airways below the pharynx may result. Inspiring through the mouth can also lead to excessive dryness of the oral cavity, which may cause oral discomfort and sores.

Nasopharynx   Air in the nasal cavities moves backward into the *nasopharynx*, which is above the soft palate. Bones and other firm tissues keep this passage open except when one is swallowing, during which the tongue pushes upward on the soft palate. The mucus, cilia, and blood in the nasopharynx further clean, moisten, and warm the air.

Pharynx   After passing through the nasopharynx, the air moves through the throat, or *pharynx*, into the opening in the voice box, or *larynx*. This opening is called the *glottis*. The pharynx is held open by the firmness of the muscles and other tissues that make up its walls.

Since food and beverages in the oral cavity that are being swallowed also pass through the pharynx, these materials can lodge in the pharynx or enter the glottis, blocking or injuring the airways.

FIGURE 5.5 Respiratory passages in the head and neck.

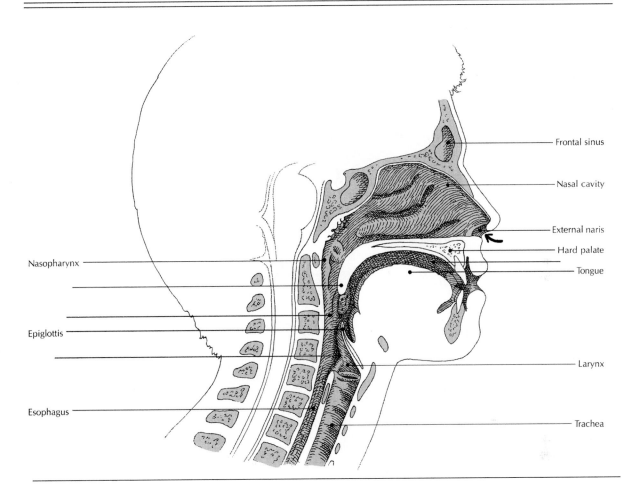

Frontal sinus

Nasal cavity

External naris

Hard palate

Tongue

Larynx

Trachea

Nasopharynx

Epiglottis

Esophagus

Two reflexes controlled by the nervous system prevent these problems.

The *swallowing reflex* occurs whenever solids or liquids are present in the pharynx behind the tongue. This reflex clears the pharynx by pushing materials down into the esophagus. At the same time, a flap called the *epiglottis* is moved over the glottis to prevent materials from entering the larynx. The epiglottis is moved off the glottis after swallowing has been completed so that ventilation can begin again (Fig. 5.6).

The *gag reflex* is caused when irritating materials enter the pharynx. This reflex causes muscles near the pharynx to close the openings into the larynx and esophagus. At the same time, muscle contractions in the abdomen raise the pressure in the esophagus and trachea to prevent materials from entering those passageways. A very strong gag reflex can result in vomiting.

Larynx, Trachea, and Primary Bronchi Air passing through the glottis moves through the larynx, down the windpipe (*trachea*), and through the two *primary bronchi* into the lower airways in the lungs. Plates and rings of springy cartilage within the walls of these airways provide support so that the airways stay open during ventilation.

Materials other than air that enter these air passages initiate the *cough reflex*. During coughing, bursts of air that are expired rapidly force foreign materials up and out of these airways.

The mucus, the cilia, and blood flow in these structures carry out further cleaning, moistening, and warming of the air. The cilia beat in an upward direction so that the mucus glides into the pharynx. Since the mucus carrying materials slides upward in a smooth continuous stream, this mechanism is called the *mucociliary escalator*. Phagocytic macrophages and immune system

FIGURE 5.6 The swallowing reflex: (a) Tongue pushes food back. (b) Soft palate elevates and epiglottis lowers to close airways. (c) Muscle contractions push food into esophagus. (d) Wave of contraction pushes food down to stomach.

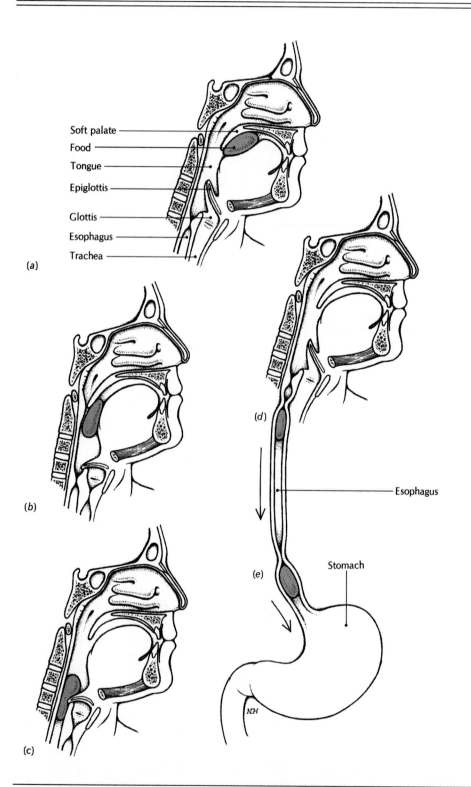

cells in the trachea and primary bronchi provide additional defense against foreign materials.

Smaller Bronchi, Bronchioles, and Alveolar Ducts   As each primary bronchus enters a lung, it branches repeatedly, forming ever more numerous smaller *bronchi* and *bronchioles* and finally microscopic *alveolar ducts*.

The walls of these airways become thinner as they branch and narrow. Cartilage in the smaller bronchi keeps them open during inspiration. There is no cartilage in the bronchioles or smaller airways. A peculiar helical structure of the collagen that coils around the airways and elastin fibers also support these smaller airways. The cartilage and fibers provide the lungs with compliance. Like the trachea and bronchi, these smaller airways are protected by the cough reflex and defense cells and condition the entering air.

Smooth muscle cells in the airway walls allow for appropriate adjustments in their diameter as the amount of ventilation needed fluctuates. The smaller airways provide most of this adaptability. The activity of the muscle is controlled by the nervous system, the endocrine system, and nearby chemicals.

As air is expired and the lungs decrease in size, the open passages in the airways become narrower. The walls of the smallest airways are so thin and weak that these airways close completely before all the air has escaped from the alveoli below them. This air remaining in the alveoli makes up part of the residual volume.

Alveolar Sacs and Alveoli   The inspired air in the alveolar ducts passes into blind cup-shaped outpocketings called *alveoli*. Most alveoli occur in clusters extending outward from slightly enlarged spaces at the ends of the alveolar ducts called *alveolar sacs*. Each cluster may look like a tightly packed bunch of plump grapes (Fig. 5.7).

There are about 300 million alveoli in the lungs. Because alveoli are hollow, filled with air, and very small and because they make up most of the lungs, dried lungs have the consistency of Styrofoam. The alveoli provide an amount of surface area equivalent to that of an area 30 feet long and 25 feet wide. The walls of the alveoli are very thin, allowing diffusion of O_2 and CO_2 between the air and the blood to occur easily.

Special cells in the alveoli secrete a material called *surfactant*. As surfactant spreads out, it coats the inner surface of the alveoli and parts of the smaller airways. The surfactant greatly increases the compliance of the lungs by reducing the attraction between the water molecules on the inner surfaces of the lungs. Without surfactant, the attraction (*surface tension*) would be so great that the alveoli and small airways would collapse. The inner surfaces would stick together tightly, making it nearly impossible for them to separate and fill with air during inspiration. These characteristics can be compared to the difference between the effort needed to inflate a new balloon that contains a powdery surfactant and the effort needed to inflate an old balloon that dried after becoming damp.

Surface tension is important because as it makes the lungs collapse, it helps increase the pressure in the lungs and therefore assists in expiration. The combination of a moderate amount of surface tension in the alveoli and the large surface area they provide makes expiration much easier.

Control Systems

Nervous System   Ventilation begins with inspiration, which requires the contraction of muscles. The nervous system signals activating these muscles originate in a region of the brain called the *medulla oblongata* and travel to the muscles through nerves. The medulla oblongata is inside the region of the skull just above the neck. The part of it concerned with respiration is called the *respiratory control center*.

The respiratory control center starts inspiration when it detects an increase in CO_2 levels or a decrease in O_2 levels in the blood flowing through it. When sensory nerve cells from the aorta and arteries in the neck detect very high levels of CO_2 or very low levels of O_2, these nerve cells also stimulate the respiratory control center. Other sensory neurons in the lungs send impulses to the respiratory control center, telling it that the lungs are in a partially collapsed condition and are ready for inspiration. Sensory nerves from muscles and joints inform the respiratory center when a person begins physical activity and will need more gas exchange.

Nerves from the lungs inform the respiratory center and a nearby part of the brain called the *pons* when inspiration is complete. The brain then

FIGURE 5.7 Lower airways and alveoli.

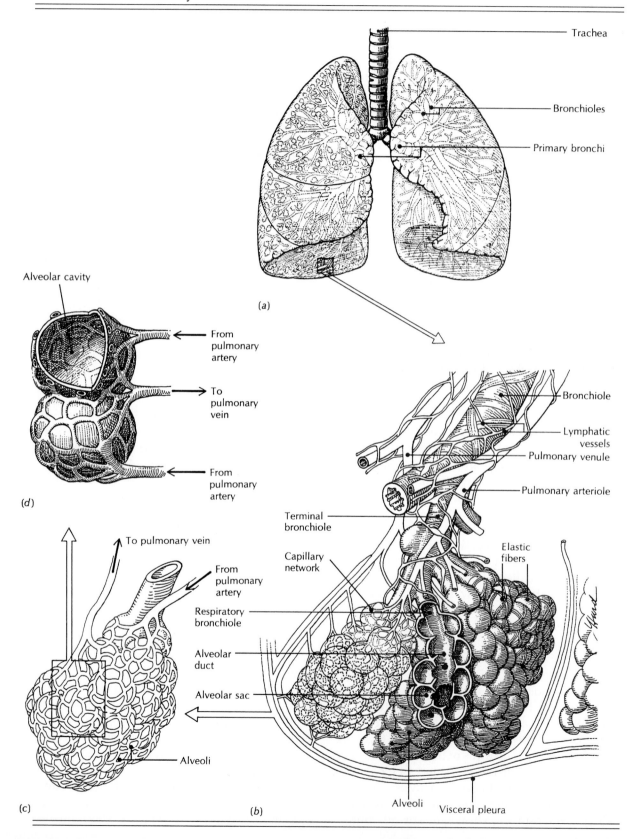

stops the impulses for inspiration. As the muscles relax, expiration begins because of elastic recoil of the thorax and lungs. The respiratory center can also send impulses to the muscles indicating that a forceful expiration is needed.

The respiratory center and the pons monitor their own impulses and are also informed by nerves from the lungs when expiration is complete. This triggers the beginning of the next inspiration. Therefore, the linking and repeating of two negative feedback systems result in rhythmic breathing.

The depth of breathing, the speed of airflow, and the respiratory rate are adjusted when the respiratory center detects that CO_2 or O_2 concentrations and the acid/base balance in the blood are beginning to wander from proper levels. The adjustments restore appropriate gas levels and acid/base balance so that homeostasis is maintained.

Ventilation is also modified by the swallowing, gag, and cough reflexes. In addition, upper regions of the brain such as the areas controlling emotions and those controlling conscious actions can influence the respiratory center. The conscious control areas allow a person to voluntarily inspire, expire, stop ventilation, or modify ventilation for actions such as talking.

The nervous system also controls ventilation by adjusting the size of the lower airways. Impulses from the *sympathetic* nervous system cause relaxation of smooth muscles in the airways, permitting them to dilate and increasing minute volume. *Parasympathetic* nerves cause the smooth muscles to contract, constricting the airways and reducing minute volume. These changes allow the rate of gas exchange to maintain homeostasis for O_2 and CO_2.

Endocrine System   Hormones from the endocrine system also help regulate ventilation. *Norepinephrine* makes a main contribution. This hormone has the same effect on the airways as do impulses from sympathetic nerves.

AGE CHANGES AFFECTING VENTILATION

Because everyone's airways are subjected to some air pollution and other environmental insults, it is difficult to know which age-related changes in airways are due to aging and which are due to

other factors. However, certain changes seem to occur in all people. These universal changes are considered age changes and thus are included in this section. Let us examine how age changes affect the five requirements for ventilation.

Open Airways

Mucus and Cilia   All airways from the nasal cavities to the smallest bronchioles produce mucus continuously. With aging, the mucus produced is more viscous and therefore more difficult to move. In addition, both the number and the rate of motion of the cilia decrease. As the clearing out of mucus slows, the accumulation of mucus narrows airways, and this inhibits ventilation. When ventilation becomes more difficult, the work of breathing increases and extra CO_2 is produced by the muscles of ventilation, making respiration less efficient. Narrower airways also reduce the rate of airflow and therefore reduce the maximum possible minute volume.

Airway Structure   Age changes in the walls of bronchioles cause them to become even narrower, amplifying the effect of mucus accumulation. In addition, the bronchioles close earlier during expiration, trapping more air in the smaller airways and in the alveoli, especially in the lower parts of the lung. One result is an increase in residual air. This causes the fresh air entering with each inspiration to be mixed with a larger amount of stale residual air, decreasing the rate of diffusion. A second result is uneven lung ventilation.

While the bronchioles become narrower, the larger airways in the lungs and the alveolar ducts increase in diameter. These changes compound the negative effects by increasing the dead space. Thus, fresh inspired air is mixed not only with more residual air in the smallest airways and alveoli but also with more dead space air. This further decreases the rate of diffusion.

The increase in tidal volume with age may help minimize the expected drop in the diffusion rate during tidal breathing by mixing more fresh air with the increasing amounts of stale air remaining in the respiratory system. The rate of diffusion remains high because the O_2 concentration in the lungs is kept high while the CO_2 levels are kept low.

Defense Mechanisms

Since age changes in the mucus and cilia cause slower movement of mucus, harmful materials such as microbes, particles, and noxious chemicals trapped by the mucus stay in the respiratory system longer. Aging also decreases the functioning of other defense mechanisms, including reflexes (see below), white blood cells, and the immune system. All these age changes cause an increase in the risk of developing respiratory infections and other respiratory problems.

Proper Pressure Changes

Muscles   As with most muscles, aging causes respiratory muscles to become weaker. The decrease in muscle strength is not enough to detract from performing tidal breathing or ventilating at moderately increased minute volumes. However, it slowly decreases the maximum pressure changes that the muscles can produce and thus decreases the maximum rate of airflow attainable.

Skeletal System   Age changes in the cartilage, bones, and joints of the thorax also reduce a person's ability to produce large pressure changes in the thoracic cavity. The cartilage attaching the ribs to the sternum becomes more calcified and stiff, and the ribs become less elastic. Age changes in the cartilage and ligaments of other joints, such as those between the ribs and the vertebrae, result in decreases in the ease and range of motion of the bones they connect.

Aging also leads to slight alterations in the positions of the bones of the chest. The chest becomes deeper from front to back, making deep inspiration more difficult. Altered posture from other changes in the skeletal system further reduces a person's ability to inspire quickly and fully.

Because of these skeletal age changes, there is a decline in the maximum minute volume attainable and an increase in the work of breathing. Older people partially compensate for these effects by relying more on diaphragmatic breathing.

Lungs   Though there are no important age changes in the elastic fibers or surfactant, other age changes in the lungs significantly affect pressure changes. For example, aging causes the coiled collagen fibers in the lungs to become somewhat limp and less resilient. Also, aging causes the alveoli to become shallower, and this reduces the amount of surface area present. The resulting reduction in surface tension decreases elastic recoil. Both of these age changes reduce the maximum rate of expiration attainable and add to the work of breathing (Fig. 5.8).

Compliance

Aging causes the coiled collagen fibers to become somewhat limp and stretch more easily. These changes increase the compliance of the lungs and tend to make inspiration easier. Note, however, that the increase in lung compliance is much less than the increase in chest stiffness caused by skeletal age changes. Thus, there is a net increase in stiffness of the respiratory system, resulting in a decreased ability to inspire.

Control Systems

Aging does not seem to affect the contributions of the nervous system to rhythmic breathing under resting conditions. However, three types of age change reduce the ability of the nervous system and endocrine system to cause adaptive changes in ventilation:

1. Neurons monitoring O_2, CO_2, acid/base balance, and muscle activity seem to become less sensitive to changes in these parameters.

FIGURE 5.8 Effects of aging on alveoli: (a) Young alveoli. (b) Old alveoli.

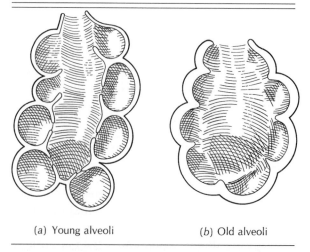

(a) Young alveoli   (b) Old alveoli

2. There may be changes in the nervous pathways through which all their impulses are sent, resulting in altered ventilation.
3. The lungs become less sensitive to norepinephrine from sympathetic nerves and the endocrine system.

These age changes result in a slower and smaller increase in minute volume when there is a decrease in O_2 or an increase in CO_2, acids, or body activity. As a result, individuals who begin vigorous activity feel out of breath and tire more quickly as they get older.

Age changes in other parts of the nervous system reduce its ability to provide the swallowing, gag, and cough reflexes that defend the respiratory system (Chap. 6). Because of these changes, it takes a greater amount of material and a longer time to start a defensive reflex. Once it begins, the response is slower and weaker.

Therefore, older individuals must avoid situations that raise the risk of choking. These include eating quickly; talking or laughing while eating; eating while lying on one's back; and eating after consuming alcoholic beverages or medications that slow the reflexes.

Consequences

Reductions in pressure changes caused by weakening of muscles, stiffening of the respiratory system, and decreased alveolar surface area combine with narrowing of the airways to produce two effects. First, they cause a decrease in the rate at which air can flow in the system. Second, they make ventilation more difficult and therefore increase the work of breathing. This reduces the amount of available O_2 and increases CO_2 in the blood.

Although aging does not change the total lung capacity, age changes affect the volumes of air that can be moved. The more rapid closing of bronchioles, together with stiffening of the system, causes a decrease in both inspiratory and expiratory reserve volume. At the same time, tidal volume increases somewhat, and the age changes cause an increase in residual volume both at rest and during increased ventilation. These changes in volumes cause the vital capacity to decrease.

These changes in respiratory volumes have two effects. First, they further decrease maximum minute volume. Second, the decrease in vital capacity, combined with the increase in residual capacity, means that less fresh inspired air is mix-

ing with more stale air remaining in the lungs. This change decreases the rate of diffusion. The problem is compounded by the increase in dead space. The age-related increase in tidal volume may help compensate for this problem during quiet breathing (Fig. 5.9).

The force of gravity on the lungs causes the lower bronchioles to close sooner than do those in the upper regions. Therefore, the lower parts have a higher proportion of the residual air than do the upper regions. This unevenness in ventilation increases with age. As seen below in the discussion of perfusion and diffusion, this further decreases the efficiency of the system. Thus, as people get older, they must ventilate more air to get the same amount of gas exchange, and this adds to the work of breathing. Breathing more deeply can partially overcome the deleterious effects of uneven ventilation. Aging also reduces the maximum respiratory rate (breaths per minute) because of age changes that slow airflow and age changes in the nervous system.

These decreases in maximum flow rate, maximum volume per breath, and maximum respiratory rate combine to cause a decrease in the maximum minute volume. Many individuals can expect their maximum minute volumes to decline by 50 percent as they pass from their twenties to very old age. This change makes a major contribution to the decrease in the maximum rate of gas exchange as people age. Age changes in perfusion and diffusion, discussed below, cause additional decreases in gas exchange.

FIGURE 5.9 Age changes in respiratory volumes.

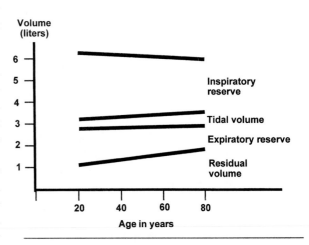

FIGURE 5.10 Pulmonary circulation and perfusion of the lungs.

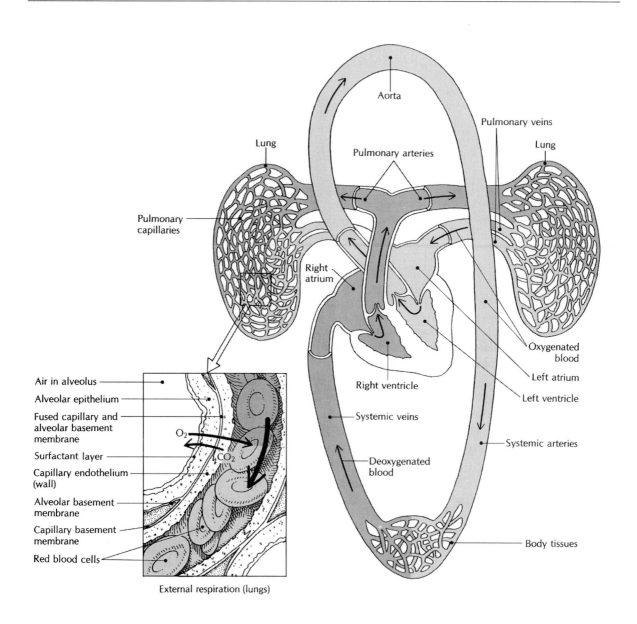

Aorta

Pulmonary veins

Lung

Pulmonary arteries

Lung

Pulmonary capillaries

Right atrium

Oxygenated blood

Left atrium

Right ventricle

Left ventricle

Systemic veins

Systemic arteries

Deoxygenated blood

Air in alveolus

Alveolar epithelium

Fused capillary and alveolar basement membrane

Surfactant layer

O_2

CO_2

Capillary endothelium (wall)

Alveolar basement membrane

Capillary basement membrane

Red blood cells

Body tissues

External respiration (lungs)

PERFUSION

Recall from Chap. 4 that perfusion is the passage of blood through the vessels of body structures. Perfusion of the lungs proceeds as follows.

The right atrium receives blood from *systemic veins* from all parts of the body except the lungs. This blood has little oxygen because the oxygen was removed and used as the blood flowed through capillaries and past body cells. Therefore,

this blood is called *deoxygenated blood*. It also has a high concentration of CO_2, which diffused into the blood from the body cells (Fig. 5.10).

Deoxygenated blood from the right atrium flows into the right ventricle, which then pumps it through the *pulmonary arteries* to the lungs. As these arteries enter and pass through the lungs, they branch into smaller vessels until they enter the thin-walled *pulmonary capillaries*. These capillaries carry the blood close to the walls of the alveoli. This allows gases to diffuse between the blood in the capillaries and the air in the alveoli. The blood then enters *pulmonary veins*, which carry blood to the left atrium. Since this blood has a high concentration of O_2, it is called *oxygenated blood*. It also has a low concentration of CO_2. This blood will pass from the left atrium into the left ventricle, which will pump it through the *systemic arteries* to all parts of the body except the lungs.

Like the rate of ventilation, the rate of perfusion must vary as a person's rate of activity, and therefore the need for gas exchange, varies. Blood flow to an area of the body can be changed by altering the cardiac output and changing the diameter of the arteries delivering blood to body structures.

Age Changes in Perfusion

There are essentially no age changes that affect the pulmonary arteries and veins. Furthermore, aging does not change cardiac output.

The reason for the minimal change in pulmonary vessels compared with other vessels in the body may be that blood pressure in the pulmonary vessels is much lower than that in the *systemic vessels*. When diseases such as emphysema cause a rise in pulmonary artery pressure, these arteries undergo changes that resemble atherosclerosis.

Though pulmonary arteries and veins remain largely unchanged by aging, the pulmonary capillaries decrease in number and accumulate some fibrous material. Whether these are true age changes or are due to the effects of air pollution is uncertain.

Normally, the reduction in perfusion due to changes in pulmonary vessels is slight. The effect on reducing gas exchange does not become apparent until the respiratory system is called on to deliver the maximum rate of gas exchange. Even then, this causes only a small reduction in maximum gas exchange.

However, heart disease and certain types of pneumonia and emphysema can reduce perfusion of the lungs. Such reductions decrease the rate of gas exchange and therefore decrease the ability of the respiratory system to maintain homeostasis of O_2, CO_2, and acid/base balance. These effects are often noticed as the sensation of being out of breath and being fatigued when one engages in vigorous physical activity.

DIFFUSION

Recall that the alveoli are the destinations for inspired air. Their great numbers and deeply curved surfaces provide a great deal of surface area.

The walls of the alveoli are only one cell thick, and the cells are flat and very thin. Thin-walled capillaries surround the alveoli. Only an exceedingly thin noncellular layer (**basement membrane**) separates the alveolar wall from the capillary wall. These structural features provide a thin surface through which gases must pass. The secretions from some alveolar cells keep their surfaces moist. Thus, the alveoli supply a large, thin, and moist surface that is ideal for the diffusion of gases. Diffusion of O_2 and CO_2 in the lungs proceeds as follows (Fig. 5.10).

The blood entering the pulmonary capillaries has a very low concentration of O_2 and a high concentration of CO_2. The air in the alveoli, by contrast, has a high level of O_2 and a low level of CO_2 because ongoing ventilation continuously refreshes the alveolar air. Therefore, O_2 diffuses from the alveolar air into the blood while CO_2 diffuses from the blood into the alveolar air.

Much of the CO_2 that diffuses into the alveoli is eliminated with the next expiration, which also delivers a new supply of O_2. Only a very small amount of the O_2 that enters the blood can be carried by the plasma. Almost all the oxygen in the blood is bound to hemoglobin molecules, which are contained in the red blood cells. Each hemoglobin molecule can bind up to four molecules of oxygen. When hemoglobin binds oxygen, the result is *oxyhemoglobin*. Decreased CO_2, increased pH, or decreased temperature of the blood increases the amount of oxygen that can be bound to each hemoglobin molecule; the converse is also true. Normally, this promotes complete oxygenation of blood in the lungs, where ventilation keeps CO_2 levels and temperature low. It also promotes greater release of oxygen in other capillaries, where body cell activities keep CO_2 levels and temperature high. These characteristics of hemoglobin are sometimes displayed as oxyhemoglobin dissociation curves.

Continuous perfusion, coupled with continuous ventilation, keeps diffusion occurring in an uninterrupted fashion. Furthermore, alterations in the rate of ventilation or perfusion can alter the rate of diffusion to meet bodily needs. Increasing ventilation (i.e., minute volume) or perfusion increases the differences in concentrations of O_2 and CO_2 between the blood and alveolar air. These changes increase the rates of diffusion and gas exchange. Reducing ventilation or perfusion has the opposite effects.

Age Changes in Diffusion

Aging causes several changes that reduce the maximum minute volume of ventilation, increase residual volume, and cause uneven ventilation in the lungs.

Age changes in the alveoli further decrease the rate of diffusion. The alveoli become flatter and shallower, decreasing the amount of surface area. The alveolar membrane that remains becomes thicker and undergoes chemical changes which further impair diffusion (Fig. 5.8).

EFFECTS FROM ALTERED GAS EXCHANGE

Biological Effects

In summary, essentially all aspects of the respiratory system involved in gas exchange are detrimentally affected by aging, resulting in a drop in the maximum rate of gas exchange. Furthermore, there is an overall decline in the efficiency of this system. Finally, the ability of the respiratory system to adjust the rate of gas exchange to meet body needs declines. These changes occur at a fairly steady rate throughout life.

As these changes occur, the maximum rate at which a person can perform physical activities declines, and a person who starts a vigorous activity such as running or climbing stairs will feel tired and out of breath sooner as age advances. Such an individual will not be able to perform at top speed for an extended period. Age changes in other systems, including the circulatory, skeletal, muscle, and nervous systems, may contribute to these decrements. The consequences of these effects can be reduced by raising one's pace gradually. Doing this provides extra time for respiratory functioning to adapt to the increased need for gas exchange. Also, going at a more moderate pace lowers the required rate of gas exchange.

Although aging causes reductions in several maximum respiratory values, these age changes are observed only when people require that the respiratory system function at maximum capacity. This system has such a great reserve capacity that the decline in maximum values caused by aging has essentially no effect on a person who engages in light or moderately vigorous activities. Thus, unless a person engages in activities such as very demanding physical work or highly competitive athletic events, age changes in respiratory functioning have little noticeable effect on his or her lifestyle. The aging respiratory system can provide adequate service in all but the most physically demanding situations.

Factors other than aging alter gas exchange. Also, much can be done to minimize age-related reductions in the ability of the respiratory system to satisfy the need for gas exchange. For example, a sedentary lifestyle further limits respiration, while regular exercise keeps the decline in maximum minute volume small. Furthermore, incorporating adequate vigorous physical activity into one's lifestyle can restore much of the decline in respiratory functioning caused by inactivity.

Another factor that adversely affects respiration is air pollution. Breathing polluted air seems to increase both the speed and severity of essentially every age change in the respiratory system mentioned thus far. People who smoke, live in areas where air quality is poor, or engage in occupations where the air contains dust, fine particles, or noxious chemicals have a much faster and greater loss of respiratory functioning. In addition, these individuals are at higher risk for developing respiratory diseases, including lung cancer, chronic bronchitis, and emphysema. Air pollutants can injure respiratory cells and tissues in several ways including physically, chemically, and through free radicals, microbes and immune responses. Radon damages lung tissues through the radiation it causes and the free radicals it induces. Breathing polluted air can be reduced by avoiding polluted areas; by not smoking; by wearing a protective mask; and by providing adequate ventilation with clean air in living and working areas.

Interactions

The biological effects of decreased gas exchange can affect other aspects of life. The nature and degree of these effects depend on the amount and importance of physical activity in a person's life. Examples of people who may be affected more dramatically include people whose chief form of recreation and social contact is competitive sports and people whose jobs involve considerable physical exertion.

Finally, changes in gas exchange can be affected by other types of age changes. For example, upon retirement, a sedentary office worker may take up a physically demanding sport, which may provide the motivation to stop smoking. The result can be a slowing and even a temporary reversal of the decline in respiratory capacity.

DISEASES OF THE RESPIRATORY SYSTEM

While age changes in the respiratory system have only a small impact on the ordinary activities of daily living, changes caused by disease can have a substantial effect. Respiratory diseases reduce a person's speed and endurance in physical activities and cause significant disability. Treatment can extend for long periods and is often expensive. Furthermore, respiratory system diseases (not including cancer of the lungs) are the fourth leading cause of death for those over age 65. If lung cancer is added, respiratory disease ranks as the third leading cause of death among the elderly.

The reasons for the high incidence of respiratory diseases among older people are similar to those for other diseases. They include reductions in defense mechanisms; more time for the development of slowly progressing diseases; and increases in the number of exposures and the total time of exposure to disease-promoting factors.

There is one factor that contributes to the development of virtually all these diseases: *air pollution*. One of the most common forms is smoking and inhaling smoke from tobacco products. Though the proportion of smokers in the population has declined, the effects of smoking among older people will be evident for many years because many older people have smoked for long periods. The rate of decline of the respiratory system slows when a person stops smoking and there may even be a period of improvement in gas exchange. However, most effects from long-term smoking are not reversible.

Other forms of air pollution include particulate matter such as dust from coal mining, woodworking, farming, and the manufacture of fabrics. Fumes and vapors such as those from painting, chemical plants, and scientific laboratories can harm the lungs. Smog, automobile fumes, and other types of air pollution associated with urban environments are also significant risk factors for lung damage.

Reducing the inspiration of air pollutants can significantly reduce both the incidence and severity of respiratory disease. Doing this will preserve much of the capacity for gas exchange by the respiratory system.

Respiratory diseases that are most common among people of advancing age include lung cancer, chronic bronchitis, emphysema, pneumonia, and pulmonary embolism. These diseases and two other abnormal conditions (sleep apnea and snoring) will be considered here. In examining these respiratory diseases and abnormal conditions, keep in mind that the ability of hemoglobin to bind oxygen is affected by CO_2, pH, and temperature. Respiratory diseases and conditions can reduce ventilation, leaving more CO_2 in the blood and more warm air in the lungs. These changes reduce the ability to oxygenate blood not only because the O_2 supply to the lungs is reduced. The elevated CO_2 reduces the pH in blood in the lungs, and the blood remains somewhat warmer. Under these conditions, the hemoglobin in blood passing through the lungs cannot pick up and hold as much oxygen. Therefore, the hemoglobin cannot transport as much O_2 to body cells.

Lung Cancer

Normally, cells reproduce when the body needs more of them; once the need is met, they stop reproducing. An example is the temporary rapid reproduction of skin cells that occurs until a cut in the skin heals. *Cancer* consists of cells that continue to divide and spread out in an uncontrolled fashion even when they are not needed. A clump of these cells is called a *tumor*.

Some forms of cancer develop from lung cells and are called *primary lung cancer*. These are the types caused primarily by smoking. Many other cancers of the lungs develop when the circulatory system moves cancer cells from another place in the body to the lungs. Cancer that moves to another part of the body is called *metastatic cancer*. Metastatic lung cancer often comes from the breasts or the reproductive system.

A person with lung cancer may have from one tumor to very many tumors. Whether the cancer is primary or metastatic, the effects on the lungs are similar. Ventilation becomes more difficult because airways get blocked when tumors grow inside them or squeeze them closed. Air volumes are reduced as alveoli become filled with cancer cells. Occasionally the cancer becomes so large or stiffens the lungs so much that they cannot inflate or deflate adequately for ventilation. Cancer cells in the alveoli may reduce diffusion by thickening or replacing the respiratory membrane between the air and the blood. Sometimes cancer will distort, squeeze, or replace the pulmonary vessels so that perfusion is reduced. Some blood vessels are weakened, causing bleeding.

Several warning signs indicate that lung cancer may be present. They include a persistent cough, coughing or spitting up blood, pain in the chest, difficulty swallowing, hoarseness, easy fatigability and the feeling of breathlessness, and a swelling of the fingertips. Any of these indicators warrant evaluation by a physician.

Though some forms of lung cancer can be cured if discovered early enough, most cases are not identified until the cancer has grown so much that it cannot be eliminated. The vast majority of cases of lung cancer result in death within a few years. The only effective "cure" is prevention: avoiding tobacco smoke and other forms of air pollution.

Chronic Bronchitis

To understand chronic bronchitis, recall that the trachea and bronchi are lined with a thin layer of mucus and that as the mucus is made, cilia sweep it up and out of the airways.

Development   If a person inspires air with an excess amount of harmful particles or noxious chemicals, the cells lining the trachea and bronchi become injured. The resulting inflammation causes those cells to make mucus much faster, and the lining of the airways becomes swollen. In addition, the beating of cilia slows. The person now has bronchitis and will begin to cough to remove the extra mucus and pollutants.

If this person breathes the pollutants frequently and continuously, the airways remain inflamed for a longer time, and the person then has *chronic bronchitis*. This condition is accompanied by extra mucus production and coughing. After a while the cilia will be damaged and may completely disappear.

Effects   The major effect of chronic bronchitis is to reduce ventilation by making the airways narrow in two ways. First, mucus accumulates because it is being produced more quickly and removed more slowly. Second, the lining of the airways swells inwardly. The effect on airflow through the trachea and bronchi is similar to the stuffed-up feeling that occurs when a head cold causes swelling and the accumulation of mucus in nasal passages.

Expiration becomes especially difficult because the lower airways normally narrow during expiration. The additional narrowing from the mucus and swelling makes them so narrow that expiration can occur only very slowly. This decreases the minute volume, and so having enough gas exchange to meet the body's needs is quite difficult. The problem is compounded because the person will begin to rely more on forced expiration, increasing the work of breathing. The effort used in coughing raises the work of breathing to levels that may leave the victim dizzy, breathless, and temporarily incapacitated.

The problem becomes very serious when the person tries to do something physically active. Fatigue and the sense of being out of breath develop quickly and are rather severe. Some individuals are disabled by this disease.

Fortunately, many cases of chronic bronchitis that have not been allowed to progress too long can be cured. The person need only eliminate breathing polluted air. Eventually, mucus production will slow and the cilia will grow back and begin to function as before.

Curing chronic bronchitis can be difficult if smoking is the source of the air pollution, however, because tobacco smoke contains addictive chemicals such as nicotine. Also, as the respiratory system begins to clear itself, coughing increases temporarily. Smokers often experience extra coughing in the morning because the clearing action began during the night, when they were not smoking. After a period of abstention, smoking seems to help because it relieves the withdrawal symptoms and stops the clearing action, and thus stops the coughing. Of course, continuing to smoke only relieves certain symptoms while the disease continues to destroy the person's respiratory system.

Besides reducing directly the performance of the respiratory system, chronic bronchitis increases the risk of infection of the respiratory system because the accumulation and slow removal of mucus allow microbes to flourish in the airways. It can also lead to emphysema, and the chronic coughing contributes to hemorrhoids. Long-term smoking is also a major risk factor for nonrespiratory diseases such as heart attack, atherosclerosis, and stroke.

Emphysema

Emphysema is a disease that involves actual destruction of some parts of the lungs. There are two main forms: *centrilobar emphysema* (*CLE*) and *panlobar emphysema* (*PLE*). Both types will be present in most people with emphysema, though one type will predominate.

Centrilobar Emphysema   Centrilobar emphysema most often develops along with or after chronic bronchitis. It involves a thinning and weakening of the smallest bronchioles and the production of much mucus. Many results are similar to those of chronic bronchitis. Additionally, the damage to the bronchioles usually results in a decrease in the number of small blood vessels in the lungs, decreasing perfusion. The reduction in blood vessels also makes it harder for the heart to pump blood through the lungs, and the over-worked heart eventually becomes weaker. If CLE continues to progress, the victim eventually dies of respiratory failure, respiratory infection, or heart failure.

Panlobar Emphysema   Panlobar emphysema is less common than CLE. Though the major cause is air pollution, some people inherit a tendency to develop this type of emphysema.

PLE causes destruction of the walls of the alveoli and alveolar sacs. The results are like a highly exaggerated version of age changes in the alveoli. Many walls between the alveoli shrink and disappear. Neighboring alveoli blend to form large air-filled spaces. The lungs change from having microscopic spaces like those found in Styrofoam to having large spaces like those in a sponge. The wall material that remains is weaker and less elastic. All these changes reduce ventilation.

With PLE, expiration becomes more difficult and more residual air is left in the lungs. As passive expiration decreases, forced expiration increases, increasing the work of breathing. Perfusion also decreases because the number of capillaries declines as the alveolar walls are destroyed. Besides reducing gas exchange, this overworks

the heart, occasionally leading to heart failure. Finally, diffusion is reduced because there is a decrease in the amount of surface area.

A complication of PLE is the partial or complete collapse of a lung. This occurs when a hollow space developing close to the lung surface bursts like a bubble. As escaping air separates the lung from the thoracic wall, the lung collapses like a balloon with a small leak. This condition is called *pneumothorax*. Proper inspiration is impossible unless the leak heals and the body absorbs the air from the thoracic cavity.

Overall Effects of Emphysema   People in the early stages of emphysema may hardly notice the decline in their ability to perform physical activities. As the disease progresses and devastates more of the lungs, gas exchange plummets. Eventually, even walking at an ordinary pace becomes a challenge. Individuals with advanced cases are so disabled that they may be unable to get up, sit up, or even roll over in bed without extreme fatigue. Mild exertion or a slight respiratory infection can cause death. Among people over age 55, emphysema is the fifth leading cause of death for men and the seventh leading cause for women.

Pneumonia

Pneumonia is actually a group of related diseases involving inflammation in the lungs. Several types reduce a person's ability to inspire. Older people are especially affected by pneumonia caused by microbes (bacteria, viruses, and fungi) and by dust and chemical vapors. Pneumonia can also result from aspirating stomach contents that have moved up into the throat.

Microbial Pneumonia   Reasons for the age-related increased susceptibility to microbial pneumonia include age changes in the functioning of the mucociliary escalator, white blood cells, and the immune system; the rising prevalence of chronic bronchitis and emphysema; and other diseases that weaken the body and make it less able to ward off infections.

Pneumonia caused by bacteria results in filling of the airways and alveoli with fluids and cells from their walls. This material blocks the airways. It usually becomes somewhat solid after 1 to 2 days. If a person is otherwise healthy and receives proper treatment, such as antibiotics, the infection can be overcome and the material will be cleared away after about a week.

Many types of bacteria that cause pneumonia leave the lungs with no residual damage. However, some forms cause serious and permanent damage that results in a reduction in respiratory functioning and can cause death. These forms are the ones most likely to occur in weakened or hospitalized individuals.

Viral pneumonia affects the walls of the alveoli, causing them to accumulate fluids and become thicker, reducing gas exchange. If a person is healthy and has a good immune system, the immune response will eliminate the virus in a few days and the lungs will return to normal functioning.

Because fungal pneumonia and tuberculosis cause death of the portions of the lungs they infect, they can be more serious than bacterial or viral pneumonia. Thus, after fungal and tubercular infections are stopped, the lungs are left with regions that no longer function. Areas affected by tuberculosis are filled in with solid scar tissue which, if calcified, can be detected on x-ray. If enough areas of the lungs are destroyed, gas exchange and activity levels are permanently reduced. More extensive damage results in death.

Unfortunately, many older individuals are not healthy and do not have strong immune systems when they get pneumonia. Weakened persons may have great difficulty combating the infection. Then the disease lasts longer and has a greater impact on the respiratory system. The proportion of individuals who survive microbial pneumonia decreases rapidly with age. Those who survive are often left weakened for long periods.

Dust and Vapors   Some individuals breathe large amounts of certain types of air pollution repeatedly for long periods, usually because of their occupations. Examples include farmers, miners, textile mill workers, sandblasters, and woodworkers. The heavy exposure and the size and chemical nature of such air pollutants cause the lungs to form large quantities of fibers and develop the condition called *pulmonary fibrosis*.

With pulmonary fibrosis, the normal amount and rate of age changes in the lungs increase dramatically, leading to a rapid decline in gas exchange. Very severely affected people will become quite disabled. Since the fibrosis is permanent,

affected individuals can recover little if any of the lost respiratory functioning even if they avoid future exposure to air pollution. The only solution is to prevent pulmonary fibrosis by avoiding its causes.

Pulmonary Embolism

Pulmonary embolism (Chap. 4) is a disease condition in which blood clots have moved to the lungs from the systemic veins or the heart. Conditions commonly promoting the formation of such emboli in the elderly include varicose veins, congestive heart failure, and immobility. The elderly are especially prone to having conditions that cause immobility. These include heart attack, stroke, hospitalization, recovery from surgery, and fractures. The effects of pulmonary embolism depend on the size and number of pulmonary emboli.

Control Errors

Two age changes involving the control of ventilation that have not yet been discussed are sleep apnea and snoring.

Sleep Apnea   *Sleep apnea* (SA) is exhibiting at least five temporary cessations of ventilation per hour or exhibiting at least 10 occasions of depressed ventilation and cessation of ventilation per hour when asleep. The incidence of sleep apnea increases with age up to age 65, after which the incidence plateaus. It is present in 4 percent of younger adults but in 25 percent to 30 percent of people over age 64. The male:female ratio for SA is approximately 3:1.

Sleep apnea may be caused by narrowing and collapsing of the pharynx, especially when in a supine position (i.e., sleeping on one's back); because the respiratory center becomes less sensitive; or because the center simply stops initiating inspiration. Then blood levels of O_2 and CO_2 change. These alterations in the blood may provide the necessary stimulation to begin inspiration again. People with sleep apnea tend to snore and to have frequent sudden awakenings with feelings of respiratory distress.

Mild sleep apnea seems to have no deleterious affect on the body. However, frequent awakenings can lead to fatigue, indications of sleep deprivation, and adverse alterations in mood and personality. Because sleep apnea causes significant fluctuations in O_2, CO_2, and blood pressure, serious cases increase the risk of heart attack and stroke. Treatments for SA include avoiding sleeping in a supine position; using masks with pumps that provide positive pressure into the respiratory system; reversal of obesity; medications; and surgical correction of the pharyngeal region.

Snoring   *Snoring*, or making loud breathing sounds when asleep, is due to partially obstructed upper airways. Approximately half of all women and well over half of all men above age 65 snore. Some individuals who snore also have sleep apnea.

Snoring causes a variety of problems. Biologically, it causes from mild to severe adverse effects on blood O_2 and CO_2 levels and on circulation. It can also contribute to high blood pressure and heart disease. Since snoring disrupts normal sleep patterns even if the person who is snoring does not awaken, it can result in fatigue and other indications of sleep deprivation.

Anyone who sleeps near a person who snores can attest to some of the social implications. Their responses to the person who snores, together with the multitude of jokes about snoring, can add to the psychological impact produced by sleep deprivation. The fatigue felt by many snorers also affects their social interactions and can impinge on their ability to carry out their jobs.

Though the causes of snoring and the role of the nervous system in snoring are not clear, research has provided methods of treatment for this condition.

SMOKING

Main consequences in the respiratory system from smoking have been mentioned. Smoking has adverse effects in other areas of the body, also. In general, smoking increases the formation of free radicals and lipid peroxides while reducing the antioxidant actions of vitamin C, vitamin E, and β–carotenes. Smoking may increase free radical damage to DNA by 50 percent. In the skin, smoking speeds up and amplifies the effects from aging and from photoaging. Smoking is associated with increased risks for most skin cancers. In the circulatory system, smoking damages the endothelium; raises blood pressure; and increases substantially the risk of blood clots, of atherosclerosis, and of their complications. Effects on these

two systems are due partly to constriction of skin vessels and reductions in blood oxygen caused by smoking. These two changes develop within minutes of initiating smoking and can last for hours, long enough to light the next cigarette. The result is continuous inadequate blood flow in the skin and elevated blood pressure. In the eyes, smoking is associated with a higher incidence of cataracts and diseases of the retina. Smoking reduces estrogen levels in women and speeds up age-related thinning of bones. Smoking doubles the problems from non-insulin dependent diabetes; suppresses normal functioning of the immune system; promotes autoimmune diseases; and is associated with higher rates of reproductive system and digestive system cancers. Cessation of smoking is associated with reduction or complete reversal of these problems and risks.

SOUND PRODUCTION AND SPEECH

The vocalizations produced by people involve words and a variety of other sounds, such as moans, grunts, whistles, cheers, and laughing. People use sound production for communication. Communication among individuals by sound and other means (e.g., visual signals) is important to a high quality of life and to survival because it is one of the three components in negative feedback systems. A common example of using vocalization as part of a negative feedback system is shouting a warning to a person in danger.

Human sound production can enhance life in other ways. Words and other vocalizations can motivate and encourage positive actions such as beginning a new career or hobby. They are also used in teaching, praising, consoling, expressing emotions, and many other human activities. And what of the beauty of a poem or song? All these are created by the sounds produced by the flow of air through airways.

Mechanisms

The respiratory system produces sound by passing air through the upper airways and the mouth. Most of the sound people make is caused when air passing through the larynx causes the vibration of two flaps of tissue called the *vocal cords* (Fig. 5.5). The sound gets louder when more air flows through the larynx.

Different muscle contractions in the larynx control the position and tension of the vocal cords and thus alter the pitch of sounds. The sounds made by the vocal cords are modified by the other upper airways, especially the nasal passages and the mouth. By changing the shape of these passages and moving the tongue, a person can create a multitude of sounds and form words.

All the actions that produce and modify sounds from the respiratory system are controlled by the nervous system.

Age Changes

Many age changes that alter inspiration and especially those which modify expiration affect sound production. Age-related stiffening of the larynx, shrinkage of the vocal cords and its muscles, and changes in the mouth are also important. Age changes in the nervous system are also important since sound production depends on the coordinated action of many muscles. Even age changes in hearing are important because the ears provide feedback information so that the sounds a person produces can be adjusted to conform to the sounds intended by that person.

Because of age changes in these areas, the voice becomes more variable in pitch and volume during speaking. Female voices often become lower in pitch, while male voices often become higher in pitch. Other common changes include increases in hoarseness, roughness, and extraneous sounds while speaking. The voice often becomes weaker, and elders have declining abilities to speak very quietly or with very loud volume. The ability to control volume declines, and the precision of word pronunciation diminishes.

Language fluency and vocabulary usually do not decline, and often increase. However, phrases and sentences often become shorter, syllables and words are repeated more often, and more words are pronounced incompletely. These trends in speaking become more prominent in stressful situations. Of course, variability among elders increases with age, and some elders retain the voice and speech of a young adult.

All these changes reduce the effectiveness of vocalization in providing communication. Additionally, some of the pleasure derived from the human voice maybe lost. As a result, the contribution of the voice to happy and healthy survival diminishes. Age changes in the voice also alter the way people respond socially to individuals who are getting older. These changes in turn affect aging individuals' responses and self-images. Therefore, biological aging of vocalization can influence nonbiological aspects of life.

6

NERVOUS SYSTEM

The nervous system is made up of three types of organs: the *brain*, the *spinal cord*, and *nerves* (Fig. 6.1). The brain and the spinal cord are referred to as the *central nervous system* (*CNS*) because they are along the midline of the body. The nerves constitute the *peripheral nervous system* (*PNS*), extending from the brain and spinal cord to the farthest reaches of the body. The functions of the brain, spinal cord, and nerves are performed by the highly specialized *nerve cells* (i.e., *neurons*) they contain.

MAIN FUNCTIONS FOR HOMEOSTASIS

The overall goal of the nervous system is to regulate the operations of parts of the body to make sure they contribute to homeostasis and a satisfactory quality of life. The nervous system regulates muscles and glands directly by sending impulses to those structures. Among the glands controlled by the nervous system are the sweat glands and salivary glands. This system regulates other parts of the body indirectly by adjusting the amounts of hormones produced by some of the endocrine glands.

Monitoring

The nervous system performs six main functions to carry out its overall goal. Three operations stem from the three steps in negative feedback systems: monitoring, communicating, and adjusting. Many of the neurons in the brain and nerves monitor conditions in and around the body. These neurons do very little if conditions are proper and fairly stable. However, they are affected by harmful conditions and are sensitive to any change in conditions. When conditions are unfavorable for the cells or when there is a change (a *stimulus*), the neurons respond by starting messages (nerve impulses) within themselves.

Communicating

The initiation of impulses by neurons leads to the second main function: communicating. The neurons carry impulses to other parts of the nervous system, where they are passed on to other neurons, which pass them to still other neurons, and so on. Thus, many parts of the nervous system are informed that a change has occurred. They are

FIGURE 6.1 The nervous system.

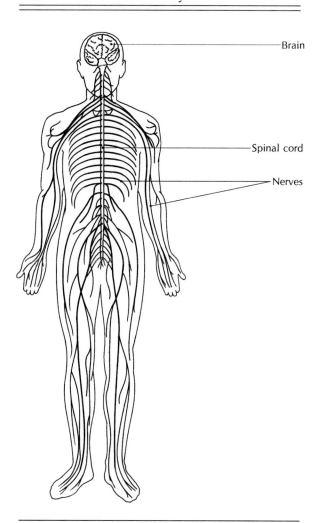

Brain

Spinal cord

Nerves

which is the third step in negative feedback. It only stimulates other parts of the body to do so.

These three functions can activate responses to promote beneficial changes as well as eliminate harmful ones. For example, when neurons in the stomach sense that it is empty and brain neurons detect that the nutrient level in the blood is low, a person feels hungry. If other neurons detect the sound of someone cooking in the kitchen while still others detect dinner aromas, the nervous system will activate muscles so that the hungry person will go to the kitchen and obtain nourishment. When the stomach has become full and blood nutrient levels begin to rise, other neurons initiate a negative response, causing the person to stop eating.

Coordinating

Making adjustments often requires the contributions of many parts of the body, and the nervous system must stimulate them so that they all work in harmony. At the same time parts of the body that can interfere with achieving the desired outcome must be inhibited from acting. The nervous system provides these stimulations and inhibitions through its fourth main function: coordinating. For example, to walk to the kitchen, a person must activate some muscles while inhibiting others in order to step forward with one foot at a time.

Remembering

When a person must adjust to a new situation, it may take quite a while for all the necessary impulses to reach their destinations, especially when the situation is complicated and the proper response requires the coordinated stimulation of many structures. Furthermore, sometimes mistakes are made and the wrong response occurs. This is when remembering, the fifth main function of the nervous system, becomes helpful.

By remembering, the nervous system stores information about past experiences that includes the recollection of a situation, the responses that were made, and the degree of success that was provided by each response. Then, when faced with the same situation, a person can avoid trial and error by remembering what to do. This procedure saves time and prevents costly mistakes. Simple examples include remembering the way home after traveling the route a few times and

also informed of the nature of the change, its extent, and where it is happening. For example, if an insect bites a person, that person feels that something is happening. He or she also knows that it is a bite rather than something soft brushing against the skin, has a sense of the severity of the bite, and knows where to scratch or hit to remove the insect.

Stimulating

Communicating leads to the third function: stimulating. In the case of an insect bite, the nervous system activates muscles in the arms to remove the source of irritation. Note that the nervous system does not actually perform the adjustment, remembering the answers to test questions using information that was studied many times.

The benefits of remembering are used on an unconscious level as well. For example, when one is practicing an activity such as walking, playing an instrument, or riding a bicycle, the nervous system remembers the sequences of muscle contractions that resulted in failure or success. With enough practice, one need only consciously start the activity. One can then continue to perform well without thinking about the activity because, like a recording, the nervous system plays the rest of the program of successful commands for the muscles.

Memory also recalls situations that led to favorable or unfavorable outcomes. When the nervous system recognizes the presence of such situations, it will alert a person to proceed or take evasive action. This is why an experienced child will reach out for candy but back away from fire.

Thinking

Remembering tends to provide the same type of successful response every time a person is in the same circumstance. The more successful the same response is in the same situation, the faster and more accurately that response will occur. However, remembering does little when a person is faced with a new situation. That person must try to find the correct response by trial and error or by mentally imagining different responses and the results they might cause. Creating mental images of new courses of action and their possible outcomes is the sixth main function of the nervous system: thinking.

Thinking depends on memory to provide initial mental images and information. In thinking, a person intelligently rearranges the remembered images and information to create new images that have not been experienced before. Many alternatives can be mentally explored in a few seconds without actually trying any of them. People are thinking when they make plans, solve problems by analysis, and create mental images of things that do not occur naturally. Thinking provides the variety of acting that many people believe separates humans from other living things.

Thinking allows people to decide the best response to a new situation quickly, accurately, and without having to risk the consequences of untested attempts. It can even produce new re-

sponses to situations that have been created by people. For example, this is how people return to the Earth from a trip to the moon.

NEURONS

Components

All the billions of neurons in the nervous system have three basic parts. The *nerve cell body* contains the nucleus of the cell along with cytoplasm and organelles (e.g., mitochondria and ribosomes) (Fig. 6.2). The nerve cell body supplies the other two parts of the neuron with the materials and energy they need. It can also pick up messages from other neurons.

One or more extensions called *dendrites* project from the nerve cell body. Each dendrite can branch up to several hundred times. Like nerve cell bodies, dendrites can pick up messages from other nerve cells. They are also the parts of the sensory cells that monitor conditions. A dendrite being activated by another neuron or by a stimulus starts nerve impulses that travel along the dendrite to the nerve cell body, which passes the impulses to the third part of the neuron: the *axon*.

Each neuron has only one axon, which extends out from the nerve cell body. Each axon may have up to several hundred branches (*axon collaterals*). The impulses that are passed to the axon travel the entire length of each of its branches. Each branch then passes the impulses to another structure. Axons can pass impulses to other neurons, muscle cells, and gland cells, although all the branches from one neuron's axon can go to only one of these types of cells.

Operations

Reception All neurons perform three main functions. *Reception* involves having impulses generated in response to environmental conditions or messages from other neurons. Dendrites and nerve cell bodies are the parts that usually perform reception (Fig. 6.3).

Conduction The second function—*conduction*—refers to the movement of impulses along the neuron to the end of the axon (Fig. 6.3). Conduction in longer dendrites and axons occurs through a special mechanism called an *action potential.*

FIGURE 6.2 Neuron structure.

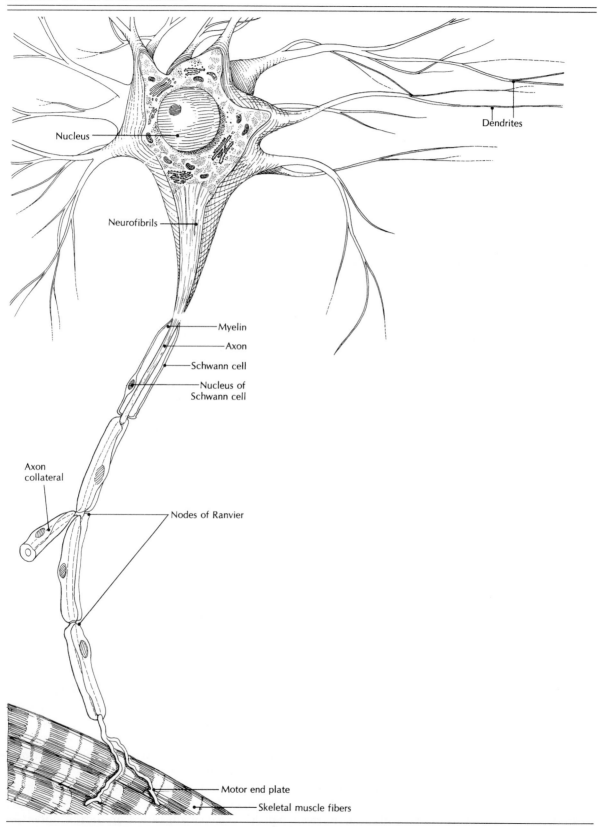

Nucleus

Neurofibrils

Dendrites

Myelin

Axon

Schwann cell

Nucleus of
Schwann cell

Axon
collateral

Nodes of Ranvier

Motor end plate

Skeletal muscle fibers

This mechanism involves several activities of the neuron cell membrane that carefully control the inward and outward movement of ions, especially sodium and potassium ions.

Transmission Once impulses have been conducted to the end of the axon, they are passed to the next structure by the third neuron function: transmission. The place where transmission occurs between neurons is called a *synapse*. Transmission to muscle cells occurs at *neuromuscular junctions*, and transmission to gland cells takes place at *neuroglandular junctions*. The process of transmission is essentially the same in all three cases (Fig. 6.3).

At a synapse, when an action potential reaches the end of an axon, it causes small packets (*synaptic vesicles*) at the end of the axon terminal to burst like blisters. These packets contain a chemical called a neurotransmitter, which is then released into the small space (*synaptic cleft)* between the neurons. Most neurons can release only one type of neurotransmitter. The neurotransmitter diffuses to the dendrite or cell body of the next neuron, where it attaches to *receptor molecules* on the cell membrane. Each type of receptor molecule is designed to bind to only one type of neurotransmitter.

Once enough neurotransmitter has been bound to the receptor molecules, the receiving neuron responds. Depending on the type of neurotransmitter and the type of neuron, the receiving neuron will be stimulated to perform reception and start its own impulses or will be inhibited from acting. The nervous system uses stimulatory transmissions to start or speed up an activity; it uses inhibitory transmissions to slow down, stop, or avoid an activity. A neurotransmitter continues to have its effect on the next cell until it is eliminated or counteracted. Neurotransmitters can be counteracted when antagonistic neurotransmitters are sent into the synapse.

Although a few synapses involve one neuron transmitting to one other neuron, synapses often have many neurons converging to transmit messages to a single neuron. The amount and length of the response by the receiving neuron depend on the balance between the amount of stimulatory and inhibitory neurotransmitters it receives at any moment from the many neurons connected to it. Thus, by changing the combinations of neurotransmitters at synapses, the nervous system

FIGURE 6.3 Neuron reception.

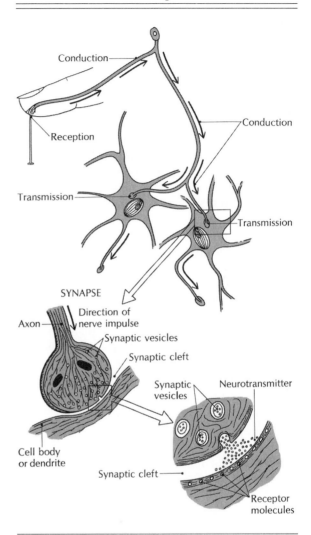

can provide exquisitely precise adjustments to its impulses and the resulting body activities. The effect of such an interplay of stimulatory and inhibitory transmitters is experienced, for example, by a person whose hands are being burned by a hot beverage but who puts down the cup slowly and carefully to avoid spilling the beverage.

The branching of axons allows for divergence. Thus, impulses in one neuron can spread to many muscle cells, gland cells, or neurons. One can experience the effects of divergence when hearing a frightening sound or noticing a flirtatious glance. The heart pounds, the breathing increases, the stomach tightens, and the legs may become weak and shaky.

Another important function of synapses is to keep order in the nervous system. Since messages can pass only from axons to the next neuron, synapses ensure that impulses move through the system only in the correct direction.

NEUROGLIA

The CNS contains *neuroglia* cells, which provide a variety of services for the neurons (e.g., support and defense). These cells do not perform reception or conduct or transmit nerve impulses. One type makes a material called *myelin*, which forms a coating on CNS axons. The myelin coating on an axon resembles beads on a string. It causes impulses to travel faster by making them jump along the neuron (Fig. 6.2). Since myelin is white, it causes the regions that contain it to become white in appearance; these areas are referred to as the *white matter* of the brain and spinal cord.

The areas of the CNS that do not have myelin possess the pinkish gray color of plain neurons; these regions constitute the *gray matter*. The gray matter is important because it contains the synapses. All the complicated nervous system functions, including coordination, remembering, and thinking, require these synapses.

SCHWANN CELLS

Neurons in the PNS are assisted by *Schwann cells*. These cells produce myelin on dendrites and axons; this myelin is structurally and functionally similar to CNS myelin (Fig. 6.2).

NERVOUS SYSTEM ORGANIZATION

Central Nervous System

Recall that there are two main subdivisions of the nervous system—the *central nervous system* and the *peripheral nervous system*—and that the two parts of the CNS are the brain and the spinal cord. The neurons in different regions of these two organs are specialized to contribute to one or more of the main functions of the nervous system. For example, certain areas of gray matter in the brain monitor conditions such as temperature and the level of CO_2, others start impulses that stimulate muscles to contract, and still other areas are for remembering. Myelinated axons in the white matter allow regions of gray matter to communicate with each other.

Peripheral Nervous System

Sensory Portion   The *sensory* portion of the peripheral nervous system contains *sensory neurons*, which monitor body conditions outside the brain and spinal cord. They also monitor conditions on the surface of the body and in its surroundings. Each type of sensory neuron is designed to monitor only one type of condition. For example, one kind responds to changes in temperature, while another is activated by pressure. Those in the nose and on the tongue respond to chemicals.

Most sensory neurons are long thin cells that extend through nerves from the regions they monitor to the brain or spinal cord. For example, sensory neurons from the fingertips extend through nerves in the arm all the way up to the middle of the back, where they enter the spinal cord. Once a sensory neuron performs reception in response to a condition, it carries impulses to communicate information about that condition to the brain or spinal cord.

Sensory neurons that do not have myelin release two substances (i.e., calcitonin gene-related peptides, substance P) at sites of wound injury. The combined effects are providing adequate inflammation while promoting healing.

Motor Portion   The *motor* portion of the PNS consists of *motor neurons* that control the activities of muscles and glands. *Somatic motor neurons* control muscles that are attached to bones. Usually there is voluntary control of these muscles, although sometimes the nervous system causes them to contract involuntarily.

Somatic motor neurons extend from the brain and spinal cord, through nerves, to muscles they control. For example, the motor neurons that enter and stimulate the muscles in the lower leg begin in the spinal cord just below the middle of the back.

Other motor neurons make up the *autonomic* portion of the PNS. *Autonomic motor neurons* control many of the functions of the integumentary, circulatory, respiratory, digestive, urinary, and reproductive systems by regulating many glands and also muscles that are usually not under voluntary control. The sweat glands and salivary glands, for example, are under autonomic control. Muscles under autonomic control include the heart and the smooth muscle in the walls of blood vessels, the bronchi, the stomach, and the urinary bladder.

Autonomic motor neurons are of two types: *sympathetic* and *parasympathetic*. Though a few structures (e.g., sweat glands, skin vessels) are controlled by only one type of autonomic motor neuron, most receive both sympathetic and parasympathetic motor neurons. In places where both types are present, one type of autonomic motor neuron stimulates the structure and the other type inhibits it. By balancing the amount of stimulation and inhibition, the autonomic nervous system can precisely control the speed and strength of activity of a structure. For example, sympathetic motor neurons increase the rate and strength of the heartbeat while parasympathetic motor neurons decrease them. By automatically adjusting the ratio between sympathetic and parasympathetic impulses, the autonomic nervous system varies the rate and strength of the heartbeat as the amount of blood flow needed by the body fluctuates.

NERVOUS SYSTEM PATHWAYS

Reflexes

The individual components of the nervous system work together to regulate the operations of parts of the body in order to maintain homeostasis. The simplest level of regulation involves a *reflex*, which is an involuntary response to a stimulus. Reflexes that use somatic neurons include blinking when something moves close to the eyes, coughing when something gets caught in the throat, and withdrawing from something that is painful. All activities controlled by autonomic neurons are reflex responses.

Many reflexes are built into the nervous system as it develops before birth. Others are *acquired reflexes* which develop when a person repeats the response every time a certain stimulus occurs. These reflexes involve the use of unconscious remembering.

A reflex occurs in basically the same way every time a particular stimulus occurs because the nervous system pathway that causes it is firmly established. Sensory neurons detect the stimulus and communicate through synapses with specific neurons in the CNS, and the CNS neurons quickly communicate with specific motor neurons. In a few reflex pathways, such as the one for the knee jerk, sensory neurons synapse directly with motor neurons. In either case the motor neurons complete the pathway by sending impulses to a muscle or gland, causing it to make the response.

Note that reflex pathways involve monitoring, communicating, and stimulating (or inhibiting). In many reflexes the adjustment caused by the response prevents or reverses the situation created by the stimulus. For example, the cough reflex removes material that enters the airways. These reflexes therefore are negative feedback systems that help maintain homeostasis. The responses produced by other reflexes contribute to homeostasis by improving conditions for the body. For example, the sight and smell of appetizing food cause a reflex that increases the secretion of saliva, which will be useful when the person begins to eat because it makes swallowing easier.

Some reflexes simultaneously use sensory impulses from several types of sense organs, such as the eyes, ears, skin receptors, and proprioceptors. Proprioceptors detect motion and tension in muscles and at joints. Some reflexes require a considerable amount of coordination by both brain and spinal cord interneurons and synapses. Some are influenced by voluntary motor impulses or by higher brain activities such as emotions and thinking, which send modifying impulses into the reflex synapses.

Reflex Pathways   The specific parts and activities in a reflex pathway must be understood to appreciate the effects of aging on reflexes. The withdrawal reflex that occurs when a sharp object jabs the bottom of the foot provides a good example (Fig. 6.4).

When sensory neurons in the skin of the left foot detect the intense pressure caused by stepping on a sharp object, their dendrites carry out (1) reception. This causes the dendrites to (2) conduct impulses up through the nerve in the leg. These impulses reach and enter the gray matter in the back of the spinal cord via the sensory neuron axons, which (3) transmit them through synapses to other neurons in the spinal cord gray matter. Since these next neurons extend from one neuron to another, they are called *interneurons*. The interneurons (4) transmit the impulses to somatic motor neurons in the front part of the gray matter of the spinal cord. The impulses are then (5) conducted down the motor axons in the nerves in the left leg to certain muscles in the thigh and calf. Neurotransmitters from the motor axons (6) stimulate these muscles to contract, causing the response of lifting the foot and thus relieving the intense pressure and protecting the foot from harm.

Proper reflex responses may require coordination in addition to monitoring, communicating, stimulating, and unconscious remembering. For example, to prevent loss of balance when lifting the foot, cooperation by a second reflex must oc-

cur. Branches of the sensory axons transmit impulses to other interneurons that cross over to the right side of the spinal cord. These crossing interneurons (7) transmit the impulses to other somatic motor neurons in the right side of the gray matter. Impulses in these motor neurons are (8) conducted down the nerves in the right leg. The impulses cause certain muscles in the right leg to contract, resulting in a straightening of the right leg at the same time that the left leg is bending and lifting the foot off the object. In this way, the right leg supports the weight of the body so that the person does not fall down.

FIGURE 6.4 Reflex pathways involving skeletal muscles.

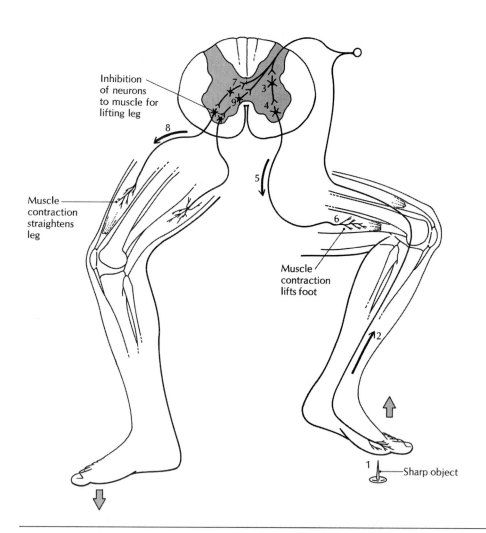

Another aspect of coordination is shown by the withdrawal reflex. As the interneurons stimulate motor neurons to the muscles that will make the appropriate actions occur, the interneurons send (9) inhibitory impulses to motor neurons controlling leg muscles that would interfere with the proper movements. This prevents antagonism among the muscles.

The reflex pathway for the withdrawal reflex is a fairly simple one. Other reflex pathways may involve interneurons that extend up or down the spinal cord or through several areas of the brain. Countless synapses may become involved before the impulses are finally transmitted to the motor neurons. Autonomic reflexes are further complicated by the synapses in the PNS. This increased complexity permits more coordination and modulation in responses. However, more complicated reflex pathways operate in essentially the same manner as simple reflex pathways.

Conscious Sensation

Though a reflex is completely involuntary and requires no conscious awareness, a person may feel the stimulus. For example, a person feels a sharp object jabbing the foot because the sensory neurons may synapse with other interneurons extending up to the brain. These other neurons help form the *conscious sensory pathways* in the nervous system.

Information from perceived sensations is used to initiate and adjust voluntary actions so that people can respond properly to conditions in their bodies and the world around them. These sensations provide information necessary for learning. Finally, conscious sensation provides much of the enjoyment that makes life worthwhile.

All conscious sensory pathways begin in the same way as do reflex pathways. That is, sensory neurons that have carried out reception conduct impulses into the CNS (Fig. 6.5). Sensory neurons that monitor regions below the head extend into the spinal cord, while those which monitor the head region pass into the brain. Once in the CNS, sensory impulses are passed to interneurons extending into the gray matter of the brain.

Impulses in each type of sensory neuron and from each part of the body are directed by synapses to the part of the brain designed to monitor that type of stimulus from that region. For example, impulses from the eyes are sent to vision

FIGURE 6.5 A conscious sensory pathway.

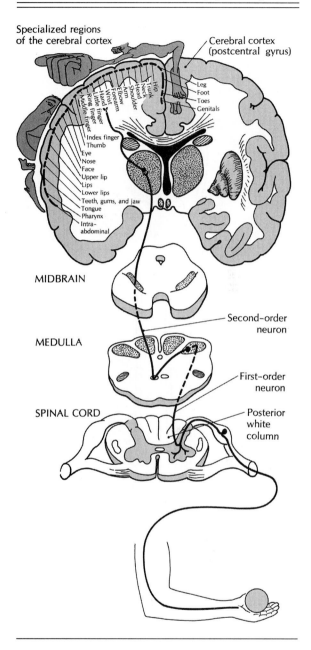

centers, while impulses from the auditory parts of the ears are sent to hearing centers. The impulses are interpreted as perceived sensations when they reach the appropriate areas of the *cerebral cortex*, a layer of gray matter on the surface of the *cerebral hemispheres*. The *postcentral gyrus* is a raised area of the cortex on each cerebral hemisphere that is concerned mostly with conscious sensations from the integumentary, muscle, and skeletal systems (Fig. 6.5). Other

FIGURE 6.6 A somatic (voluntary) motor pathway.

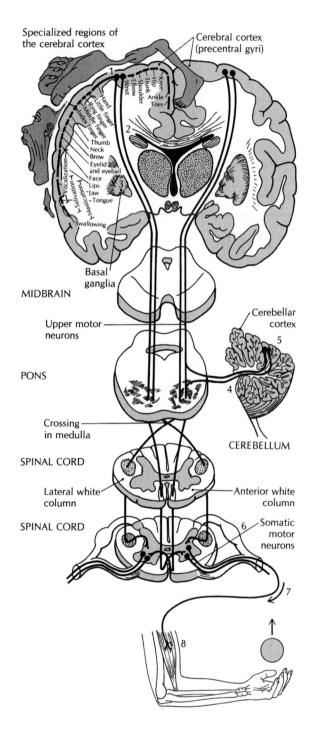

Specialized regions of
the cerebral cortex

Cerebral cortex
(precentral gyri)

Hip
Knee
Trunk
Ankle
Shoulder
Toes
Elbow
Wrist
Hand
Little finger
Ring finger
Middle finger
Index finger
Thumb
Neck
Brow
Eyelid
and eyeball
Face
Lips
Jaw
Tongue

Vocalization
Salivation
Mastication
Swallowing

Basal
ganglia

MIDBRAIN

Upper motor
neurons

Cerebellar
cortex

PONS

Crossing
in medulla

CEREBELLUM

SPINAL CORD

Lateral white
column

Anterior white
column

SPINAL CORD

Somatic
motor
neurons

regions of the cortex are used for the special senses, such as vision, hearing, and smell.

Voluntary Movements

In many situations a person voluntarily chooses to move or not move in response to stimuli. One can also choose the type and degree of motion to make. For example, if someone calls, a person can choose to answer or not answer. If that person answers, the response may include a variety of motions or sounds. If the response is vocal, the sounds produced may be loud or soft, enunciated quickly or slowly, and projected with different intonations.

In addition to deciding whether to move in response to conscious stimuli, one can decide whether to take action on the basis of internal thought processes. Again, the type and degree of motion are usually up to the individual. Thus, one need not be called in order to decide to move or say something. A person may spontaneously decide to start a conversation or simply to sing.

Voluntary movements allow a person to take what he or she judges to be an appropriate action to optimize conditions in a given situation. Unlike reflex responses, voluntary movements allow freedom to select among many options rather than forcing a person to respond in a particular way.

Whether voluntary motion is initiated by stimuli or by thought processes in the brain, the nervous system pathway causing the motion is the same. It is called the *somatic motor pathway* because it controls voluntary muscles. The somatic motor pathway begins in a band of the cerebral cortex running down the side of each cerebral hemisphere. Each band is called a *precentral gyrus* (Fig. 6.6).

Each region of a precentral gyrus is designed to control the voluntary muscles in one area of the body. To move, a person (1) starts impulses from the area of the precentral gyrus that controls the muscles for the part of the body to be moved. Impulses from the precentral gyrus begin to (2) travel down the brain through white matter. As the impulses descend, they pass through areas of gray matter, where they are modified as they move through the gray matter synapses. In this way, the motion is performed at exactly the speed, strength, and distance chosen. Several important areas of gray matter that modify the motor impulses are called the (3) *basal ganglia*, located inside the cerebral hemispheres. In general, the basal ganglia dampen motor impulses so that motions are not exaggerated.

Descending motor impulses are also channeled through the gray matter of the (4) *cerebellum*, which lies behind and below the cerebral hemispheres. Its gray matter forms a wrinkled coating called the (5) *cerebellar cortex*. The synapses in the cerebellar cortex modify the impulses so that the resulting motion starts and stops smoothly, at the proper time, and within the desired distance. The cerebellar cortex also adds impulses to ensure that all muscles that can assist in the motion are stimulated appropriately. The additional impulses activate muscles that move in the same direction and muscles that hold other parts of the body still or prevent loss of balance. At the same time the cerebellar cortex blocks impulses that would cause muscle contractions antagonistic to the desired action.

The cerebellar cortex continues to work throughout the time during which the desired action is occurring. It monitors the motion that is occurring and, if the motion is not exactly what was intended, provides impulses to muscles that can correct the error. With practice, the cerebellar cortex improves its ability to adjust the action, leading to increasing skill at performing that action. Similar control activities occur in the part of the cerebral cortex in front of the precentral gyrus.

Other synapses in the descending somatic motor pathways modify impulses to a lesser degree. Finally, the impulses reach (6) synapses to the dendrites and cell bodies of the somatic motor neurons. These synapses are the last places where the impulses can be modified. Once within the somatic motor neurons, the impulses leave the CNS and travel along the (7) motor axons in the nerves.

Upon arriving at the ends of the motor axons, the impulses cause the (8) release of the neurotransmitter *acetylcholine*. Like neurotransmitters in synapses, acetylcholine binds to the receptor molecules on the cell membranes of muscle cells. Once enough acetylcholine is bound, the muscle cells initiate the steps that lead to contraction, producing the desired action. Enzymes from the muscle cells then destroy the acetylcholine, and the cells relax until the next nerve impulses arrive.

The brain improves the efficiency of this process by sending some impulses to somatic motor neurons just before the person attempts a motion. Some of these anticipatory impulses are sent to motor neurons controlling the muscles that will contract, making them more sensitive to the main impulses telling the muscle to contract. The result is that when the motion should occur, the correct muscle moves faster and stronger while muscles that oppose its motion are inactivated.

Higher-Level Functions

The nervous system is also involved in activities that produce conscious remembering, thinking, interpretations, emotions, and personality traits. All these higher-level functions take place in the brain.

The neuron pathways that produce these activities are poorly understood. There seem to be complicated interactions among several areas of the brain for each activity. Also, each activity seems to influence and interact with the others. However, many of the areas of the brain that are involved with these higher-level functions have been identified, and some of the details of their operations have been discovered.

AGE CHANGES IN SENSORY FUNCTIONING

Age changes that affect the sensory neurons are important because by providing monitoring and communication, these neurons initiate reflexes and start or influence many voluntary actions, memories, thoughts, and emotions. Therefore, alterations in sensory functioning can affect homeostasis and the quality of life.

Aging causes a gradual decline in sensory functioning as a result of a reduction in the numbers of several types of sensory neurons, a decline in the functioning of the remaining sensory neurons, and changes within the CNS. The following section concentrates on changes in PNS sensory neurons other than those involved in vision, hearing, and other inner ear functions.

Skin Receptors

In the skin there is little change in either the number of sensory neurons for touch that are associated with hairs or the number of pain receptors. However, touch receptors called *Meissner's corpuscles*, which are not associated with hairs, and pressure receptors called *pacinian corpuscles* decrease in number and become structurally distorted. In addition, the capsule in each pacinian corpuscle becomes thicker. Further reductions in sensations from the skin seem to result from a weakening of the action potentials that conduct impulses to the CNS. Alterations in action potentials may be due to age changes in neuron cell membranes or thickening of the myelin that surrounds many sensory neurons.

The age changes in Meissner's corpuscles and pacinian corpuscles lead to a decreased ability to (1) notice that something is touching or pressing on the skin, (2) identify the place where touch or pressure is occurring, (3) distinguish between being touched by one object and being touched by more than one at the same time, and (4) identify objects by touching them. In addition, some skin sensory neurons require more time to respond to stimuli; this may contribute to the declining ability to feel vibrations, particularly those with higher frequencies. An age-related increase in impulse speed in some sensory neurons may partially compensate for these changes.

In addition to the effects of age changes on sensory neurons, the monitoring of conditions in and on the skin may be altered by changes in the thickness of the skin and the subcutaneous layer; the quality and distribution of hair; the ability of the CNS gray matter to respond to and interpret impulses from sensory neurons; and psychological status. Because of these factors, the effects of aging on the perception of temperature and pain are ambiguous.

Decreases in the ability to detect, locate, and identify objects touching or pressing on the skin result in decreases in the ability to respond to those objects. As a consequence, harmful objects may be encountered more frequently, more severely, and for longer periods. There is also a decline in the ability to perform precise actions that depend on good sensory input, such as moving the lips when forming words and manipulating small objects with the fingers. Reductions in skills may lead to problems in certain professions and loss of satisfaction with hobbies. Furthermore, reduced sensation means reduced pleasure from favorable physical contact, and this can have psychological and social consequences. Since sensory neurons associated with pain release substances that promote wound healing, age-related decreases in these neurons or in processing impulses from them may contribute to the age-related slowing of healing.

Sense of Smell

Aging causes decreases in the number of sensory neurons for smell. These neurons are called *olfactory neurons* and are high in the nasal cavities. Aging also causes deterioration of the pathways that carry olfactory impulses through the brain. All these changes cause a decline in the ability to detect and identify aromas. The degree of change is difficult to measure, however, because of the influence of changes in other brain functions (e.g., memory, emotional state) and of previous experiences. Furthermore, the degree of change seems to be highly variable among individuals.

Since much of what is commonly referred to as flavor is actually aroma, age changes in the sense of smell reduce the pleasure derived from eating and can contribute to malnutrition. Reduced olfaction also means a reduced ability to detect harmful aromas such as toxic fumes and dangerous gases. Finally, a declining ability to notice offensive odors can lead to socially embarrassing situations.

Sense of Taste

The sense of taste accounts for only four of the sensations that many people call flavors; all other flavors are due to the sense of smell. The four taste flavors are salt, sweet, sour, and bitter. Aging seems to cause slight decreases only in the ability to detect salty and bitter substances. The amount of change is highly variable among individuals, and the ability to detect salt declines the most.

Even in the oldest individuals, the *threshold* levels for these four taste sensations are well below the levels in ordinary foods. The threshold for a stimulus is the lowest level of that stimulus which causes a response. If the threshold for tastes approaches the values found in foods, adding more of the ingredient that produces the flavor can compensate for this age change. Therefore, unlike the sense of smell, age changes in the sensory neurons for taste normally do not have a significant effect on food selection or diet. Of course this may not be true for persons with medical problems such as high blood pressure because these individuals may be on restricted diets that prohibit the use of flavorings such as salt. It may also be untrue for individuals who smoke because smoking greatly reduces taste sensations.

A main reason for the small age change in the sense of taste may be the lifelong ability of these sensory neurons to reproduce rapidly and thus replace taste receptors lost to aging or injury (e.g., from hot foods).

Other Sensory Neurons

Other types of sensory neurons that seem to have reduced functioning because of aging include those which monitor blood pressure in arteries; materials in the throat; thirst; amount of urine in the urinary bladder; amount of material in the rectum (the end of the large intestine); and positions, tensions, and lengths of the joint structures, muscles, and tendons. Additional decrements in these sensory functions may derive from changes in the ability of the organs being monitored to stretch and from alterations in the ability of the CNS to respond to sensory impulses.

Corresponding outcomes from these decreases in sensory functioning include high blood pressure; dehydration; swallowing and choking problems; urinary incontinence; constipation or bowel incontinence; and reduced control and coordination of voluntary movements.

AGE CHANGES IN SOMATIC MOTOR FUNCTIONING

Somatic Motor Neurons

Important age changes in somatic motor neurons involve their numbers, action potentials, and transmission sites. The first two changes are similar to those we have noted in sensory neurons.

Number   There is a decrease in the number of motor neurons, and this reduces the number of cells that can be stimulated in a muscle. Therefore, the maximum strength of contraction that muscles can produce declines. In the lumbar region of the spinal cord, which controls muscles in the lower half of the body, as many as 50 percent of the somatic motor neurons are lost by age 60. Muscle cells that lose their motor neurons degenerate completely because they are no longer stimulated.

The resulting decrease in muscle strength can be minimized by increasing the strength of contraction provided by muscle cells that retain their motor neurons. This effect can be achieved on a

short-term basis by increasing the amount of stimulation by the surviving motor neurons. However, using this strategy puts extra strain on the stimulated muscle cells. It also can produce the feeling that one must work harder to perform a strenuous activity which formerly was not difficult. Over the long term much of the strength of each muscle can be retained through programs of physical training and ordinary activities that require very strong muscle contractions.

Action Potentials   The second age change in motor neurons is a slight decrease in the speed of action potentials in their axons. The amount of slowing is different in different neurons. The changes in speeds caused by aging increase the original differences in speed found among young neurons. As a result, when an aging muscle is supposed to contract, the burst of impulses sent to it by the motor neurons arrives over an increasingly long period. Therefore, the contractions of muscle cells are spread out over a longer period.

Slower action potentials in motor neurons may result from age changes in motor neuron cell membranes, myelin, or blood vessels within the nerves. Aging causes some myelin in peripheral nerves to separate from its axons. Damaged myelin is removed by macrophages, and its replacement occurs more slowly with age. Age changes in blood vessels were described in Chap. 4. These changes reduce blood flow in the nerves and therefore decrease the supply of nutrients and the elimination of wastes.

Alterations in muscle contraction resulting from slower action potentials and the spreading of muscle cell contractions include slower contraction, lower peak strength of contraction, and slower relaxation. Age-related decreases in anticipatory impulses increase these changes. All these alterations reduce the maximum amount of strength a muscle can produce when it performs very quick movements.

Neuromuscular Transmission   The third age change is a substantial decrease in the speed of transmission from motor neurons to muscle cells. This decline may be from the formation of irregularities at the ends of aging motor axons. Slower transmission results in further delay in starting a motion.

All three age changes mean that activities that require strong and/or fast actions cannot be performed as well. This can have a significant impact on individuals whose careers or recreational activities depend on such actions. For other people, modifying or changing strategies to achieve their goals can help compensate for the slow decline in strength and speed.

AGE CHANGES IN AUTONOMIC MOTOR FUNCTIONING

Aging of the autonomic motor neurons has not been as well studied as aging of other parts of the nervous system because of difficulties in distinguishing such changes from other age-related changes. Therefore, little can be said with confidence about the effects of aging on autonomic motor neurons. However, some aspects of the aging of these neurons are coming to light. In general, aging seems to have little effect on their ability to regulate body functions under normal conditions. This is due in part to overall slow loss of sympathetic motor neurons in the spinal cord (i.e., 5 percent to 8 percent per decade). Additionally, sympathetic motor neurons compensate for some age changes by modifying their dendrites and axons throughout life. However, when conditions become unfavorable, the autonomic neurons controlling certain structures have difficulty causing adequate adjustments to preserve homeostasis.

Autonomic Motor Neurons

An apparently inadequate autonomic response occurs when older people stand up or remain standing for long periods. Normally, sympathetic neurons prevent a substantial drop in blood pressure by stimulating the heart and causing constriction of many blood vessels. The ability of the sympathetic neurons to cause these adjustments decreases in many people. The resulting low blood pressure when one is in an upright position—*orthostatic hypotension*—can cause dizziness, light-headedness, and fainting. This is a major cause of falls and physical injury (e.g., fractures). Orthostatic hypotension does not occur in all older individuals, and some cases result from abnormalities in the circulatory system.

Aging of autonomic neurons can lead to elevated blood pressure as well as low blood pressure. Normally, parasympathetic impulses slow and weaken the heartbeat to keep blood pressure down while a person is at rest, when a person ends vigorous physical activity, and during each inspiration. Aging causes this parasympathetic function to decline and therefore diminishes the ability of these neurons to prevent blood pressure from exceeding the proper levels.

Age changes in autonomic neurons may also contribute to a decrease in the ability to adjust to extremes in temperature. Normally, sympathetic impulses cause blood vessels in the skin to constrict when a person is getting cold; this helps stabilize body temperature by reducing the rate of heat loss. With increasing age, there is a decrease in such constriction. Thus, older individuals are at greater risk of developing hypothermia. This age change may be due largely to age changes in blood vessels.

Another age change that may be due in part to aging of autonomic neurons involves erection of the penis. Normally, erection occurs when parasympathetic neurons cause dilation of blood vessels in the penis during sexual arousal, increasing blood flow into the penis and causing it to enlarge and become stiffer. With advancing age, these processes occur more slowly and to a lesser degree. These age-related changes may be due to reduced parasympathetic functioning or to age changes or disease in penile vessels. Parasympathetic control of other blood vessels is not changed by aging.

Another age change believed to result from aging of autonomic nerves is a decrease in the responsiveness of the pupil. Normally, sympathetic nerves stimulate muscles in the iris that cause dilation of the pupil and parasympathetic nerves stimulate muscles in the iris that cause constriction of the pupil. Balancing these autonomic influences results in letting enough light enter the eye for vision while preventing the entry of excess light, which can hinder vision and damage the eye. With advancing age, there is a decrease in the amount of pupillary dilation and slower constriction of the pupil, which reduces adaptation by the eye. Both changes may be caused by changes in the autonomic neurons or in the iris.

Finally, there is a decrease in the number of neurons controlling the movements of the esophagus during swallowing. Normally, when solids or liquids enter the esophagus from the throat, these materials are pushed down to the stomach by a wave of muscular contraction in the esophagus. The contraction is initiated by the swallowing reflex and is coordinated by a group of motor neurons (*Auerbach's plexus*) in the esophagus. With aging, the number of neurons in Auerbach's plexus decreases. Swallowing becomes more difficult because the wave of contraction starts later, is weaker, and is less well coordinated. Sometimes the esophagus fails to empty completely, resulting in considerable discomfort.

Sympathetic Neurotransmitters

Sympathetic functioning is also affected by changes at neuromuscular and neuroglandular junctions. Sympathetic neurons become especially active when conditions become unfavorable and homeostasis is threatened or when such a threat is suspected or anticipated. The effects of sympathetic activity include increases in heart functioning, blood pressure, and perspiration as well as dilation of the airways. At the same time, sympathetic neurons inhibit certain activities, including digestion, urine production, and the functioning of the reproductive organs. Overall, these effects are adaptive and beneficial because they channel more of the body's energies into actions that help the individual overcome or escape danger. The combination of effects caused by the sympathetic neurons is often referred to as the *fight-or-flight response*, which is part of the body's reaction to stress.

Most sympathetic motor neurons use *norepinephrine* as a neurotransmitter at neuromuscular and neuroglandular junctions. At the direction of sympathetic neurons, norepinephrine is also produced and secreted into the blood by a gland called the adrenal medulla (Chap. 14). Norepinephrine from the *adrenal medulla* increases the intensity and duration of the effects of sympathetic norepinephrine.

Aging affects blood levels of norepinephrine in three ways:

1. The concentration of norepinephrine in the blood of resting individuals rises.

2. When a stressful situation is encountered, the level of norepinephrine increases faster.

3. Once the stress has passed, the level of circulating norepinephrine returns to its resting concentration more slowly.

There seem to be two reasons for the higher levels of norepinephrine in older individuals. One may be the stiffening of arteries (Chap. 4). The other seems to be a compensatory response for an age-related decline in the effectiveness of norepinephrine in some organs. This decline may be due to age changes in receptor molecules (e.g., lungs) or in reactions within cells (e.g., heart).

In conclusion, although the effects of age changes in autonomic neurons are not unimportant, such changes are few compared with the number of autonomic functions that seem to be unaffected by aging. Autonomic neurons can provide proper regulatory impulses to most of the structures they control regardless of age or the degree of stress placed on the body.

AGE CHANGES IN REFLEXES

Since aging causes many detrimental changes in sensory and motor neurons as well as in myelin, it produces deleterious effects on the reflexes that use those structures. Some of these effects were mentioned in the sections on sensory, somatic, and autonomic neurons. The decrease in number and the decline in sensitivity of certain sensory neurons mean that more stimulation is required to start many reflexes. It takes more time for the response to begin because reception takes longer and action potentials are weaker and slower. Changes in action potentials, together with decreases in the number of motor neurons and the effectiveness of certain neurotransmitters, cause the response to be weaker and of longer duration.

Age changes in the structures that surround the sensory neurons, such as the skin and blood vessels, further alter reflexes by preventing sensory neurons from properly detecting stimuli. Reflex responses are also reduced by age changes in the glands and muscles producing the responses and in the skeletal system.

Reflexes also seem to be detrimentally affected by age changes in the CNS. It has been observed that the more complicated the pathway in the CNS, the more dramatic the effect of aging on reflexes. In addition to reflexes occurring more slowly and weakly, there is a decline in the amount of coordination provided by the CNS in complicated reflex responses. Reflex contraction of large muscles is a good example.

The simplest muscle reflexes in the body are those which help maintain posture. These *stretch reflexes* or *deep tendon reflexes* use few synapses and no interneurons. A stretch reflex is initiated when a muscle is stretched, as occurs when a person's posture begins to change because of slumping, an external force causes a joint to bend, or an object hits a tendon. When the impulses in the reflex pathway reach the muscle that has been stretched, it contracts to restore the body to its original posture. The knee-jerk reflex is an example of a stretch reflex. Such simple reflexes become weaker but only slightly slower with age. The degree of weakening in different individuals is highly variable. The degree ranges from virtually no change in the strength of the response to essentially total loss of the response. However, many cases of very weak or absent stretch reflex responses result not from aging but from abnormal or disease conditions such as traumatic injury, atherosclerosis, arthritis, and diabetes mellitus.

In contrast to stretch reflexes, reflexes that maintain balance while one is standing in place require the proper timing of a sequence of many muscle contractions. Keeping one's balance while there is movement of either the body or the surface on which a person is standing requires an even more complicated series of muscle contractions. Though the same sensory and motor neurons involved in stretch reflexes may be used, many interneurons and synapses in various parts of the brain and spinal cord are involved in these pathways. Sensory inputs from the eyes, ears, and skin may assist in these reflexes.

Complex reflexes such as those which maintain balance show a substantially greater slowing with age than do simple muscle reflexes. Aging also causes disturbances in the coordination required for such reflexes. For example, there is a change in the sequence in which the muscle contractions occur during these reflexes and an increase in the number of antagonistic muscle contractions. In comparison to simple reflexes, some of the additional slowing and much of the decline in coordination seen in complex reflexes seem to be due to age changes in the synapses and interneurons in the CNS.

Interestingly, some age changes in the CNS seem to involve adjustments in reflex pathways that compensate for diminished sensory functioning, muscle strength, skeletal system functioning, and confidence in one's ability to maintain balance. This can be observed in the age change in gait. Part of walking involves voluntary activity, but many of the muscles used for walking are controlled by acquired reflexes. Older individuals walk with smaller steps, at a slower pace and with the feet more widely spread. Such a gait minimizes the risk of losing one's balance. Gradu-

ally modifying voluntary actions and reflexes to walk in this manner seems to reduce the demands on the muscles, joints, and reflexes needed to maintain balance.

In summary, reflexes undergo several age changes. They require more stimulation to be activated, and it takes longer for a response to begin. The response is weaker, takes longer to occur, and shows less coordination. These changes are caused by alterations in both the PNS and the CNS. With more complicated reflexes, aging of the CNS makes a larger contribution to alterations in reflexes than do age changes in the PNS. As aging diminishes the functioning of reflexes, it reduces their ability to provide automatic, fast, and accurate responses to changes in internal and external conditions and therefore to maintain homeostasis.

AGE CHANGES IN CONSCIOUS SENSATION AND VOLUNTARY MOVEMENTS

As with reflexes, aging affects conscious sensation and voluntary movements because of age changes in sensory neurons, motor neurons, myelin, and CNS neurons and synapses. Since conscious sensation and voluntary movement use even more CNS synapses and interneurons than are used in reflexes, age changes in the CNS have a greater impact on these activities.

The results of PNS and CNS age changes on conscious sensation include a declining ability to detect, recognize, and determine levels of stimuli. These decrements make selecting and performing appropriate voluntary actions more difficult, inhibit learning, and diminish enjoyment from experiences.

The ability to maintain homeostasis and the quality of life is decreased further because aging of nerve pathways used for voluntary movements causes such movements to become slower, weaker, less accurate, and less well coordinated. Since these changes occur gradually, individuals are able to make adjustments in their activities and minimize the undesirable effects.

AGE CHANGES IN THE CNS

Correlations between the alterations in reflexes, conscious sensation, and voluntary movements and age changes in the structure and functioning of the CNS are not well understood. The reasons for this ambiguity include (1) the necessity of studying brains obtained from autopsies, which have undergone variable degrees of postmortem changes, (2) the difficulty in determining how much, if any, disease was present in the brain or in other organs, and (3) the paucity of psychological or behavioral information about the people whose brains are studied. However, as these correlations become clear, it may become possible to influence the decreases in nervous system functioning caused by aging.

Spinal Cord

In the white matter, there is an age-related decrease in the motor neurons, especially of motor neurons that control somatic motor functions. These neurons carry anticipatory impulses and main impulses from the brain to lower somatic motor neurons in spinal cord gray matter. Within the gray matter, the average loss of motor neurons is approximately 25 percent during adulthood and into very old age. The rate is highly variable, and may be two to three times higher in some individuals. There seems to be a preferential loss of somatic motor neurons. This corresponds with the loss of motor units in muscles (see Chapter 8).

Brain

Dimensions Many studies report that there is a decrease in the size and weight of the brain as age increases. The fluid-filled cavities inside the brain enlarge, the raised ridges (*gyri*) on the surface shrink, and the grooves (*sulci*) between the gyri become wider (Fig. 6.7).

How much of the age-related shrinkage of the brain is due to aging and how much is due to diseases such as atherosclerosis has not been determined. One reason for the overall shrinkage may be a decrease in the number of neurons in several areas of the brain. The cause of this neuron death is not known, and there is no indication that what is considered to be a normal amount of overall shrinkage has any effect on brain functioning.

Numbers of Neurons Some parts of the brain show a substantial decline in the number of neurons, and this may affect specific functions. In the cerebrum, these parts include areas that control voluntary movements, areas for vision and

FIGURE 6.7 Structure of the brain.

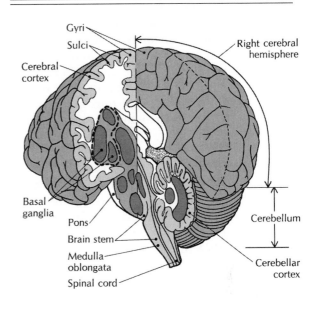

hearing, and possibly areas involved in other conscious sensations. Other parts of the cerebral cortex seem to lose few if any neurons. The cerebellar cortex, which coordinates muscle movements and controls many complicated muscle reflexes, and the basal ganglia, which are also involved in modifying muscle actions, lose many neurons.

The regions of the brain other than the cerebral hemispheres and the cerebellum are refered to as the **brain stem**. The only regions of the brainstem that seem to lose neurons because of aging are the **nucleus of Meynert**, which produces acetylcholine for short-term memory, and the **locus coeruleus**, which produces norepinephrine and helps regulate sleep.

It has been suggested that neuron losses in these areas contribute to age-related detrimental changes in the functions to which they contribute: voluntary movements, conscious sensation, muscle reflexes, memory, and sleep. However, there is no conclusive evidence that localized loss of brain neurons caused by aging has any effect on the functions performed by the areas that incur neuron loss.

One reason why neuron loss may have no effect is that the remaining brain neurons can branch and produce many new synapses. The new neuron pathways created by the new synapses may compensate for the decrease in neurons.

Second, there may initially be many more neurons in the brain than are needed, and these additional neurons may constitute a reserve capacity. Third, the loss of neurons may actually improve the brain by eliminating neurons that are not used or have made errors. The brain may be able to recognize and eliminate unused or undesirable neuron pathways and thus improve its efficiency. This process may constitute part of the development of wisdom.

Neuron Structure and Functioning   Many neurons remaining in the aging brain undergo several age changes. For example, the cell membranes of brain neurons become less fluid and stiffer. These changes may contribute to age-related alterations in brain functioning by altering reception, conduction, and transmission. Second, internal membranes (e.g., endoplasmic reticulum) become irregular in structure, and many neurons have an accumulation of lipofuscin. The effects of abnormal amounts of lipofuscin are not known.

A third change in brain neurons is the formation of *neurofibrillar tangles*. Normally, neurons contain long thin structures called **neurofibrils.** These structures are present in the cell bodies and extend down the axons. Neurofibrils seem to be important in the movement of neurotransmitters from their sites of production to the ends of the axon. The formation of tangled neurofibrils may mean that not enough neurotransmitter is reaching the end of the axon; this could result in a decrease in or an elimination of transmission by neurons with neurofibrillar tangles. The result would be a decrease in the functioning of synapses.

Synapses   Because most research on changes in brain synapses has been directed toward alterations caused by disease, the effects of aging are not well understood. For example, there may be dozens or even hundreds of different neurotransmitters in the brain, and much confusion and contradictory information exist about age changes in these. All that can be said at this point is that aging seems to cause decreases in some neurotransmitters in some areas of the brain. There are few or no cases where the amount of a neurotransmitter increases with aging.

Information about age changes in the number of synapses in various brain areas is also incomplete. It is known that the number of synapses in an area may increase or decrease depending on how much use is made of that area. Neurons that are heavily used increase their number of synapses by growing new axon branches or new dendrites and dendrite branches (*dendritic spines*). The ability of neurons to do this decreases with age. There is also evidence that at least in some areas, neurons that get little use reduce their number of dendrites or dendritic spines and thus decrease the number of synapses in those areas of the brain.

The interpretation of information about changes in the number of synapses is complicated because the effectiveness of synapses depends not only on their numbers but also on changes in their neurotransmitters and in the exact neuron pathways that are gained or lost. For example, many inefficient synapses may be replaced by a few efficient ones, resulting in an improvement rather than a decline in functional capacity.

Adding to the confusion is the fact that synapses undergo age changes in structure as well as number. For example, though there is a decrease in the number of synapses in the precentral gyrus, the remaining synapses become broader. This may mean that these synapses work better and therefore compensate for those which are lost.

Perhaps the best-known age change in synaptic structure is the buildup of the protein called *amyloid*. A mass of amyloid in a synapse is called a *plaque*. As with other age changes, different amounts of plaques develop in different areas of the brain. It is believed that plaques decrease the functioning of synapses. Normally, however, a person does not form enough plaques to alter brain functioning to a detectable degree.

AGING OF OTHER BRAIN FUNCTIONS

Memory

The process of consciously remembering information is referred to as *memory*. Memory is a very complicated process that is not well understood. Though certain areas of the brain, such as the hippocampus, are especially important, many areas in the cerebral cortex and other brain regions act cooperatively to provide memory.

Memory can be divided into two broad types: *short-term memory* and *long-term memory*. Information that is stored in short-term memory is retained for brief periods (seconds or minutes). The brain may be temporarily storing this information by continuously repeating the impulses containing the information, and the information is forgotten as soon as the impulses fade away. The information is also easily forgotten if a person is distracted by different information that sends other impulses through the neurons.

It is possible to increase the time information is stored in short-term memory by keeping the impulses going. This can be done by repeating the stimulus over and over, just as a person can keep a wheel spinning by giving it a push now and then. This technique is used when people remember a telephone number for a short period by repeating it until the number is dialed.

Long-term memory can store information for many years. For example, remembering an incident from childhood requires the use of long-term memory. Apparently, information is stored in long-term memory when impulses produce physical changes in the neurons processing the information. The more times impulses about an incident pass through the neurons, the greater the chances that they will cause the physical changes. This is why a person studies material over and over to remember it for a test.

Two types of changes are believed to occur in neurons that store information in long-term memory. In one case, new molecules are produced in the neurons. Alternatively or additionally, the synapses in the nerve pathway are altered. In either case the impulses for the information travel much more easily. Then a small stimulus can trigger the neurons to produce the same impulses, resulting in the person consciously remembering the information.

Memory can also be classified according to the types of information stored. *Incidental memory* involves remembering information or skills that were self-taught. *Procedural memory* involves recalling how to perform a process or series of steps. Both types may include explicit memory and implicit memory. *Explicit memory (declarative memory)* involves remembering specific facts that a person tried to learn so they could be remembered. *Implicit memory* involves remembering specific facts that a person did not try intentionally to learn so they could be remembered. For example, a person may be unaware of learning procedures, processes, motor skills, or vocabulary by experience. *Episodic memory* involves recalling the times and places events happened. The events are mentally separated and oriented correctly regarding their proper time, sequence, and locations of occurrence. *Working memory* involves holding information at or close to the level of consciousness so it can be used in cognitive processing, such as solving a problem or planning a complex activity.

Age Changes in Memory Aging causes a decline in short-term memory in most people. The rate of decline varies highly between individuals. This may be due in part to differences in the rate of age changes within the nervous system, but it is caused by other factors to a greater degree. These factors include differences in general health, diet, presence of specific diseases, past patterns of mental activity, motivation, and diverse psychological, social, and economic parameters. So many features affect memory that it is impossible to predict which changes have occurred or will occur in a particular individual.

On the average, the decline in short-term memory is gradual and slow until approximately age 60 and then becomes ever more rapid, especially after age 70. However, the total amount of loss in memory functioning in a normal individual is relatively slight regardless of age. In many cases changes in memory can be noticed only in carefully controlled experimental situations and because people develop compensatory strategies, such age changes usually do not affect ordinary activities significantly.

The greatest decline in short-term memory occurs for information that is presented quickly and verbally. Information about completely unfamiliar things also becomes much harder to remember. Older people have more difficulty recalling information than simply recognizing it. For example, questions that require an older person to supply the answer are harder than those which require the person to select the correct answer from among several incorrect ones. To help elderly people remember, information should be presented slowly, in an organized manner, using relevant and concrete examples and visual aids. People are better able to recall information when cues such as notes and mnemonic devices are used and when they are allotted additional time to study and respond. It is also helpful to make adjustments to compensate for deficits in vision and hearing.

The reasons for the decline in short-term memory are not understood but may include age changes in the number of neurons, the number or structure of synapses, and the amounts of different neurotransmitters present in memory pathways.

Long-term memory seems to be largely unaffected or to improve as people get older.

Age changes in incidental memory and procedural memory depend upon whether they use explicit memory or implicit memory. Explicit memory decreases with aging, especially when the facts have to be learned quickly or they must be remembered quickly. Aging has much less effect on explicit memory when more time is used to learn or to remember facts. Implicit memory shows little decline when elders unknowingly experience or are given prompts related to the passed information, such as being placed in a familiar setting. Implicit memory shows the greatest age-related decline when a person tries intentionally to remember. Because of different age-related changes in these two types of memory, elders largely retain ability perform even complex procedures they have practiced, but they may have difficulty explaining how to carry them out. Episodic memory also decreases with age. Failure of episodic memory results in erroneously remembering widely separate events as having occurred together or being unable to connect related events.

Working memory decreases with aging. Therefore, while the ability to remember specific information does not decline much, the ability to use multiple pieces of information in complex cognitive activity declines significantly. This may result from age-related reductions in effectively selecting, retrieving, and processing information consciously.

Elders can increase their memory functions through educational and training programs about memory. Memory training programs may emphasize specific memory techniques. Examples of such techniques include using written notes; mentally repeating information often; organizing material into large meaningful blocks rather than many unrelated details; making up sentences or words where letters (e.g., first letter in each word, letters of the words) stand for the items being re-membered; mentally picturing information, images, or processes; putting information into a story, rhyme or a song; sketching pictures or diagrams; finding experiences in life that are relevant or related to the information. Factors that help learning information include studying when energy levels are high, but not after eating a large meal; avoiding large quantities of aspartame artificial sweetener (e.g., diet beverages); avoiding distractions when learning; getting restful REM sleep.

Other memory training programs take less direct approaches. Sometimes using cognitive restructuring to promote positive expectations in memory performance produces greater and more lasting beneficial effects on memory. This may result from using practical techniques in only similar situations, while cognitive restructuring techniques are often used in diverse situations.

Knowledge of the associations between memory and aging are important for improving outcomes from training programs for elders. For example, modifying job training programs to accommodate age changes in memory becomes more important as the numbers and ages of older workers increase.

Thinking

Like memory, thinking occurs entirely within the brain, but it is an even more complicated and less well understood process. Thinking includes problem solving, planning, and other activities that may be called intelligence. Intelligence may be divided into two categories. *Crystallized intelligence* involves using cognitive skills with familiar learned activities. *Fluid intelligence* involves using cognitive skills in new situations. Examples of fluid intelligence include learning novel problem solving, motor activities, or reasoning. It involves more flexibility in dealing with situations. No attempt will be made here to explain how the brain performs thinking.

Age Changes in Thinking   As with age changes in short-term memory, there is on the average a slow and gradual decline in thinking to approximately age 60; the rate of decline increases more each year after that, especially after age 70. Note, however, that the loss of thinking ability is relatively slight regardless of age and that changes can be noticed only through careful testing. The small amount of change, coupled with the use of compensatory strategies, usually means that there is not a significant effect on ordinary normal activities. There is much variability among individuals in regard to age-related changes in thinking because of variations in aging of the nervous system and differences in other factors that affect thinking. As a result, no one can anticipate how aging will affect an individual's ability to think. Some individuals show no changes in thinking, and up to 10 percent of older people show an increase in thinking ability. This increase seems to be due to continued use of thinking, ongoing education, or good economic status. Among those whose thinking declines with aging, thinking becomes slower and changing one's train of thought becomes more difficult.

Aging has little effect on crystallized intelligence, and many people show age-related increases. Fluid intelligence usually shows age-related decreases. Men show earlier decline in crystallized intelligence; women show earlier decline in fluid intelligence. Deterioration in the ability to solve problems and make decisions quickly and accurately is most evident when these processes require the consideration of many factors.

Vocabulary and Conversation

Language functions rely heavily upon memory and intelligence. There is little or no age-related change in knowing the meanings of words, though vocabulary may increase throughout life. Age-related changes in conversation include using more short and simple sentences; sentence fragments; pronouns and less specific terms; vague adjectives; vague references to time and place. Working vocabulary, ability to find the right word, and adherence to one topic decline. These changes increase as background distractions increase (e.g., noise, motion). Comprehension of conversations decreases as the content of a conversation becomes more complex; more disjointed; more novel; faster; and with increased distractions. These

age-related changes usually do not prevent elders from carrying on meaningful conversations. The changes seem to result from age-related changes in memory and in cognition, including changes in methods of processing verbal information.

Supporting Memory and Intelligence

Factors that reduce age-related decreases in memory and intelligence and often improve these functions include good health; exercise; past and continuing education; activities requiring complex mental functions; self-determination and self-direction; and a sense of self-efficacy. Estrogen therapy in postmenopausal women improves some aspects of memory and cognition including short-term verbal memory, abstract reasoning, logical thinking, and overall cognitive functioning. Using proper prevention, intervention, and cognitive training programs for elders help to sustain and improve memory and intelligence as age increases.

Personality

Personality includes many facets, including levels of anxiety, depression, self-consciousness, vulnerability, impulsiveness, hostility, warmth, assertiveness, gregariousness, and emotions.

Age Changes in Personality Personality undergoes changes up to about 30 years of age, after which most of its aspects are extremely stable. However, major upsetting events in a person's life, such as a major illness, may significantly alter one's personality.

Personality greatly influences the choices made throughout life, particularly in matters related to education, exercise, diet, and health care. All these parameters influence the length and quality of life. Also, personality is a major determinant of an individual's ability to adapt to changing circumstances. Since personality becomes stable, the nature of its contribution to the ability to adjust remains about the same throughout life. Therefore, knowledge of personality can be useful in predicting an individual's future ability to adapt to the new life situations that develop with aging.

Sleep

The effects of aging on sleep are of great interest. One reason for this is the perception that older individuals are sleepier during the day. Second, there is evidence that compared with wakeful (daytime) values, body functions are different during sleep and at night. To fully understand aging, the body must be studied in the sleeping as well as the wakeful state.

Age Changes in Sleep   As people get older, several changes in sleep usually occur. Complaints about sleep difficulties rise from 15 percent among young adults to almost 40 percent among elders. With aging, more time is needed to fall asleep, there are more awakenings during the night, and wakeful periods are longer. Reasons for the increased number of awakenings include a higher incidence of indigestion, pain (e.g., arthritis), rhythmic leg movements, sleep apnea, and circulatory problems (e.g., irregular heartbeat). Some individuals have more awakenings because a decline in the capacity of the urinary bladder requires them to void urine more often. The rise in the level of norepinephrine may also contribute since norepinephrine increases alertness. The increase in awakenings is greater in men than in women. Even though sleep becomes more fragmented, the total amount of time spent asleep in each 24-hour period remains about the same because more time is spent in bed as age increases.

Changes occur in the type of sleep as well as in its continuity. While there is increasing variability among people as they get older, there is an average increase in the time spent in stage 1 sleep, the least restful of the five types. The existence and significance of age changes in amounts of stage 2 and stage 3 sleep are uncertain.

While asleep, people switch between stage 4 sleep and *rapid eye movement* (*REM*) sleep every 80 to 100 minutes. These are the most restful stages of sleep. There is an age-related decline in the amounts of time spent in stage 4 sleep and REM sleep, although the decrease in REM sleep becomes substantial only in very old age.

It is difficult to determine how much or which of the changes in sleep are due to aging of the brain and which are due to other age-related factors, such as having diseases, taking more medication, being past menopause, having different daily routines because of retirement, having more freedom for daytime napping, and experiencing altered social situations such as death of a spouse or a move to a different home or institution.

Sleep can be improved by keeping to a schedule; adhering to bedtime routine; creating an environment conducive to sleep (e.g., quiet, dark); exercise; treating medical problems and sleep apnea; entraining circadian rhythms with bright light therapy; biofeedback training; and mental relaxation techniques. Things to avoid include daytime naps; stimulants (e.g., caffeine) late in the day; strenuous activity shortly before bed; using the bed and bedroom for work, worrying, or solving problems; medications that adversely affect sleep (e.g., diuretics at bedtime); and chronic use of sedatives, hypnotics, and other sleep inducers.

The effects of age-related changes in sleep include a reduction in the quality of sleep and alterations in the time when it occurs during each 24-hour period. These effects probably explain why more people feel sleepy during the day as they get older. However, this is not a normal part of aging. When daytime sleepiness interferes with regular activities, it should be considered abnormal and warrants further diagnosis. The presence of age-related increases in abnormal sleepiness has contributed to the stereotype of the older person who nods or falls asleep at inappropriate times.

Biorhythms

Many activities in the body show regular cyclic fluctuations or *biorhythms*. One of these is a daily biorhythm that repeats itself approximately every 24 hours. It is aptly called the *circadian rhythm*, meaning "approximately daily rhythm." Perhaps the most obvious manifestation is the cycle of sleeping and being awake. Another well-known biorhythm is the menstrual cycle in women, which recurs approximately every 28 days. Faster cycles include the cardiac cycle and the breathing cycle. People also exhibit annual rhythms that accompany seasons of the year.

In the body, the circadian rhythm is controlled primarily by the brain. When light entering the eyes causes impulses to be sent to the brain, many of the impulses reach a brain area called the *suprachiasmatic nucleus* (*SCN*). The SCN is in the *hypothalamus*, located between the basal ganglia

(Fig. 6.7, Fig. 14.1). Impulses from the SCN travel an indirect route to the *pineal gland* of the brain. The pineal gland is in the crevice between the cerebral hemisphere and the cerebellum (Fig. 6.7, Fig. 14.1). The pineal gland secretes the hormone *melatonin*. When less light enters the eyes, more impulses travel from the SCN to the pineal, causing more melatonin secretion. More light entering the eyes causes the opposite effect. Both the intensity and wavelengths of light influence its effects on melatonin secretion.

Since usually more light enters the eyes during the day and light decreases during the evening, remaining very low during the night, melatonin secretion increases during the evening and remains low during the day. Melatonin influences many body functions including the SCN, and it produces some manifestations of the circadian rhythm. However, even with no light entering the eyes, the SCN causes melatonin to be secreted in a circadian rhythm. The SCN is the main regulator of the body's circadian rhythm. The circadian rhythm is influenced by other factors including environmental cues, physical activity, and eating.

Body circadian rhythms include sleep: wakefulness; stages of sleep; lowering of body temperature, blood pressure, and urine production at night; and oscillations in blood levels of many substances including hormones (e.g., melatonin, glucocorticoids, growth hormone, testosterone, estrogen, progesterone). Oscillations of these hormones cause manifestations of the circadian rhythm (see Chap. 14). The importance of maintaining normal circadian rhythms is evident when they are disrupted. Examples include "jet lag," working night shifts, or having sleep: wakefulness cycles disrupted by environmental irregularities (e.g., nighttime noise).

Aging causes changes in circadian rhythms. Many changes begin during the third decade and increase after that through old age. In general, manifestations of the circadian rhythm have lower peak intensities. Examples include difficulty falling asleep; poorer sleep quality; more urine production at night; and lower peak hormone levels.

The circadian rhythm tends to shorten, and most manifestations begin up to one hour earlier in the 24-hour day. However, phase shifts are unequal, and some manifestations of the circadian rhythm occur later rather than earlier. The result is an age-related loss of synchrony among manifestations of the circadian rhythm. Perhaps the most obvious troublesome consequence is the age-related deterioration of the sleep: wakefulness cycle accompanied by deterioration of sleep quality.

Age changes in circadian rhythms may be due to a combination of age changes in the brain and the eyes. Weak or disrupted circadian rhythms can be brought toward normal by regulating exposure to bright light, by voluntarily regulating routines (e.g., physical activity), and by carefully timed melatonin supplementation.

In general, there are only small age changes in seasonal rhythms. Exceptions include levels of clinically important substances in the blood (e.g., creatinine, urea, urate, blood proteins).

Understanding and accounting for age-related changes in circadian rhythms and seasonal rhythms are important because circadian rhythms influence patient evaluations and effects of medications. The changes should also be considered in research studies so that measurements are taken at proper times of the 24-hour day.

CONCLUSION

In spite of age changes, the normal nervous system can help maintain homeostasis and sustain a satisfactory quality of life for many decades. However, as with other body systems, the comfort derived from these conclusions may diminish when one considers the frequency and effects of nervous system diseases that increase with age.

DISEASES OF THE NERVOUS SYSTEM

Strokes

Strokes are the third leading cause of death among people over age 65, accounting for approximately 9 percent of all these deaths. Beginning at a rate of less than 6 percent at age 65, the percentage of deaths from strokes rises steadily as age increases, surpassing 12 percent for those over age 85.

Heart disease accounts for 4.5 times as many deaths, and cancer, which is the second leading cause of death among the elderly, accounts for more than twice as many deaths among people above age 65. While the death rate from cancer has remained stable for many years, the death rates from strokes and heart disease have declined steadily since about 1960. These declines are probably due in large part to better prevention of atherosclerosis and better diagnosis and treatment of strokes and heart disease.

Many people who have a stroke survive. Therefore, not only the percentage of deaths but also the overall incidence of strokes increases with age, especially after age 65. About 4.5 percent of those between 65 and 75 years of age have a stroke, and the rate among those over age 75 is about 7.3 percent. Strokes occur more frequently in men than in women and much more frequently in blacks than in whites. Those who survive are often left with serious lifelong disabilities.

Causes and Types   To understand how and why strokes occur, some additional information about the brain must be understood. Most of these facts are also true of the heart.

Brain neurons are always very active and therefore need a constant supply of energy. This energy is obtained by breaking down glucose in processes that consume oxygen and thereby prevent the formation of lactic acid and other harmful waste products. Flowing blood delivers the glucose and oxygen to the brain. If the supply of glucose or oxygen drops, the brain neurons will be injured or killed. A low oxygen supply for a few seconds will cause the neurons to malfunction, and a very low oxygen supply for several minutes can result in neuron death. The brain can adjust the amount of blood flow it receives by signaling the heart to adjust cardiac output, directing blood vessels throughout the body to adjust blood pressure, and constricting or dilating its own blood vessels.

Blood being pumped to the brain by the left ventricle passes first through part of the aorta and then through the arteries (carotid and vertebral arteries) that lead up the neck and into the skull (Fig. 6.8). Blood can be felt pulsing through the carotid arteries on either side of the neck. Branches from these ascending arteries carry blood over the brain's surface and deep into the brain.

Strokes occur when blood flow to and through the brain is disrupted. Because of the sudden and devastating effects on the brain, the victim may appear to have been struck with a heavy blow, hence the name "stroke." Since strokes affect the brain and are almost always caused by abnormalities in the blood vessels or heart, they are also

called *cerebrovascular accidents* (*CVAs*).

The most common circulatory system problem resulting in strokes is atherosclerosis. As in all arteries, atherosclerosis in brain arteries reduces blood flow by causing them to become narrow, rough, and stiff. Recall that roughness leads to thrombus and embolus formation, blocking blood flow, and that stiffness prevents an artery from dilating when necessary. Blood flow to the brain can also be reduced by emboli formed on a myocardial infarction that causes roughness of the inner lining of the heart. Additional causes of blocked brain arteries include emboli or pieces of plaque that break free from the wall of an artery leading to the brain. Coronary artery disease may decrease blood flow by reducing heart functioning. Since all these strokes prevent adequate blood flow in the brain, they are called *ischemic strokes*. *Ischemia* means inadequate blood flow, and about 80 percent of all strokes are ischemic strokes.

Thrombus formation, narrowing, and stiffening in brain arteries develop gradually, and there is some time for enzymes in the blood to dissolve some of the thrombus and for blood vessels to compensate for the reduced flow by dilating. Sometimes this restores blood flow sufficiently so that even though neurons are injured, they survive. Additional neuron injury occurs when blood flow is restored because the increase in O_2 combined with injured cells causes an increase in free radical production and damage. Injured neurons can repair themselves and regain their normal functions.

Ischemic strokes from emboli tend to produce greater injury and more neuron death because they cause blood flow to be stopped suddenly and completely. Even if the blocked artery dilates, the embolus is likely to slide farther along until it gets stuck at the next narrowing. Furthermore, since a rough spot in the heart or in an artery may continue to produce emboli, many brain regions may be affected and many strokes can occur in succession.

After a stroke, the neurons that were killed are not replaced since neurons cannot reproduce. However, the remaining neurons may form new dendrites and synapses to compensate for the dead neurons. The surviving neurons may be trained to take on some of the jobs previously performed by the killed neurons.

Atherosclerosis of brain arteries also causes strokes in another way. Arteries weakened by ath-

FIGURE 6.8 Blood vessels that supply the brain.

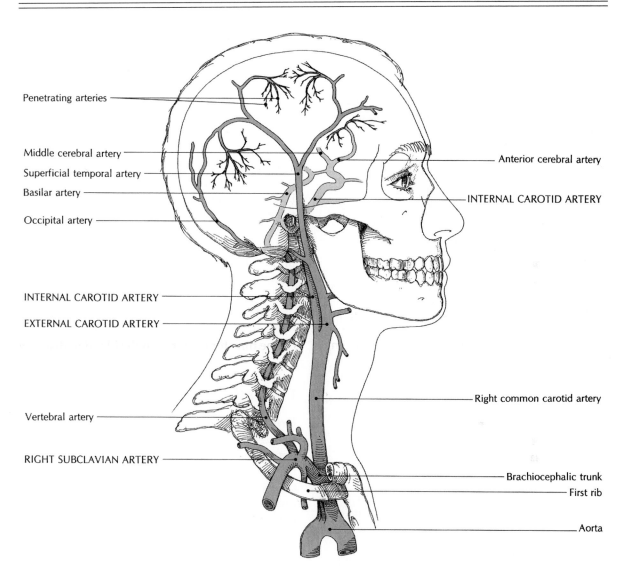

Penetrating arteries

Middle cerebral artery

Superficial temporal artery

Basilar artery

Occipital artery

INTERNAL CAROTID ARTERY

EXTERNAL CAROTID ARTERY

Vertebral artery

RIGHT SUBCLAVIAN ARTERY

Anterior cerebral artery

INTERNAL CAROTID ARTERY

Right common carotid artery

Brachiocephalic trunk

First rib

Aorta

erosclerosis can rupture and bleed, causing *hemorrhagic strokes*. These constitute the remaining 20 percent of all strokes. Since they are often associated with high blood pressure, hemorrhagic strokes are also referred to as *hypertensive hemorrhagic strokes*.

When an artery in the brain ruptures, the region it supplies no longer receives adequate blood flow because some of the blood is leaking. Neurons near the site of the rupture are injured as blood sprays on them and pushes them apart and aside. The hemoglobin that leaks out of the red

blood cells further injures these neurons. Part of the injury is from free radicals produced in the presence of iron in hemoglobin.

Hemorrhagic strokes cause additional brain damage because as more blood leaks from the artery, it increases the pressure inside the skull. This condition begins to damage neurons in all parts of the brain. The pressure also tends to compress vessels, reducing blood flow to many parts of the brain. If the blood pushes the brain far out of position, more neurons will be torn and crushed and more blood vessels will be squeezed shut.

Pressure within the skull increases further when inflammation in the injured areas causes the brain to swell, and all areas of the brain can be injured. If the brain regions controlling the heart, respiration, or blood pressure are affected, the victim's life is severely threatened.

Since hemorrhagic strokes can injure many parts of the brain, they are more serious than ischemic strokes and are much more likely to cause death. About 80 percent of all hemorrhagic strokes in people with high blood pressure are fatal.

Signs and Symptoms   Malfunctioning of the brain starts as soon as a stroke begins. Depending on which regions are injured and the severity of the damage, the malfunctions are apparent as any of a wide variety of signs and symptoms. Some more common ones are tingling, numbness, and muscle weakness or paralysis in one or more parts of the body. These alterations often occur on only one side of the body. Other frequently encountered changes include loss of balance or muscle coordination, altered vision, difficulty speaking, mental confusion, and diminished or lost consciousness.

Sometimes the signs and symptoms disappear in a few seconds or hours. Strokes of this type are called *transient ischemic attacks* (*TIAs*) because they result from a brief decline in blood flow and the injured neurons recover quickly. TIAs frequently occur over and over in exactly the same way because a thrombus forms in the same place in a brain artery. Though TIAs may appear to be unimportant, they are often followed by more serious strokes.

The signs and symptoms of other strokes last for days or weeks and subside very gradually, if at all. Strokes of this type are referred to as *reversible ischemic neurological deficits* (*RINDs*) because the brain is able to regain some of its functions.

The third type of stroke is called a *completed stroke* because the signs and symptoms develop quickly and show no improvement.

Treatments   The best way to reduce the effects of strokes is prevention. Since most strokes result from atherosclerosis, this entails reducing the risk factors for atherosclerosis. This process should begin as soon as possible and continue throughout one's life (Chap. 4).

Other ways of reducing the risk of a stroke in individuals of advanced age include reducing high blood pressure, treating blood disorders, and avoiding exhaustion. When a person seems to be having a stroke, medical attention should be obtained immediately to minimize possible complications.

Treatments for strokes involve reducing the risk of having another stroke and may include medications or surgery. During and after medical treatment, steps should be taken to provide psychological and social support for the stroke patient and the members of his or her family. Physical therapy and rehabilitation often help the patient improve or compensate for functions detrimentally affected by the stroke.

Many stroke patients are disabled for the rest of their lives. The disabilities not only adversely affect their ability to care for their physical needs but also may impinge heavily on their self-image, mental health, interactions with others, and ability to support themselves economically. The cost of treatments and care may add substantially to the economic difficulties.

Dementia

Dementia is a broad category of diseases, all of which involve a serious decline in memory accompanied by a major decline in at least one other mental function. Three other criteria must be met before a person can be said to have dementia. First, the person must be affected to such an extent that he or she has significant difficulty carrying out normal activities and interacting with other people. Second, these difficulties must be present on a continuing and long-term basis rather than sporadically. Third, they must be caused by an identifiable physical abnormality or at least must not be caused by an identifiable mental illness such as depression. Functions that are often reduced in patients with dementia include abstract thinking; speaking, reading, and writing; making judgments; solving problems; identifying common objects; and performing simple voluntary tasks.

The number and rate of cases of dementia are increasing because the number of older people and the proportion of the population made up of older people are growing. In addition, since better diagnostic tests are being developed and the social stigma attached to the diagnosis of dementia is declining, more cases are being identified and reported. However, incidence rates and death rates are only estimates because of difficulty with diagnosis and other diseases can mask the presence of dementia. Also, dementias contribute to other causes of death, leaving cases of dementia unreported as the cause of death.

The incidence of dementia increases with the age of a population. The incidence rate rises exponentially, meaning the greater the age, the faster the rate of incidence rises. Very few cases occur in people below age 60, and less than 2 percent of all people between the ages of 60 and 65 have dementia. The percentages approximately double for every five years above age 65, so that more than 30 percent of those over age 85 suffer from dementia to some degree. Overall, between 16 and 24 percent of the population over age 65 suffer from mild dementia and up to 8 percent of those over age 65 have severe dementia. Among those over age 65, the number of people with dementia is greater than the number who have strokes. For adults, the death rate from dementias approximately doubles with each decade of life until age 90, when the death rate begins to plateau.

There are more than 60 different types of dementia. Some forms are reversible, including dementia caused by medications; drugs; alcohol; anemia; malnutrition; CNS infection; malfunction of the thyroid gland or adrenal glands; and malfunction of organs such as the liver and kidneys. Some forms of dementia are irreversible, including the forms associated with Alzheimer's disease and Parkinson's disease and those caused by strokes, heart failure, repeated head injury, AIDS, and Huntington's disease.

Some individuals have more than one type of dementia. Others have dementia along with nervous system disorders such as delirium and depression. As a result, a definitive diagnosis of dementia is quite difficult to obtain. At present, cases can be diagnosed with about 90 percent accuracies.

The many causes of dementia occur as follows: 10 percent to 20 percent from atherosclerosis or, occasionally, another circulatory system disease: at least 55 percent from Alzheimer's disease; 8 percent from Parkinson's disease; 4 percent from head trauma; 12 percent from a mixture of these causes; and 6 percent from other causes. Approximately 70 percent of cases after age 60 are caused by Alzheimer's disease.

Multi-Infarct Dementia

Dementia caused by circulatory disease is often called *multi-infarct dementia* because it results from having many areas of the brain die from inadequate blood flow. Free radicals also cause damage. The amount of infarction usually increases over an extended period because the victim has one stroke after another or because arteries remain nearly completely blocked. Therefore, multi-infract dementia becomes progressively worse. Sometimes one large stroke will leave the patient with dementia. Since almost all cases of multi-infarct dementia result from atherosclerosis, taking steps to prevent atherosclerosis reduces the risk of multi-infarct dementia.

Alzheimer's Disease

Alzheimer's disease (*AD*) is named for Alois Alzheimer, who first described the disease in 1907. The rate of occurrence of Alzheimer's disease doubles every five years after age 60 up to age 90. The earliest cases occur at about age 40. However, less than 1 percent of those under age 65 have AD, compared with up to 20 percent of those over age 80. Overall, 10 percent to 15 percent of people over age 64 have Alzheimer's disease, and it affects approximately 50 percent of those over age 84. Alzheimer's disease occurs more frequently in women compared with men.

There are now four million people with AD. The number is expected to reach nine million by AD. 2040. Alzheimer's disease is now the fourth leading cause of death in the U.S., causing 100,000 deaths per year. AD is becoming more important, as death rates from cardiovascular disease and strokes continue to decline, the elder population continues to increase, and the proportion of very elderly people increases. Older statistical tables do not list AD as a major cause of death among older people because widespread and accurate diagnosis of AD has occurred only in recent years. By AD. 2020, costs from AD are expected to exceed costs from heart disease and cancer. Costs come from physicians, health care providers, social workers and in-home care givers; diagnostic procedures and medications; hospitalizations and nursing homes; special apparatus, diets, and living accommodations; and loss of income and productivity. These costs bear on families, insurance companies, and society as a whole. Non-economic costs include social costs (e.g., disrupted family life, isolation, increased conflicts) and personal costs (e.g., stress, fatigue, psychological detriments such as depression and anger, reduced quality of life). These costs increase synergistically as the disease progresses and as other disorders develop.

Types Alzheimer's disease can be subdivided into two types. One type is *early onset AD* or *familial AD (FAD)*. Onset occurs before age 65, usually during the sixth decade of life. The second is *late onset AD* or *senile dementia of the Alzheimer's type (SDAT)*, with onset usually after age 60. SDAT, also called *sporadic Alzheimer's disease*, is the most common form of AD.

Causes Though the causes of most AD cases are not known, as many as 50 percent are probably caused by genetic abnormalities since AD tends to run in families. Other factors must be involved because when one identical twin develops AD, the other may not develop the disease (see Genetics of AD, below). A main difficulty in finding causes of AD is that no animals are known to develop AD or conditions very similar to AD. Therefore, research is limited.

Some scientists propose that AD is not a disease but is part of normal aging. They point out that all aging brains develop the same physical changes found in brains from AD victims, though to a lesser degree. Perhaps like other age changes, AD develops in everyone, though at different rates. They suggest that if people lived long enough, everyone might eventually develop AD.

Though the causes of some forms of AD remain unknown, risk factors have been identified. The greatest risk factor is increasing age. Other risk factors include having relatives with AD; suffering head trauma (e.g., boxing); being exposed to aluminum; having high blood cholesterol; having low education; and for women, being postmenopausal. Factors that seem to reduce the risk for AD include education; taking anti-inflammatory medications (e.g., steroids, ibuprofen); smoking; and for postmenopausal women, taking estrogen supplements.

Effects The effects of Alzheimer's disease develop in a steady and fairly predictable sequence. At first there is a decrease in short-term memory. Because the change is gradual and resembles the normal decrease, it is not uncommon for normal individuals to fear that they have Alzheimer's disease when the ability to remember begins to decline. Conversely, individuals with Alzheimer's disease may attribute their memory impairment to aging.

With AD, however, memory function declines to such an extent that affected individuals have considerable difficulty performing ordinary daily activities such as preparing food, dressing, and shopping. Patients with AD become disoriented with respect to location and have trouble learning new information. Early in the disease some patients begin to have trouble with language skills such as speaking. Perhaps because of fear of some of these changes or because of the disease itself, personality changes such as irritability, hostility, and agitation may appear. Often affected individuals tend to withdraw from social contact.

As AD progresses, loss of short-term memory becomes severe enough to dramatically decrease the ability to learn information or new skills, solve problems, and perform the ordinary tasks of daily living or working. Abstract thinking and making judgments become increasingly impaired. Language functions such as speaking, reading, and writing decline. Affected individuals become easily disoriented not only in terms of where they are but also with regard to time and date. Confusion occurs easily and frequently. Many patients wander away from home and become lost. Long-term memory, including recognizing familiar people, may also diminish.

Major personality changes that commonly accompany these more advanced effects of Alzheimer's disease may include high levels of agitation, paranoia, hostility, and aggressiveness. These patients may have verbal and physical outbursts of anger or other emotions. They may strike out violently. These changes make cooperation and acceptable interactions with others difficult. For many people, social withdrawal becomes more intense.

By this stage affected individuals require a great deal of care. They need to be bathed, dressed, and fed. Their behavior must be monitored so that they do not engage in destructive actions or wander off. Eventually the care must extend for 24 hours a day. The changes in personality and behavioral traits caused by AD make providing such care emotionally draining on family members. Families that cannot provide adequate care are faced with the financial burden of paying others to provide it. All these problems intensify as the disease progresses.

In the most advanced stages of Alzheimer's disease patients lose essentially all memory and intelligence capabilities. Performing any task and talking with others become impossible. Apparently, there is a complete loss of awareness of one's surroundings. Bladder and bowel incontinence develop. The nervous system seems to forget how to stimulate muscles so that walking, eating, and other voluntary motions dwindle and finally cease. Curiously, long-lasting muscle spasms may occur. The victim becomes bedridden and paralyzed. The final result of Alzheimer's disease is death, which is caused by complications from immobilization. The complications may include infections of the skin and respiratory systems, thrombus and embolus formation, malnutrition, and respiratory failure.

Though this sequence of events occurs in most patients with Alzheimer's disease, individual cases vary considerably. For example, changes in personality may be the first noticeable indication that something is wrong. In other cases, problems with speaking may occur early in the disease or not until most of the other effects have developed.

There is also much variation in the time that passes from the diagnosis of AD until death occurs; this period may range from 2 to 20 years. The average length of time from diagnosis to death is eight years. More rapidly progressing and serious cases are correlated with an earlier age of onset. Alzheimer's disease almost always progresses at a steady rate. There is never a period of improvement.

Diagnosis Diagnosing Alzheimer's disease by observing changes in behavior is difficult until the disease has progressed into more advanced stages because at first these changes seem to be normal fluctuations. Only specific tests can detect early abnormalities in mental status. Repeating tests every few years to detect changes associated with AD may help detect AD at earlier stages.

Making a definitive diagnosis of Alzheimer's disease remains difficult even after the recognition of abnormal behaviors because similar behavioral changes can be caused by many other factors (e.g., medications, depression, altered social situations) and by other diseases of the nervous system or other systems. Furthermore, the simul-

taneous presence of other types of dementia can mask the presence of AD. Researchers continue developing other diagnostic procedures including tests at the chemical, genetic, and cellular through system levels. Being able to detect and diagnose AD earlier could lead to developing effective treatments.

Eventually, after all other possible causes of the behavioral signs and symptoms have been ruled out, a clinical diagnosis of Alzheimer's disease can be made with an accuracy of over 90 percent. Only an autopsy examination of the brain can determine conclusively that a person had Alzheimer's disease.

Changes in the Brain A brain from an Alzheimer's patient can be identified because it has two characteristics: an excessive number of senile plaques and neurofibrillar tangles. A third important finding is a low level of the neurotransmitter acetylcholine. These features are especially prevalent in brain areas involved in memory. The functioning of synapses in these areas may be hampered because the neurons produce inadequate neurotransmitters; the tangles may prevent enough neurotransmitters from reaching the ends of the axons; and the plaques may block transmission at synapses.

Plaques and Tangles *Senile plaques (SPs)* are round microscopic masses having various mixtures and densities of materials. They are at or near synapses. SPs usually contain a protein called *beta-amyloid (β-A)*, dead neurons and neuroglia cells, pieces of synapses, and fibrous material called *neurofibrillar tangles (NTs)*. Neurofibrillar tangles are composed of one or two protein fibers twisted into a helix. Much of the protein is *tau protein (T-protein)*. NTs also contain other materials including enzymes, inflammatory molecules, β-A, a lipoprotein called *apolipoprotein E (APOE)*, and carbohydrate/protein complexes. NTs also form in neuron cell bodies, axons, and dendrites.

SPs and NTs appear first in the hippocampus region, which is near the center and bottom of the cerebral hemispheres. The hippocampus has a major role in memory functions. Later, SPs and NTs appear in wider areas near the bottom of the hemispheres. Later still they appear in upper regions of the cerebral cortex. Eventually, all regions of the cerebral cortex develop SPs and NTs. Neuron connections to the nucleus of Meynert also develop many SPs and NTs, and SPs form in the cerebellum. The final distribution of SPs and NTs in brain areas corresponds to the sequence in which they appear. Areas showing SPs and NTs first develop the highest densities of them.

As SPs and NTs form, neurons are damaged and die, and synapses are destroyed. Scientists do not know if SPs and NTs form and then cause damage to neurons or if neurons damage occurs first, causing SPs and NTs to develop. Neurons that interconnect other neurons (i.e., association neurons) are affected much more than sensory neurons and motor neurons.

As AD progresses, brain vessels also change. Small vessels accumulate much β-A in their middle layer. Vessels become twisted, shrunken and broken, which reduces blood flow in the brain. The cerebral hemispheres shrink dramatically. Some scientists believe that reduction in blood flow causes the SPs, NTs, and other neuronal and synaptic changes in the AD brain.

Beta-amyloid   Many cells in the body produce *amyloid protein*. There are more than 10 types of amyloid protein. The type called *β-amyloid (β-A)* is found in AD. Its function is unknown. Beta-amyloid may be produced by neurons and by blood vessels. It is produced when an enzyme breaks a protein called *amyloid precursor protein (APP)*, which extends across cell membranes. Breaking normal APP produces a small amount of soluble short β-A. In AD, APP is abnormal. When it is broken by enzymes, much abnormal long β-A is produced and released from the cell membrane.

The abnormal "sticky" β-A binds easily to APOE and to τ-protein, forming many SPs quickly. The abnormal β-A increases free radicals, inflammation, cell membrane damage, and neuron apoptosis. Excess glycation of proteins also occurs. All these processes seem to promote each other synergistically. Finally, APP itself binds to t-protein and to APOE, suggesting that it can contribute to the formation of NTs and SPs.

The causes, method of formation, sources of β-A and NTs, and sequence in which materials are deposited are unknown.

Tau Protein   Brain cells produce other proteins called *tau proteins (τ-proteins)*. Their functions are unknown, though they seem to promote mi-

crotubule formation. The brain contains at least six types of τ-protein, and their proportions vary from childhood through adulthood. Abnormal modifications of τ-proteins (e.g., glycation, adding phosphate groups) cause τ-proteins to help form NTs.

APOE   Many cells produce apolipoprotein E (e.g., brain, liver, adrenals). Most brain APOE comes from neuroglia cells and macrophages. Though neurons do not produce APOE, it enters them. APOE helps move cholesterol and other lipoproteins from cell to cell and through cell membranes. APOE also seems to help in neuron development and repair.

Brain APOE has different forms including APOE-ϵ3 and APOE-ϵ4. APOE-ϵ4 seems to promote the formation of SPs and NTs. The mechanisms are not clear, but they may involve disruption of neuron membranes; formation of free radicals; excess accumulation of β-A; and the formation of abnormal microtubules in neurons. Interactions between the β-A and the abnormal microtubules seem to result in SPs and NTs.

Presenilins   The last groups of brain proteins to mention are the *presenilins*. Two important forms of presenilin in the brain are *presenilin-1 (PS-1)* and *presenilin-2 (PS-2)*, which are membrane proteins. Their functions are unknown.

In summary, AD may be caused by or promoted by abnormal protein formation; chronic inflammation; inadequate blood flow; free radical damage from brain proteins, metal ions, damaged endothelium, or neurotransmitters; decreased *FR defenses; mitochondrial malfunctioning; reduced insulin sensitivity; immune responses; or abnormal apoptosis of neurons. Regardless of the causes or mechanisms, the results are the same; too many SPs, too many NTs, too much neuron death, and too much loss of synapses.

Genetics of AD   There are several genes that promote different types of AD. Though these genes are in different chromosomal locations, have effects at different ages, and may act by different mechanisms, they all produce the same outcomes in the brain and the same manifestations of AD. Some genes that promote or modify AD have not been identified. One or more of these genes may contribute to a form of AD that begins after age 70. These latter genes may be on chromosomes 12 or 3.

Three genes for one type of familial Alzheimer's disease (FAD) are on **chromosomes 21**. The mutated forms of the genes cause the production of abnormal "sticky" β-amyloid, resulting in 7 percent of AD cases and 25 percent of FAD cases. Age of onset is between ages 45 and 65, with most cases developing before age 60. The mutations are present in approximately 19 families. An individual with only one copy of one of the mutated genes has a 100 percent chance of developing AD because each mutated gene is a dominant gene.

Certain forms of a gene on **chromosome 19** promote SDAT. The gene has three forms (i.e., three alleles), each of which contains the genetic information for producing **APOE-ε**. One form codes for **APOE-ε4**, one form for **APOE-ε3**, and one for **APOE-ε2**. Since a person has two copies of chromosome 19, each person has two of these genes. The pair of genes may be in any combination (i.e., ε4:ε4, ε4:ε3, ε4:ε2, ε3:ε3, ε3:ε2, ε2:ε2). In the general population, the genes are found in the proportion ε4:ε3:ε2::14:78:8.

The genes are codominant, meaning that each produces its form of APOE-ε regardless of which other forms of the gene are present. Having two ε4 genes provides the highest risk from APOE genes and makes the AD occur at earlier ages. The risk for developing AD is eight times higher in people with two ε4 genes than in people with two ε2 genes. However, people with two ε4 genes do not always get AD.

The different combinations of APOE-ε genes provide decreasing risk of getting AD and increasing average age of onset in the order ε4:ε4 (age 68), ε4:ε3 (age 71), ε3:ε3 (age 74). Still, age of onset shows great variability with any of these combinations. Very few people with even one ε2 gene develop AD.

The APOE-ε gene influences other problems. Having an ε4 gene increases the age-related decline in cognitive functions even if AD does not develop. The ε4 gene also promotes amyloid formation in blood vessels, so people with the ε4 gene are at higher risk for developing atherosclerosis. Having an ε2 gene reduces the risk of atherosclerosis.

Chromosome 14 has the gene for PS-1, and *chromosome 1* has the gene for PS-2. Nearly 50 percent of FAD cases are associated either with mutations in the APP gene on chromosome 21 or a presenilin gene. Nearly 70 percent of cases of FAD are associated with mutations in the PS genes. Mutations in either presenilin gene increase the risk of developing AD, apparently because abnormal PS-1 and abnormal PS-2 increase the production of "sticky" β-A.

The PS-1 mutation is known to occur in nearly 50 families. The PS-2 mutation is known to occur in descendants from certain German families (i.e., Volga Germans). For people with the PS-1 mutation, average age of onset is in the fifth decade, but cases develop as early as age 30. The PS-1 mutation also promotes late onset SDAT. For people with the PS-2 mutation, the average age of onset is higher than with the PS-1 mutation, but onset may occur before age 30.

Treatments   There are no effective treatments to slow, stop, or reverse the effects of Alzheimer's disease. Therapies being investigated include antioxidant supplements; anti-inflammatory drugs; medications that increase brain acetylcholine (e.g., tacrine); medications that slow atherosclerosis or reduce blood clotting; and for women, estrogen supplements. Until effective treatments are found, all that can be done is reduce the signs and symptoms and maintain as much functioning as possible. In the early stages of the disease memory aids such as notes and verbal reminders help. Various medications can alleviate the behavioral and psychological problems. Maintaining social contacts and providing emotional support for the patient and his or her families are important components in a complete treatment program.

As the disease progresses, outside help from support groups and social agencies is usually required. Day care centers can relieve the burden of full-time care by family members. Attention must be paid to preventing complications such as malnutrition and infections. Finally, full-time institutionalization may be necessary.

Parkinson's Disease

Though the incidence of **Parkinson's disease** is less than half that of strokes or Alzheimer's disease, it remains a leading disease of the nervous system among older Americans. Its rate of occurrence is extremely low before age 50, but the rate increases gradually after that until about age 75; after that age it diminishes steadily. About 2 percent of those over age 50 will develop Parkinson's disease. This disease occurs with equal frequency in men and in women and among people of different races.

Causes   The cause of true, or *primary*, *Parkinson's disease* is unknown, and it does not tend to run in families. Scientists suspect the involvement of free radicals and reduced blood flow. Many cases of what appear to be Parkinson's disease actually result from abnormalities such as CNS infections, atherosclerosis, brain tumors or other brain diseases, head injury, toxins, and medications. These cases are called *secondary parkinsonism*.

Effects   The development of Parkinson's disease is shown primarily by changes in the control of muscle contractions. These changes usually occur in the same sequence. At first, ongoing movements of the fingers and hands occur. The movements of the fingers give the appearance that the victim is rolling pills between the fingers.

Tremors of the hand, arm, and leg muscles often develop next. The movements are rhythmic, with alternating contractions between muscles that bend the joints and muscles that straighten them. Four to eight contractions occur each second. The tremors are greatest when the person is awake but resting. They diminish during voluntary movements and stop when the patient falls asleep.

Further progress of the disease causes muscle stiffness and difficulty moving rapidly and smoothly. As control of muscle contraction diminishes further, the patient may find it impossible to complete a motion once it has been started. For example, a person who is walking may suddenly stop in the middle of taking a step. Ordinary motions occur ever more slowly. Performing ordinary tasks and job-related activities becomes difficult or impossible.

As normal contractions of muscles continue to diminish, facial expressions disappear. The voice becomes soft and loses inflection. Weaker, slower, and fewer contractions of leg, trunk, and arm muscles cause walking to occur more slowly and with shuffling of the feet, a stooped posture, and little swinging of the arms. Muscle contractions for swallowing and breathing also weaken and slow.

Declining muscle control and muscle activity causes drooling. Constipation is not uncommon because patients are less active and have weaker contraction of the abdominal muscles that normally help with bowel movements.

Gradually, coordination of muscles declines to such an extent that the person has trouble with balance. Not only do these patients tend to fall more frequently, they make little or no effort to slow or stop themselves as they are falling.

Parkinson's disease often produces effects other than those involving control of muscles. During the night patients tend to wake up and have difficulty going back to sleep. They become restless and begin to wander about. Because of declining muscle coordination and balance, they are at great risk of physical injury from falls. The interrupted sleep also causes these patients to be sleepy during the day.

Psychological changes may begin at any stage in the disease. Many patients experience depression, loss of interest in activities, and other mood changes. These psychological alterations seem to be caused partly by the disease itself and partly by the awareness of its effects. Reductions in very short-term memory are common. Parkinson's disease causes dementia in over 15 percent of patients.

While the sequence of changes caused by Parkinson's disease is fairly regular and progresses steadily, the rate of change varies greatly from one person to another. A few cases reach extreme conditions in as few as five years, although most cases progress more slowly, so that severe disability is delayed for many years.

Nervous System Changes   The mechanism by which Parkinson's disease affects muscle control is fairly clear. Recall that impulses controlling voluntary movements are modified as they descend through the somatic motor pathway. Some areas of modification are in the basal ganglia inside the cerebral hemispheres (Fig. 6.7). The normal impulse modifications occurring in the basal ganglia actually result from the interplay among several neurotransmitters in the basal ganglia. Acetylcholine tends to increase the impulses and thus increases muscle contractions. *Dopamine (DOPA)* and another neurotransmitter (gamma-aminobutyric acid) tend to dampen the impulses and the movements they cause.

In Parkinson's disease a major decline in the amount of DOPA in the basal ganglia creates an imbalance among the antagonistic transmitters. This imbalance causes impulses and the muscle contractions they produce to become excessive and uncontrolled. Hence, muscle contractions occur. Neurotransmitter imbalances also cause the other effects of this disease.

Diagnosis   Parkinson's disease is accompanied by a decrease in certain CNS chemicals that are used by the brain to manufacture dopamine. Dopamine is a neurotransmitter that is present in inadequate amounts in patients with Parkinson's disease. Because this and the other effects of the disease are somewhat different from those of other diseases and the effects develop in a fairly regular sequence, Parkinson's disease can be diagnosed accurately.

Treatments There is no cure for Parkinson's disease and no way to slow its progress. However, its effects can be greatly diminished by administering *levodopa* because this chemical boosts brain production of DOPA. Dosages must be carefully monitored and adjusted during the disease to minimize adverse side effects such as increased uncontrolled movements. Since increasing the level of DOPA seems to be so important, attempts have been made to implant into the brains of Parkinson's disease patients tissues that produce DOPA. Pieces of adrenal medulla and pieces of brain regions from aborted human fetuses have been used. Transplants of adrenal medulla have not yet produced satisfactory results. However, experiments using fetal brain tissue have resulted in dramatic and long-term improvements in muscle control in individuals having severe cases of Parkinson's disease. As the controversial and experimental techniques employing fetal brain tissue improve and become more standardized, they may gain widespread acceptance and use.

Other medications can relieve certain signs and symptoms sometimes. However, the specific types and amounts of substances used to treat Parkinson's disease vary from case to case because individuals have such varied responses to these medications and because their responses change as the disease progresses.

Besides medications, treatment of Parkinson's disease should include physical therapy to help sustain the movements used in ordinary and occupational tasks. Speech therapy and psychological support are also important components of a treatment plan.

Dementia with Lewy Bodies

Dementia with Lewy bodies is a newly classified type of age-related dementia. It has been identified in nearly 20 percent of the brains from people who died after developing any dementia. Lewy bodies are round masses of clumped microfilaments in neurons. They occur in all areas of the brain. This type of dementia also shows amyloid deposits.

7

EYES AND EARS

The sensory function of the nervous system was explained in Chap. 6. Information about the eyes and ears was not included there because these sense organs are more specialized and complicated than other sense receptors.

THE EYES

The eyes assist in the monitoring function of the nervous system by receiving information about conditions on the surface of the body and the region surrounding it. This information is used to make proper responses that promote healthy survival and maintain the quality of life.

Unlike most sense receptors, the eyes can also provide information about conditions far from the body and thus enable a person to respond to harmful factors long before they affect the body. Examples include avoiding an oncoming car and seeking shelter from an approaching storm. A person can also efficiently and effectively seek out, move toward, and utilize helpful factors at some distance from the body. For example, the eyes are of inestimable value in obtaining and preparing food and in any project that involves obtaining and assembling parts.

In addition, information provided by the eyes contributes greatly to learning, and much communication occurs through the eyes. Finally, much of the beauty of the world can be appreciated and enjoyed because of the eyes.

Image Formation and Vision

To perform these functions successfully, each eye must produce accurate images of objects by focusing some of the light that reflects from or is given off by those objects. Focusing light involves bending it so that by the time it reaches the retina in the rear of the eye, the light has the exact organization it had before leaving the object. For example, a group of people in a room can recognize each other because light from a lamp strikes them and reflects in all directions from each one. Some of the reflected light from each person enters the eyes of each of the other people, where it is focused into accurate, recognizable images (Fig. 7.1).

If enough light enters the eye, it stimulates certain neurons to initiate impulses. The impulses are then processed by other neurons in the eye and by neurons in the brain. The *vision center*, the part of the cerebral cortex at the back of each

FIGURE 7.1 Image formation and the pathways used for vision.

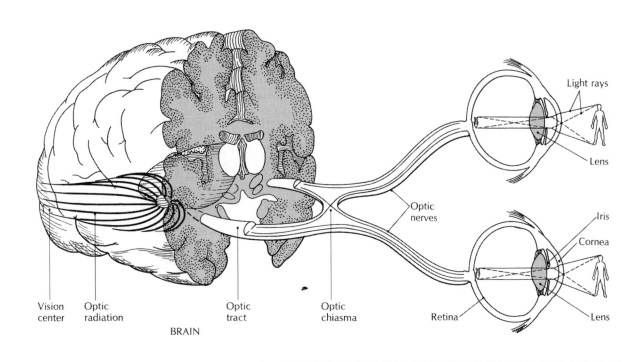

Vision center | Optic radiation | Optic tract | Optic chiasma | Optic nerves | Light rays | Lens | Iris | Cornea | Retina | Lens

BRAIN

cerebral hemisphere, plays a major role in processing these impulses. The final result is the conscious perception of the images, a phenomenon called *vision*.

Deterrents to Clear Vision

Clear vision is not always possible for three reasons. First, the eye may not be able to focus light well enough for an accurate image to be formed. This can occur if the light is too disorganized before it enters the eye. For example, the water droplets in fog and the irregular surfaces of the glass used in rest room windows cause this effect. Inadequate focusing of light can also occur if parts of the eye are improperly shaped or malfunction because of nearsightedness or farsightedness.

The second reason involves having improper amounts of light strike the receptors in the eye. With insufficient light, not enough impulses are produced for accurate perception to occur, and so errors in vision are made. For example, objects in a very dimly lit area may be misidentified, may

seem to move, or may go completely unnoticed. By contrast, when excess light hits the receptors, they may be overly stimulated or damaged.

The third reason is that neurons in the eye and brain may be unable to properly process the impulses sent to them. For example, a person with normal eyes may be blinded because of a stroke.

FORMING IMAGES

Conjunctiva

Each eye is nearly spherical in shape. The front surface consists of a smooth thin layer, the *conjunctiva*, which covers the part of the eye exposed to air (Fig. 7.2). The conjunctiva also extends away from the edge of this region to form a lining on the inner surfaces of the upper and lower eyelids. It is transparent, and so it does not absorb or block any of the light that strikes the eye.

The conjunctiva secretes a fluid that helps prevent eye damage from drying and lubricates the eye so that the eyelids slide over it easily. Small

FIGURE 7.2 Structure of the eye.

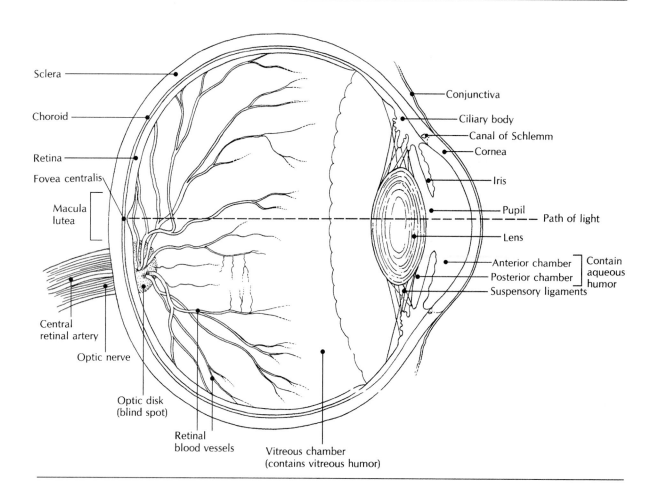

blood vessels in the conjunctiva that are visible over the white part of the eye help nourish the cornea.

Age Changes   Aging results in a small decrease in the smoothness of the conjunctiva, causing light entering the eye to be slightly disorganized and scattered. These changes make focusing the light into a clear image more difficult. The conjunctiva also diminishes slightly in transparency, causing it to absorb and block the passage of some light.

A more important age change is a gradual decline in the amount of fluid the conjunctiva secretes. In some individuals fluid production is so low that the eyelids can no longer glide smoothly over the eye. The irritation and inflammation of the conjunctiva that result can be quite uncomfortable. Artificial tear solutions can relieve this discomfort in most individuals.

Cornea

Immediately behind the central region of the conjunctiva is another transparent structure, the *cornea* (Fig. 7.2). Because the cornea is fairly thick and curved, it bends (*refracts*) the light that passes through it.

Age Changes   Aging results in a gradual decrease in the transparency of the cornea. This change is usually great enough to block a significant amount of light. Light at the blue end of the spectrum is blocked more than is light toward the red end; therefore, seeing blue objects or objects lit with blue light is reduced preferentially. The blue end of the spectrum has light of shorter wavelengths than does the red end.

Aging of the cornea also increases the degree of scattering of light that passes through the cornea.

Much of the scattered light still reaches the retina, but since it strikes the retina in the wrong places and in a disorganized way, it causes the viewer to see bright areas in the wrong places in the field of view. This phenomenon is usual noted when one looks at outdoor objects on a bright but hazy day. An extreme example of glare can be created by looking at bright lights at night through a window covered with drops of water.

With aging, the cornea becomes flatter, reducing the amount of refraction it can cause and making it difficult to see close objects clearly. The curvature of the cornea also becomes irregular so that light from certain parts of the field of view is not focused properly. This condition, called *astigmatism*, decreases the clarity of images from some parts of the field of view. Eyeglasses or contact lenses can often compensate for astigmatism. Finally, the sensitivity of the cornea to pain from pressure on the eye diminishes, which in turn decreases a person's ability to ward off eye injury from external pressure.

Iris and Pupil

A short distance behind the cornea is the *iris* (Fig. 7.2), which is shaped like a phonograph record or compact disk. Pigments in the iris give the eye its color.

The hole in the center of the iris is called the *pupil*. The pupil allows light to pass from the front of the eye into the rear region. Muscle cells in the iris regulate the size of the pupil. These cells are controlled reflexively by autonomic motor neurons.

In the presence of bright light, some muscle cells in the iris constrict the pupil to reduce light entering the eye; this helps protect the eye from being damaged by excess light. The pupil is also constricted when a person looks at an object close to the eye. This helps form a clear image by blocking out stray light.

When light is dim, other muscle cells in the iris dilate the pupil. This allows enough light into the eye to adequately stimulate the receptor neurons.

Age Changes   With aging, the number and strength of the muscle cells that cause dilation of the pupil diminish and the thickness and stiffness of the collagen fibers increase. As these processes continue, the size of the pupil for any light intensity decreases with each passing year, starting at age 20. The result is a continuous decline in light available to form images.

Age changes in the cells and fibers in the iris may also slow the rate at which the pupil dilates when changing from bright to dim light. This effect, combined with age-related slowing of pupillary constriction, retards pupillary adaptation to changing light intensities.

Ciliary Body

The outer edge of the iris is attached to a thickened ring of cells called the *ciliary body* (Fig. 7.2), which contains muscle cells that regulate the curvature of the lens. As in the iris, these muscle cells are controlled by reflexes. The ciliary body also secretes a fluid called *aqueous humor*, which is discussed below.

Age Changes   An important age change in the ciliary body is the slowing of its secretion of aqueous humor. The significance of this change is discussed below.

Lens and Suspensory Ligaments

A short distance behind the iris and the pupil is the transparent *lens*, which is nearly round (Fig. 7.2). However, because the lens is elastic, its shape can be changed. The lens increases in thickness throughout life.

A ring of thin fibers called *suspensory ligaments* radiates outward from the lens much as spokes radiate from the center of a bicycle wheel. These ligaments reach and attach to the ciliary body much as spokes connect to the rim of a wheel.

Since the lens is a thick curved structure, it refracts the light passing through it. Unlike the cornea, however, the curvature of the lens can change so that the lens can refract light to a greater or lesser degree (Fig. 7.3). Alterations in refraction are important because light coming to the eye from close objects must be refracted more than is light from farther objects. Therefore, to focus light from close objects, the lens becomes more rounded so that it bends the light more. The lens becomes flatter and bends light less to help the eye form a clear image of a distant object.

Age changes   As the lens ages, it is altered in four ways. One alteration is a decline in transparency to all colors of light, especially blue light. The markedly declining transparency of the lens blocks more light than does the reduction in the transparency of any other part of the eye. This change begins during the third decade of life and increases exponentially with age.

The second alteration is the development of opaque spots, which begins during the fifth decade. Usually these *opacities* are toward the periphery of the lens. Lens opacities block the passage of light and cause a great increase in light being scattered. They account for more scattering of light and more glare than do age changes in any other eye component.

If many opacities form close to the center of the lens, vision is greatly impaired and the condition is called *cataracts*. While cataracts are categorized as a disease, their development is part of aging of the lens. Everyone who lives long enough will eventually develop cataracts.

The third alteration of the lens is a reduction in its ability to refract light. This change results from accumulation of damaged proteins plus age-related formation of abnormal proteins. These two changes offset the age-related increase in lens thickness and curvature. With diminishing refractive power, it becomes increasingly difficult to see close objects clearly. Individuals who are near-sighted during youth may benefit from this age change since their ability to see distant objects improves.

Many individuals have lenses so flat that they cannot focus light even from far objects. This condition may eventually begin to improve because thickening of the lens increases its ability to refract light.

The fourth age change is a decrease in elasticity, which may result partly from an increase in the cross-linkages among collagen fibers. As the lens loses elasticity, its shape changes more slowly when it adjusts to near or distant objects. Declining elasticity actually begins some time before age 10 and continues at a steady rate until about age 50. The decline in elasticity decreases the amount of curvature the lens can achieve, and objects must be farther away from the face to be seen clearly. Thus, the smallest distance from the eye at which an object can be seen clearly—the *near point of accommodation*—increases.

This increase is usually not noticed until about age 40 because most objects used in daily living are located beyond the near point. After age 40, the reserve capacity of the lens for elasticity has dwindled sufficiently and the near point becomes large enough to interfere with ordinary activities such as reading and writing. This condition, called

FIGURE 7.3 Focusing light from (a) close objects and (b) distant objects.

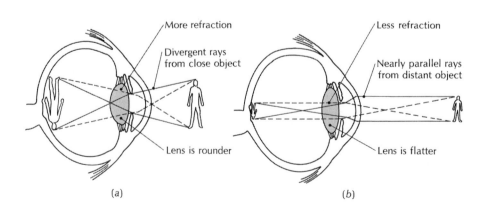

presbyopia, involves farsightedness caused by stiffening of the lens. Difficulties and limitations caused by presbyopia can be reduced by wearing eyeglasses or contact lenses.

The increase in the near point is rapid from about ages 40 to 50, but the rate of change slows during the sixth decade of life. By age 60 there is usually no further increase in the near point because the lens has lost all ability to change its curvature.

Because adjusting the curvature of the lens requires the use of the aqueous humor and vitreous humor, a more complete explanation of how the curvature of the lens is adjusted and how this process is affected by aging follows the description of those two humors.

Aqueous Humor

As mentioned above, the ciliary body produces a liquid called *aqueous humor* (Fig. 7.2). The aqueous humor flows forward from the ciliary body, passes through the pupil, and is removed from the eye by a special tube located around the edge of the cornea.

Since aqueous humor is produced and removed continuously, its steady flow delivers nutrients and removes wastes. These services are important for the cornea and lens, which have no blood vessels.

The aqueous humor fills the cavity between the cornea and the lens. The fluid provides a slight outward pressure that helps keep the cornea curved outward so that it refracts light properly. By causing the entire eye to bulge outwardly, this pressure also helps provide tension on the suspensory ligaments. The tension helps to hold the lens in place and pull it into a slightly flattened shape.

Age Changes As a person ages, the rate at which aqueous humor is produced declines. This change reduces the rate at which the cornea and lens are serviced. There is also a decrease in the amount of aqueous humor present, and this may contribute to the flattening and irregular curvature of the cornea that develop with aging.

Vitreous Humor

Another eye humor, the *vitreous humor*, fills the cavity in the eye behind the lens (Fig. 7.2). The transparent vitreous humor is composed of a soft gel that has the consistency of partially solidified gelatin. The center of the gel becomes liquefied early in childhood. Thereafter, the liquefaction slowly and continuously spreads outward.

The vitreous humor is held in place by a ring of attachment on the front edge of the retina and an attachment point at the back of the eye where the optic nerve begins.

The vitreous humor produces an outward pressure so that, like the aqueous humor, it puts tension on the suspensory ligaments. The soft consistency of the vitreous humor allows it to protect eye structures by absorbing shock. It also holds the retina and choroid layers in place by pushing against them.

Age Changes With aging, alterations in the chemistry of the vitreous humor make it lose transparency and cause more scattering of light. Blocking and scattering of light are increased by small areas of vitreous humor that become opaque and often grow large enough to be visible in the field of view. These areas, which appear as objects of varying sizes, shapes, and textures, are called *floaters* because they are seen to move, especially when the eye moves. Though floaters are not dangerous, they decrease the quality of the images that are formed, are often distracting, and can obscure parts of the field of view. Some floaters are pieces of vitreous humor that have broken off from the main mass and float between the vitreous humor and the retina.

Chemical changes caused by aging of the vitreous humor also make it decrease in size and shrink away from the retina. These alterations reduce the amount of support provided for the retina. At the same time, more of the central region is becoming liquefied, making the vitreous humor move about when the eye moves. As the vitreous humor shifts, it pulls on the retina, especially during rapid eye or head movements, and causes the person to perceive flashes of light called *flashers*.

The presence of a few flashers may be distracting but is of little importance. However, as the vitreous humor ages, the tension it places on the retina increases. If the tension becomes great enough, the person may perceive many flashes and part of the field of view may become darkened. These symptoms constitute a warning that the vitreous humor may have detached the retina from the back of the eye. If not treated and corrected immediately, detachment of the retina usually causes some degree of blindness.

ACCOMMODATION

Having described the aqueous and vitreous humors, we now move to a more complete description of the process of adjusting the curvature of the lens. This process is called *accommodation*.

Since the lens is normally in a slightly flattened condition because of pressure from the aqueous and vitreous humors, it is normally set for focusing light from distant objects (Fig. 7.4). For the lens to become more rounded so that it can focus light from near objects, the tension on the suspensory ligaments must be decreased; this is accomplished by having the ring of muscle cells in the ciliary body contract and pull the edges of the ciliary body inward. When this happens, the normal elasticity of the lens causes the lens to spring back to its round shape.

To flatten the lens again to focus on distant objects, the muscle cells in the ciliary body relax, allowing the lens to be pulled back to its slightly flattened shape. When the amount of contraction of the ciliary body is regulated, the curvature of the lens can be adjusted. Then light from the object a person is viewing is in focus. Other objects at different distances appear to be less clear.

FIGURE 7.4 Accommodation to (a) distant objects and (b) near objects.

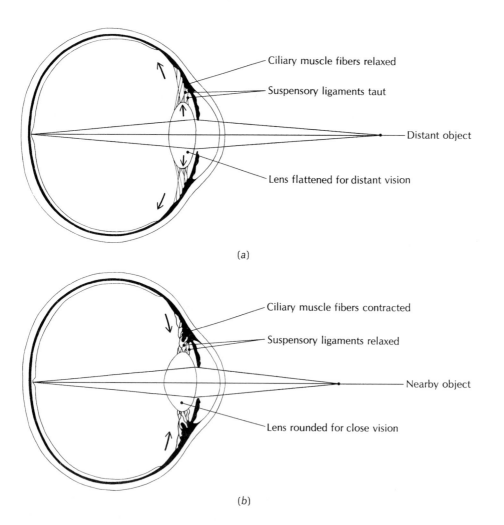

Ciliary muscle fibers relaxed

Suspensory ligaments taut

Distant object

Lens flattened for distant vision

(a)

Ciliary muscle fibers contracted

Suspensory ligaments relaxed

Nearby object

Lens rounded for close vision

(b)

Age Changes

Age changes in the ciliary body and the lens gradually reduce and finally eliminate the ability to adjust the shape of the lens. Therefore, the range of distances over which the eye can focus diminishes considerably.

A person can partially compensate for diminishing accommodation by using eyeglasses with different refractive powers for viewing objects at different distances. Of course, it is inconvenient to change glasses when the distances being viewed change frequently, as occurs when a person checks prices while shopping or takes notes during a lecture. The expense of two or more sets of eyeglasses is also a significant factor.

These problems can be largely overcome by using eyeglasses with bifocal or trifocal lenses, which have regions with two or three different refractive powers, respectively. Thus, one set of glasses can be used over a wide range of distances simply by looking through the appropriate region of the multipowered lenses. Unfortunately, each region of the lens has a limited field of view, and such lenses are sometimes considered to have undesirable cosmetic effects. They are also expensive.

RETINA

We have seen that light from an object being viewed passes through the *optic media* (conjunctiva, cornea, aqueous humor, lens, and vitreous humor) and the pupil so that it is focused and its intensity is adjusted. In this way, clear and accurate images are formed on the retina. The *retina* is a thin layer that lines the rear portion of the inner cavity of the eye. It extends back from the edge of the ciliary body and thus partially surrounds the vitreous humor (Fig. 7.2).

Layers and Regions

The retina consists of two main layers of cells. The inner layer is called the *sensory retina* (Fig. 7.5). It is in contact with the vitreous humor and contains several layers of neurons. The neurons in the deeper region are called *photoreceptors* because they respond to light by starting impulses in the form of action potentials. The neurons in the surface region, closest to the vitreous humor, use their synapses to process these impulses. The impulses are then passed to the *optic nerve*, which begins near the back of the retina at a spot called the *optic disk*. This nerve carries the impulses to the brain.

Blood vessels in the retina nourish the neurons that process impulses; these vessels pass through the optic disk. The photoreceptors are nourished by vessels behind the retina in the *choroid* layer, as is the outer layer of the retina, the *pigmented epithelium*.

The photoreceptors in the sensory retina are of two main kinds. The *cones* are clustered together in a small circular region at the very back of the eye. This region is in line with the center of the cornea and the lens and is called the *macula lutea* or simply the **macula**. In the center of the macula is a slightly depressed area that contains a very high concentration of cones. This central area is called the *fovea centralis* or simply the *fovea*.

The outer layer of the retina is called the pigmented epithelium because is it a darkly colored thin layer of cells. A noncellular membrane (*Bruch's membrane*) lies behind the pigmented epithelium (Fig. 7.5). The pigmented epithelium and apparently Bruch's membrane regulate the exchange of materials between the choroid and the photoreceptors of the sensory retina.

Cones

There seem to be three types of cones, and each type is most sensitive to light of a different color, either blue, green, or red. If only one type of cone is stimulated, the person sees the corresponding color, while stimulation of various combinations of cone types allows a person to see many other colors and shades of color. This principle explains how people can see many colors on a television screen that employs patterns of red, blue, and green dots to form images.

Cones initiate impulses because of specific chemical reactions that occur when light strikes pigmented molecules within them. Cones are not particularly sensitive to light; therefore, relatively high intensities of light must strike them before they initiate impulses.

When a person views a scene, the image of what is in the center of the field of view is focused on the fovea. Since the fovea has a very high concentration of cones, it allows the person to see the

FIGURE 7.5 Structure of the retina and associated eye components.

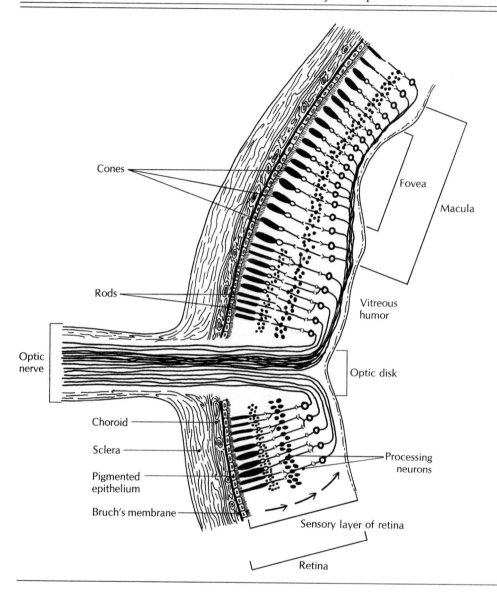

object in the center of the scene in color and with maximum *visual acuity* (the amount of detail that can be seen).

Objects immediately surrounding the center of the field of view are focused on the outer regions of the macula. Since the cones are less concentrated there, these objects are seen with less visual acuity but still are seen in color.

Rods

While the regions of the retina immediately surrounding the macula contain some cones, most

of the photoreceptors are of the second type, called *rods* (Fig. 7.5). The concentration of rods diminishes steadily at greater distances from the edge of the macula, and the front edge of the retina has few rods. Therefore, images from very peripheral objects are focused on retinal areas with low concentrations of photoreceptors and are seen with little visual acuity.

Rods are of only one type. Images focused on rods are seen as black and white images. Though rods do not permit us to perceive color, they allow us to see in dim light because they are more sensitive to light than are cones. As in the cones, the rods produce impulses when light causes specific chemical reactions within them.

Age Changes

Cones Beginning at age 40, cones decrease in length and many are lost. This decrease in number is greatest in the fovea. The cones that remain in the retina widen to fill the spaces left by degenerated cones.

Since the level of visual acuity and color vision are directly related to the concentration of cones in the retina, the declining number of cones with aging contributes to a gradual drop in visual acuity and a diminishing ability to distinguish colors.

Rods Aging causes little if any change in the number of rods. However, the rods lengthen, causing them to be bent into irregular shapes to fit within the confines of the sensory retina. These age changes further diminish an aging person's ability to see in dim light.

Processing Neurons The sensory retina loses not only photoreceptors but also neurons in its inner layer that process impulses from the photoreceptors. The loss is greatest in the neurons that serve the macula. The declining number of neurons processing impulses probably decreases the quality of perceived images and the ability to interpret those images.

Pigmented Epithelium and Bruch's Membrane Cell numbers and cell functioning also diminish in the pigmented epithelium of the retina. Furthermore, the exchange of materials between the sensory retina and the pigmented epithelium decreases, resulting in diminished servicing of the photoreceptors by the pigmented epithelium.

Age changes in Bruch's membrane seem to further inhibit the exchange of materials required by the cones and rods. Changes in this membrane also cause a reduction in capillaries in the choroid. These capillaries normally provide nutrients for the cones and rods and carry away wastes.

Free radicals in the eye A unifying factor contributing to several age-related changes in the eye is damage from free radicals (*FRs) and reactive oxygen species (ROS), especially H_2O_2. Free radicals in the eye are produced by radiation (e.g., UV light); atmospheric oxygen; air pollutants; normal metabolism (e.g., mitochondria); and impulse generation.

The FR damage may be from a combination of age-related increases in FRs and ROS in the eye; age-related decreases in FR and ROS defenses in the eye; and accumulation of damage from FRs and ROS due to slow turnover of some eye components. Slow turnover seems especially relevant in the lens and the retina. The damage from FRs and ROS may lead to cataracts, glaucoma, decreasing sensitivity of the retina to light and to color, and age-related maculopathy.

OTHER EYE COMPONENTS

Choroid

The *choroid* lies in contact with Bruch's membrane on the outer surface of the retina (Fig. 7.2). As has been mentioned, it contains many blood vessels that nourish and remove wastes from the photoreceptor layer in the sensory retina.

The choroid also contains a large amount of black pigment that absorbs light that has passed through the retina. Therefore, reflection of the light back through the retina is prevented. If the light were reflected, it would pass through the vitreous humor and strike other parts of the retina, causing foggy or blurry images.

Age Changes With advancing age, the blood vessels in the choroid become irregular, decreasing their ability to service the retina. Other age changes in the choroid include thickening, weakening, declining elasticity, and the formation of irregularities. These changes cause irregularities in the retina and therefore interfere with the focusing of light on the retina. These changes also allow the choroid to be more easily torn as a result of trauma.

Sclera

The outermost layer of the eye is the *sclera*, which is the white part (Fig. 7.2). It extends from the edge of the cornea at the front of the eye around to the optic nerve at the back.

The sclera consists of a tightly woven mat of collagen fibers. Its strength allows it to support eye structures and protect them from trauma. The sclera also serves as a firm attachment point for the external muscles of the eye, which allow a person to turn the eye in its socket.

Age Changes   The structure and functioning of the sclera are virtually unaffected by aging, though it becomes somewhat yellow and develops translucent areas that appear as darkened spots. These color changes may have some cosmetic impact.

External Muscles

Six external muscles attach to each eye (Fig. 7.6). The arrangement of these muscles allows a person to turn the eye in any direction. Turning the eyes serves several purposes. First, it permits the eye to be pointed directly at a stationary object. Second, it allows the eye to keep a moving object centered in the field of view. Maximum visual acuity is achieved when the eye follows the object smoothly. Third, it permits the scanning of a wide field of view.

Rotating the eyes in their sockets also allows both eyes to be aimed at the same object simultaneously so that a person does not have double vision. This goal cannot be achieved by any other means. For example, since the eyes are positioned

for distant viewing when at rest, they must be rotated inward toward the nose (converged) when a person wishes to view a close object. By contrast, they must be rotated outward (diverged) when the person wants to view a distant object again.

To observe divergence and convergence, have a volunteer hold a pencil away from his or her face at arm's length. Then watch the volunteer's eyes as he or she looks first at the pencil, then at a distant wall, and then back at the pencil.

Eye movements produced by the external eye muscles are controlled voluntarily at some times and occur under reflex control at others. For example, a person may voluntarily contract the muscles to move the eye to the right and left. By contrast, convergence and divergence, which are caused by contraction of the same muscles, usually occur reflexively.

Age Changes   The ability of the external eye muscles to turn the eye smoothly declines with age. This may be due to age changes in the muscles or to alterations in the nerve pathways that control the muscles.

The decline in the smoothness of eye movements greatly decreases visual acuity both when

FIGURE 7.6 The eye in its orbit, with external eye muscles and fat tissue.

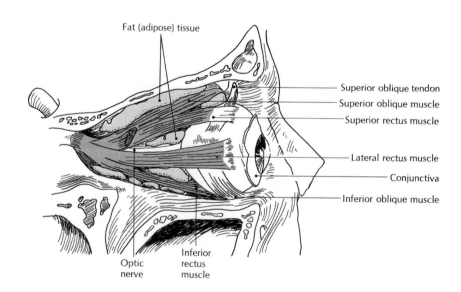

one is viewing moving objects and when one is viewing stationary objects while moving. Decline in vision becomes greater as the speed of movement increases. The reduction in visual acuity caused by this age change results in difficulties in activities such as participating in sports, driving a vehicle, and watching motion pictures.

Adipose Tissue

The space between the eye and the surrounding skull contains much fat tissue (*adipose tissue*). This tissue cushions the eye from trauma, supports the eye in position, and helps maintain its round shape. The skull bones forming the eye socket also protect and support the eye.

Age Changes   The amount of adipose tissue around the eye decreases with aging. This allows the eye to sink deeper into its socket, causing a cosmetic change in the face and a decrease in the size of the field of view. The decrease in fat tissue also reduces the support the eye receives; therefore, the eye may become misshapen. Depending on the nature of the change in eye shape, the position and functioning of several structures, including the cornea, vitreous humor, retina, and choroid may be affected. Any alteration in these structures will adversely affect vision.

Eyelids

The upper and lower eyelids each consist of a fold of skin (Fig. 7.7). Part of the conjunctiva covers the inner surface of each eyelid. A row of hairs is present along the edge of each eyelid.

The front of the eye must be kept moist because it consists of living cells. If the cells on the surface of the eye were to die, as do those on the surface of the skin, light could not be properly focused. The eyelids keep the eye moist by secreting fluid from the conjunctiva, spreading the conjunctival and lacrimal fluids over the eye surface, and closing to prevent evaporation of moisture. The eyelids also protect the eye from traumatic injury and prevent exposure to dust, noxious chemicals, microbes, and excess light.

The functioning of the eyelids depends on their positioning and movement, both of which are determined mostly by the contraction of muscles. Closing the eyelids is accomplished by contracting a ring of facial muscle that surrounds the eye.

Opening the eyelids occurs through contraction of a muscle attached to the upper eyelid. Contraction of other facial muscles can also influence the position of and movement of the eyelids.

Contraction of muscles affecting the eyelids occurs under voluntary control at some times and under reflex control at others. For example, a person may blink any time he or she chooses, but blinking also occurs reflexively. Closing the eyes when falling asleep or sneezing is another example of reflex control of the eyelids.

Age Changes   Age changes in the skin of the eyelids are the same as those that occur in other parts of the facial skin. The discomfort and possible injury resulting from the decrease in secretion by an aging conjunctiva were described earlier in this chapter. Finally, age changes in the muscles of the eyelids are the same as those that occur in other voluntary muscles.

Lacrimal Gland

The upper outer portion of the eye socket contains a *lacrimal gland,* which secretes *lacrimal fluid.* The eyelids spread this fluid over the front of the eye and the fluid moves gradually toward the corner of the eye near the nose. There it drains through an opening into lacrimal ducts and is eventually carried by ducts to the nasal cavity.

FIGURE 7.7 Eyelids and the lacrimal apparatus.

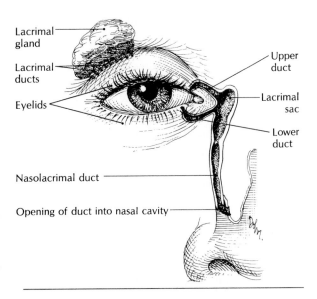

Lacrimal gland

Lacrimal ducts

Eyelids

Upper duct

Lacrimal sac

Lower duct

Nasolacrimal duct

Opening of duct into nasal cavity

Besides moistening the eye, lacrimal fluid washes away irritating objects, chemicals, and microbes. It also prevents infections by killing certain bacteria through the action of a chemical called lysozyme.

Age Changes The production of lacrimal fluid decreases with aging. Many people produce enough regardless of age, but production becomes so low in some individuals that they develop dry and irritated eyes. The use of artificial tear solutions significantly reduces these problems, just as it does when secretion by the conjunctiva is inadequate.

TWO EYES

Though each eye can perform all the functions carried out by the other, there are two advantages to having two eyes. The first is that a person's field of view is wider when viewing with both eyes at the same time. This feature provides more information about the environment and improves a person's ability to respond to the surrounding conditions.

Depth Perception (Binocular Vision)

The second advantage of viewing with two eyes is that a person can perceive the distance between himself or herself and the objects being viewed. This is known as *depth perception* or *binocular vision*. It is very important in activities where a person needs to make contact or avoid making contact with objects in the environment, as occurs when people reach out to grasp something or drive a vehicle.

Depth perception is possible because the eyes are a few inches apart. Because of this separation, the images of objects at different distances from the eyes are seen in slightly different positions relative to each other. This difference can be demonstrated by holding one hand out at arm's length and looking at a distant scene first with one eye and then with the other. It will be noted that the hand blocks out different parts of the scene when the scene is viewed with each eye. During impulse processing, when viewing occurs with both eyes open, the brain interprets the differences in the impulses from the eyes as depth or distance.

Age Changes Aging causes a decrease in depth perception. Some of this decline occurs because aging does not affect both eyes equally. As the quality of images formed in one eye becomes substantially different from the quality of those formed in the other, the brain has increasing difficulty comparing the impulses from the two sets of images. Any alteration in the ability to aim both eyes at the same object has the same effect. The brain may even completely block its interpretation of impulses from one eye and provide vision that is equivalent to seeing with only the other eye. Another reason for decreased depth perception is aging of the retinal and brain neurons that process the impulses originating in the cones and rods.

AGE CHANGES IN VISION

Light Intensity

As people age, brighter lighting must be present if they are to see as much as they did when they were younger. The main reason for this is that the eye allows less of the light striking the conjunctiva to reach the retina. Less light can pass through the eye partly because of the declining transparency of the conjunctiva, cornea, lens, and vitreous humor. The decreasing size of the pupil also blocks the passage of a substantial amount of light. With less light reaching the retina, fewer cones and rods are stimulated enough to send impulses in response, and the photoreceptors that do respond send fewer impulses.

Brighter lighting is also needed to compensate for age changes in the cones and rods. These changes reduce the sensitivity of the retina to light, causing an additional decrease in the number of impulses being produced for a given amount of light. Aging of the processing neurons of the retina may further reduce the number of impulses that leave the eye. Since fewer impulses go to the brain, it is less able to perceive clearly and accurately what is being viewed. Aging of neurons in the visual pathways and processing areas of the brain seems to add to the problem.

The deficit in retinal impulses is probably greatest in the macula region because that region contains cones. Recall that cones need bright light to be stimulated and that they and the processing neurons for the macula are more affected by aging than are the rods and their processing neurons. Therefore, aging causes a decline in the ability to distinguish colors. Since age-related decreases in transparency preferentially affect short wavelengths, the perception of violet, blue, and green decrease the most.

The need for brighter lighting causes vision problems in aging individuals whenever the field of view is lit dimly. This is especially apparent when they move from a brightly lit environment to a dimly lit one or when the lighting consists mostly of light with shorter wavelengths, such as violet and blue light.

Aging individuals also see less well when light intensity changes rapidly, because the pupil adjusts to such changes more slowly. Rapidly interspersed periods of dim light may produce alternating periods of inadequate lighting and excess lighting. Such conditions may be encountered during the day when one is driving through patchy shade (e.g., wooded areas) and at night when one is driving past bright lights (e.g., oncoming traffic, electric lights in roadside signs).

Quality of Light

Vision in old age is also affected by the quality of light. This phenomenon occurs because more of the light entering the eye is scattered. Age changes in the cornea, the vitreous humor, and especially the lens contribute significantly to light scattering.

Since much of the scattered light strikes the retina in the wrong places, the retina produces scattered impulses, which cause glare. Depending on the degree of aging of the eye and the conditions present, the amount of glare may range from being barely a nuisance to obscuring objects completely.

As age increases, external conditions that intensify problems from glare include having brightly lit objects or bright lights against a dark background (e.g., driving at night); having bright light strike the eye at an angle (e.g., multiple bright lights); and viewing light with shorter wavelengths, such as blue light. Sunlight causes glare because it contains much short-wavelength light. Yellow light produces less glare. Aging also slows recovery of good vision after being exposed to glare.

Visual Acuity

Even when light is bright and glare is minimized, there is a decrease in visual acuity with aging. Almost all the decrease for close objects is due to stiffening of the lens, which causes a great reduction in the refractive power of the eye. Other age changes that contribute to declining refractive power include flattening of the cornea and lens and the diminishing ability of the ciliary body to move. Close visual acuity is decreased to a lesser degree by reductions in the sensitivity to contrast caused by changes in the neuron pathways. Since most of the decline in close visual acuity is caused by a decrease in refractive power, the use of eyeglasses that increase refractive power can restore much of the lost acuity.

Recall that the decline in near visual acuity caused by aging is usually not noticed until about age 40. Most of the deterioration of close visual acuity is completed by age 60 because by then the lens has become completely stiff. However, since the neuronal changes and flattening of the cornea and lens continue for many years, close visual acuity may continue to decline gradually after age 60. By contrast, in some individuals the continued thickening of the lens and the resulting increase in its refractive power may restore some close visual acuity that has been lost.

Visual acuity for distant vision also declines with aging. However, distant visual acuity does not begin to decline until about age 45, after which the decline is usually steady but slow. The deterioration of distant visual acuity is so slow that more than 50 percent of those who reach age 80 have fairly good distant vision as indicated by scores of 20/40 or better on a Snellen eye chart.

Most of the decline in distant visual acuity is caused by age changes that reduce the amount of light that passes through the eye. Since only a small proportion of the decline is caused by focusing problems, eyeglasses and contact lenses can compensate for only a limited amount of the loss. Thus, unlike near visual acuity, distant visual acuity declines steadily even if corrective lenses are used.

Declining close and distant visual acuity is noticed whether a person is observing stationary or moving objects. However, the decline for moving objects is much greater than that for stationary ones because viewing moving objects involves additional processes that also undergo age changes.

One of these processes is smooth movement of the eyes to keep the object being viewed centered in the field of vision. Recall that with aging, the movement of the eyes becomes less smooth, causing the image of a moving object to move irregularly on the retina. The brain has difficulty interpreting the increasing complexity of the resulting retinal impulses.

Another relevant process is the rapid recovery of each neuron every time it is stimulated. This is needed because moving objects produce rapid changes in the light striking the photoreceptors. Since aging causes slowing of the neurons involved in vision, neuron recovery lags behind changes in the moving images. The result is a lengthening of afterimages.

An *afterimage* is a lingering perception that an image is present even though the image in the eye has changed or disappeared. For example, a light that is flickering rapidly may be perceived as emitting a steady light because the individual images blend into one continuous image. Another effect of long afterimages is the apparent blurring of an object as it moves faster. This effect is easily observed by watching a fan as it begins to spin.

As aging lengthens afterimages, blurring of moving objects increases and therefore visual acuity for moving objects decreases. Detecting small movements in nearly stationary objects also becomes more difficult. These results adversely affect a person's performance in walking and driving, recreational and occupational activities, and even spectator activities.

Depth Perception (Binocular Vision)

A person's eyes usually do not undergo age changes at the same rate. As the eyes become more different from each other, they send increasingly different impulses to the brain. These differences, together with the declining ability of the brain to interpret them, result in a decrease in binocular vision, as was discussed previously.

Field of View

The width of the field of view (i.e., peripheral vision) decreases gradually from 170 degrees to 140 degrees. Also, there is an age-related decline in the ability to notice a specific object in a field of view. This latter change is more evident in the periphery of the field of view and with increasing complexity of the items in the field of view.

Optimizing Vision

Though age changes that affect vision cannot be prevented, the limitations and disabilities they cause can be minimized by (1) providing adequate lighting, (2) reducing sources of glare, (3) enlarging items, (4) increasing contrast, (5) positioning objects at greater distances from the eyes, (6) slowing the motion of moving objects, and (7) using eyeglasses or contact lenses. These steps can help prevent accidents and injuries while preserving much of an aging person's activity, independence, and quality of life.

DISEASES OF THE EYES

In many people, one or more diseases compound the adverse effects of aging of the eyes. The incidence of such diseases increases with age, so that the risk for those over age 75 is 2.5 times greater than the risk for those between ages 50 and 65. Some of these diseases are little more than a nuisance, while others drastically reduce the quality of vision and in severe cases cause blindness.

Minor Diseases

Lacrimal Fluid The production of lacrimal fluid may decline so far that the eyes become dry, causing irritation. The application of wetting solutions can largely relieve this problem. Other individuals experience what seems to be an excessive production of lacrimal fluid because the fluid pours out of the eye, forming tears.

In most cases the cause of the tearing is not excess production of fluid but some factor that prevents the fluid from draining properly into the nasal cavity. One common cause of such tearing is excessive weakening of the circle of muscle surrounding the eyelids. When it is weak, the muscle allows lacrimal fluid to accumulate behind the eyelid until it overflows onto the face. Tearing may also be caused by infection or other factors that block the lacrimal ducts that lead from the eyes to the nasal cavity.

Eye Muscles Weakening of the eye muscles may cause problems other than tearing. For example, inadequate contraction of the circular muscle surrounding the eyelids may cause drying of the eye if the eyelids are not closed completely during sleep. Also, weakening of the muscle that raises the upper eyelid causes drooping of the lid. Besides producing the appearance of drowsiness, a drooping lid may obstruct part of the field of view.

The circular eye muscle sometimes contracts

excessively, causing the lower eyelids to turn inward. The edge of the lid and its eyelashes then scrape on the eye, producing irritation. If this condition is severe and is not corrected, the scar tissue that forms can drastically reduce vision by blocking and scattering light.

Cataracts

Cataract formation is the most common serious eye disease among the older population. However, its impact in the United States has been lowered considerably by the high rate of success in treating this disease surgically. Cataract surgery is the most common surgical procedure in the U.S. There are one million surgeries per year, accounting for 12% of annual Medicare expenses.

Cataracts are the most common age-related eye disease and the main cause of blindness in "third world" locations.

Recall that aging causes the formation of opacities in the lenses. Free radicals and glycation are main contributing factors. At first opacities usually develop toward the periphery of the lens, where they scatter light and cause glare, but eventually they form toward the center of the lens. Central opacities cause glare and decrease visual acuity because they scatter and block light. An affected individual is said to have cataracts when central opacities reduce visual acuity substantially.

Main risk factors for cataract formation in descending order of importance are increasing age; exposure to UV-B light; and topical or internal corticosteroids. Other risk factors include being female; having diabetes mellitus; smoking; family history; low socioeconomic status; malnutrition or low levels of AOXs (e.g., vit A, vit C, vit E, carotenoids); dehydration; and eye trauma or internal eye inflammation. *Diabetes mellitus* is a disease involving hormone imbalances and abnormally high levels of a sugar called *glucose* in the blood. Opacities in the lenses of diabetics seem to be caused by the conversion of glucose to another sugar, *sorbitol*, which accumulates in the lenses. Estrogen supplements in postmenopausal women seem to reduce their risk of developing cataracts.

When cataracts are treated by surgical removal of the affected lens, the ability of the eye to refract light is restored by implanting an artificial lens or by wearing eyeglasses or contact lenses. However, not all individuals with cataracts can

or should undergo eye surgery.

Age-Related Macular Degeneration

The second most common serious eye disease among the elderly is *age-related macular degeneration (AMD)*, also called *senile macular degeneration*. It accounts for about 25 percent of visual loss among those under age 80 and about 40 percent among those over age 80. As the name indicates, this disease causes deterioration of the retina. Though the cause of AMD is not known, factors that increase the likelihood of its occurrence include being of advanced age, having high blood pressure, having family members with the disease, and having atherosclerosis.

The mechanism by which AMD causes retinal damage is not clear, but it seems to involve FR damage and formation of lipid peroxides (LPs). It seems that age changes in the pigmented epithelium and Bruch's membrane begin to occur in excess near the edge of the macula. These changes substantially reduce the passage of nutrients from the choroid to the sensory retina, leading to degeneration of the cones. There is no treatment for the disease at this stage.

If the disease progresses no further, there will be decreased visual acuity in the center of the field of view. However, in about 10 percent of cases blood leaks from choroid vessels and passes between the pigmented epithelium and the macular region of the sensory retina. Phagocytic cells from the choroid also invade the retina. These changes cause more severe degeneration, and about 90 percent of these advanced cases result in macular blindness.

The onset of AMD is indicated by gradual deterioration of vision in the central region of the field of view. Sudden rapid distortion of vision in this area is a warning sign that bleeding and rapid macular degeneration are occurring. Professional help should be sought at once. Treatment with lasers can slow or stop further vessel damage and bleeding in about 50 percent of advanced cases. If treatment is not obtained or is unsuccessful, macular blindness is likely.

Since AMD affects only the macula, peripheral vision is not appreciably altered. When central vision is affected, activities such as driving and reading become difficult or impossible.

Glaucoma

Glaucoma is the third leading serious eye disease among older people. It causes diminished vision because the high pressure that develops from accumulation of aqueous humor inside the eye damages the retina and optic nerve. The pressure inside the eye is called *intraocular pressure (IOP)*. About 9 percent of people over age 65 have one of the three types of glaucoma.

Open-Angle Glaucoma   About 80 percent of all patients with glaucoma have *open-angle glaucoma*. Its risk factors include being of increasing age, having relatives with glaucoma, being black, and being male.

The cause of and the mechanism producing high IOP are unknown. Pressure in both eyes is affected. When the pressure remains somewhat high for extended periods, it slowly damages the retina and optic nerve. The damage may be noticed by the affected individuals as a gradual shrinkage in the width of the field of view. Unfortunately, this narrowing may go undetected until permanent injury has occurred. If left unchecked, open-angle glaucoma leads to total blindness.

Fairly simple procedures that are normally part of a complete eye examination can detect the presence of open-angle glaucoma long before significant eye damage has occurred. These procedures include measurement of IOP and visual examination of the optic nerve using an *ophthalmoscope*. The combination of the two tests identifies about 80 percent of all affected people.

Though there is no cure for open-angle glaucoma, the IOP can be controlled and the effects of the disease can be prevented with appropriate medications.

Angle-Closure Glaucoma   *Angle-closure glaucoma* or *narrow-angle glaucoma* gets its name from the accompanying abnormally narrow space between the lens and the cornea. It accounts for about 10 percent of all cases of glaucoma.

With angle-closure glaucoma, the IOP may rise quite high within a matter of minutes or hours. Affected individuals often experience eye pain, headache, nausea and vomiting, halos around lights, and blurred vision. If this is not treated within 2 to 3 days, permanent eye damage and blindness can result. Treatment usually involves surgery that reestablishes normal drainage of the aqueous humor.

Secondary Glaucoma   The remaining 10 percent of glaucoma cases involve *secondary glaucoma*. This condition is characterized by elevated IOP resulting from another disorder such as diabetes, tumor, or disease of the vessels in the eye. Each type of secondary glaucoma is treated according to its specific cause.

Diabetic Retinopathy

Diabetic retinopathy (DR) means "disease of the retina associated with diabetes." The form of diabetes involved is diabetes mellitus. While DR is not nearly as common as cataracts, AMD, or glaucoma, it is the third leading cause of blindness in the United States and the most prevalent cause of blindness among those with diabetes mellitus. The incidence of DR among diabetics increases with the length of time since the onset of diabetes. For example, only 7 percent of those who have had diabetes for less than 10 years have developed DR, while more than 62 percent of those who have had it for more than 15 years have developed DR.

Diabetic retinopathy seems to develop because some of the extra glucose in the blood is converted to sorbitol, which weakens retinal capillaries so much that they are overly dilated by normal blood pressure. Some weak spots in the capillaries bulge outward as aneurysms that leak and bleed. Injured vessels and hemorrhaged blood can be seen when the inside of the eye is examined with an ophthalmoscope. Fluids and blood from the leaking dilated capillaries injure the retina.

As more blood flows through the dilated capillaries, unaffected capillaries receive less blood flow and eventually shrink and close. The areas of the retina that were served by these shrunken capillaries are injured because they receive inadequate blood flow. Such ischemic areas can be seen with an ophthalmoscope. As the number of abnormally dilated and shrunken retinal vessels increases, more of the retina is adversely affected and vision deteriorates.

In more advanced cases of DR, retinal vessels near the optic nerve grow into the vitreous humor. Since these proliferating vessels are weak, they rupture easily, especially when the vitreous humor moves. Hemorrhaging near or into the vitreous humor often leads to blindness from secondary glaucoma or from detachment of the retina from the choroid.

The best treatment for diabetic retinopathy is to maintain blood sugar levels within the normal range. Once vessels become dilated, the progressive destruction of the retina can be slowed by destroying the dilated vessels with laser surgery.

THE EARS

Much of what was stated about the eyes at the beginning of this chapter is also true of the ears. The ears help monitor conditions and provide information not only about the body, but also about conditions far from the body. Their functioning helps preserve homeostasis, contributes to learning, aids in communication, and provides enjoyment.

The ears accomplish all this by detecting four different types of stimuli. One type consists of vibrations within a certain range of frequencies, which are called sound vibrations. *Hearing* is the conscious perception of sound vibrations. The second type is the pull of *gravity*. A person's awareness of gravity is perceived as a sense of the position of the head and results in the person's knowing whether the head is right side up, upside down, or tilted. The third type of stimulus is *change in the speed of motion of the head*. For example, a person feels the speed of the head change when a vehicle in which he or she is riding speeds up or slows down. The fourth type of stimulus is *rotation* of the head. For example, a person in a vehicle senses when the vehicle makes a turn even if the speed remains the same.

Aging adversely affects the ability of the ears to detect all four types of stimuli. The perception of these stimuli is further hampered by age changes in the CNS that interfere with the processing of impulses sent by the ears. Beginning with the sense of hearing, we will examine how aging alters the monitoring of each type of stimulus and see the impact these alterations have on people.

HEARING

External Ear

Most sound vibrations that are heard travel in air or, occasionally, water before reaching the ear. The visible part of the ear, the *pinna* or *auricle*, collects the vibrations and directs them into the *ear canal*, which is about 2.5 cm (1 inch) long. These two components make up the *external ear* (Fig. 7.8).

Air in the ear canal carries vibrations to a thin flexible membrane covering the inner end of the ear canal. This membrane is called the *eardrum* or *tympanic membrane* because it resembles the membrane on a drum. The eardrum marks the beginning of the *middle ear.* Air vibrations cause the eardrum to vibrate.

The lining of the outer two-thirds of the ear canal contains modified apocrine sweat glands called *ceruminous glands*. These glands secrete a semisolid waxy material called *earwax* or *cerumen*. Hairs are also located in the outer part of the ear canal.

The sticky cerumen and the hairs trap small particulate matter and insects, preventing such items from reaching the eardrum and injuring it or interfering with its vibrations. The cerumen also keeps the eardrum pliable so that it can vibrate without cracking. Cerumen slowly moves to the outer opening of the ear canal, where it is easily removed by wiping or washing.

Age Changes   Aging of the external ear has some cosmetic impact. The pinna becomes thicker, longer, broader, and stiffer. Hairs on the pinna and within the ear canal become more visible because they thicken and lengthen. Aging of the skin on the pinna is similar to aging of other parts of the facial skin.

Each ceruminous gland produces cerumen at the same rate regardless of age. However, the overall rate of production decreases because the number of ceruminous glands slowly decreases. The cerumen may also become thicker in consistency. Therefore, it takes longer to move to the end of the ear canal and becomes even firmer. Age-related reductions in skin elasticity and adipose tissue cause the ear canal to sag. Therefore, cerumen tends to accumulate within the ear canal. The problem is increased when attempts to remove the cerumen with cotton swabs or other objects push it deeper into the canal.

A buildup of cerumen in the ear canal inhibits the passage of vibrations to the eardrum and thus diminishes a person's ability to hear. Removal of accumulated cerumen, which should be done by properly trained individuals, restores normal functioning of the ear canal. Except for increasing the likelihood of cerumen retention, aging of the external ear has no effect on hearing.

FIGURE 7.8 Structure of the ear.

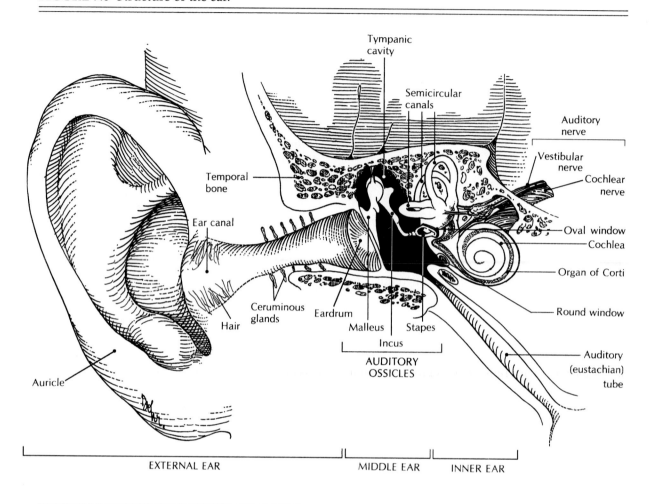

Middle Ear

Sound vibrations causing the eardrum to vibrate are passed to a small bone attached to the inner surface of the eardrum, the *hammer* (*malleus*). The vibrations then pass in turn through two other bones, the *anvil* (*incus*) and the *stirrup* (*stapes*). The stirrup passes the vibrations to another thin flexible membrane, the *oval window*, which marks the end of the middle ear and the beginning of the *inner ear*.

The bones of the middle ear provide a system of levers that amplify the sound vibrations passing through them. Amplification is adjusted by altering contraction of the small muscles attached to the eardrum and stirrup. These muscles are controlled reflexively.

The space surrounding the three bones of the middle ear is filled with air so that the bones can vibrate easily. To prevent bulging of the eardrum, the air pressure within the middle ear must be adjusted so that it always equals the air pressure in the ear canal. This is done by allowing air to pass between the middle ear and the nasal cavity through the *eustachian tube*.

Age Changes   As age increases, the eardrum becomes slightly stiffer and the joints between the hammer and anvil and the anvil and stirrup become calcified and deteriorated. However, these age changes do not have a significant effect on hearing. Age changes in the muscles and the eustachian tube are also of no consequence in hearing.

Inner Ear

The vibrations of the oval window are passed to a liquid called *perilymph* in the inner ear (Fig. 7.9). The vibrations travel through the perilymph in the part of the inner ear that detects sound vibrations. This section is called the *cochlea* because it has a spiral shape like that of a cockle or snail shell.

Vibrations in the perilymph pass through a flexible membrane (vestibular membrane) within the cochlea and enter another liquid called *endolymph*. The vibrating endolymph causes the vibration of another flexible membrane within the cochlea, the *basilar membrane*. This membrane protrudes inward from the wall of the spiraling cochlea much as a spiral staircase protrudes from the inner wall of a building.

Sound vibrations with the highest frequency or pitch cause vibration of the beginning region of the basilar membrane. This region is comparable to the bottom steps of a spiral staircase. Sound vibrations of lower frequency or pitch cause more distant regions of the basilar membrane to vibrate. The lower the frequency of the vibrations, the farther along (higher on the spiral staircase) the membrane vibrates.

The basilar membrane bristles with rows of neurons, called *hair cells*, which are sensitive to vibrations. The rows of neurons make up the *organ of Corti*. Vibration of the basilar membrane agitates the hair cells of this organ, causing them to initiate impulses.

All the components of the inner ear may also be set into motion by vibrations reaching them through the skull bones. Scratching one's head, chewing on crunchy food, and clicking one's teeth together are a few ways to produce skull bone vibrations. These vibrations also initiate impulses.

The impulses from the hair cells are passed to other neurons in the ear that carry them to the brain. Auditory centers in the brain process and interpret the impulses, and the result is hearing.

Age Changes　Several changes are known to occur in the inner ear as people get older, but a variety of other factors are also involved in producing these changes. These factors include the amount of fat and cholesterol in the diet, genetic factors, noise, and atherosclerosis. Identifying which factors besides aging cause each of the following observed changes is not yet possible, but these changes will be referred to as age changes here because they seem to occur to some extent in all people.

As age increases, there is shrinkage of the mass of small blood vessels servicing the cochlea and producing endolymph. The resulting decline in nourishment may be partly responsible for changes in the organ of Corti. The reduction in endolymph production diminishes the passing of vibrations through the cochlea, resulting in a decreasing ability to hear all frequencies of sound.

There are decreases in the numbers of several types of cells, including the hair cells in the organ of Corti, the cells that support that organ, and the neurons that carry impulses to the brain. The organ of Corti becomes flattened and distorted, most often at the beginning of the basilar membrane. The net result is an exponential decline in the ability to hear; since most age changes occur at the beginning of the basilar membrane, hearing loss of high-frequency sounds is usually greatest.

Localizing Sound

While each ear can provide the same information about sound, the use of both ears at the same time allows a person to detect an additional feature: the direction of the source of the sound.

Since the ears are on opposite sides of the head, sounds originating closer to one side of a person reach the ear on that side with greater intensity and the brain receives more impulses from that ear. By comparing and interpreting the differences in impulses, the brain gives the person a sense of the direction from which the sound originated. This process is called *localization of sound*.

When the source of a sound is in line with the center of the body, each ear receives the sound with equal intensity. The brain often has difficulty localizing such sounds because the impulses from both ears are similar.

Age Changes　Aging does not affect a person's ears equally, and the ability to hear with one ear may decline much more than does the hearing ability of the other ear. The brain will receive fewer impulses from the cochlea of the more affected ear regardless of the source of the sound. Since the localization of sound depends on comparing the

FIGURE 7.9 Pathway of sound vibrations and structure of the cochlea.

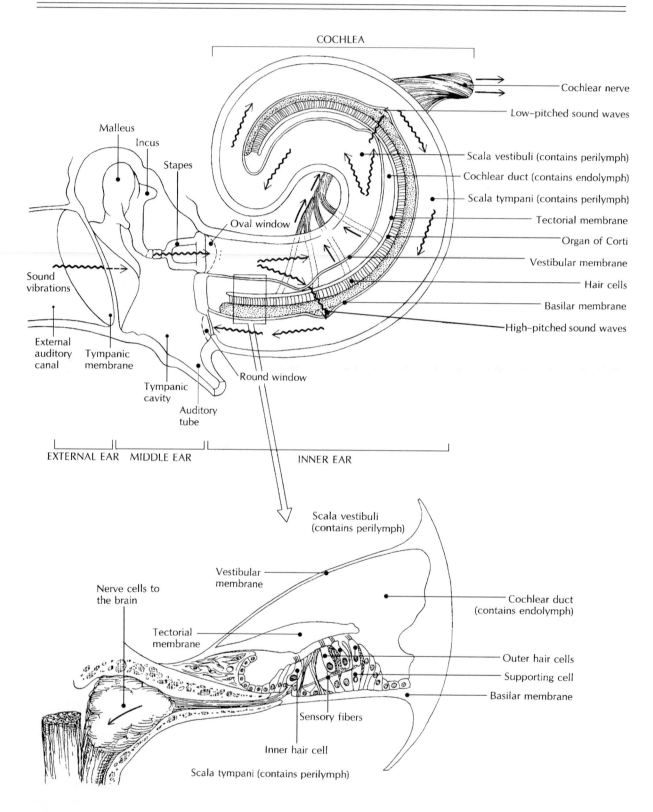

differences in impulses from the ears, greater hearing loss from one ear causes errors that can lead to disorientation.

Central Nervous System

Age changes in the central nervous system cause further hearing impairment because the ability of the brain to process and interpret impulses from the cochlea is adversely affected. These effects are noticed as increased difficulty understanding sounds that contain echoes or background noise, sounds that change quickly, and speech that is broken up or has syllables or words missing.

DISORDERS IN HEARING

Presbycusis

Aging causes different degrees of hearing loss. Individuals who have lost a great deal of hearing because of aging are said to have *presbycusis*, which is the third leading chronic condition among those over age 65. Only arthritis and high blood pressure occur more frequently among the elderly.

The incidence of presbycusis increases exponentially with age. About 12 percent of those between ages 45 and 65 are affected. This increases to more than 24 percent for those age 65 to 74 and may reach 39 percent for those over age 75. The incidence among those in institutions is considerably higher, reaching 70 percent.

Both the percentage of cases and the seriousness of presbycusis are greater for men than for women. It is thought that greater hearing loss among men may be caused by noise associated with occupations traditionally held primarily by men. Presbycusis becomes progressively worse as the age of an affected person increases. This condition tends to run in families.

Presbycusis seems to result when the previously mentioned age changes in the cochlea and in neurons and brain areas involved in hearing occur to an unusually severe degree. Though some of these changes cause hearing loss of all sound frequencies, most cause hearing loss predominantly of higher-frequency sounds.

Effects   Presbycusis reduces the ability of the ears to alert a person to desirable factors and harmful factors in his or her surroundings, hinders learning and communication, and reduces the enjoyment sound provides. Since presbycusis affects high-frequency sounds most, it substantially reduces the ability to hear the quality of music called brilliance. Presbycusis makes understanding speech difficult primarily because many consonant sounds in words are high-frequency sounds. High-pitched voices become especially difficult to understand.

The increasing difficulty in understanding speech has diverse effects on people with presbycusis. For example, these people may fail to respond or respond incorrectly when others speak to them. Other people may begin to believe that the person is becoming demented. Both the people with this condition and others may feel that they are being ignored. Individuals with presbycusis sometimes believe that others are trying to deceive them or talk about them in demeaning ways and often tend to withdraw from social contacts. Depression and paranoia are not uncommon outcomes.

Presbycusis may also alter speaking ability because affected individuals can no longer hear their own voices well. People with presbycusis may become annoying because they sometimes talk very loudly.

Prevention and Compensatory Techniques
While presbycusis is at least partially due to aging, it seems that chronic exposure to loud sounds increases both the likelihood of developing it and the severity of the problems it causes. Therefore, avoiding exposure to loud sounds, such as loud music and noise from tools and machinery, may help prevent presbycusis. When loud sounds are unavoidable, earplugs or other protective devices should be used.

Though there is no cure for presbycusis, people can be helped to compensate for it. Hearing can be increased by raising the volume of sound, such as by talking louder, or by using amplifying hearing aids.

The probability that a particular person with hearing loss will benefit from the use of a hearing aid depends on several factors. These include the type of presbycusis or other condition (e.g., eardrum injury, middle ear infection) that produced the hearing loss, the nature and severity of the hearing loss, and the ability of the person to use a hearing aid. The evaluation for hearing aid use is best done by a qualified audiologist.

Reducing background noise, echoes, and the speed of speaking can also improve the understanding of what is heard. Individuals with presbycusis can be trained to do lipreading and use other visual cues to increase their ability to understand those who are speaking. Speakers can help by getting the listener's attention; facing the listener; using more gestures and facial expressions; speaking slowly and clearly; repeating; keeping their mouths visible; and asking for feedback to confirm the listener's understanding.

Tinnitus

Tinnitus is the perception of sound when there is no sound external to the person. Sometimes the perceived sound originates from within the body; in other cases there is no sound whatever. Until 1990, tinnitus was among the ten most common chronic conditions among the elderly. It has now been displaced to a lower rank because of the higher reported frequencies of visual impairment and varicose veins.

Tinnitus can result from many causes including obstruction of the ear canal; abnormality of any of the parts of the middle ear, inner ear, or eustachian tube; infection in the ear or CNS; tumor in the ear or CNS; high blood pressure; atherosclerosis; diabetes mellitus; hormone imbalances; malnutrition; migraine headaches; medications; and toxic chemicals.

In many cases of tinnitus the sound being perceived is little more than a nuisance, although sometimes it becomes quite distracting. Tinnitus may alter or prevent normal sleep.

Eliminating tinnitus requires elimination of the cause. Individuals who cannot be cured can sometimes be helped by using low levels of other sounds as a distraction or cover for the tinnitus. Other treatments include dietary modifications (e.g., vitamin supplements, reduce caffeine), and surgery.

DETECTING OTHER STIMULI

Gravity and Changes in Speed

Recall that the second and third types of stimuli detected by the ear are gravity and changes in speed. Two chambers in the inner ear are specialized to detect these stimuli. The *saccule* is con-

nected to the beginning of the cochlea (Fig. 7.10). The *utricle* is connected to the saccule. Both chambers are filled with endolymph.

One region of each chamber contains a patch of nerve cells protruding into the endolymph. Each patch is called a *macula*, the same name used for a patch of cones in the retina. Attached to the ends of the nerve cells is a gel containing heavy crystals called *otoliths*.

When gravity pulls on the otoliths, it causes them and the gel to shift. This causes a binding of the nerve cells to which the otoliths and gel are attached. Speeding up or slowing down of the head also causes a shifting of the otoliths and gel, resulting in bending of the neurons. Changes in speed produce this effect because the otoliths tend to keep moving at the same speed because of inertia.

The effects produced by gravity and inertia on the otoliths and nerve cells can be observed if a weight is attached to the end of a flexible rod such as a fishing pole. When the rod is pointed straight up, it will be fairly straight. If the rod is tilted or moved from side to side, gravity or inertia will cause it to bend.

Bending the nerve cells in the macula causes them to initiate impulses. Since the macula in the saccule is at a different angle from the macula in

FIGURE 7.10 The inner ear.

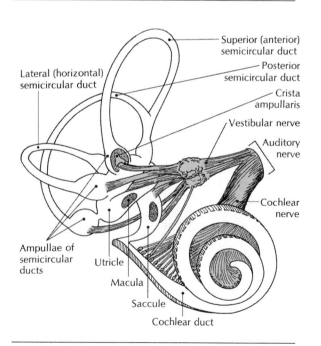

the utricle, some neurons will always be bent. Depending on the angle of the head or its direction of acceleration or deceleration, the shifting and bending occur in different directions, causing different impulses to be sent.

The impulses are passed to neurons that carry them to the brain, where they are processed and interpreted. The result is the perception of head position relative to gravity or the sensation of speeding up or slowing down. The person can then adjust his or her position or motion voluntarily to suit the situation.

Some ear impulses are sent to the cerebellum, causing reflexive contraction of muscles automatically to maintain posture and balance. Maintaining posture and balance allows a person to interact with the environment effectively and helps prevent falling.

Rotation

Head rotation is monitored by three curved tubes—the *semicircular canals*—which are connected to the utricle. These canals are filled with endolymph. One end of each canal is enlarged, forming an ampulla (Fig. 7.11). Each *ampulla* contains a patch of inwardly protruding neurons called the *crista ampullaris* (Fig. 7.11).

Each of the semicircular canals is positioned at right angles to the others. Therefore, rotation of the head in any direction in three-dimensional space will cause one or more of the semicircular canals to shift relative to the endolymph, causing an apparent swirling of the endolymph. The shifting and apparent swirling bend the neurons in the crista ampullaris. This effect on the neurons is like the bending of aquatic plants in a current or of tall trees or grass in a breeze.

Bending the neurons causes them to initiate impulses. These impulses are passed to other ear neurons that send them to the brain, where they are processed and interpreted. The result is the perception of turning or rotation. The person can then adjust his or her movements voluntarily. Some impulses are sent to the cerebellum to initiate reflexive muscle contractions that maintain balance.

Age Changes (Gravity, Changes in Speed, and Rotation)

With aging, there is a decrease in the number of

FIGURE 7.11 A semicircular canal and the detection of rotation.

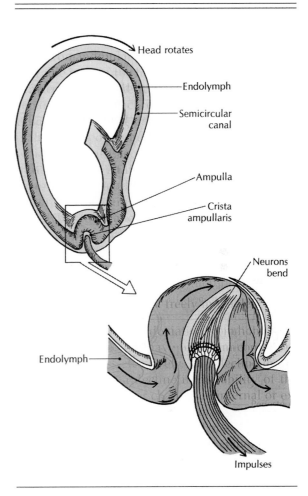

sensory cells in the saccule, the utricle, and the three ampullae. The number of neurons that carry impulses from the ear to the brain also decreases. The cells supporting the sensory neurons show degeneration. Aging also results in a steady decrease in the size and number of otoliths.

The combined effects of all these changes cause a decreased sensitivity of the ear to gravity, changes in speed, and rotation of the head because fewer impulses are sent to the brain. Also, age changes reduce the ability of the brain to process and interpret the impulses and initiate voluntary and reflexive responses.

Properly responding to gravity and changes in speed is hampered further because the saccule loses more otoliths than does the utricle. The brain may be confused when the saccule sends an unusually low number of impulses relative to the utricle.

Dizziness and Vertigo

A severely diminished ability to detect and respond to gravity, changes in speed, and rotation of the head produces two types of sensations. One is *dizziness*, which is the sensation of instability. Affected individuals feel that they are unable to maintain their posture, body position, or balance. The other is *vertigo*, the feeling that either the body or the surrounding environment is spinning when no spinning is actually occurring.

Vertigo is usually experienced under one of four conditions: holding the head stationary, rotating the head, changing head position relative to gravity (e.g., sitting up from a recumbent position), and walking. Some relief can be obtained by avoiding conditions that initiate or amplify vertigo and by moving more slowly.

Both dizziness and vertigo caused by age changes in the ears are unpleasant. They can also be dangerous because they increase the risk of losing one's balance and falling.

Several other factors increase the incidence and seriousness of injuries from falling as age increases. Aging reduces the information provided by the eyes, the skin receptors, and the proprioceptors in muscles and joints. This information normally helps the ears in maintaining posture and balance. Voluntary and reflexive movements used to stop or slow a fall are reduced because of age changes in the nervous system, the muscles, and the skeletal system. Many disorders and a variety of medications further hamper the functioning of all these organs and systems. Because so many factors are involved in promoting loss of balance and falling, diagnosis of the causes of individual cases of falling is complicated and requires a broad perspective on the body.

Another factor that makes falls more serious for the elderly is thinning of the subcutaneous fat, leading to less cushioning of the body. Furthermore, age-related weakening of the skin, bones, blood vessels, and other structures make them more susceptible to injury from a fall. Aging also lengthens the time needed for recovery from an injury.

The risks from loss of balance and falls can be reduced by providing handrails, better lighting, stable floor surfaces, and uncluttered walking areas.

8

MUSCLE SYSTEM

There are three types of muscle in the human body. The most abundant type is called *skeletal muscle* because virtually all these muscles are attached to the bones of the skeletal system. Skeletal muscle makes up the more than 600 muscles in the body, most of which are close to the surface of the body, between the integumentary system and the bones (Fig. 8.1). Many muscles bulge when they contract; therefore, they are visible and can be felt as firm lumps under the skin. This chapter is concerned mainly with skeletal muscle.

Cardiac muscle is found exclusively in the heart (Chap. 4). *Visceral muscle*, or *smooth muscle*, is found within organs in many body systems.

MAIN FUNCTIONS FOR HOMEOSTASIS

The muscle system performs three functions that help maintain homeostasis: movement, support, and heat production.

Movement

The movement produced by muscles allows a person to carry out the last step in negative feedback systems: making an adjustment to a change in conditions. Movement is used to get away from impending danger (e.g., fire, falling objects), escape from unfavorable conditions (e.g., intense sunlight), and eliminate wastes and unwanted materials (e.g., carbon dioxide, splinters).

Movement is also important in taking positive actions. It allows a person to move toward, obtain, and use items and conditions that promote the welfare of the body and quality of life. These needs include basic physical needs (e.g., food, water, shelter) and other needs (e.g., social interactions, recreational activities). Movement allows people to rearrange their environment and construct and repair useful and decorative artifacts to suit human requirements and desires.

Support

The muscle system provides support when muscle contractions prevent the movement of a part of the body. Support maintains proper positional conditions of parts of the body so that they function well. For example, muscle contractions can maintain an upright posture. This activity includes holding the bones in place and

FIGURE 8.1 Skeletal muscles in the muscle system.

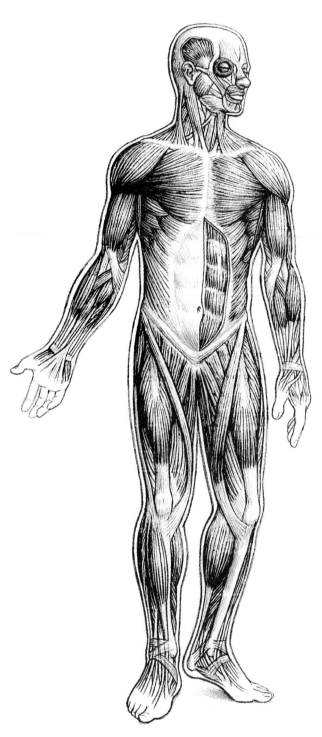

preventing the protrusion of the organs in the lower trunk. With proper posture, circulation is improved because blood vessels are open rather than pinched shut, and respiration is assisted because the lungs have room to inflate easily. Holding the head up positions the eyes for viewing the surrounding environment.

Heat Production

Heat production is essential for maintaining a proper and fairly stable body temperature because most people live in environments that are cooler than normal body temperature. Therefore, the body is always losing heat to the environment, just as any warm object or substance, such as warm food, loses heat and becomes cool. However, if the body is allowed to become cool, this will make the rate of its chemical reactions too slow to sustain life functions (e.g., heartbeat, respiration, brain activities) and perform effective negative feedback responses. Therefore, to prevent cooling of the body, the amount of heat loss must be balanced by an equal amount of heat production.

Heat is produced by many chemical reactions in the body, but the muscle system is the main heat producer. One reason for this is that the muscle system is one of the largest systems, usually accounting for one-third or more of body mass. Second, the muscle system is one of the most active systems in the body. When a person is awake but resting, this activity involves steady muscle contractions (muscle tone) that help maintain posture. This system is especially active and produces much more heat when a person is forcefully contracting muscles during vigorous exercise. Still, the muscle system performs many chemical reactions even when the muscles are relaxed; this is why a sleeping person remains warm.

AGE CHANGES VERSUS OTHER CHANGES

The functioning of the muscle system depends on the nervous, circulatory, and respiratory systems. As a result, many age-related changes in this system derive from age-related changes in the other systems, which vary greatly among individuals. Furthermore, alterations in exercise received by muscles quickly and dramatically affect the muscle system. Though there is an average age-related decline in exercise, changes in the total amount of exercise of the body and of each muscle vary greatly both from time to time and from one person to another. Two consequences of these variables are that they add to the age-related increase in heterogeneity among people and make it quite difficult to identify true age changes in the muscle system. Therefore, in this chapter the causes of each age-related change will be noted.

MUSCLE CELLS

The muscles are composed primarily of muscle cells, which perform the three functions of the muscle system. Other materials in each muscle include nerve cells, collagen and elastin fibers, fat, and blood vessels (Fig. 8.2).

Structure and Functioning

The main activity of muscle cells is *contraction*, which produces both the force needed for movement and support and most of the heat derived from the muscle system. Muscle cells have many specializations that permit them to perform contraction.

Muscle cells are very long and thin, reaching lengths of up to several centimeters. Usually, these cells are as long as the muscle in which they are contained. Because of their shape, muscle cells are also called *muscle fibers* (Fig. 8.2).

Cell Membrane (Sarcolemma)   The muscle cell membrane (*sarcolemma*) is modified in three ways (Fig. 8.3). First, the spot on the membrane that receives stimulatory messages from a somatic motor neuron is highly convoluted. This modified area (*motor end plate*) apparently provides more surface area and receptor molecules to receive and respond to molecules of acetylcholine from the somatic motor neuron. Second, the cell membrane can carry messages in the form of *action potentials*, just as axons do. Third, the membrane has many penetrating indentations (*T tubules*), which deliver action potentials deep within the cell.

Myoglobin, Oxygen, and Energy   The muscle cell cytoplasm (*sarcoplasm*) contains a protein called *myoglobin*, which is found only in muscle cells and causes muscles to appear red in color.

FIGURE 8.2 Components of a muscle.

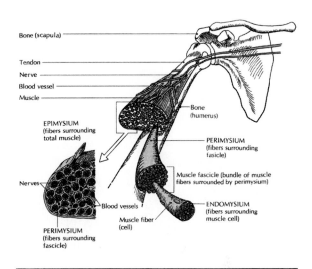

Myoglobin attracts oxygen from the blood into the muscle cells and stores oxygen. As soon as a muscle cell uses some of the oxygen, its myoglobin quickly attracts more (Fig. 8.4).

The muscle cell uses the oxygen to obtain energy from sugar and other nutrient molecules. As long as the cell has enough oxygen, it can obtain much energy from nutrients while producing only carbon dioxide (CO_2) and water as waste products. The CO_2 and water are easily removed from the cell and can be eliminated from the body by the respiratory and urinary systems, respectively.

When a person engages in vigorous activity, the amount of oxygen required to produce the energy needed by a muscle cell often rises above the supply of oxygen to the cell. The cell can continue to work because some energy can be obtained by breaking nutrients down partially. One of the main waste products from this process is *lactic acid* (Fig. 8.4), which tends to accumulate in muscle cells and causes them to become acidic. A result of lactic acid accumulation is weakening of the muscle cell's contractions. The affected person experiences fatigue in the forms of muscle weakness and muscle pain. The person also feels out of breath.

If activity decreases, the circulatory and respiratory systems can again deliver oxygen to the muscle cell faster than oxygen is consumed. The extra oxygen is used to complete the breakdown of lactic acid into CO_2 and water; this not only eliminates the lactic acid but also makes a large

FIGURE 8.3 Components of a muscle cell.

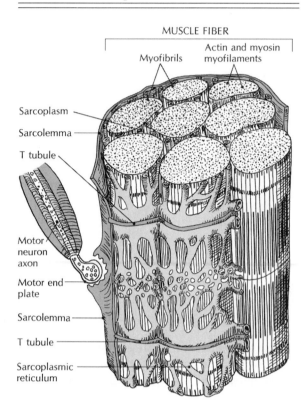

FIGURE 8.4 Obtaining energy in muscle cells.

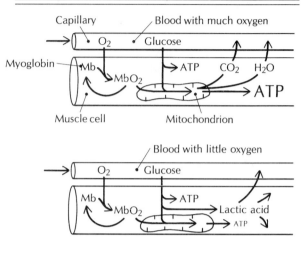

called *sarcoplasmic reticulum* and regulate the movement of calcium ions needed for contraction.

The region in the cell surrounded by each tube of sarcoplasmic reticulum contains an array of tiny fibers called *myofilaments*. Clusters of thick myofilaments alternate with clusters of thin myofilaments (Fig. 8.5). The alternating clusters overlap to form units called *sarcomeres*, which extend from one end of the cell to the other like links in a chain. Each chain of sarcomeres is surrounded by sarcoplasmic reticulum and is called a *myofibril*.

When the cell is stimulated and action potentials pass over the sarcolemma, the sarcoplasmic reticulum releases calcium, which causes the thick myofilaments to pull on the thin myofilaments and slide farther among them. The pulling and sliding cause the muscle cell to become shorter; contraction has occurred. The contraction applies a pulling force to the bone or other structure to which the muscle is attached, and the structure is either moved or held in place.

Recall that energy for contraction comes from the breakdown of nutrient molecules. Only some of the energy released from these molecules is converted into movement of the myofilaments; the remainder is converted into heat. This is why muscles produce so much heat when they contract.

Types of Muscle Cells   The proportions of muscle cell components are different among muscle cells, so the cells have different characteristics. *Type I fibers* contract more slowly and can work longer

amount of energy available to the muscle cell. The affected person's sensation of fatigue subsides and he or she may claim, "I have caught my breath." The oxygen used to eliminate the lactic acid produced by vigorous exercise is called the *oxygen debt*.

Much of what has just been said is also true of cardiac muscle cells. For example, when cardiac or skeletal muscle cells accumulate lactic acid, they become weak. However, unlike cardiac muscle cells, skeletal muscle cells are rarely seriously injured or killed by lactic acid. These cells can continue to work with lactic acid present as long as the acid concentration does not become too severe. Still, exceedingly high levels of lactic acid will prevent skeletal muscle cells from contracting.

Contraction   The membranes of the endoplasmic reticulum within muscle cells are arranged in the form of lacy tubes that extend over the length of the cell (Fig. 8.3). These membranes are

FIGURE 8.5 Myofilaments and the process of contraction: (a) Contraction begins. (b) Contraction ends. Sarcomeres and cells are shorter.

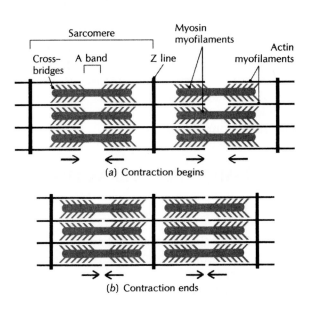

(a) Contraction begins

(b) Contraction ends

before becoming fatigued. *Type IIA fibers* contract more quickly and resist becoming fatigued, also. *Type IIB fibers* contract quickly, but they become fatigued quickly. *Type IIC fibers* are intermediate between Type IIA and Type IIB. Type IIA and Type IIB fibers are most important for fast and powerful movements. Different muscles have different combinations of these types of muscle cells, and the combinations change gradually during adulthood.

Age Changes in Muscle Cells

Internal Components As muscle cells age, the convolutions in the motor end plate decrease and the sarcolemma becomes smoother. The resulting decrease in surface area diminishes the ability of the muscle cell to be stimulated by the motor neuron. Other changes in the sarcolemma cause the action potentials that lead to contraction to become weaker, slower, and more irregular. Because of the changes in the action potentials, the cell takes longer to begin to contract and is less able to recover from one contraction and prepare for the next. Age-related slowing of calcium release and retrieval by the sarcoplasmic reticulum contribute to these effects.

The large-scale results of these cellular age changes include a longer time to respond when a person wants to move suddenly and a diminished ability to perform rapidly repeated movements such as playing fast music on a piano. Muscle composed of aging cells also has a weaker maximum strength when used for activities requiring rapid and very strong contractions, such as grasping a handrail to stop a fall.

Another change in aging muscle cells is a decrease in the substances used to supply energy for contraction (ATP, creatine phosphate, glycogen). Much of this change seems to be caused by a decrease in exercise rather than by aging. Lack of exercise also seems to cause most of the decrease in the enzymes that extract energy from nutrients. There is even a decrease in the number and size of mitochondria, which perform most of the energy extraction. Many remaining mitochondria have been damaged, so they are less efficient and produce more free radicals (FRs). Some muscle cells seem to accumulate damaged mitochondria and become sources of FR damage to surrounding cells. All these changes leave the cells with less energy, especially for tasks requiring a prolonged effort.

The final substantive change inside muscle cells is a decrease in the number of sarcomeres within the myofibrils. This tends to cause the cells and the muscles they compose to become shorter and have a reduced distance through which they can move. The affected person experiences stiffness and diminished freedom of movement. The loss of sarcomeres also reduces the strength of the cells and muscles.

Cell Thickness Since muscle cells that get little exercise lose parts of their internal components, they decrease in thickness. This shrinkage is prevalent among the elderly because of the general reduction in physical activity as people age. Regularly exercised muscles show little change in cell thickness until age 70 or beyond. Even then, there is only a slight thinning of cells in muscles that receive plenty of exercise. Therefore, reduction in exercise rather than aging is the main cause of muscle cell thinning and much of the consequent decrease in muscle thickness and strength that usually accompanies advancing age.

Cell Number Most of the decrease in the thickness of muscles with aging is caused by the death of muscle cells. Up to half the muscle cells in a muscle may be lost by late old age. This loss occurs in exercised muscles and in muscles receiving little use. Lost muscle cells are not replaced by new ones because except in very unusual circumstances, adult muscle cells cannot form new muscle cells.

Type II fibers become thinner and are lost faster than Type I fibers. The ratios of loss are different for different muscles. Some muscles may lose Type II fibers more than twice as fast as they lose Type I fibers. Type IIB fibers are lost faster than Type IIA fibers. Some of the loss of Type II fibers may be from conversion to Type I fibers. Most age-related decreases in strength and speed result from thinning and loss of Type II fibers.

In muscles receiving much regular strenuous exercise, the space left by the lost cells may be largely filled by the remaining cells. This occurs because muscle cells pulling against heavy loads on a regular basis adapt by synthesizing more internal components. The additional components increase the thickness and strength of these cells, which encroach on the vacant areas. As a result, the decline in thickness and strength of exercised muscles is slow.

Aging muscles that receive little strenuous exercise have the spaces left by lost cells filled with fibrous tissue and fat. Such muscles become thinner and considerably weaker as time passes.

Cell Repair   Though muscle cells are unable to reproduce, they can repair themselves after an injury. One common cause of injury is contracting against a load much heavier than that normally encountered by muscle cells. This type of injury can be sustained when a person who normally lifts objects weighing less than 30 pounds tries to support a 60-pound object.

A muscle containing muscle cells injured by an excessive force, such as lifting a heavy object, is weakened and causes the sensations of muscle soreness and stiffness. If the muscle is rested, the injured muscle cells will repair themselves within a few days and the soreness and stiffness will subside. As was mentioned above, the cells will adapt to the heavy demands previously placed on them by becoming thicker and stronger. They will then be more resistant to injury caused by excessive loads. For muscle cells receiving regular strenuous exercise, the ability to repair injury and re-

cover from such weakness and soreness is not altered by aging.

It is uncertain whether muscle cells that receive little exercise can repair themselves as quickly as exercised muscle cells do. Still, muscle cells in exercised or unused muscles retain the ability to adapt to heavier loads by manufacturing internal components. Thus, the thickness and strength of muscles can be increased by strenuous exercise regardless of age. However, muscle cells in older individuals make the compensatory increase in thickness more slowly.

NERVE-MUSCLE INTERACTION

Motor Units

Recall from Chap. 6 that skeletal muscle cells are stimulated to contract by nerve cells called *somatic motor neurons*. The axon from each motor neuron branches as it passes through its muscle. Some motor neurons have only a few branches, while others have several hundred.

Each branch from a motor neuron axon ends on a muscle cell, and each muscle cell receives a branch from only one motor neuron (Fig. 8.6). Thus, the muscle cells in a muscle are organized into groups, with all the cells in each group being controlled by one motor neuron. The combination of one motor neuron and all the muscle cells it controls is the functional unit of the muscle and is called a *motor unit*.

When an impulse travels down a motor neuron, it passes along every branch of its axon. Therefore, every muscle cell in the motor unit is stimulated to contract; it is not possible to cause only some of these cells to contract.

The strength of each contraction is determined by which motor units and how many motor units are activated at a given time. Since more varied combinations of numbers of muscle cells can be selected in muscles with small motor units, a person has more control over the amount of strength provided by each contraction in such muscles. It is more difficult to select precise levels of strength from muscles with large motor units because the muscle cells contract in larger groups. The difference in the degree of control is similar to the difference between the ability to pay an exact amount when one has many one dollar bills and small

FIGURE 8.6 Motor units.

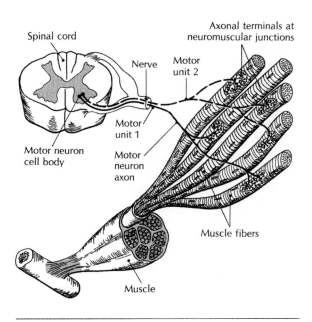

change and the ability to pay an exact amount when one has only bills of large denominations.

Changes in Motor Units

Since motor units change in many ways as people get older, the functioning of a muscle also changes. Some of these changes and their consequences were described in Chap. 6.

One change is an exponential decrease in the number of motor neurons. The loss may reach 50 percent by age 60. This is the main reason for the decrease in the number of muscle cells because a muscle cell degenerates and dies if it does not receive stimulation from a motor neuron. As more motor neurons and their muscle cells are lost, the maximum strength of contraction the muscle can produce diminishes.

Fortunately, many surviving motor neurons produce additional axon branches that connect to the orphaned muscle cells. These adopted muscle cells survive and function. This compensatory process helps slow the decline in the strength of the muscle. Note, however, that the size of the remaining motor units increases. This means that there is a decrease in control of the strength of each contraction. This may be one reason people have a reduced ability for fine movements as age increases. Also, Type II fibers are often "adopted" by motor neurons from Type I fibers. This alter-

ation speeds up the conversion of Type II fibers to Type I fibers.

A second age change is a slowing in the passage of impulses to muscle cells. There is a variable amount of slowing among the motor neurons controlling a muscle. As mentioned in Chap. 6, three alterations in the overall contraction of the muscle result. First, it takes longer for the muscle to reach its peak strength of contraction. Second, the peak amount of strength is lower. Third, the entire contraction takes more time. These alterations further reduce the maximum amount of strength a muscle can produce and make it more difficult to perform very quick movements.

Another change in motor neurons is a decrease in the frequency with which impulses are sent to the muscle. Normally, a motor neuron sends a volley of impulses in rapid succession so that the muscle cells contract rapidly. A rapid series of contractions—*incomplete tetany*—provides a fairly smooth and strong contraction that can be maintained for a long time. Since age changes in muscle cell action potentials decrease the frequency at which muscle cells can contract, reducing the frequency of neuron impulses may be compensatory. Sending impulses faster than the muscle cells can respond would be wasteful of neuron energy and neurotransmitter materials.

Other Nerve-Muscle Interactions

Several other changes in the nervous system alter the operation of muscles as people age. Recall from Chap. 6 that age changes in sensory neurons, synapses used by reflex pathways, and other areas of the central nervous system involved in controlling voluntary movements all affect adversely the ability of the muscle system to maintain homeostasis and the quality of life.

BLOOD FLOW IN MUSCLES

Muscle cells depend on the circulatory system to supply oxygen, nutrients, and other needed materials and to remove wastes. Service of muscle cells diminishes somewhat even when no disease of the circulatory system is present. Exactly how much of this decrease is due to age changes and how much is due to a reduction in exercise is not known.

Reasons for the reduced ability of the circulatory system to meet the needs of muscle cells include the decrease in the density of capillaries among muscle cells and age changes in capillary structure. These changes may contribute to the decline in the maximum rate of working and the faster onset of muscle fatigue as people advance in age. These and other age changes in and diseases of the circulatory system that can affect the muscle system were discussed in Chap. 4.

CHANGES IN MUSCLE MASS

The many changes at the cellular and microscopic levels in the muscle system combine to reduce the thickness of each muscle and therefore the total amount of muscle mass. Serious loss of muscle mass is called *sarcopenia*. On the average, sarcopenia begins during the third decade. The rate of loss is low at first, but the rate increases with age, rising quickly after age 50. Muscle mass may decrease as much as 50 percent by age 80. This increasingly rapid loss seems to be due primarily to the decline in physical activity that usually accompanies advancing age. Most of the loss of muscle mass and thickness is due to the loss of muscle cells rather than to thinning of the cells.

Effects of Mass on Strength

The decline in muscle mass produces several effects, one of which is a decline in muscle strength. This loss of strength is related to the total thickness of a muscle since the amount of strength per unit of cross-sectional area of muscle cells remains fairly stable regardless of age. However, the reduction in muscle strength that accompanies aging is only partially due to thinning of the muscles. Other important factors include changes in muscle cell structure and functioning and increases in fat and fibrous material among the muscle cells. Changes in factors outside the muscle system (e.g., nervous system, joints, motivation) also play an important role in the decline in strength with age.

In general, strength peaks during the third decade and declines little during the fourth decade. Age-related decrease in strength becomes more rapid and significant during the fifth decade. The decline in strength becomes faster as age increases after that. However, the decline in muscle strength varies considerably from person to person and from muscle to muscle. There is variation with respect to the age at which a substantial reduction in strength can first be detected and the rate at which strength declines afterward. Muscles used for quick strong contractions show a greater decline in strength than do muscles used to maintain posture or perform other actions requiring long-lasting mild contractions.

It seems that the most important reason for heterogeneity in loss of strength is the increased variability in the amount of strenuous exercise performed regularly by each person and each muscle. For example, individuals whose daily routines include gripping objects or tools lose grip strength slowly, but these individuals may have fairly rapid loss of leg strength if their activities include little use of the legs.

The amount of strength lost over a period of years may impair an individual's ability to carry out ordinary activities such as shopping, gardening, cleaning, climbing stairs, and breathing heavily during exertion. It becomes increasingly difficult to continue in certain lines of employment, such as those requiring lifting or moving heavy loads. It may be necessary to forsake strenuous recreational activities such as sailing. Still, many aging individuals can tolerate declining strength by using methods requiring less brute strength, substituting power tools and appliances for muscle power, and enlisting aid from others.

The unevenness in loss of strength among different muscles creates an additional problem in the form of reduced coordination. This occurs because the balance in strength among the muscles used to perform an action is altered. An important effect of dwindling strength and decline in coordination is an increase in the risk of falling. Reduced and unbalanced muscle strength also modifies posture. Detrimental outcomes from deteriorating posture may include biological effects (e.g., restricted ability to inhale, impingement of bones on nerves), social and psychological effects of altered appearance, and economic effects from the need to obtain different clothing or furniture.

Other Effects

The reduction in muscle mass accompanying aging can have effects other than changes in strength. A change in body proportions can have social, psychological, and economic consequences for the reasons noted above related to altered posture. Another effect is the need to modify one's diet. With less muscle mass, there is a decrease in the basal metabolic rate and a consequent decrease in the amount of calories needed per day. Though the diet consumed by most aging people should contain fewer calories, it should be richer in protein to maintain the remaining muscle mass while preventing an undesirable gain in weight. The declining metabolic rate, along with a relative decrease in the proportion of body mass composed of lean muscle, also necessitates adjustments in the doses of medications.

MUSCLE SYSTEM PERFORMANCE

Reaction Time and Speed of Movement

All the changes in the muscle system already mentioned, combined with aging in the nervous, circulatory, respiratory, and skeletal systems, lead to other noticeable alterations in the actions produced by muscles. One is an increase in the time needed to begin a voluntary motion in response to a stimulus (*reaction time*). For example, it takes longer for a driver to move his or her foot from the gas pedal to the brake when a traffic signal turns red. Most of the increase in reaction time is caused by slowing of the processing of impulses in the central nervous system.

Note that by definition, reaction time ends when the person begins to move. The time from the beginning of a motion to the end of that motion also increases with age. This second alteration, a decrease in the *speed of movement*, is caused by decreasing muscle strength.

Both the increase in reaction time and the decrease in speed of movement make the performance of rapid movements difficult. As can be seen in the example of driving, these changes increase certain risks. There is also an increased risk of falling. The probability of sustaining greater injury from a fall also rises because it takes longer to grasp a handrail or piece of furniture or to change body position to break or cushion the fall.

Longer reaction times and slower movements also make it more difficult to perform rapidly repeated movements such as those used in playing fast music or dancing. The effects of these changes become greater when individuals attempt more complex or less familiar movements. As with declining strength, changes in reaction time and speed of movement occur faster as age increases.

Skill

A third aspect of muscle activity that changes with age is skill in performing tasks. Though changes in reaction time and speed of movement have profound adverse effects on a person's skill in performing novel activities, they have much less of an impact on activities that have been performed routinely for many years. Skill in well-practiced actions can even improve with age if repetition of the movements involved continues.

Practice also reduces the frequency of errors in performing an intended movement and selecting sequences of movements to complete complicated tasks. New strategies are formulated, and the efficiency of energy use improves with practice. Therefore, experienced older individuals may perform better than do younger individuals in activities requiring both strength and speed.

Stamina

The advantage of experience can be overshadowed by a gradual drop in stamina. *Stamina* may be defined as the ability to perform vigorous activity continuously for more than a few seconds. The effect of dwindling stamina on overall muscle system performance is proportionately greater than is the effect of the age-related decrease in speed of movement. Stamina declines faster as age increases.

The decrease in stamina is manifested in four ways. First, there is a decline in the maximum rate at which vigorous activities can be performed. For example, the maximum speed at which a bicycle can be ridden diminishes. Second, the length of time such activities can be performed without stopping to rest becomes shorter. This decrease in *endurance* is evident whether a person is working as fast as possible or at a rate somewhat lower than the maximum rate. As will be explained below, important causes of the reduction in endurance include a more rapid accumulation of lactic acid in muscles and a faster and more intense onset of discomfort at a given rate of vigorous activity.

The third indication of reduced stamina is a lengthening of the time needed to recover after ending an activity such as running. For example, it may take longer for respiration and heart rate to return to resting values. One reason for the increase in recovery time is the faster accumulation of lactic acid caused in part by a decline in the efficiency of movement. Another factor is a slowing in the rate at which the heat produced by muscle contraction is released from the body.

The fourth indication of dwindling stamina is a rise in muscle stiffness and soreness experienced hours or days after a vigorous activity has ended. Lactic acid buildup also seems to be a main reason for this indication.

The decline in the maximum rate of performing physical activity has been studied intensively. Therefore, this age-related change will be discussed in detail below.

˙Vo2max   The maximum rate at which a person can use muscles to perform an activity is commonly determined by measuring the rate at which that person uses oxygen while engaging in an activity at the fastest rate possible. The maximum rate of working is expressed as the Vo_2max. A person's ˙Vo_2max is the amount of oxygen used per kilogram of body weight per minute while a person is exercising at the fastest rate attainable. Exercises commonly used for determining ˙Vo_2max include riding a stationary exercise bike and walking or running on a treadmill. ˙Vo_2max is also called *aerobic capacity*.

˙Vo2max declines with age. The decline begins at about age 20 for men and about age 35 for women. These are average values, however. As with many other age-related changes, there is great variability among individuals of the same age in regard to actual ˙Vo2max values and the rate of decline in ˙Vo2max values.

A main reason for differences in the levels and rates of change of ˙Vo2max is variation in the amount of exercise a person gets. For example, the ˙Vo2max for people who have a rather sedentary lifestyle drops about twice as fast as does the ˙Vo2max of individuals whose jobs, home lives, and recreational activities include large amounts of physical activity. Also, ˙Vo2max begins to decline faster when a person's activity decreases. By contrast, when a person's participation in regular vigorous exercise increases, the decline may be delayed and become slower or even be temporarily reversed. Still, some reduction in ˙Vo2max eventually occurs in all people, including individuals who engage in highly demanding physical activities throughout life. When vigorous physical training continues, ˙Vo2max declines 5

percent per decade.

Much of the decline in ˙Vo2max is due to the age-related decrease in total muscle mass combined with a relative increase in the proportion of body fat. The rate of oxygen consumption of each kilogram of muscle may be the same despite age.

Many other factors seem to contribute to the decline in ˙Vo2max. One factor is a reduction in the ability of muscles to extract oxygen from blood. Other factors include changes and diseases that limit the functioning of the circulatory, respiratory, and skeletal systems. A person may be affected by more than one factor, and many people are affected by most or all of them. Therefore, it is extremely difficult to identify how much of the decline in ˙Vo2max is due to aging of the muscle system rather than to other factors.

Consequences of Lowered ˙Vo2max   Since ˙Vo2max is an indicator of the maximum rate at which a person can perform activities, a small decline means a drop in the maximum rate at which a person can run, climb stairs, and carry out other vigorous activities. Individuals with lowered values tend to stop physical activities sooner because of the discomfort such activities induce. As ˙Vo2max decreases further, limitations in less demanding activities, such as walking briskly, become evident. When very low values are reached, individuals may have trouble walking slowly or even getting up from a chair or bed.

Since a substantial decline in ˙Vo2max adversely affects the performance of all types of physical activity, it can reduce a person's effectiveness and participation in occupational, recreational, and social activities. When ˙Vo2max becomes very low, the performance of ordinary daily activities needed to maintain a person becomes difficult or impossible. Examples include shopping, dressing, and bathing. Serious losses in the sense of independence and other negative psychological consequences often develop. Undesirable alterations in one area can cause detrimental effects in other areas, leading to a synergistic spiral of decline.

As mentioned previously, the decline in ˙Vo2max can be slowed or even reversed when an adult of any age begins a program of exercise or includes vigorous activity in his or her daily life. Individuals with relatively high ˙Vo2max values need to engage in activities with high intensity and frequency to derive beneficial alterations in 'Vo2max. People whose 'Vo2max is fairly low can slow the decline or increase this parameter with less strenuous activities. For many, substituting muscle power for convenience can achieve real gains. For example, parking farther from stores and walking to reach them or climbing stairs rather than using an elevator can significantly increase a person's amount of exercise.

STAYING PHYSICALLY ACTIVE

Many age-related changes in the muscle system are caused or greatly increased by a decrease in physical activity. Conversely, many of these adverse changes can be greatly slowed or even negated by continuing to engage in regular exercise. It is also possible to delay, slow, reduce, or prevent many undesirable changes and diseases in other body systems by living a physically active lifestyle. In general and within reasonable limits, the more exercise a person gets, the greater the benefits.

Specific Effects

Many effects from maintaining a high level of physical activity are listed in Table 8.1.

Muscle Mass Besides retaining the strength to perform both heavy and ordinary tasks, maintaining muscle mass helps stabilize body proportions.

It also reduces detrimental changes in the ability of the hormone insulin to regulate blood sugar and certain metabolic activities in the body (Chap. 14). The effects of ongoing exercise on the nervous system help slow both the increase in reaction time and the decline in speed of movement.

'Vo2max The impact of ongoing exercise on slowing the decline in 'Vo2max is so great that very active elderly people have values equal to or greater than those of sedentary individuals of about age 30. However, no amount of exercise can completely stop the decrease in 'Vo2max as age increases. Therefore, younger people who exercise will have values greater than those of older individuals who get in the same amount of exercise.

Overview

In considering the beneficial effects of years of physical activity, it is important to realize that these benefits are obtained only by persons who continue to lead active lives. People who are very active or athletic during youth but then become sedentary for many years lose most of the benefit they acquired in their previously active lives.

Getting regular exercise throughout life has been shown to increase life expectancy, possibly because exercise reduces the risks of certain causes of death. Exercise has not been shown to increase maximum longevity. Finally, exercise undoubtedly improves the quality of life. Long-

---

TABLE 8.1 EFFECTS OF MAINTAINING A HIGH LEVEL OF PHYSICAL ACTIVITY THROUGHOUT LIFE

Muscle system
  Slower decline in energy molecules (ATP, creatine phosphate, glycogen), oxidative enzymes, muscle cell thickness, number of muscle cells, muscle thickness, muscle mass, muscle strength, blood supply, speed of movement, stamina, endurance, 'Vo2max
  Slower increase in fat and fibers, reaction time, recovery time, development of muscle soreness

Nervous system
  Slower decline in processing impulses by the CNS
  Slower increase in variations in speed of motor neuron impulses

Circulatory system
  Maintenance of lower levels of LDLs and higher HDL/cholesterol and HDL/LDL ratios
  Decreased risk of high blood pressure, atherosclerosis, heart attack, stroke

Skeletal system
  Slower decline in bone minerals
  Decreased risk of fractures and osteoporosis

term exercise enables an individual to participate more fully and with greater pleasure in many more life activities. Years of regular exercise also markedly reduce the risk of developing many disabling diseases. Those who exercise and still develop a disease are often less affected.

STARTING OR INCREASING EXERCISE

Clearly, elderly individuals who have been involved in vigorous physical activity throughout their lives benefit from such a lifestyle. Young people who adopt active lifestyles can expect to reap the same benefits when they become elderly. Furthermore, people of any age who have lived sedentary lives and begin to get exercise and those who have been getting only low or moderate amounts of exercise for many years and increase their exercise can improve their well-being. We will now examine outcomes in older people who begin vigorous exercise or substantially increase their level of physical activity.

Effects

Many effects on older people who begin or increase physical activity are listed in Table 8.2.

Circulatory System   The rise in maximum cardiac output is evident within a few days to weeks of initiating an exercise program. The more intense the exercise program, the sooner a significant increase in maximum cardiac output appears. This rise begins to be reversed within days of ending the exercise program. The final maximum cardiac output of those leaving an exercise program will be about the same as that which existed when the exercise program began. Altering blood lipoprotein levels requires a decrease in body fat along with the effects of the exercise.

Respiratory System   There is disagreement about whether increasing exercise increases respiratory volumes and speed of airflow, but long-term participation in exercise programs slows the decline in respiratory functioning. Therefore, in the long

---

TABLE 8.2 EFFECTS OF STARTING OR INCREASING EXERCISE

Circulatory system
Increases cardiac output, cardiac efficiency, HDL/cholesterol and HDL/LDL ratios, blood vessel diameter, muscle capillaries
Slows decline in heart functioning
Decreases resting blood pressure, blood pressure during exercise, heartbeat abnormalities

Respiratory system
Increases clearance of mucus and respiratory efficiency
Slows decline in respiratory functioning and closing of airways

Nervous system
Increases formation of new axon branches to orphaned muscle cells, speed of impulse processing by the CNS, balance, short-term memory, sleep, mental abilities (possibly)
Decreases variability in speed of action potentials in motor neurons and risk of falling

Muscle system
Increases oxidative enzymes, stored glycogen, capillary numbers, blood flow, uptake of oxygen from blood, cell thickness, muscle strength, muscle mass, speed of movement, stamina, endurance, Vo_2max
Slows decline in efficiency of movement and increase in recovery time

Skeletal system
Increases ease of movement, range of movement, joint flexibility
Slows bone demineralization
Decreases risk of falling and sustaining fractures

Endocrine system
Increases glucose tolerance and sensitivity to insulin
Slows decline in growth hormone

run elderly individuals who exercise will eventually have better respiratory system functioning than they will if they remain sedentary.

The respiratory system changes caused by a proper exercise program are of special importance to persons who have chronic obstructive pulmonary diseases (COPDs) such as chronic bronchitis and emphysema.

Nervous System   The mechanisms by which strenuous exercise increases strength in older people are different from those in younger people. At younger ages the increase in strength from training with heavy weights is caused almost exclusively by thickening of the muscle cells rather than alterations in the nervous system. Perhaps the change in mechanisms for increasing strength is a way the aging body partially compensates for a decreased ability of the muscle cells to adapt to lifting or moving heavy loads.

Muscle System   The gain in strength achieved by older individuals is proportionately the same as that which younger adults attain with the same type of exercise. For example, consider an older person who has had little exercise for many years and a younger adult who has had the same level of activity. The older person will not be as strong as the younger adult because the older adult has been losing strength for a longer period. If both individuals begin an exercise program designed to double the strength of an adult, both will end the program with twice the strength they had when they started. Of course, since the younger person was stronger at the start of the program, he or she will be the stronger person at the end. However, an older person who participates in such a program can become stronger than a younger adult who remains sedentary.

Older individuals whose exercise is not strenuous enough to cause an increase in strength still benefit because they at least have a slower decline in strength. Therefore, after long-term involvement in physical activity these individuals will be stronger than they would have been if they had remained sedentary. They will also have retained more total muscle mass. Keeping a high muscle mass helps the functioning of insulin.

Alterations in Vo2max caused by increased exercise are similar in three ways to exercise-induced changes in strength. First, the increase in Vo2max attained by an older person is propor-tionately the same as that achieved by a younger adult who starts with the same capability and participates in the same exercise program. Second, elderly people who increase their physical activity enough can develop values that are greater than those of much younger adults who remain sedentary. Third, elderly people whose increase in exercise is not enough to produce an increase in Vo2max will still benefit because even small increases in physical activity slow the decline in Vo2max. Therefore, these individuals will eventually have a greater Vo2max than they would have had if they had remained sedentary.

There is an important difference between the effects of exercise on alterations in conditions such as blood lipoproteins, body composition, percent body fat, functioning of insulin, and strength and the effects on alterations in Vo2max. Though vigorous activity is needed to effect substantial changes in the first five parameters, for very sedentary older people even low levels of easy activities such as walking can substantially increase Vo2max. The resulting improvements can restore the ability of very sedentary elderly individuals with extremely low Vo2max values to perform the ordinary activities of daily living.

All the exercise-induced alterations in the nervous and muscle systems just described combine to produce several other benefits in the elderly. These benefits include the perception that less effort is needed to perform demanding tasks; improved mood and sense of well-being; improved social interactions; and increased independence.

Skeletal System   Aging of the skeletal system raises the risk of sustaining fractures by causing demineralization of bones and reducing the ease of movement and range of motion of joints. Some forms of the joint disease called *arthritis* exaggerate these changes.

No one knows the best exercise program for slowing or reversing bone demineralization caused by aging or osteoporosis; different programs may be effective for different individuals. Also, different individuals can tolerate different amounts and types of exercise. The causes of these differences include physical condition, presence of diseases, lifestyle, and motivation.

The possible benefits of slowing bone demineralization and deterioration of joint functioning through a large increase in strenuous or vigorous physical activity must be weighed against the added risk of injury. Some more common problems include; increased risk of fracture of the bones in the spinal column caused by lifting or holding heavy loads; increased risk of fracturing hip, leg, or arm bones by falling; traumatic injury to the bones, muscles, ligaments, and tendons in the lower legs from walking or running on hard surfaces or with improper footwear; and injury to the joints from excessive movements or forces, including impact forces.

Endocrine System   Exercise leading to a decrease in body fat significantly improves glucose tolerance and insulin sensitivity in elderly people who have a reduced glucose tolerance or non-insulin-dependent diabetes mellitus (NIDDM). Individuals who improve their glucose tolerance and insulin sensitivity have a substantially reduced risk of developing the complications associated with diabetes.

These beneficial effects of exercise begin to develop within days or even hours after a person increases the level of vigorous physical activity. Improvement increases as body fat decreases. However, the beneficial effects of the exercise begin to dwindle within a few days of ending involvement in the exercise program. Therefore, to sustain the benefits of exercise, a person with reduced glucose tolerance or NIDDM must engage in the exercise at least once every three days.

By contrast, individuals with type I (insulin-dependent) diabetes mellitus (IDDM) have very unstable blood sugar levels. Therefore, the amount of exercise they get must be carefully monitored and adjusted according to factors such as the severity of the disease, diet, and insulin treatments.

Other Effects   The effects of increasing exercise mentioned up to this point are related to specific body systems, but elderly people who increase their exercise seem to benefit in many other ways. These other benefits include helping to maintain normal body weight by improving nutrition and using more calories; increasing independence by generally slowing the onset of disability and physical limitations; and helping to enhance psychological health by improving mood and sense of well-being while reducing boredom, anxiety, and stress.

The physical and psychological effects of exercising also increase older individuals' ability to remain productive and economically self-sufficient. Their social situation is bolstered by the additional social interactions obtained through exercise programs and through an enhanced ability to participate in other activities in the community. Therefore, while increasing exercise has not been shown to lengthen life, it dramatically improves the quality of life for older individuals.

EXERCISE RECOMMENDATIONS

Having reviewed how exercise benefits the elderly, we will now consider information and suggestions that have been found important in achieving these benefits.

First, as with other mechanisms that maintain homeostasis, adjustments made in body systems to each form of exercise represent attempts to minimize or prevent disturbances in internal conditions. Furthermore, adjustments and improvements made by the body are specific to the demands placed on it. For example, if conditions in leg muscles are significantly disturbed by lifting heavy loads, those muscles will become stronger and therefore will be less affected when the loads are lifted a few days later. By contrast, if conditions in leg muscles are disturbed by an activity involving many repeated actions that do not require much strength, such as walking briskly for a long distance, the adjustments in the body will increase stamina for walking but will have little effect on muscle strength.

Set Goals

The first step in preparing to increase exercise is to establish specific goals. Then activities can be selected that will cause the body to make the adjustments needed to achieve those goals. For example, if an increase in the range of motion of the arms is desired, activities requiring movement of the arms over wide angles can be selected. If increasing the grip strength of the hands is a goal, activities using strong grasping should be undertaken. As more goals are identified, a greater number and variety of exercises or activities must be employed.

In a more general way, if exercise is being used to improve the functioning of the circulatory and respiratory systems, a number of activities demanding faster blood flow and increased respiration can produce the desired results. Examples include walking or riding a bicycle at a fast pace, running, and swimming.

Evaluate and Individualize

When one is deciding on exercises, attention must be paid to the condition of the person who is participating in the exercises. Careful attention to an elderly person's physical condition is particularly important because of the higher incidence of disease and the increased heterogeneity among older people. At this point, a qualified professional should perform a physical examination and evaluation of the participant and the information obtained should be used to determine the appropriateness of the anticipated activities. At least one follow-up examination and evaluation should be performed several days or a few weeks after the individual has taken up the new level of activity. Data from the first and later examinations and evaluations should be used to determine what changes are occurring because of the increased exercise and to suggest improvements in the activities.

Plan a Program

Once appropriate activities have been selected and the condition of the participant has been ascertained, decisions about the intensity and the length of exercise can be made. The time allotted for exercise should include time for warming up and cooling down. The frequency with which the activity will be performed can also be established. A healthy person should exercise at least 30 minutes each session, with sessions occurring at least every three days.

Generally, starting it with a fairly low level of intensity and a short duration of activity is best, especially for very sedentary or frail individuals, who can achieve substantial benefits from relatively low levels of exercise. Also, such individuals are more likely to sustain injuries or other adverse effects from a sudden increase in physical activity. Beginning with low levels of exercise also helps prevent negative attitudes by minimizing the discomfort caused by an increase in exercise.

Consideration might also be given to the number of weeks or months during which the participant expects to perform the activity. Exercise programs lasting only a few weeks produce little ben-

efit, while those lasting several months or longer yield significantly better results. Longer-lasting programs are especially important for older people since the rate of improvement caused by exercise decreases with age. Sustained participation in the exercise program is aided by using positive feedback and other motivational strategies, such as combining exercise sessions with social activities. Since the exercise program should last for an extended time, it is important to provide for proper nutrition.

As the exercise program continues, the intensity or duration of each exercise or the frequency with which it is performed each week should be increased gradually so that the participant continues to improve. If the exercise is not increased, the participant's level of physical fitness will soon stabilize, and boredom may become a problem if there is no variation. Furthermore, the psychological benefits of exercise become most apparent once a high intensity and greater frequency have been achieved.

Minimize Problems

Although the benefits are directly proportional to the intensity, duration, and frequency of exercise, care must be taken not to exceed reasonable limits. As levels of exercise increase, so do the risks of overheating, being physically injured, and developing complications from existing diseases.

Very strenuous activities present a special danger to those with atherosclerosis because people tend to hold their breath while pulling or pushing with great force. Blood pressure rises to a very high level during such maneuvers, placing a great burden on the heart and arteries. A heart attack, a stroke, or damage to the retina or vitreous humor can result. When the straining ends, there is a sudden drop in blood pressure, placing additional burdens on the heart and sometimes causing dizziness, fainting, and falling. These problems can be largely avoided by minimizing exercises requiring great strength and maximizing activities involving free movement of parts of the body.

Consider Alternatives

When one is discussing exercise, focusing it on activities whose primary purpose is exercise is easy (e.g., aerobics, weight lifting, jogging). Using such activities and the many available exercise machines and devices provide means of carefully regulating the amount of exercise obtained and measuring physical status and improvements in physical fitness.

While some individuals enjoy such purposeful exercise, others find it unpleasant, expensive, or unavailable. These individuals can still obtain plenty of beneficial exercise through activities with other primary purposes. Examples include recreational activities such as dancing, sports, and hiking and activities related to occupations requiring physical work. Activities performed in caring for one's home and family, such as gardening and mowing a lawn, shopping, and doing laundry, can provide opportunities to get healthful exercise. Choosing to walk or climb stairs rather than riding can add substantially to the amount of beneficial exercise obtained.

Achieving the benefits of increasing exercise often involves nothing more than substituting muscle power for motor power. What may be needed first, however, is replacement of the notion that using minimal physical effort means living well with the realization that only through regular physical exercise can an older person achieve "the good life."

DRIVING MOTOR VEHICLES

Driving accidents increase in numbers and in rates as the age of the drivers increase. For example, elderly drivers have twice as many accidents per mile compared with younger drivers. Elder drivers who have accidents are more likely to sustain serious or fatal injury than are younger drivers. At the same time, the number and proportion of elder drivers are increasing, and they will continue to increase at faster rates for the next few decades. By 2024, drivers over age 64 will make up 25 percent of all drivers.

Potential accident situations often require making quick and coordinated responses in new situations. Therefore, elder drivers can be very safe drivers until they meet surprising or complicated situations that demand quick reactions in unfamiliar situations. Examples of problematic situations for elders are intersections. In such demand-

ing situations, age changes in muscle strength, speed, reaction time, and coordination contribute to an age-related decrease in driving ability. Age changes and age-related abnormalities in other body systems also contribute to reduced driving ability. These changes and abnormalities are described in other chapters.

Neurological parameters that change very little with normal aging include implicit memory of driving skills and making automatic coordinated responses. Other age-related changes that do not have a major impact on driving ability in elders include modest decrements in vision and in hearing. The small effects on driving safety among elders from decrements in vision and in hearing may be due to elders avoiding driving at times and conditions where these decrements are important (e.g., nighttime, bad weather, heavy traffic, high speeds, time pressures). In general, overall cognitive ability has little to do with driving skills until cognitive abilities become severely reduced. Therefore, people in early stages of dementia can still be good drivers.

Important neurological factors that limit driving ability for elders include age-related decreases in avoiding distraction; changing attention quickly; responding quickly in unfamiliar situations; noticing, accurately identifying and responding to sudden changes in the visual field; noticing and responding to a novel change in the environment; and distinguishing between important and unimportant items in the visual field.

Driving is very important to elders for many reasons including mobility; independence; a sense of self-efficacy; and a sense of self-worth. Loss of driving often causes major psychological, social, and economic problems for elders. Demands to provide alternative means of transportation increase (e.g., family, friends, community groups, private companies, governmental agencies).

Recommendations that can accommodate these diverse factors include providing reliable, practical and regular tests for evaluating elder drivers; providing education and training to maintain or improve elders' driving skills and safety; and developing alternative transportation for elders as they give up or lose their driving privileges.

9

SKELETAL SYSTEM

The *skeletal system* has two main components. The first consists of the *bones*, which are the organs of the system and number approximately 206 in the adult body (Fig. 9.1). Some individuals have a few additional small bones, sometimes in the skull, which provide additional strength. The second component consists of the *joints*, which attach bones to each other and contribute to the proper functioning of the system (Fig. 9.1).

MAIN FUNCTIONS FOR HOMEOSTASIS

The bones and joints work together to maintain homeostasis in two main ways. One is by minimizing changes in the body's internal conditions; this is accomplished by providing support and protection from traumatic injury. The second way is by helping to restore errant conditions to proper levels. The skeletal system contributes to this goal by helping the muscles move, storing minerals, and producing blood cells.

Support

The support provided by the skeletal system is important for the same reasons mentioned in Chap. 8 about the muscle system. Recall that some structures of the body are not strong enough to hold themselves up but must be held in position to work properly. For example, the spinal cord, which extends down the back from the base of the head to slightly above the waist, is quite soft and very flexible (Fig. 9.2); it cannot stand on end by itself.

If the spinal cord is bent sharply or excessively, the resulting injury can inhibit its impulse conduction and result in permanent paralysis. To prevent such disastrous alterations in the position of the cord, it is encased within the vertebral column. The vertebral column is composed of a row of ring-shaped bones that are firmly attached to each other by joints that allow only slight movement. Therefore, the skeletal system holds the spinal cord in proper position while allowing it to bend an acceptable amount and in a smooth curve.

Similarly, without support from the skeletal components in the thoracic region, the lungs would collapse like leaky balloons and a person would be incapable of breathing. However, the joints among the thoracic bones permit limited

FIGURE 9.1 The skeletal system.

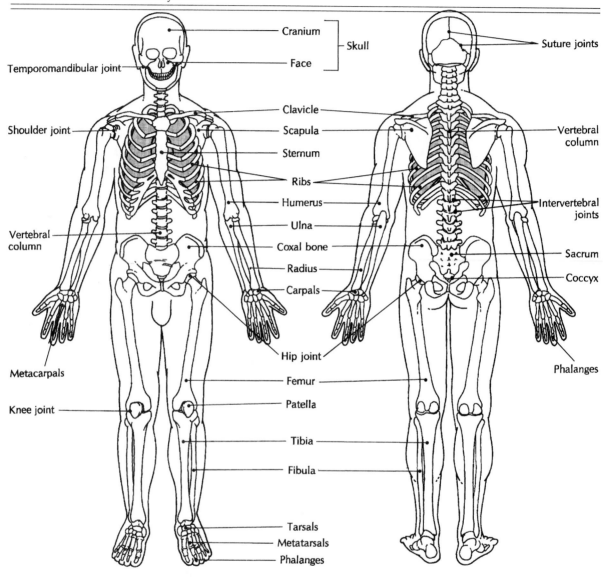

flexibility of the thorax, allowing for breathing (Fig. 9.1).

Protection from Trauma

The weak, delicate nature of many parts of the body requires that they be protected from injury. Recall that fat in the subcutaneous layer of the integument contributes to such protection. The skeletal system also provides protection from trauma.

The soft and delicate nature of the spinal cord requires not only that it be given support but also that it be protected from pressure and sharp objects. Even a slight squeezing of the cord could crush its nerve cells: Nerve impulses would be blocked, and the victim would be paralyzed. However, the spinal cord is rarely damaged because it is shielded by the vertebrae that encircle it (Fig. 9.2).

Consider also the brain, lungs, heart, and bone marrow. Each is essential for life, is easily damaged, and, like the spinal cord, is encased within bones. Through this arrangement, these organs are kept safe from crushing, tearing, cutting, and other forces that may be encountered.

Movement

Movement is a second homeostatic function shared by the muscle and skeletal systems. Recall that movement is one key means by which the body makes adjustments when it detects that a change in conditions is about to take place or has already taken place. By moving, the body can attain what is desirable and avoid what is detrimental. Since muscle cells are the only cells that can furnish motion, one might ask what role the skeletal system plays in movement. The answer to this question has two parts.

The first part is that the skeletal system provides stable anchoring points for muscles (Fig. 9.3). These points are needed to make the force generated by muscle contraction effective. If muscles were not firmly attached to other structures, they would slide about inside the body when they contracted, and no helpful actions would be performed.

The second part of the answer is that the skeletal system acts as a set of levers to modify the motion provided by the muscles (Fig. 9.3). This converts the simple shortening of muscles into the multitude of varied movements that people perform to maintain themselves. Therefore, people can perform bending, twisting, turning, and lengthening actions as well as shortening ones. All these actions can be observed when one watches people perform ordinary tasks such as household chores. A skeletal lever is also able to increase or decrease the distance, speed, and force obtained from the contraction of a muscle so that they better suit the task to be performed. For example, the muscles of the leg can move only a few inches. However, since these muscles are attached to the long bones of the leg, a person can quickly jump far out of the way of an oncoming vehicle.

FIGURE 9.2 Skeletal support for the spinal cord.

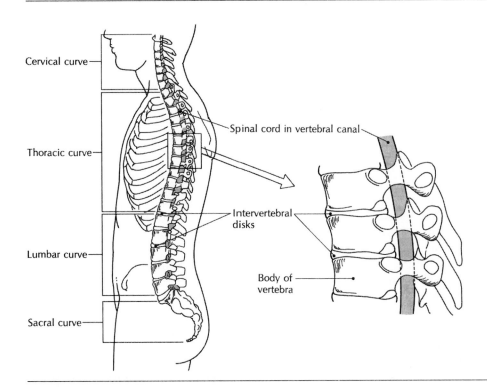

FIGURE 9.3 Skeletal components as anchors and levers.

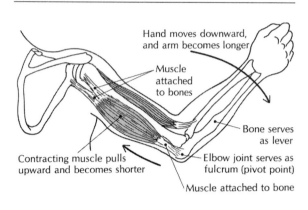

Hand moves downward, and arm becomes longer

Muscle attached to bones

Bone serves as lever

Contracting muscle pulls upward and becomes shorter

Elbow joint serves as fulcrum (pivot point)

Muscle attached to bone

Mineral Storage

Besides helping to maintain homeostasis in mechanical ways, the skeletal system helps in maintaining homeostasis of certain chemicals through mineral storage.

Minerals are needed so that the body cells can perform properly. For example, calcium is necessary for muscle contraction and nerve impulses and is also important in regulating the speed of many cell activities. Phosphorus is used in the processes that supply energy in all cells, and it is a main ingredient of cell membranes. Each cell must be supplied with the correct level of each mineral at all times. An overabundance can injure or poison cells, and a deficiency can make a cell function abnormally or prevent it from functioning.

The skeletal system helps maintain homeostasis of minerals through two activities. First, extra minerals are taken out of the blood by the bones when their concentration begins to rise above the proper point. This situation often occurs after one has eaten calcium-rich foods such as dairy products. Later, the level of minerals in the blood begins to drop below the proper concentration because they are being used by cells and are being lost in urine and perspiration. Then the bones put back into the blood just enough of the minerals they had stored so that the body cells always have enough. The skeletal system can also store toxic minerals such as lead.

Blood Cell Production

The skeletal system helps prevent changes in the amount of cellular components of the blood (red blood cells, certain types of white blood cells, platelets) by producing them whenever their numbers drop too low. Blood cell production is the one function of the skeletal system that is not performed by the bone material; rather, it is accomplished by specialized **bone marrow** tissue within the bones (Fig. 9.4).

Red blood cell production is increased when oxygen in the blood begins to drop. With more red cells, the oxygen is restored to normal levels. As long as oxygen levels remain high, red cell production is slow. Thus, the skeletal system helps keep red blood cell numbers and oxygen levels proper and fairly stable.

White blood cells play a variety of roles, including defending against infection and controlling the inflammatory response. Platelets prevent the loss of blood by helping it clot. The mechanisms controlling the bone marrow so that levels of white blood cells and platelets remain within a normal range are not clearly understood.

The production of blood cells occurs only in red bone marrow. In adults, this marrow is found within the bones of the head, the trunk, the arm bones at the shoulder, and the leg bones at the hip. The rest of the marrow is yellow bone marrow, which stores fat molecules. Yellow marrow is converted to active red marrow when the body needs an extra supply of blood cells and is converted back to yellow marrow once the need has been met.

BONES

Though the bones are of many sizes and shapes, they are all built of the same types of materials. Each component is in the same position relative to the others, and each contributes to one or more of the five functions of the skeletal system.

Bone Tissue Components

The bony material (**bone matrix**) in bones contains three types of cells. **Osteoblasts** are cells that produce bone matrix by first secreting fibers made of collagen and then coating the fibers with mineral materials. In this way, the osteoblasts build bone, repair damaged or broken bone, and place

FIGURE 9.4 Structure of a bone.

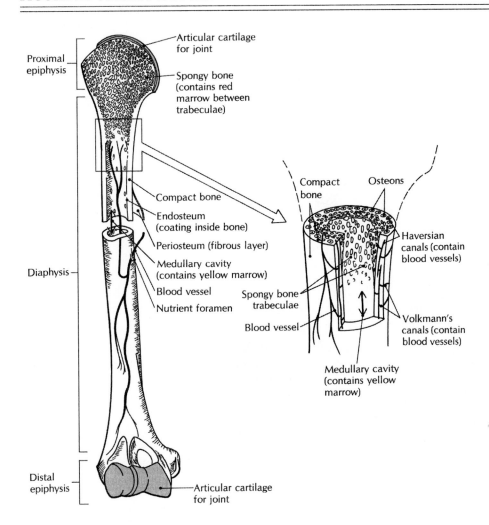

extra amounts of certain minerals found in the blood into bones for storage (Fig. 9.5). Osteoblasts also activate osteoclasts (see below) by secreting signaling materials (e.g., IL-6) (see Chapter 15).

As the osteoblasts work, many of them become surrounded and entrapped by the mineralized material they produce. These cells become quiescent and are called *osteocytes*. They will remain in a retired state unless a severe condition such as a fracture develops in the bone.

Osteoclasts dissolve some minerals in the matrix whenever the concentration of these minerals in the blood drops too low, restoring blood mineral concentrations to normal levels. This action is regulated by many factors (e.g., IL-6, hor-

mones). Osteoclasts also remove unwanted bone material when bones have to be remodeled or repaired. Osteoblasts often fill the vacated areas with new bone matrix at a later time.

The activities of the three types of bone cells are carefully controlled by a variety of hormones and other substances so that minerals are simultaneously being added to and removed from the bones (see Chapter 14). Since the speed of these two processes is not always equal, the bones may gain minerals at one time and lose them at another. These control mechanisms usually assure preservation of homeostasis in the body.

Since minerals cannot be easily deposited into bone matrix unless fibers are present, the fibers

FIGURE 9.5 Bone tissue: compact bone and trabecular bone.

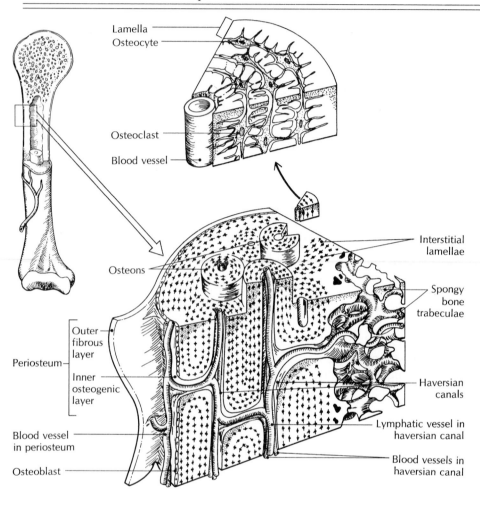

Lamella
Osteocyte
Osteoclast
Blood vessel
Osteons
Outer fibrous layer
Periosteum
Inner osteogenic layer
Blood vessel in periosteum
Osteoblast
Interstitial lamellae
Spongy bone trabeculae
Haversian canals
Lymphatic vessel in haversian canal
Blood vessels in haversian canal

in bone matrix play a role in mineral storage. The fibers make up about one-third of the bone matrix in young adults; the remainder consists of mineral salts. The fibers hold the minerals together and keep them from cracking. Therefore, the reinforcing fibers contribute to three other functions of the skeletal system: support, protection from trauma, and movement.

About 90 percent of the minerals in the matrix consist of calcium and phosphorus. The presence of minerals provides the means by which the skeletal system stores minerals. The minerals also make the bone hard and rigid. These two properties allow the bone to provide good support and protection from trauma. The hard and rigid bones also provide secure anchoring points and levers to aid the muscles in performing varied movements.

Bone Tissue Types

Cortical Bone   Bone cells produce two types of bone tissue. One type forms the outer layer of the bone (*cortical bone*). This bone tissue is made up of many long thick tubes of bone matrix called *osteons* (Figs. 9.4 and 9.5). The osteons are welded tightly together with bone matrix; therefore, this type of tissue is also called *compact bone*. Old osteons are always being gradually dissolved and replaced with new osteons.

Trabecular Bone The second type of bone tissue is inside the bone. It consists of small pieces of bone matrix called *trabeculae*; therefore, this type of bone tissue is called *trabecular bone*. The trabeculae are of varied shapes, including needles, chips, and flakes, which are fused together by bone matrix. However, unlike the osteons in cortical bone, the irregular arrangement of the trabeculae leaves many open spaces. Since this arrangement provides a structure resembling that of a sponge, this is also called *spongy bone*. However, since trabecular bone is quite hard, rigid, and rough rather than soft and spongy, it might be better named coral bone.

Other Tissues

Bones contain more than just bone tissue. For example, the hollow shell formed by cortical bone and the arrangement of trabeculae in spongy bone leave much space inside a bone, which is filled with bone marrow (Fig. 9.4). The outer surface of compact bone is covered by a tough layer of fibrous material called the periosteum, which serves as a place of attachment for ligaments and tendons. Additionally, where the surface of a bone meets another bone to form a joint, the bone may have a coating of cartilage, which may help join the bones or help them move. Finally, bones have blood vessels and nerves to serve the parts already mentioned. Therefore, bones are complex, dynamic, living parts of the body that change continuously. Some alterations are reversible physiological changes, such as removing and adding minerals, while others are biological age changes.

AGE CHANGES IN BONES

The matrix and the cartilage in bones seem to undergo most age changes; the functional capacity of bone cells and bone marrow seems to remain largely unchanged regardless of age. Some bone cells may function slower with age, though this appears to be due to changes in the control signals they receive. If the bone cells are stimulated, as when a fracture occurs, they resume rapid functioning. Exceptions are osteoblasts in the endosteum, which covers the inner surfaces of bones (Fig. 9.4) These osteoblasts have an age-related decrease in sensitivity to stimulation by vitamin D. Sensitivity may decline by 60 percent by age 50. This change contributes to age-related thinning of bones.

Bone Matrix

Age changes in bone matrix are complex and are far from being understood. This is partly due to the variety of factors that influence bone matrix. The factors include genes; amount of exercise; nutrition; levels of hormones; amount of exposure of skin to sunlight; levels of chemicals in the blood; and the functioning of skin, intestines, and kidneys.

Proteins and Minerals With aging, the balance between the amount of protein and the amount of minerals in bone matrix shifts in favor of the minerals. Therefore, bones become more rigid, brittle, and likely to fracture.

Quantity The quantity of bone matrix decreases with aging because matrix formation becomes slower than matrix removal. This decline may begin in some individuals at age 20, and by age 30 most people are losing bone matrix. It seems that by age 35 everyone has begun to lose bone at a substantial rate.

Structure At first, only trabecular bone is removed. In this type of bone, all the trabeculae become thinner and weaker (Fig. 9.6a). Trabeculae can become thicker and stronger again if osteoblasts are stimulated to replace the missing matrix. This often occurs when a person who has been sedentary begins to exercise. However during aging, some trabeculae disappear completely and cannot be replaced. The weakening of the bone at that spot is permanent. Also, as the matrix joining trabeculae dissolves, some trabeculae become disconnected from the others and can no longer contribute significantly to bone strength.

The decline in cortical bone is not detected until about age 40, and the loss is quite slow until age 45. The rate of loss then begins to increase significantly, though it remains approximately half the rate of trabecular bone loss. The loss of cortical bone takes place on the inside of the bone only, and the process proceeds outward toward the surface. Therefore, the layer of cortical bone becomes thinner while the overall width and length of the bone remain the same.

During the removal and replacement of cortical bone matrix, old mature osteons are gradually dissolved and shrink while new osteons are formed next to them (Fig. 9.6b). Small portions of the old osteons often remain behind. As years pass, the number of osteon remnants and the number of new osteons increase. As a result, the number of points of fusion among the osteons increases, causing the matrix to become weaker. At first the new osteons that form fill all the space left by the old ones. However, as a person gets older, the new osteons fail to fill these spaces completely and the number of gaps between the osteons increases. This change in structure also weakens the matrix.

Effects of Menopause Most experts agree that the rates of loss of trabecular and cortical bone matrix in women are increased by menopause, the time when menstrual cycles cease. Menopause usually takes place between ages 45 and 55. As it occurs, the production of the hormone estrogen by the ovaries is greatly reduced. Actually, since estrogen production probably drops gradually in the years just before menopause, the effects of declining estrogen begin before menopause.

FIGURE 9.6 Age changes in bone tissue: (a) Trabecular bone. (b) Cortical bone.

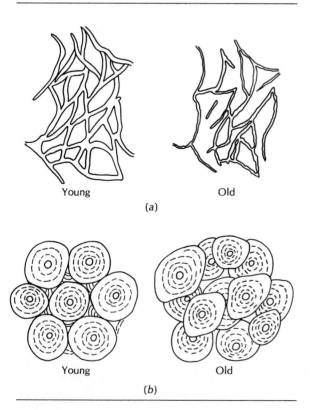

Young   Old

(a)

Young   Old

(b)

The combined effects of aging and menopause result in a loss of 15 percent to 20 percent of trabecular bone in the 10 years after menopause. This is two times to three times faster than the rate of loss in women before menopause or the rate of loss in men. As a result, very old women may have only half the amount of trabecular bone they had at age 25. Men have lost only two-thirds as much trabecular bone during the same period.

Cortical bone loss also accelerates because of menopause, with a loss of 10 percent to 15 percent of the cortical bone in the decade after menopause. This is a threefold to fourfold increase over the rate of loss before menopause. The rate of loss eventually slows down so that by age 70 it has dropped to the same rate found in men of that age. As with trabecular bone, a very elderly woman probably has less than half the amount of cortical bone she had during her twenties. Again, bone loss in a man of the same age is one-third less.

Therefore, because of menopause, a very elderly woman can expect to have considerably less bone material than does a man of the same age. This difference in the amount of bone material is usually made greater because men generally have more bone matrix than do women when bone loss begins.

Variability in Loss Trabecular bone loss begins earlier and occurs faster than does cortical bone loss. Since some parts of the skeleton have a higher proportion of trabecular bone, different regions have different rates of decreases in matrix. We will examine two important examples.

First, vertebrae are composed mostly of trabecular bone. Therefore, they begin to lose bone sooner and lose more bone than do bones in the arms and legs, which contain mostly cortical bone. Because of this, there is a higher incidence of vertebral fractures among the elderly, especially elderly women.

The second example involves the bone in the thigh called the femur (Fig. 9.1). The upper part of the femur, which joins with the pelvis at the hip, contains a high percentage of trabecular bone, while the long central shaft is made up almost entirely of cortical bone. Therefore, the upper end loses bone earlier and faster than does the shaft. This contributes to the higher rate of hip fractures as age increases.

Consequences Although age-related alterations in the composition, quantity, and structure of bone matrix result in weakening of the bones, the degree of weakening normally is not great enough to reduce substantially the reserve capacity of the bone matrix. Unless a very heavy load or strong force is applied, the bones are able to provide support and protection for the body as long as a person lives. Of course, large forces from accidents and severe falls cause fractures more frequently as a person ages. Since women end up with less bone matrix than men do, they are at higher risk for fractures.

Fractures are painful, hinder or prevent normal activities, and can lead to serious complications such as infection. Treatment of fractures is often quite expensive. Though elderly individuals who develop a fracture face the same problems, the adverse effects multiply with increasing age because healing of a fracture proceeds slower as one gets older. Slower healing can mean prolonged immobility, which increases the risk of complications, such as bedsores, blood clots, and pneumonia. Prolonged immobility also leads to faster loss of matrix, which in turn increases the risk of developing another fracture.

To assure that a normal skeleton serves a person well, it is important to compensate for the weakened condition of the aging skeleton. One way to do this is to avoid abusing the skeleton. Since falling is among the most common causes of skeletal abuse resulting in fractures among older people, reducing falls is of prime importance. A second way to assist the skeleton is to reduce the loss of bone matrix.

Minimizing Loss of Matrix Though the cause of bone matrix loss with aging is not known, much has been discovered about factors that modify the rate of loss. Many of these factors are easily regulated; therefore, much can be done to reduce the loss of bone matrix. In so doing, individuals can make significant contributions to the strength of the skeleton and its ability to serve them.

The condition of a person's bone matrix depends on how well it is treated throughout life. Much more benefit can be derived from taking steps to assist in building and maintaining bone matrix through early and late adulthood rather than just in old age, after a considerable amount of matrix has been lost. While the loss of bone matrix can be slowed at any age, little of the bone matrix and bone strength that are lost early in adulthood can be replaced later in life.

TABLE 9.1 RECOMMENDATIONS FOR BUILDING AND MAINTAINING BONE MATRIX AND REDUCING THE RISK OF OSTEOPOROSIS

Keep physically active: Nearly any activity that puts body weight on the feet helps.

Consume enough calcium: Eat calcium-rich foods. Take calcium supplements if necessary.

Maintain adequate levels of vitamin D: Eat foods containing vitamin D. Spend some time in the sun each day. Take vitamin D supplements if necessary.

For women, maintain adequate levels of estrogen: Avoid activities or operations that stop menstrual cycles. Use estrogen replacement therapy when necessary.

Avoid excessive consumption of alcohol.

Avoid smoking.

Avoid excess caffeine in the diet: Regular coffee, tea, and many carbonated beverages contain caffeine.

Do not consume excess phosphates: High levels of phosphates are present in many carbonated beverages.

Choose antacids that contain calcium, not aluminum.

Use corticosteroids only when needed: Corticosteroids are commonly used to reduce pain and inflammation.

Avoid consuming excess insoluble fiber in the diet: Wheat bran and certain bulk laxatives contain high levels of insoluble fiber. Excess fiber is usually a problem only for people who already consume low levels of calcium.

Numerous steps can be taken to build a large reserve of bone matrix before age 35 and minimize decreases in bone matrix at any age (Table 9.1). Each of the steps in the table contributes to one or more of the following: (1) maintaining high calcium levels, (2) stimulating matrix production, and (3) inhibiting matrix removal. Of course, elderly individuals and persons with known diseases should seek qualified professional advice before changing their normal activities.

OSTEOPOROSIS: A BONE DISEASE

Though normal bone matrix retains much of its strength throughout life, many individuals develop a disease called *osteoporosis*, which causes substantial reductions in the quantity and strength of bone matrix. Affected individuals develop bone fractures quite easily even when carrying out ordinary daily activities or simply walking or sitting.

Osteoporosis means "bones with pores." This name is appropriate since osteoporosis causes matrix production to be much slower than matrix formation, leaving bones full of holes. Affected bones become thin, hollow, and fragile.

Type I, or Postmenopausal, Osteoporosis

There are two types of osteoporosis. *Type I osteoporosis* is also called *postmenopausal osteoporosis* because it usually affects women after menopause. This occurs because the lowered estrogen levels after menopause seem to be the main factor in bone deterioration. Type I osteoporosis rarely occurs in men since the level of the hormone testosterone in men does not decline dramatically with aging.

Postmenopausal osteoporosis affects trabecular bone more than it affects cortical bone and thus leads to fractures in regions of the skeleton that consist largely of trabecular bone. These regions include the vertebrae, the neck of the femur near the hip, the radius near the wrist, and the humerus near the shoulder (Fig. 9.1).

Type II, or Senile, Osteoporosis

Type II osteoporosis is also called *senile osteoporosis* because it usually affects people of advanced age, especially those over age 60. The late appearance occurs because senile osteoporosis affects cortical bone more that trabecular bone. Type II osteoporosis occurs twice as frequently in women as in men.

Though senile osteoporosis affects many areas of the body, most of the resulting fractures occur in the neck of the femur. Other fractures occur in the radius near the wrist, the humerus near the shoulder, the tibia (shin bone) near the knee, and the pelvis (Fig. 9.1).

Incidence

Osteoporosis affects more than 24 million Americans, and this number is rising as the number of elderly people increases. Approximately 80 percent of those with osteoporosis are women, and as many as 60 percent of all women over age 60 have osteoporosis. In the three decades following age 50 and based on World Health Organization standards, the incidences of osteoporosis among women are 15 percent, 25 percent, and 40 percent respectively. A substantial percentage of women classified as not having osteoporosis have serious thinning of bone. The incidences among men are 33 percent less.

Effects

Each year osteoporosis causes more than 1.5 million fractures. Almost half are fractures of vertebrae, while about 20 percent are hip fractures.

Vertebral Fractures Seven of eight vertebral fractures occur in women. The incidence of these fractures in women rises quickly and continuously, beginning soon after menopause. Vertebral fractures occur in about two-thirds of all women over age 65.

Vertebral fractures caused by osteoporosis are usually *crush fractures*, or fractures in which the supporting part of a vertebra, called the body, becomes so weak that it collapses (Fig. 9.7b). When this happens, the upper part of the vertebral column settles down on the part below the fracture. These fractures often happen spontaneously or when a person is lifting a heavy object.

Crush fractures produce serious problems. Extreme pain often occurs because the nerves extending out from the spinal cord become pinched where they pass between the collapsed vertebrae. The misalignment of the vertebrae limits mobil-

FIGURE 9.7 Vertebrae and slightly movable joints: (a) Young vertebrae and joints. (b) Crush fractures. (c) Age changes. (d) Intervertebral joints with osteoarthritis.

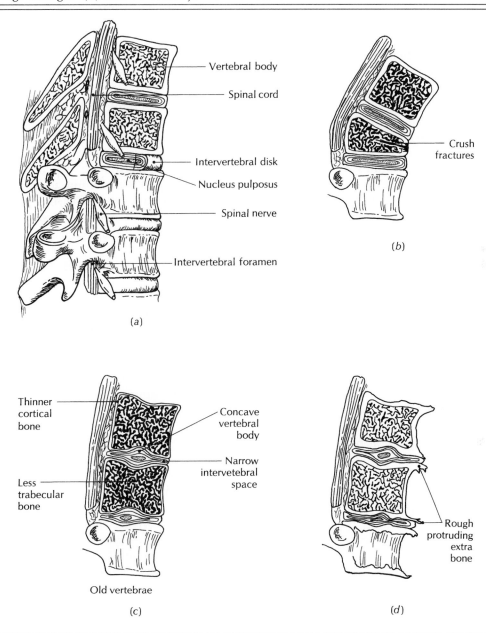

(a)
- Vertebral body
- Spinal cord
- Intervertebral disk
- Nucleus pulposus
- Spinal nerve
- Intervertebral foramen

(b)
- Crush fractures

(c) Old vertebrae
- Thinner cortical bone
- Less trabecular bone
- Concave vertebral body
- Narrow intervetebral space

(d)
- Rough protruding extra bone

ity, and the settling down of the vertebrae results in a decrease in height and a hunched-over or humpbacked posture. These people may have a drastically altered appearance, and their clothing often does not fit properly. The poor posture also produces complications in other systems, such as difficulty breathing and poor circulation. All these changes have an impact on the ability of affected individuals to care for themselves. Their social interactions change, they frequently have lower self-esteem and suffer from depression, and they may encounter problems in performing occupational tasks. These effects also affect the people in their families and communities.

Hip Fractures   The incidence of hip fractures is low until about age 60, after which the rate of occurrence increases gradually each year until about age 70. Then the rate of occurrence rises much more quickly each year. One-third of all women and one-sixth of all men who reach age 90 will have had a hip fracture due to osteoporosis.

Like vertebral crush fractures, many hip fractures occur spontaneously. Others often result from falling. It is sometimes difficult to tell whether a person fell because a hip fractured or fractured a hip because of a fall.

Fractures of the hip caused by osteoporosis lead to significant problems that are different from those resulting from vertebral fractures. Over half the individuals who suffer a hip fracture lose the ability to walk without assistance. Many people with hip fractures no longer can perform normal daily activities such as bathing and dressing without help. Between 15 and 25 percent of individuals with a hip fracture need to enter an institution for extended care. Most of these patients will never be able to return to living in the community. As with vertebral fractures, all-encompassing shifts in lifestyle are imposed on people who suffer hip fractures due to osteoporosis, and hip fractures have an impact on people associated with hip fracture patients. Finally, 20 percent to 30 percent of those who have osteoporosis-related hip fractures die within one year as a result of complications such as pneumonia and blood clots.

Causes

The cause of osteoporosis remains unknown, but the changes in bone matrix brought about by Type I osteoporosis in women result primarily from drastic reductions in estrogen. Low levels of estrogen result in profound changes because estrogen helps build and maintain bone matrix through numerous complex mechanisms. Type I osteoporosis occurs relatively infrequently in men, but when it does occur, it seems to follow abnormally large decreases in testosterone.

Type II osteoporosis results primarily in an age-related decrease in vitamin D activation by the kidneys. In addition, the aging intestines seem to become less sensitive to vitamin D and less able to respond to bodily needs for calcium. The result is a declining supply of calcium to body cells, leading to breakdown of bone matrix. Since the intestines do not absorb enough calcium, the matrix that is destroyed is not replaced.

Diagnosis

Diagnosing osteoporosis is very difficult, and in almost all cases individuals do not find out that they have osteoporosis until they have suffered a fracture. The diagnosis is difficult because the appropriate tests are dangerous (e.g., radiation, surgery), time-consuming, expensive, and difficult to interpret. In addition, to truly determine that a person has osteoporosis, that person should be tested regularly every few years to track the rate of loss of bone matrix.

Modifiable Risk Factors

The best way to protect oneself from the effects of osteoporosis is to build as much bone matrix as possible before age 35 and keep the deterioration of matrix as slow as possible thereafter (Table 9.1). The first four items in the table seem to be the most important. Following these recommendations will minimize the modifiable risk factors for osteoporosis and diminish the incidence of fractures and the destructive immobility that usually follows.

Lifelong involvement in weight-bearing activities such as walking and running is one of the best ways to minimize the effects of normal bone demineralization and decrease the chances of developing osteoporosis. However, any increase in weight-bearing exercise by sedentary older individuals will be helpful. Any strenuous activity will slow the demineralization process.

Intrinsic Risk Factors

While all people should take the appropriate steps to protect themselves from osteoporosis, this is especially important for those who are intrinsically at high risk. These high-risk categories include being female, having early menopause, having the ovaries removed, being white or Asian, having fair skin, having relatives with osteoporosis, being very thin, having kidney disease, having thyroid or parathyroid gland disease, having an intestinal disease that inhibits calcium absorption, and having chronic bronchitis or emphysema. Of course, belonging to more than one category places a person at even greater risk. Individuals at very high risk may benefit from diagnostic testing by qualified professionals.

Treatments

Though prevention is the key to success in battling osteoporosis, individuals who have already lost much bone matrix because of this disease can be helped to some degree.

Strengthening Bone Many treatments have been shown to slow the loss of bone matrix in at least some individuals. Often these treatments are based on the recommendations in Table 9.1.

For some individuals, such as very sedentary older women, vigorous exercise can reverse the process of demineralization and increase the amount of minerals in bones subjected to heavy loads. The length of time over which more minerals will be added to the bones depends on the nature of the exercise regimen. Regardless of the nature or duration of the activity, however, demineralization of the bones will eventually resume. If a high level of physical activity is continued, demineralization occurs at a slow rate. If the exercise is stopped, demineralization soon recurs at a rapid rate.

Many researchers believe that the exercise-induced addition of minerals to bones or the slowing of bone demineralization decreases the risk of fractures among the elderly. Exercise such as walking a mile three times a week has been shown to reduce the risk of fractures in many older individuals. Though exercise alone may be beneficial, the most successful treatment plans incorporate several recommendations. For example, an exercise program may be combined with dietary supplements of calcium and vitamin D.

Since changes in bone matrix occur slowly, treatment programs should be continued for many years. Because of the heterogeneity and the higher incidence of diseases among the elderly, a complete assessment of an older person should be performed before a treatment program is initiated.

The most successful treatment programs for postmenopausal women usually include estrogen replacement therapy, and most women can be helped by such therapy. However, there is a small risk from complications such as the formation of blood clots and the development of uterine or breast cancer. Women at high risk for these complications probably should avoid estrogen therapy. For other women, the risks are very small when estrogen is administered with certain types of progesteronelike hormones. Women receiving such treatments have much higher life expectancies than do women who do not receive them. Not only are the effects of osteoporosis reduced, but with the proper progesterone the risk of other diseases, such as atherosclerosis, is reduced. To be most effective, estrogen therapy should begin at menopause and continue for up to 10 years afterward.

The use of other substances for treating osteoporosis is increasing. Biphosphonates (e.g., alendronate, etridonate) reduce the risk of vertebral fractures, and are most effective when combined with estrogen. Biphosphonates can cause inflammation of the esophagus and stomach. Fluoride has little effect on the femur and results in brittle matrix. Calcitonin is expensive and causes painful calcium deposits.

Avoiding Injury A second aspect in treating osteoporosis patients involves minimizing traumatic injuries that may make weakened bones fracture. One way of doing this is to refrain from putting strain on the skeleton by, for example, lifting heavy objects. Perhaps an even more important factor is reducing the risk of falling, one of the most common causes of such injury among the elderly. Falling causes so many fractures that it has become an area of specialized research.

The risk of fractures and other injury from a fall increases with age. Reasons beyond the decreased strength of bones include weaker muscles and slower reflexes to break the fall and reduced subcutaneous fat and muscle mass to absorb the shock.

Falls are more common as age increases, partly because of the increased occurrence of diseases such as atherosclerosis, stroke, and parkinsonism. Other contributing factors include poor vision, age changes in the ears, muscle weakness, joint stiffness, altered gait, slow reflexes, and certain medications.

Much can be done to reduce the incidence of falls, including providing adequate lighting and grab bars, avoiding slippery surfaces such as wet or highly waxed floors, removing obstacles such as throw rugs, and wearing well-fitted shoes.

After a Fracture   Various combinations of approaches are employed to help people who have sustained a fracture. Surgery may be performed to quickly mobilize the individual because surgical repair compensates for the slow healing of bones in older people. Rapid mobilization reduces further bone and muscle deterioration and the risk of blood clots, pneumonia, and psychological problems such as depression. Physical therapy and the use of support devices such as a back brace or cane can also help restore a person to activities. Medications are often prescribed to reduce pain.

In summary, four weapons are used to treat osteoporosis: reducing the loss of matrix; replacing matrix to strengthen bones; preventing injury; and assisting in the recovery from fractures. Currently, all except replacing bone matrix can be successfully employed.

JOINTS

Functions

The bones that constitute the skeletal system are held together by special structures called *joints*. By furnishing strong attachments, joints contribute to the support provided by the skeletal system. They also protect the body from traumatic injury by absorbing shock and vibration in two ways. One way is by allowing the bones to move somewhat, which permits the skeletal system to yield to sudden physical forces. The other way derives from the cushioning provided by the fluids and resilient cartilage found in many joints. Because of these two features, joints can prevent much of the damage to delicate parts of the body caused by jolting forces from activities such as running and jumping. Assistance with movement derives, of course, from the various movements of bones permitted by the joints. The joints do not help the skeleton store minerals or produce blood cells.

Immovable Joints

There are three main types of joints in the body. An *immovable joint* consists of tough collagen fibers that bind bones tightly together. The unyielding strength of the collagen, together with the tight fit of the bones, essentially eliminates shifting of the bones. Among the immovable joints are the suture joints between skull bones (Fig. 9.1). These joints keep the shieldlike skull bones in place to support and protect the brain.

Age Changes   As people age, the collagen fibers between the bones at immovable joints are coated with bone matrix, and so the space between the bones gets narrower. Eventually the bones may fuse together. Thus, immovable joints improve with aging because they get stronger.

Slightly Movable Joints

The second type of joint is the *slightly movable joint*. There is a layer of cartilage between the bones joined by these structures. Some of these joints have ligaments, which help hold the bones together. Ligaments are cablelike structures consisting primarily of collagen fibers.

There are two kinds of slightly movable joints. One kind contains *hyaline cartilage*, which is a smooth, slippery white substance with the consistency of hard rubber. Slightly movable joints with hyaline cartilage join the ribs to the sternum (Fig. 9.1). The limitation of movement in these joints allows the ribs to support and protect the lungs, the heart, and other organs in the chest cavity while permitting enough movement of the ribs to allow breathing.

In the other kind of slightly movable joints, *symphysis joints*, the bones are separated by a pad of *fibrocartilage*. This type of cartilage is also smooth, slippery, and resilient. It is stronger than hyaline cartilage because it contains many more thick collagen fibers. The fibers add toughness by binding the rubbery cartilage matrix together.

Symphysis joints are located where greater strength is needed, such as between the bodies of the vertebrae (Fig. 9.7a), where the *intervertebral disks* of fibrocartilage permit limited and smooth bending of the vertebral column. Together with the vertebrae, the disks support the weight of the body. Each disk contains a soft center called the *nucleus pulposus*. The nucleus pulposus helps with support and is important in shock absorption. The intervertebral joints also contain strong ligaments to hold the vertebrae together while allowing a limited amount of bending and twisting of the spine.

Age Changes   With aging, the hyaline cartilage in slightly movable joints becomes stiffer because of a decrease in water and an increase in hard and rigid calcium salts within the cartilage. The fibers in the ligaments develop more cross-links with age, causing the ligaments to become stiffer and less elastic. The combination of these age changes reduces the movement allowed. For example, such stiffening in the chest area makes breathing more difficult.

Aging causes the fibrocartilage disks in symphysis joints to lose water and gain calcium. These changes may contribute to age-related stiffening of the joints and a decrease in the movement permitted by the vertebral column. The nucleus pulposus becomes weaker and somewhat crumbly, decreasing its ability to provide support for the body and cushioning for the spinal cord and head.

The center region of each vertebral body weakens with aging (Fig. 9.7c). The weight of the body then forces the central part of each intervertebral disk to expand into the body of the vertebra, forming a concave region. This alteration in structure seems to place more of the weight of the body onto the outer edge of the intervertebral disk, compressing it somewhat. The net result is a decrease in the height of the body with aging.

Decreasing height with age also has other causes. There is a thinning of the cartilage in other joints, such as the knees and hips. The weakening of muscles and a decrease in muscle tone often lead to poorer posture, further reducing overall height. The rate of loss of height is slow at first and becomes more rapid with age. Like all collagen, the collagen in the ligaments of the intervertebral joints becomes shorter, stiffer, and less elastic. These changes further reduce the mobility of the vertebral column.

Overall, age changes in symphysis joints reduce the ease and range of motion that these joints provide. This hampers bending and twisting of the vertebral column and makes performing activities such as tying shoes, picking up objects, and dancing more difficult or less enjoyable. The loss is not great enough to be a serious threat to homeostasis. Moreover, as will be discussed below, the decline in movement of symphysis joints can be minimized and even reversed by exercise.

Freely Movable Joints

The third type of joint is the *freely movable joint*, which is the most common type. These joints make up virtually all the joints in the arms, legs, shoulders, and hips; the joints between the ribs and the vertebrae; and the joints between the vertebrae except the joints between vertebral bodies. The joint between the lower jaw and the skull, the temporomandibular joint (TMJ), is the only freely movable joint in the head.

The bones joined by freely movable joints are separated from each other by a narrow space called the *synovial cavity* (Fig. 9.8). This cavity is surrounded by a thin *synovial membrane*, which constantly secretes a fluid (*synovial fluid*) into the cavity. At the same time, the membrane removes the old fluid. Synovial fluid contains water and protein molecules. This mixture is somewhat thick and very slippery, allowing the bones to slide over each other easily. It also absorbs some shock sustained by the joint.

FIGURE 9.8 Structure of freely movable joints.

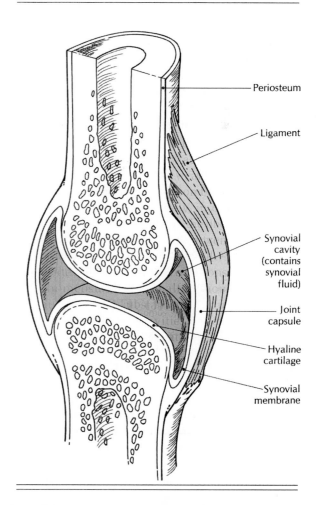

Periosteum

Ligament

Synovial cavity (contains synovial fluid)

Joint capsule

Hyaline cartilage

Synovial membrane

The end of each bone is covered by a layer of hyaline cartilage that is very smooth and somewhat resilient. Since the cartilage is lubricated by the synovial fluid, it is very slippery. The synovial fluid also supplies nutrients to the cartilage. The slippery cartilage permits easy movement and cushions the bones and the parts of the body they support.

Surrounding the synovial membrane is the thick sleevelike *joint capsule*, which consists mostly of flexible strong collagen fibers. The joint capsule helps bind the bones together and encases the synovial membrane for support. The flexibility and slight elasticity of the capsule allow the bones to move freely, though over a limited range.

Outside the joint capsule and extending from one bone to the other are cablelike *ligaments*, which also consist mostly of collagen fibers. Like joint capsules, ligaments bind the bones together and allow limited movement of the joint.

Motion allowed at a joint is also limited by the shapes of the bones and by muscles and tendons. While the joints must allow the bones to move easily and over enough distance to meet the needs of the body, limiting motion is important in preventing injury to muscles, nerves, and blood vessels. Excessive joint motion (e.g., joint dislocation) stretches, twists, and pinches these soft structures.

Age Changes   With aging, there is an increase in the amount of fibrous material in the synovial membrane and pieces of cartilage may form in it. These changes make the membrane stiffer and less elastic. The membrane also loses some of its blood vessels so that it is less able to produce and remove synovial fluid. Though there is disagreement about which age changes take place in the synovial fluid and the cartilage on the ends of the bones, it is generally agreed that these changes are slight and have little effect on the functioning of the joint.

More important than these age changes are changes in the joint capsule and ligaments. Because of an increased formation of cross-links among their fibers, these structures become shorter, stiffer, and less able to stretch. These changes make it more difficult to move and reduce the range of movement of the joint. Both changes cause the initiation of movement and the speed of movement to occur more slowly. This results in a reduction of the ability to maintain balance and take action to minimize the force of impact from a fall or another traumatic event. Thus, the aging of freely movable joints substantially reduces the ability of the skeletal system to provide cushioning and movement, resulting in increased injuries and diminished performance of activities.

The functioning of freely movable joints begins to decline at age 20. The joints move less easily and over less of a range as time passes. The decrease in blood vessels in joint structures results in slower healing of injured joints.

All these changes occur very gradually but unremittingly. It seems that only part of the reduction in functioning is due to age changes. Some change may be due to the accumulated effects of the small but repeated injuries sustained by joints during ordinary activities. Distinguishing true age changes from these other changes is difficult.

The progressive decrease in mobility caused by aging in both freely movable and slightly movable joints can be slowed by keeping physically active. Exercises that involve bending, stretching, and turning minimize the restraining effects caused by shortening of the collagen fibers. Some mobility that has been lost over time because of inactivity can be regained by initiating exercises that stretch and increase the flexibility of restrictive joint components such as ligaments. Exercise also seems to increase circulation to the joints. Such exercises reduce the risk of fractures and contribute to better balance, greater independence, and improved psychological well-being. However, when people of advanced age engage in new exercises, care should be taken to avoid injuring the joints.

DISEASES OF JOINTS

The problems caused by age changes in the joints are often compounded by a disease called *arthritis*, which means "joint inflammation." The name was chosen because arthritis results in injury and pain in the joints.

The incidence of arthritis increases with age. More than half the cases occur in people over age 65. In fact, arthritis is the most common disease among the elderly and is second only to heart disease in causing older people to visit a physician.

Different cases of arthritis vary greatly in severity. In some individuals the symptoms are so mild as to be barely noticed. At the other extreme, arthritis can cause excruciating and unremitting pain as well as deformity and crippling incapacitation. This disease results in more limitation in activity and more disability than does any other chronic illness. Only heart disease causes people to spend more days in bed.

There are more than 100 types of arthritis, and different forms are prevalent at different stages of life. A person may have two or more forms at the same time. The two types discussed below are the forms most frequently encountered in the elderly.

Osteoarthritis

Osteoarthritis is by far the most common type of arthritis in adults. It causes more than half of all cases of arthritis. Approximately 75 percent of people reaching age 75 will have OA in at least one joint. Most cases of osteoarthritis occur in women.

The cause of osteoarthritis is still unknown, there is no method of prevention, and there is no cure. It usually progresses continuously, though the rate of progress differs among individuals. Main risk factors for OA include injury to joints, inadequate treatment of injured joints, and extreme overuse of joints.

Osteoarthritis often affects weight-bearing joints, including the knees, the hips, and the intervertebral joints in the lower (lumbar) region of the vertebral column. The joints in the cervical vertebrae and those in the fingers are also frequent sites of this disease.

Effects   When osteoarthritis attacks freely movable joints, it causes breakdown of the cartilage between bones and the cartilage becomes rougher and softer and cracks. The cartilage becomes weaker and thinner because its cells are removing cartilage faster than they are replacing it. Because of these changes, the cartilage loses the ability to cushion and lubricate the ends of the bones, diminishing the operation of the joints.

So much cartilage may be removed that the hard ends of the bones bump and rub against each other. This contact can sometimes be heard and felt when a person moves. The bones respond to the resulting abuse by producing extra bone matrix at the joint. When this buildup occurs in arthritic finger joints, it may be observed as enlargements of the joints.

The new bone matrix produced is rough and sometimes jagged, and it abrades the softer tissues in the area, causing pain. As the bone matrix grows, it protrudes farther, making movement of the joint more difficult and reducing its range of motion because the edges of the bones bump against each other.

Other changes from osteoarthritis often reduce the action of the joint even further. The injured synovial membrane becomes more irregular, thick, and stiff. It may bind the bones abnormally by adhering more tightly to them. Pieces of cartilage and bony spurs sometimes break off from the bones and become lodged within the joint.

Osteoarthritis of the symphysis joints in the vertebral column causes the same type of extra bone formation that occurs in freely movable joints (Fig. 9.7d). This extra bone may cause pain by irritating surrounding tissues or pressing on nerves attached to the spinal cord, reduce ease of movement, and reduce the range of motion permitted by the joints.

Treatments   Treatment of osteoarthritis is aimed at slowing its progress and reducing the pain and disability it causes. Affected individuals can be taught how to perform activities in ways that minimize abuse of diseased joints. The use of canes and other devices that support some body weight helps in this regard. Mild exercise reduces stiffness and loss of range of motion, and a variety of medications relieve pain.

Severely diseased joints may be repaired surgically. Often the diseased joint is removed and replaced with an artificial one. Total hip replacement is a common example. Since replacement of intervertebral joints is impossible, surgeons may eliminate the joint by fusing the vertebrae above and below the joint. Though this procedure prevents further motion at the joint site, it relieves the pain and deformity that usually accompany osteoarthritis of the spine.

Rheumatoid Arthritis

Rheumatoid arthritis (*RA*) has a much lower frequency of occurrence than does osteoarthritis. Only about 1 percent of all adults have RA. Two of every three RA patients are women. Most cases begin between the ages of 30 and 40, and the number of cases increases with age.

Effects Rheumatoid arthritis usually attacks the freely movable joints of the wrists and hands as well as those in the ankles and feet; the joints closest to the ends of the fingers are spared. It sometimes affects other joints, including the shoulders, elbows, and knees. Like osteoarthritis, RA causes pain and loss of joint mobility. Unlike osteoarthritis, it often produces many other effects, including weakness, fatigue, and damage to organs such as the heart, blood vessels, lungs, nerves, skin, and eyes. This widespread damage occurs because RA can attack fibrous materials everywhere in the body. Another peculiarity of RA is that it goes into temporary remission in some individuals.

Though the cause of RA is unknown, the method by which it destroys joints is understood. The root of the problem lies in the immune system, which mistakenly identifies normal connective tissues as being foreign to the body. The immune system then reacts in its usual manner by trying to eliminate the "foreign" substances. In so doing, it kills and removes these normal connective tissue materials. This reaction is called *autoimmunity* since the body is attacking itself.

In joints, the immune system kills and removes cartilage, which is replaced with a unique type of scar tissue, called a *pannus* (Fig. 9.9). The pannus releases enzymes that destroy more of the cartilage. The immune system also removes bone material and the synovial membrane. As more normal tissues are eliminated, the pannus enlarges and spreads into the joint, substituting for normal components. All these activities cause considerable pain and joint swelling, and proper functioning of the joint becomes impossible.

As the joint weakens, the bones shift out of position. Sometimes, the pannus becomes calcified and stiff. Progressive calcification sometimes results in fusion of the bones. In addition, the ordinary scar tissue produced at the joint shrinks as time passes, pulling the bones farther out of alignment and locking them into abnormal positions. Thus, the joint becomes distorted and immovable. The result is crippling deformity, a condition most easily seen in the hands and feet.

Treatments There is no way to prevent or cure RA. The goals of treatment are the same as those for osteoarthritis: slowing the progress of the disease and minimizing pain and disability. Various medications may be prescribed to inhibit the immune system and relieve pain. Mild exercise helps maintain joint mobility. A variety of other treatment modalities may be initiated. Unfortunately, not all individuals respond well to these treatments.

FIGURE 9.9 Effects of rheumatoid arthritis on joint structure: (a) Normal joint. (b) Cartilage replaced with pannus. (c) Pannus and immune reaction remove cartilage and bone. (d) Bones fused by calcification of pannus.

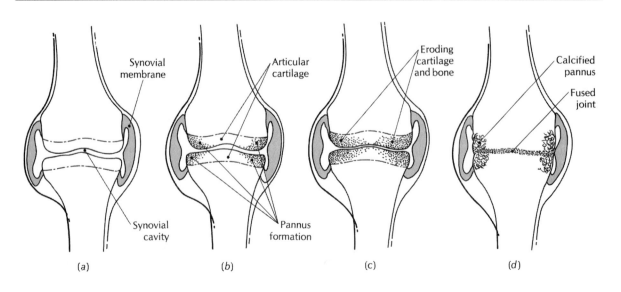

10

DIGESTIVE SYSTEM

The digestive system consists of a long tube extending from the mouth to the anus and the accessory structures attached to that tube. The tube has several names, including the *GI tract*, the *gastrointestinal tract*, and the *alimentary canal*. The regions of the GI tract include the *mouth*, *oral cavity*, *pharynx*, *esophagus*, *stomach*, *small intestine*, and *large intestine* (Fig. 10.1). These regions are specialized for certain aspects of digestion. The accessory structures that assist the functions of the GI tract include the *salivary glands*, *liver*, *gallbladder*, and *pancreas*.

MAIN FUNCTIONS FOR HOMEOSTASIS

Supplying Nutrients

The digestive system has six main functions. One function is to supply nutrients needed in several ways by body systems. For example, nutrients are needed for building and maintaining structures such as bones, producing substances such as glandular secretions and neurotransmitters, and supplying energy to power the operations of all body systems. To maintain homeostasis, the nutrient supply must be steady so that the body cells have adequate amounts of each required material at all times while not being exposed to excessive amounts of any nutrient.

Converting Foods to a Usable Form Five processes are involved in supplying nutrients at proper and fairly steady levels. One is converting foods to a usable form. This is necessary because although foods provide most of the nutrients later supplied by the digestive system, they are usually not in a form that can be used by the body. For example, it is impossible to swallow a whole apple, have chunks of meat float through the blood vessels, or have a piece of candy enter a brain cell. To pass through the circulatory system and into cells, foods must be converted into small molecules that are dissolved. The main aspects of this conversion process include mechanically breaking large pieces of food into small pieces by chewing; adding saliva to moisten food and dissolve small molecules; and chemically breaking large nutrient molecules into smaller ones by using enzymes in digestive juices.

FIGURE 10.1 The digestive system.

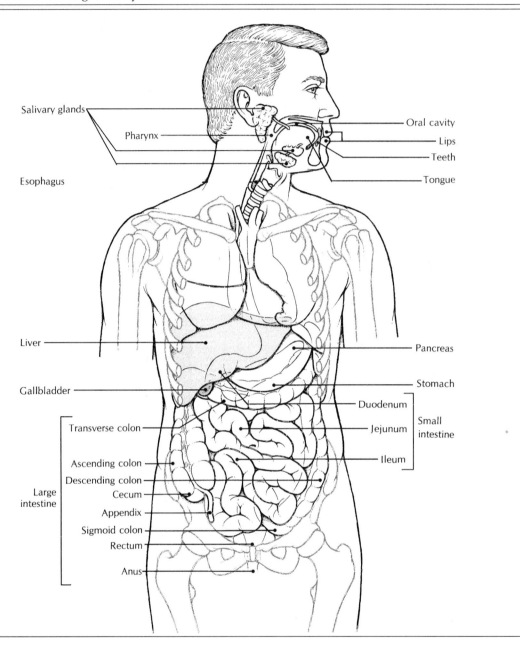

Salivary glands

Pharynx

Oral cavity

Lips

Teeth

Esophagus

Tongue

Liver

Pancreas

Gallbladder

Stomach

Duodenum

Transverse colon

Jejunum

Small intestine

Ascending colon

Descending colon

Ileum

Large intestine

Cecum

Appendix

Sigmoid colon

Rectum

Anus

Absorption Once nutrients are in a usable form, the digestive system moves them from the GI tract into the circulatory system. This step is called *absorption*. The blood and lymph distribute the absorbed nutrients to other body regions.

Manufacturing Certain Materials Some nutrients cannot be obtained from foods in adequate amounts. The digestive system compensates for some of these deficiencies by manufacturing cer-

tain nutrients. For example, only a portion of the substance called vitamin K, which helps form blood-clotting materials, is obtained from foods. The rest is produced by bacteria in the large intestine. Also, certain types of foods have inadequate amounts of the amino acids needed to repair muscle cells and build red blood cells. The liver can manufacture some of these amino acids so that they are supplied at proper and fairly steady rates even when they are not eaten regularly.

Storing and Converting Excess Nutrients Though certain foods have inadequate amounts of some nutrients, they often have a great abundance of others. Furthermore, people usually eat only occasionally during the day and eat only a few types of food. Therefore, the GI tract periodically absorbs large quantities of certain nutrients. To prevent body cells from receiving excessive amounts of these nutrients and becoming deficient in others, the liver performs the fourth and fifth steps in supplying nutrients: storing excess nutrients until they are needed and converting excess nutrients into other nutrients that are in low supply in the foods eaten.

Eliminating Toxins and Wastes

A second main function of the digestive system is eliminating toxins and wastes. Though most materials absorbed by the GI tract are beneficial, some are harmful to body cells (e.g., alcohol, detergents, industrial solvents, inappropriate medications, bacterial wastes). Body cells are protected from exposure to many of these substances because the liver eliminates them. The liver also eliminates toxic substances produced by body cells, such as ammonia and bilirubin. It converts harmful materials such as alcohol and ammonia into useful or harmless substances and secretes others (e.g., bilirubin) into the GI tract in a liquid called *bile*. Many substances in bile pass through the GI tract and are eliminated during a bowel movement.

Other Functions

Various parts of the digestive system have other functions, including helping with voice and speech (mouth region), storing blood (liver), regulating certain components in the blood (liver), and producing hormones (GI tract and pancreas).

THE CHEMISTRY OF DIGESTION

Recall that large nutrient molecules must be broken into smaller ones before they become usable. These large molecules include large *carbohydrates*, all *proteins*, large *lipids*, and all *nucleic acids* (Chap. 2). Chemical breakdown is necessary because these molecules cannot usually pass through cell membranes such as those lining the GI tract. Therefore, unless these molecules are broken down into smaller ones inside the GI tract, they cannot be absorbed into the blood. Even if they could enter the blood, they could still not pass through the cell membranes of body cells.

Using Water and Enzymes

The digestive system accomplishes chemical breakdown of large molecules by using water molecules to split them into their constituent parts. Large carbohydrates such as starch are split into simple sugars, proteins are split into amino acids, lipids such as fat are split into fatty acids and glycerol, and nucleic acids are split into nucleotides.

The digestive system assists this splitting through special molecules called digestive *enzymes*, which are secreted as part of the digestive juices. Digestive enzymes seem to work by properly positioning water molecules at the chemical bonds linking constituents and by applying pressure or tension to the water and nutrient molecules. As a result, the water molecules split and the fragments are used to separate the constituent parts of the nutrient molecules. This process is called hydrolysis.

Each enzyme molecule can act over and over again, splitting molecule after molecule without being destroyed. However, each type of enzyme can help split only one type of nutrient molecule. Therefore, many different enzymes are needed to split the many different types of nutrient molecules that are ingested.

Only large nutrient molecules need to be hydrolyzed. Small nutrients such as water, vitamins, minerals, simple sugars, and small lipid molecules (e.g., cholesterol) can be absorbed simply by being dissolved in digestive system fluids.

AGE CHANGES VERSUS OTHER CHANGES

Like the respiratory system, many parts of the digestive system are in more or less direct contact with substances and conditions from the external environment. Examples include a wide variety of foods prepared and eaten in diverse forms, physical trauma, dangerous chemicals, extreme temperatures, high internal pressures, and unusual and noxious substances in unregulated concentrations. Therefore, it is difficult to distinguish age changes from changes produced by regularly ingested materials. However, certain changes are observed in the digestive systems of essentially all people in the United States. These changes will be presented as age changes. Alterations in the digestive system that occur in only some people and alterations believed to be caused by dietary factors or factors besides aging will be identified where possible. Many of these alterations are discussed in the sections on abnormal changes.

ORAL REGION

The *oral cavity* extends from the lips and mouth to the back of the tongue, where the *pharynx* begins (Fig. 10.1). A moist membrane called the *oral mucosa* lines the inside of the oral cavity and covers the tongue. The oral cavity also contains the *teeth*. The *salivary glands* are connected to the oral cavity and secrete *saliva* into the cavity through tubes called *salivary ducts*. Many *muscles* for moving the mouth, the cheeks, the tongue, the lower jaw, and the region leading into the pharynx are also present around the oral cavity.

Oral Mucosa

The *oral mucosa* is similar in structure to the epidermis on the surface of the face except that it is thinner and does not have a surface layer of keratin. Like the epidermis, it serves as a barrier against microbes, chemicals, water, and physical trauma. However, because the oral mucosa is thin and lacks keratin, it is easily damaged and penetrated. For example, medications such as nitroglycerine pass through it quickly and easily.

The lining of the oral cavity also provides information about materials that enter the mouth, including their size, shape, texture, temperature, and chemical composition. Special neurons that detect various chemicals are located in the taste buds on the tongue. Impulses from these neurons provide the taste sensations of salt, sweet, sour, and bitter.

A third function of the oral mucosa is the production of a watery secretion that moistens the oral mucosa and foods. Moistening food dissolves some of its molecules, which can then stimulate the taste neurons on the tongue. This process also prepares food for absorption. The secretion from the oral mucosa also lubricates food, making it easier to swallow.

Age Changes   The lining of the oral cavity undergoes age changes that are similar to those that occur in the epidermis. For example, it heals more slowly. However, the oral mucosa can normally perform its functions rather well throughout life. (Age changes in taste perception are discussed in Chap. 6.)

Teeth

There are 32 *teeth* in the oral cavity. Sixteen of them form the upper row, which is attached to the upper jaw, and the others form the lower row attached to the lower jaw.

The exposed surface of each tooth is covered with a cap (*crown*) that is made of very hard *enamel* (Fig. 10.2). Internal to the enamel is a firm layer called the *dentin*. The dentin surrounds the soft innermost material—the *pulp*—which contains nerves and blood vessels serving the tooth. Some pulp nerve cells extend into the dentin.

The lower part (*root*) of the tooth is embedded in the jawbone and is composed of only dentin and pulp. It is surrounded by a layer of *cementum* and an outermost *periodontal membrane*, which attach the tooth to the jaw. Soft tissue called the *gum* surrounds each tooth where it projects from the jawbone.

The role of teeth in supplying nutrients is to cut, tear, and grind food into small pieces. Small pieces of food mix more easily with fluids, fit more easily into the GI tract, and are better exposed to digestive enzymes. Teeth also help in pronouncing words and affect the appearance of the face.

Age Changes   Though aging may have little effect on tooth enamel, the enamel may become stained from foods. It also becomes thinner with age because of normal wear from chewing hard materials. Faster thinning of the enamel results from frequently eating very acidic foods; habitually grinding the teeth, which often accompanies emotional tension; and excessively brushing the teeth, particularly with a stiff toothbrush. If enough enamel is lost, the underlying dentin may become exposed, and since the dentin contains nerve cells, the tooth may become sensitive to touch or extremes in temperature.

As age increases, the dentin is slower to repair itself when injured and often enlarges inwardly as the amount of pulp decreases. The loss of nerve cells in the pulp reduces the sensitivity of the teeth; this increases the risk of developing more serious tooth decay since it reduces a person's

FIGURE 10.2 Tooth structure.

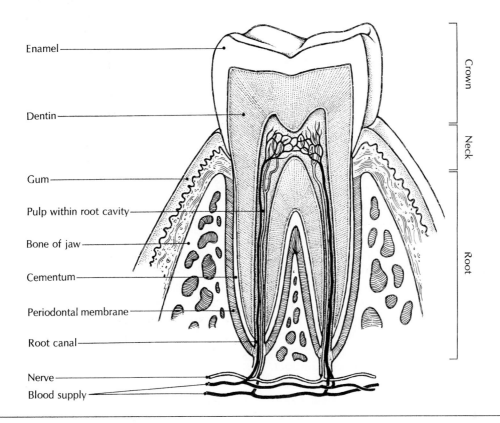

Enamel

Dentin

Gum

Pulp within root cavity

Bone of jaw

Cementum

Periodontal membrane

Root canal

Nerve

Blood supply

Crown

Neck

Root

ability to detect tooth problems. However, reduced pulp sensitivity also lessens the discomfort from dental procedures.

Though the cementum becomes thicker with aging, there is an overall weakening of the attachment of the teeth to the jaw, and age changes in bone cause the jaws to weaken. At the same time the gums recede, exposing more dentin, and so bacteria are better able to invade the base of the teeth and the spaces between the teeth and the jaw. The combination of these changes increases the incidence of disease around the base of the tooth (*periodontal disease*).

Periodontal disease is a risk factor for atherosclerosis. The mechanism by which periodontal disease contributes to atherosclerosis is not known.

Salivary Glands

The main *salivary glands* occur as three pairs of glands: the parotid, submandibular, and sublingual glands. Saliva produced by these pairs of glands passes through salivary ducts to reach the oral cavity. Some additional saliva is produced by small groups of cells and by individual cells in the oral mucosa. The production of saliva by the salivary glands is controlled by the autonomic part of the nervous system. Production is slow under resting conditions but can become rapid and profuse when food is present in the oral cavity.

Saliva is a watery solution that contains a mixture of minerals and proteins. The water in saliva functions like the secretion from the oral mucosa and helps to remove food from the teeth. Therefore, it also reduces bacterial growth and delays the onset of cavities. The minerals and proteins in saliva help preserve the mineral content of the enamel by neutralizing acids and replacing lost minerals. Some proteins inhibit the growth of certain types of bacteria and fungi. Finally, one salivary protein (salivary amylase) serves as an enzyme that helps break starch molecules into *maltose*. Maltose consists of two glucose molecules linked together.

Age Changes Though aging results in structural changes in the salivary glands, age changes do not significantly affect the chemical content of saliva produced by the main salivary glands. Also, aging causes no important changes in the amount of saliva produced either at rest or after stimulation by food.

Muscles

The muscles of the mouth and oral cavity are skeletal muscles under voluntary control. A few of these muscles open and close the mouth; others move the cheeks, tongue, and lower jaw. The movements of these muscles assist in eating, drinking, and speaking. Still other muscles in the tongue, the region near the back of the oral cavity, and the pharynx are important in swallowing.

Swallowing involves the coordinated action of many muscles. First, food other than liquids is formed into a mass. The food mass and liquids are then pushed to the back of the oral cavity and into the pharynx by the tongue. When the mass of material in the pharynx has become large, a reflex causes muscles above the mass to contract. Recall from Chap. 5 that the swallowing reflex ensures that the pharynx is continually cleared of food (Fig. 5.6). At the same time, the reflex causes muscles below the mass to relax so that the opening through the pharynx and into the esophagus enlarges. The remaining reflexive muscle contractions in the region of the pharynx cause the larynx to elevate, covering the opening into the larynx with a flaplike structure called the *epiglottis*. All these muscle contractions can be set into motion by voluntary contraction of muscles in the oral region and upper pharynx even if there is little or no material in the pharynx.

As the muscle contractions above the food move farther backward and downward, the food is pushed into the esophagus. At about this time the continued downward movement of the contraction wave above the food and the continued relaxation of muscles below it cause the food to be pushed all the way down the esophagus and into the stomach. The wave of contraction down the esophagus is called *peristalsis*.

Age Changes Muscles in the oral region undergo the same types of age changes as do all skeletal muscles. These changes, together with age changes in the nervous system, cause a slight weakening and reduced coordination in their functioning. There is a tendency to chew food less and swallow larger pieces. Under normal conditions these changes have no important effects on eating or speaking. However, when a person is in a stressful situation, they increase the risk of choking because large pieces of food do not pass through the pharynx as easily. Choking also may occur because food has entered the larynx, which may not be completely closed. If food and other materials being swallowed enter the respiratory system through the larynx, blockage of airways, pneumonia, and other respiratory problems may develop.

Bones and Joints

The principal bones of the oral region are the upper and lower jawbones, which support many oral structures. These bones are especially important in supporting the teeth. The lower jawbone is attached to the skull at the *temporomandibular joint* (*TMJ*). Proper operation of the TMJ is important for chewing, swallowing, and speaking.

Age Changes Though the jawbones and the TMJ undergo age changes similar to those in other bones and joints, these changes are so small that the functioning of these components is not affected.

Abnormal Changes

Oral Mucosa Many older people have difficulties with the oral mucosa. The reasons for these problems include atherosclerosis in arteries serving the oral region, dentures, medications, and many age-related diseases. The results include weakening, injuries and sores, and slower healing. Some medications and diseases affect the oral mucosa because they cause drying by lowering saliva production. Others alter the sense of taste. These undesirable abnormal changes can have adverse effects on nutrition and personality traits (e.g., increasing irritability).

Teeth The combined effects of all age changes in teeth and the passage of time increase the risk and incidence of spots of tooth decay called *cavities* (*caries*). While many new cavities develop, many are formed where previous cavities were filled by a dentist. With advancing age, new cavities occur more in the roots of the teeth than in the crowns.

Periodontal disease and cavities are a major source of diverse problems for the elderly. First, the pain from these conditions can reduce chewing. Less chewing leads to attempts to swallow larger pieces of food, and this in turn raises the chances of choking and developing indigestion. Difficulty with chewing also reduces the variety of foods eaten and promotes the selection of foods with little fiber. Malnutrition and constipation are common consequences of such choices. Oral discomfort can also affect speaking, emotions, and personality traits.

Second, tooth disease spreads infection from the teeth to other parts of the body. Third, tooth disease alters taste and can produce unpleasant taste sensations. Fourth, obtaining professional help to treat tooth disease is costly. Finally, an altered appearance from diseased teeth can affect a person's social interactions and self-image and cause considerable embarrassment. All these problems are made worse by the loss of teeth. The use of dentures can only partially compensate for functional changes resulting from tooth loss. Also, dentures are a main cause of injury, irritation, discomfort, and infections in the oral mucosa.

The higher rates of periodontal disease and cavities with age are the main causes of the high incidence of tooth loss among the elderly. On the average, people over age 65 have lost approximately 11 percent of their teeth. About 65 percent of those over age 65 have lost all their teeth in either the upper or the lower jaw, and about 40 percent have lost all their teeth. This number has declined from 50 percent over the last three decades because of better dental care and, possibly, the introduction of fluoride into drinking water. However, with more elderly people retaining more teeth longer, there has been an increase in the incidence of periodontal disease and cavities.

There are several ways to reduce or prevent dental diseases among elderly people and the younger people who will become the elderly of the future. Examples include drinking fluoridated water, especially during youth; getting regular professional dental care; and following a program of good dental hygiene. Good dental hygiene includes avoiding sweets, avoiding sugary beverages such as soft drinks, rinsing the mouth with water after eating, and brushing and flossing the teeth frequently. Since dental diseases at older ages usually result from an accumulation of ef-

fects during one's lifetime, it is important to start good dental practices during childhood and continue them throughout life.

Salivary Glands Though aging has no important effects on the functioning of the salivary glands, a number of conditions that occur more frequently at older ages reduce saliva production. Such conditions include reductions in fluid intake, infections of the salivary ducts, diseases such as diabetes mellitus, certain medications, and radiation therapy.

Inadequate saliva production and the resulting oral dryness can lead to (1) discomfort, (2) difficulty speaking, (3) bad tastes in the mouth, (4) lowered taste perception, (5) increased risk of cavities, periodontal disease, and oral infections, and (6) difficulty swallowing dry and solid foods. The dietary modifications that may result, such as selecting only soft moist foods, can lead to malnutrition and constipation.

Muscles The functioning of the oral muscles can be adversely affected to a substantial degree by abnormal changes in or diseases of the nervous system. For example, muscles around the mouth may become so weak that the mouth has a drooping appearance and drooling occurs. When nerve cells controlling other muscles are affected, speaking may be altered and swallowing may occur abnormally.

Swallowing abnormalities are not common among fairly healthy elderly people. However, up to 50 percent of elderly people in institutions may have trouble swallowing. Serious consequences of swallowing problems that result from improper muscle functioning include choking, pneumonia, and death. Such consequences occur more often when liquids are being swallowed because liquids can slip into the pharynx and larynx before reflexive muscle contractions close the opening into the larynx. Since difficulty swallowing is an abnormal condition that can lead to serious consequences, affected individuals should seek qualified medical diagnosis and treatment.

Bones and Joints Serious alterations in the jawbones and TMJ are also caused by abnormal conditions that increase in frequency with age. First, loss of teeth usually results in shrinkage of the jawbones. As these bones shrink, dentures fit less well and the appearance of the face changes. Second, the functioning of the TMJ can be substantially reduced by arthritis. In some individuals adverse psychological changes also lead to pain and malfunctioning of the TMJ.

ESOPHAGUS

The *esophagus* is a tube that transports materials from the pharynx to the stomach (Fig. 10.1). During swallowing, peristaltic muscular contractions in the esophagus push materials into the stomach. Coordinated contractions of the muscles are reflexively controlled by a network of nerve cells (*Auerbach's plexus*) in the wall of the esophagus. Since this network extends from the esophagus to the end of the large intestine, it can coordinate many functions throughout the GI tract.

Age Changes

Aging causes esophageal peristalsis to become slightly slower and weaker. This change seems to be caused by aging of neurons in Auerbach's plexus. The result is a small increase in the frequency with which materials from the stomach move up into the esophagus and cause discomfort, which may be experienced as heartburn.

Abnormal Changes

Esophageal Rings and Webs Though the esophagus normally functions well throughout life, many older people develop abnormalities such as the formation of *esophageal rings and webs*. These growths project inward from the wall of the esophagus and partially block the passage through the esophagus, causing difficulty swallowing.

Strictures A second abnormality is the formation of *strictures*, which are rings of scar tissue that develop from repeated injury to the esophagus. One cause of such injury is repeated passage of stomach contents into the esophagus. A stricture blocks the esophagus because the collagen in the scar tissue gradually shrinks, resulting in a narrowing of the passage through the esophagus and difficulty swallowing.

Sliding Hiatal Hernia A third structural abnormality of the esophagus is *sliding hiatal hernia*. In this condition, the connection between the esophagus and the stomach slips above the dia-

phragm rather than remaining in its normal position below the diaphragm. The incidence of sliding hiatal hernia increases with age, and up to 70 percent of those over age 70 develop this disease. Most cases result from alterations in esophageal muscles and decreased elasticity of the diaphragm.

Other Abnormalities A fourth cause of abnormal esophageal functioning is diabetes mellitus. Diabetes substantially slows peristalsis in the esophagus and all other parts of the GI tract. Other abnormalities that disturb esophageal functioning include nervous system diseases (e.g., strokes), alcoholism, medications, and cancer.

Effects and Complications Abnormalities in the esophagus can result in a variety of esophageal malfunctions. For example, peristalsis may not begin during swallowing, or it may be very slow or uncoordinated or occur with spasms. Each of these situations or partial blockage of the esophagus will inhibit the movement of materials down the esophagus and into the stomach. Two results are mild to severe discomfort and difficulty eating. Food selection may be limited, and completing a meal may take an inordinate amount of time. In addition, medications that fail to travel through the esophagus quickly can injure the esophagus. Finally, esophageal malfunction can allow stomach contents to flow upward and into the esophagus, a process called *gastric refluxing*. Since the stomach contains strong acids, this can cause pain, ulcers, and bleeding in the esophagus as well as esophageal strictures. Sometimes stomach contents may enter the respiratory passages through the larynx, causing hoarseness, inflammation of the respiratory system, or death.

Prevention and Treatment The frequency and severity of the adverse effects of esophageal malfunctioning can be reduced in several ways. The head and trunk can be kept slightly elevated so that the force of gravity assists in swallowing and helps prevent gastric refluxing. Avoiding large meals or obesity results in the same benefits because pressure in the abdomen is kept low. Other methods to reduce gastric refluxing include (1) avoiding foods and medications known to increase stomach acid and gastric refluxing, (2) using medications that promote esophageal clearance, coat the esophagus, or reduce gastric refluxing,

(3) using antacids to reduce stomach acidity, and (4) undergoing surgical correction of structural abnormalities.

STOMACH

The stomach is like a large sac (Fig. 10.1) whose walls can stretch to store large amounts of food. Food is normally prevented from moving back into the esophagus by proper functioning of the esophagus and proper pressures in the thoracic cavity. Contraction of a ring of muscle (the *pyloric sphincter*) at the lower end of the stomach temporarily prevents food from moving into the small intestine.

Secretion and Absorption

The inner lining of the stomach is a thick layer containing many secreting cells. Some of these cells secrete *hydrochloric acid* (*HCl*), and others secrete pepsin. When HCl and pepsin combine, they cause the rapid breakdown of large protein molecules, which are usually split into short chains of amino acids. HCl also kills bacteria and other microorganisms that have been swallowed. A third secretion from the stomach lining consists of *intrinsic factor*. Upon reaching the small intestine, intrinsic factor promotes the absorption of *vitamin B12.* This vitamin is important in the production of red blood cells. The lining of the stomach also absorbs water and small molecules that have become dissolved (e.g., simple sugars, salts, alcohol, certain medications). These materials enter the blood.

Movements

The middle layer of the stomach wall contains sheets of *smooth muscle.* Rhythmic contractions of this muscle churn the food and stomach secretions. The churning thoroughly mixes all materials and aids absorption by bringing dissolved materials into contact with the stomach lining.

Once the stomach contents have been adequately liquefied, the muscular contractions of the stomach become strong peristaltic waves. At the same time the pyloric sphincter relaxes some-

what so that a portion of the stomach contents is pushed into the small intestine. The pyloric sphincter then closes, and stomach churning and absorption continue until the small intestine is ready to receive more material. The functioning of the smooth muscle and the pyloric sphincter is controlled by the autonomic nervous system and Auerbach's plexus.

Age Changes

Aging causes small changes in the structure and functioning of the stomach, including a slight thinning of the stomach lining, a small decrease in HCl production, a possible decline in pepsin and intrinsic factor secretion, and a minimal slowing of stomach emptying. These changes are usually so slight that they do not prevent the stomach from performing its routine functions. However, they can alter the absorption of some medications and the functioning of the small intestine.

Abnormal Changes

Though the normal stomach functions well regardless of age, several abnormal and disease conditions become more frequent and severe with age.

Atrophic Gastritis *Atrophic gastritis* results in an excessive thinning of the stomach lining. The causes of many cases of this abnormality are unknown, but many other cases result from the immune system attacking the stomach.

Atrophic gastritis results in inadequate production of HCl and intrinsic factor. The consequences include poor protein digestion, alterations in the number and types of bacteria in the GI tract, and poor absorption of vitamin B12. Finally, atrophic gastritis is a risk factor for stomach cancer.

Poor protein digestion can lead to indigestion and malnutrition. The alterations in bacteria can also adversely affect nutrition. The reduction in vitamin B12 absorption leads to a significant reduction in red blood cell (RBC) production. The number of RBCs in the blood eventually becomes too low, and the person becomes anemic. *Anemia* caused by inadequate production of intrinsic factor is called pernicious anemia. The effects of *pernicious anemia* include sleepiness and persistent fatigue. These symptoms are not part of aging.

Atrophic gastritis can be treated with medications to relieve gastric discomfort and vitamin B12 supplements to prevent anemia.

Acute Gastritis   A second age-related stomach disorder is short-term stomach inflammation (*acute gastritis*). Reasons for the increased incidence of acute gastritis include reductions in the resistance of the stomach to environmental insults; increases in stomach infections due to lowered stomach acid production; and increases in the use of medications such as analgesics to relieve pain. Many analgesics (e.g., aspirin, steroids, nonsteroidal anti-inflammatory drugs) irritate the stomach.

The main problem arising from acute gastritis is the discomfort it causes. When attacks are frequent or severe, affected individuals may eat less, lose weight, and develop malnutrition.

Most cases of acute gastritis can be prevented by avoiding specific foods and medications. Taking antacids can relieve the symptoms in some situations.

Peptic Ulcer   In a *peptic ulcer*, stomach acid and enzymes cause cells lining the GI tract to die and peel away, leaving a pit in the wall of the tract (Fig. 10.3).

The occurrence of peptic ulcers in the esophagus has been mentioned in connection with gastric refluxing. These ulcers also occur in the stomach (*gastric peptic ulcers*) and the beginning of the small intestine (*duodenal peptic ulcers*). Although duodenal peptic ulcers are more common than gastric ones among younger adults, gastric peptic ulcers become higher in frequency among the elderly.

Gastric peptic ulcers often result from weakening of the stomach lining. Common causes include bacteria (*H. pylori*) or having unusually high levels of anti-inflammatory steroids in the blood. The elderly are more likely to have such elevated steroid levels since many medications used to relieve pain and inflammation contain them. Even nonsteroid pain relievers such as aspirin can increase the risk of gastric peptic ulcers.

A gastric peptic ulcer causes considerable pain. Usually the pain becomes worse shortly after eating because more stomach acid is produced then. Although the pain is generally of lower intensity at advanced age, other complications are usually more serious. If scar tissue forms, it can shrink

FIGURE 10.3 Peptic ulcers in the stomach and duodenum.

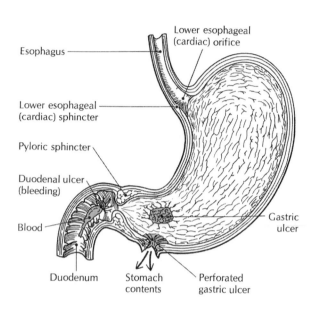

and narrow the stomach, leading to partial obstruction. Peptic ulcers that bleed slowly lead to anemia, while those that bleed profusely can cause sudden death. When an ulcer becomes very deep, the stomach may perforate, allowing its contents to leak into the abdominal cavity. The consequences may include severe pain, bleeding, digestion of neighboring organs, a severe drop in blood pressure, and death.

The incidence and severity of gastric peptic ulcers can be reduced by avoiding risk factors such as ulcer-promoting medications. Treatments include antibiotics or avoiding foods and medications that promote stomach irritation and ulcer formation. Conversely, medications such as antacids can promote healing or retard worsening of the ulcer. Some ulcers require surgery.

SMALL INTESTINE

The small intestine is a tube approximately 6 meters (20 feet) long and 2.5 cm (1 inch) in diameter that extends from the pyloric sphincter to the beginning of the large intestine (Fig. 10.1). It fits into the expansive region below the stomach by being coiled and bent.

Secretion

Like the stomach, the small intestine has a fairly thick inner lining that secretes digestive juices. Materials in the intestinal secretions include water and a variety of digestive enzymes. The water dissolves small molecules and, with the enzymes, breaks down large nutrient molecules. The breakdown of nutrients is aided by secretions from the liver and pancreas.

Absorption

The small intestine is also the section of the GI tract where most nutrients are absorbed. As in the stomach, churning aids absorption by bringing materials into contact with the inner absorptive surface. Absorption by the small intestine is especially efficient because its lining has a series of inward foldings and many microscopic fingerlike projections (villi).

The inward foldings and villi permit rapid absorption by increasing the surface area that is in contact with dissolved nutrients. Absorption by the villi is aided by the presence of many capillaries and lymph vessels, which allow nutrients to enter the circulatory system quickly. Most absorbed nutrients enter the blood vessels, though fat enters the lymph vessels.

The absorption of three nutrients by the small intestine requires special assistance. First, *iron* can be absorbed effectively only if the stomach has treated it with adequate amounts of HCl. Iron functions in the production of red blood cells. Second, *vitamin B12* can be absorbed in adequate amounts only if the stomach provides the small intestine with enough intrinsic factor. Vitamin B12 is also important for RBC production. Finally, *calcium* absorption requires the presence of activated vitamin D, which also allows the small intestine to increase the efficiency of calcium absorption when calcium in the diet or the body falls below desirable levels. Having adequate calcium is necessary for several functions, including maintenance of strong bones, muscle contraction, and nervous system activities.

Movements

As digestion and absorption continue, the contents of the small intestine are moved forward periodically by peristalsis. The basic actions and control mechanisms for peristalsis are similar to those in the stomach. By the time the contents have reached the end of the small intestine, almost all useful nutrients have been fully digested and absorbed. The remaining indigestible substances, wastes in bile from the liver, bacteria, and much water are pushed into the large intestine.

Age Changes

Aging seems to have little effect on the structure and functions of the small intestine. The age changes that have been observed, such as alterations in villi, apparently do not have important effects on intestinal functioning.

Lactase Secretion   One exception is a gradual decrease in the secretion of *lactase*, which splits *lactose* into two simple sugar molecules. Lactose is found in milk and many foods made from milk.

The decline in lactase varies from person to person with respect to time of onset and severity. Because of genetic factors, several groups (e.g., blacks, Asiatics, people of Mediterranean descent) have a significant decrease in lactase secretion during childhood or adolescence while most white people retain adequate lactase secretion well into adulthood. However, lactase secretion eventually becomes quite low in many older individuals. When it becomes too low, much of the lactose consumed in milk and milk products is not broken down. Certain types of bacteria in the intestine then use the undigested lactose for their own nutrition, resulting in much intestinal gas production. The gas can cause considerable discomfort or temporary disability. Such individuals are said to have *lactose intolerance*.

Many people with lactose intolerance avoid its consequences by abstaining from milk and foods containing milk. However, since dairy products are a major source of calcium, this can lead to calcium deficiency, which is a main risk factor for osteoporosis.

There are ways to avoid the adverse effects of lactose intolerance while still consuming milk and milk products. One way is to consume milk or milk products that have had lactose converted to other substances by bacterial action or lactase additives. Examples include certain types of yogurt and hard cheeses. Another way is to take lactase supplements. People who cannot use these methods should consume nondairy foods containing high levels of calcium, such as green leafy vegetables, canned fish, and calcium-supplemented orange juice.

Absorption   A second exception is a decline in the ability of the small intestine to absorb vitamins A, D, K and zinc. These decreases become important only for individuals whose diets contain low levels of these nutrients. The consequences of these deficiencies include skin and vision problems; weak bones; slow blood clotting; and decreased healing, immune function, and taste sensation, respectively.

Although the small intestine retains most of its absorptive power, its ability to absorb certain nutrients is adversely affected by other changes that often accompany aging. These changes include reduced production of HCl and intrinsic factor by the stomach and declining levels of active vitamin D.

Low HCl production reduces the absorption of iron and calcium and alters the numbers and types of bacteria that grow in the small intestine. As the bacteria change, the ability of the small intestine to absorb many nutrients declines. Individuals with marginal diets or severe HCl deficiencies are likely to develop iron or calcium deficiencies as well as other types of malnutrition. Individuals with very low intrinsic factor production, such as those with atrophic gastritis, and people with minimal vitamin B12 intake are likely to have vitamin B12 deficiency and the resulting anemia.

Recall that vitamin D is produced in a series of steps and is finally activated by the kidneys. The amount of active vitamin D in the body usually decreases with age because of several factors. These factors include less absorption of dietary vitamin D by the small intestine; less exposure of the skin to sunlight; less vitamin D production by skin cells; less activation of vitamin D by the kidneys; and inadequate intake of vitamin D in the diet.

The dwindling levels of active vitamin D and a gradual decline in the ability of the small intestine to respond to vitamin D cause calcium absorption by the small intestine to decline. Older people with very low levels of vitamin D and those with a poor dietary intake of calcium are at high risk of developing calcium deficiency and osteoporosis.

Abnormal Changes

Peptic Ulcer   As was mentioned earlier, one abnormality in the small intestine is a duodenal peptic ulcer (Fig. 10.3). Factors contributing to duodenal peptic ulcers include bacteria (*H. pylori*), emotional stress, excess caffeine consumption, and excess stomach acid production. The incidence of this condition does not change with age.

Unlike gastric peptic ulcers, the pain associated with duodenal peptic ulcers usually subsides after eating and intensifies when the stomach is empty. This pattern probably results because movement of acidic stomach contents into the small intestine is slowed when the stomach contains food. Other than the pain, the effects and complications of duodenal peptic ulcers are similar to those of gastric peptic ulcers.

Duodenal peptic ulcers can be prevented by avoiding the contributing factors. Treatment strategies are similar to those for gastric peptic ulcers, though the specific medications used and other details of the treatment may differ.

Secondary Abnormalities   The functioning of the small intestine also is adversely affected by many abnormal changes and conditions that are more common among the elderly. Examples include poor diet; infections; poor circulation; diseases of the skin, stomach, liver, gallbladder, pancreas, or small intestine; hormone imbalances; medications; and surgery. When any of these factors lead to detrimental changes in the small intestine, serious malnutrition can result.

LARGE INTESTINE

The large intestine, which is about 1.5 meters (5 feet) long, extends from the end of the small intestine to the end of the GI tract (Figs. 10.1 and 10.4a). Most of the large intestine is referred to as the *colon*. The last section of the colon has an S shape and is called the *sigmoid colon*. The sigmoid colon leads into the final segment of the large intestine, the *rectum* (Fig. 10.4b). The rectum, which is several centimeters long, leads into a very short passage called the *anal canal*. This canal ends at the *anus*, which is the opening from the large intestine to the outside of the body. The appendix is a fingerlike projection jutting out from the colon just below the junction of the small intestine and the colon.

Secretion and Absorption

A smooth layer of cells lines the inner surface of the large intestine. Many of these cells secrete a

FIGURE 10.4 The large intestine: (a) Regions. (b) Rectum and anus.

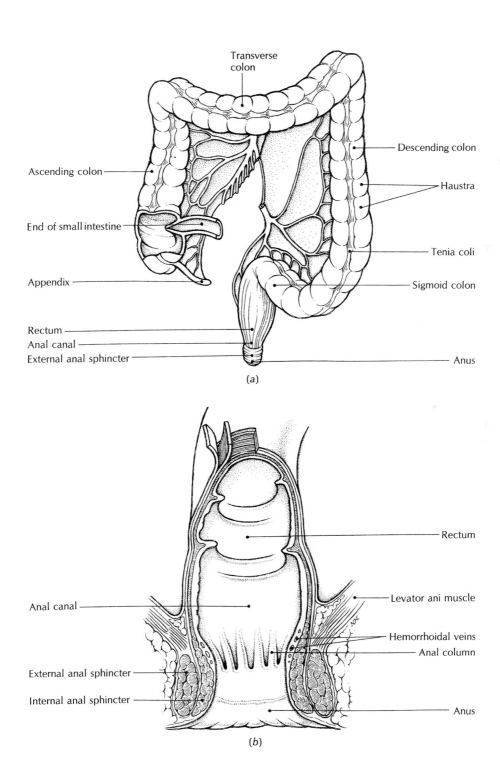

Transverse colon

Descending colon

Ascending colon

Haustra

End of small intestine

Tenia coli

Appendix

Sigmoid colon

Rectum
Anal canal
External anal sphincter

Anus

(a)

Rectum

Anal canal

Levator ani muscle

Hemorrhoidal veins
Anal column

External anal sphincter

Internal anal sphincter

Anus

(b)

lubricating mucus that aids the passage of materials. The materials in the large intestine are called *feces*. Many other lining cells absorb water, minerals, and vitamin K from the feces. Much of the vitamin K is produced by bacteria that normally reside in the feces. The absorption of most of the water and minerals from the feces is essential for maintaining adequate supplies of these substances for body cells. Vitamin K is needed to form blood-clotting materials. Absorbed substances pass into the blood.

Movements

Within the wall of the colon, some smooth muscle lies in three narrow ribbonlike bands (teniae coli). These bands run parallel along the length of the colon and are nearly evenly spaced around its circumference. The remainder of the smooth muscle lies in rings spaced little more than 2.5 cm (1 inch) from each other. This leaves regularly arranged patches of the colon wall with little muscular support. Since these patches (haustra) are weaker areas, they bulge outward.

As in the stomach and small intestine, smooth muscle in the colon churns materials and moves them along by peristalsis. The activity of this smooth muscle is encouraged by exercise and other physical activity. The muscle has diminished activity when a person is sedentary.

The churning action assists in the absorption of useful substances by mixing the feces and bringing them into contact with the lining of the large intestine. The absorption of water causes the feces to change in consistency from a liquid to a soft semisolid. As more indigestible material continues to enter the large intestine from the small intestine, the total amount of feces gradually increases.

The accumulation of feces causes the pressure inside the large intestine to increase, and the intestine is stretched outward. These changes cause neurons in the intestinal wall to activate large-scale peristalsis (*mass peristalsis*) throughout the length of the large intestine. Eating a meal or having stimulatory chemicals in the large intestine can also reflexively activate mass peristalsis, which pushes feces into the rectum.

Defecation

Initially, the feces are prevented from passing out of the rectum through the anus by two rings of muscle in the anal canal. The inner ring (*internal sphincter*) is made of smooth muscle and is controlled by reflexes (Fig. 10.4b). The outer ring (*external sphincter*) is made of skeletal muscle and is under voluntary control. These muscular rings cause retention of feces because they are usually in a state of contraction, which closes the passage through the anal canal and the anus. The internal sphincter provides about 85 percent of the force needed to retain feces, and the external sphincter provides the remaining 15 percent.

As the quantity of feces in the rectum increases, the rectum is stretched outward. When it stretches nearly as far as it can, further influx of feces causes pressure in the rectum to increase dramatically. Sensory neurons in the rectum are stimulated, and the person begins to perceive the need to pass the feces. The stretching and rise in pressure also cause rectal neurons to produce reflexive relaxation of the internal sphincter muscle, and there is an increase in the perception that elimination of feces is needed. Then only the external sphincter prevents the exit of feces through the anus. It is easier for the external sphincter to retain feces with a more solid consistency than to retain less solid or more watery feces.

At this point the individual can prevent the release of feces by voluntarily continuing contraction of the external sphincter. Alternatively, the person can cause the external sphincter to relax. Then the pressure in the rectum will begin to push feces out of the body through the anal canal and the anus. This action is often accompanied by reflexive peristalsis in the large intestine and voluntary contraction of the abdominal muscles. Since these additional contractions increase pressure in the large intestine, they assist in pushing feces through the anus.

As the rectum empties, the stretching and pressure in the rectum subside. Then reflexive contraction of muscles in and around the anal canal eliminates the feces still in the canal. Finally, the internal sphincter contracts again, and elimination of feces stops. Elimination of feces is often called a *bowel movement* or *defecation*.

Voluntary contraction of the external sphincter can delay defecation for an extended period. If a sizable mass of feces is retained in the rectum, the rectal muscles relax, allowing the rectum to stretch outward and the pressure to drop. As a result, the urge to defecate subsides until more feces enter the rectum and the pressure rises again. However, control of the external sphincter is retained only as long as the pressure in the rectum remains low or moderate. Very great pressure in the rectum causes uncontrollable reflexive relaxation of the external sphincter, and defecation occurs.

Appendix

The *appendix* is a small and unimportant part of the large intestine (Figs. 10.1 and 10.4*a*). Though its size and shape vary considerably from person to person, typically the appendix looks like a slender finger approximately 5 to 7 cm (2 to 3 inches) in length. The narrow passage within the appendix has one opening, which leads into the colon, and the wall of the appendix contains many lymph nodes.

In humans the appendix has no role in digestion. Normally, no materials from the intestine enter the appendix. Like all other lymph nodes, its lymph nodes help prevent the spread of infection and serve as part of the immune system.

Age Changes

With aging, the lining of the large intestine becomes somewhat thinner and less mucus is secreted. These changes may be due partially to arteriosclerosis in small arteries that serve the large intestine. There is also a slight shrinkage of smooth muscle in the large intestine. Usually the extent of these changes is not great enough to alter the functioning of the large intestine significantly.

Movements An age change in the colon that may produce important consequences in some older individuals is a decrease in the responsiveness of the smooth muscle to neuron messages. This tends to delay the onset of peristalsis, and more time is needed for feces to pass through the large intestine.

When feces spend more time in the large intestine, more water is absorbed and the feces become firmer and more difficult to move. Bowel movements occur less frequently, and defecation becomes more difficult. If periods between bowel movements become abnormally long or defecation requires excessive straining, the abnormal condition called *constipation* has developed.

Defecation Aging and other age-related factors cause three significant changes in the rectum. The other factors include surgery in the rectal area, forced bowel movements, and childbearing and reduced estrogen levels. One resulting change in the rectum is an increase in the amount of fibrous tissue in the walls, which reduces the amount of rectal stretching as feces enter. A second change is a decline in the strength of contraction of the internal sphincter. The third change is a decrease in the motor neurons controlling the external sphincter.

Each of these changes reduces the ability of the sphincter muscles to delay defecation until it is appropriate. Though inappropriate fecal elimination can happen to anyone, having such occurrences on a regular or frequent basis is abnormal. When inappropriate elimination of feces happens at least once a month, the condition is called *fecal incontinence*.

Appendix Age changes in the appendix (e.g., narrowing, increasing fibers, decreasing lymphatic tissue) have no known effect on the body.

Abnormal Changes

Constipation Though age has little effect on the structure and functioning of the large intestine, many aging people develop one or more abnormalities. One of the most common conditions is *constipation*. Generally, constipation occurs when bowel movements occur less than three times a week. However, since there is great variation in the frequency of bowel movements among individuals, better indicators are having an unusual decrease in the frequency of bowel movements and needing to strain excessively during defecation.

Although the incidence of constipation rises with age, it is difficult to estimate its actual frequency because of preconceived notions about how frequently bowel movements should occur and the use of *laxatives*. (A laxative is a substance that is ingested to promote bowel movements.) However, regardless of the confusion over the precise incidence, constipation is a problem for many elderly people. A variety of factors can contribute to the onset of constipation, including inadequate fluid intake, inactivity, delayed defecation, inadequate or excess fiber intake, and laxatives.

One of the most abundant types of dietary fiber is cellulose, which is contained in fruits and vegetables. Bran also contains much cellulose. Cellulose is especially important because it binds much water and thus makes feces large in volume but soft in consistency. Large soft feces stimulate peristalsis and are easy to move and eliminate by defecation.

Laxatives come in a variety of forms and work in different ways. Among the different types are fiber supplements, oils, magnesium salts, and intestinal stimulants.

Constipation can lead to significant discomfort and many complications. The discomfort may include indigestion, a bloated feeling, pain and embarrassment from intestinal gas, and difficulty defecating. One complication is an increased risk of future constipation. This occurs because the sensitivity of rectal sensory neurons declines when the rectum is stretched for long periods. With reduced rectal sensitivity, the urge to defecate and the reflexes involved in defecation are suppressed and defecation is delayed. Additional complications include diarrhea, fecal incontinence, diverticulosis, hemorrhoids, and cancer, which are discussed below. Still other complications include obstruction of the large intestine, formation and absorption of toxins, infections, difficulty urinating, irregular heart functioning, and heart attack.

Since constipation is not an age change and since its causes are largely known and controllable, much can be done to prevent its occurrence. Preventive measures are based on minimizing factors that contribute to constipation. Special consideration is required to ensure that toilet facilities and assistance are readily available to those who are disabled.

Treatments for people with constipation are derived from these preventive measures. They include gradually increasing fluid intake, physical activity, and dietary fiber. Regular bowel movements can be promoted by scheduling regular visits to the toilet. Laxatives should be used only when other treatments fail since regular laxative use promotes constipation and can cause other serious side effects, such as dehydration, vitamin deficiencies, and mineral imbalances.

Some cases of constipation can be relieved by using suppositories containing substances that stimulate mass peristalsis. These suppositories are inserted into the rectum through the anus. Other cases of constipation can be relieved by enemas. If feces are especially hard and difficult to pass, it may be necessary for a physician to remove them through the anus.

Diarrhea   A second abnormality of the large intestine that is important in elderly populations is *diarrhea*. Diarrhea is characterized by the presence of more than three relatively liquid and voluminous bowel movements a day. The causes of diarrhea are listed in Table 10.1.

One of the most serious consequences of diarrhea is an abnormally high loss of water and minerals. When body fluid levels drop, maintaining adequate blood pressure and blood flow to vital organs such as the heart, brain, and kidneys becomes difficult. These organs begin to malfunction and fail. Disturbances in mineral homeostasis can also adversely affect the heart and brain and disrupt muscle functioning.

Three factors make these outcomes from diarrhea especially likely among the elderly. First, many elderly people already have reduced blood flow to the heart, brain, and kidneys because of atherosclerosis. Second, aging reduces the ability of the kidneys to maintain adequate fluid levels. Third, the decrease in thirst sensation caused by aging reduces the urge to replace fluids by drinking. Individuals who take diuretic medications that increase fluid output by the kidneys or that lower blood pressure have an even greater risk of developing circulatory failure.

Another important consequence of diarrhea is temporary loss of the ability to retain feces when defecation is inappropriate. An individual of any age who has diarrhea may have difficulty retaining feces because of their liquid consistency and large volume and because of the strong urge to

TABLE 10.1 CAUSES OF DIARRHEA

Antacids containing magnesium

Toxins produced by bacteria in spoiled food or
   contaminated water

Laxatives

Certain antibiotics and other medications

Constipation

Toxins produced by bacteria in the small intestine

Viral infections in the GI tract

Diseases of the large intestine (e.g., ulcerative colitis)

Emotional disturbances

Cancer of the large intestine

Cancer of other body organs (e.g., pancreas)

release them. Older individuals have even greater difficulty because of age changes that weaken and reduce control of the anal sphincters. Any other condition that reduces conscious control of muscles or delays access to toilet facilities further increases the risk of unwanted defecation when a person has diarrhea. Examples of such conditions that are more prevalent among older individuals include CNS disorders (e.g., Alzheimer's disease) and physical disabilities (e.g., fractures, arthritis). Repeated incidents of diarrhea can cause fecal incontinence (see below).

Many cases of diarrhea can be prevented by avoiding factors that cause diarrhea. A main aspect of treatment is replacing lost fluids and minerals. However, care must be taken to avoid introducing excess fluids, which can overwork the heart and cause fluid accumulation in the lungs. When diarrhea is caused by certain types of bacteria, treatment may include antibiotic therapy. Good nutrition and anti-inflammatory medications can speed recovery from diarrhea. Medications that slow motility of the large intestine should be avoided since they increase the risk of retaining feces and absorbing toxins.

Fecal Incontinence A third abnormal condition of the large intestine is *fecal incontinence*, which can be defined as having inappropriate elimination of feces at least once a month.

Recall that fecal incontinence can be a complication of constipation. This occurs because mass peristalsis finally becomes strong enough to force feces into the rectum. Age changes in the rectum and anus that reduce the ability to retain feces voluntarily contribute to fecal incontinence from constipation and to the age-related increase in fecal incontinence from other causes. Among the factors that cause fecal incontinence are constipation; diarrhea; physical injury to the colon, rectum, or anus; nervous system abnormalities; diabetes mellitus; disability; and psychological disturbances.

Since many of the causes of fecal incontinence develop slowly and progressively and since each aging person may encounter more of these causes as time passes, fecal incontinence usually develops gradually and becomes worse over time.

The consequences of fecal incontinence are diverse and devastating. If feces are present on clothes or bed linens, they may also be in contact with the skin for prolonged periods, irritating the skin and causing infections. Even if these problems are avoided, people with fecal incontinence often suffer severe social disruption, social isolation, and psychological upheaval. The magnitude of the disturbances is sufficient to make fecal incontinence (together with urinary incontinence) the second leading cause of institutionalization among the elderly. Only disorders of the nervous system such as strokes and dementia rank higher as a cause of institutionalization.

Since fecal incontinence is an abnormal condition and since its causes are well known, much can be done to prevent, reduce, or stop its occurrence. The main strategy in each case is to identify the specific causes and reduce or eliminate them. Suggestions for preventing or stopping constipation and diarrhea were presented above. Individuals with structural irregularities of the colon, rectum, or anus may be helped by corrective surgery. Many individuals whose fecal incontinence results from disorders affecting the nervous system can be helped by biofeedback training combined with exercising the external sphincter muscle to increase its strength. Biofeedback training and muscle strengthening can also be helpful for individuals who are incontinent for reasons other than nervous system disorders. Such training increases the individual's awareness of the need to defecate and control of the external sphincter. Strengthening the external sphincter enables the individual to retain feces until defecation is desired. A combination of training and muscle strengthening is required for significant and long-lasting progress in reducing or stopping fecal incontinence.

Another key factor in preventing or reducing fecal incontinence is making toilet facilities easily accessible or available at appropriate times by scheduling visits to the toilet at specific times or intervals. Attempting to defecate shortly after eating is often effective because filling the stomach leads to reflexive mass peristalsis and a high probability of having a successful defecation shortly afterward. Using dietary modifications to regulate intestinal functioning can also increase control of defecation. Examples include regulating water intake and reducing the intake of foods such as beans, cabbage, and cauliflower, which cause intestinal gas production. Control of defecation can be further increased by using medications to regulate intestinal functioning or by modifying the use of medications for other disorders.

Many aspects of preventing or reducing the incidence and effects of fecal incontinence depend on the actions of those who care for affected individuals. While ample care is important, the extra attention and social interaction that usually accompany the extra care may perpetuate or increase the problem by unintentionally providing positive reinforcement for incidents of incontinence.

Diverticulosis and Diverticulitis *Diverticulosis* is characterized by the presence of deep outpocketings (*diverticula*) in the wall of the large intestine. Usually, each diverticulum has a narrow opening leading to an expanded outer region. Most diverticula occur in the sigmoid colon.

Diverticulosis is present in about 30 percent of people over age 60, its incidence increases to about 50 percent of those over age 70, and it may occur in 60 percent of those over age 80. The number of diverticula increases with age.

Though 80 to 85 percent of people with diverticulosis have no ill effects from this disorder, diverticula in the remaining 15 to 20 percent become inflamed. When this occurs, the condition is called *diverticulitis*. The longer a person has diverticulosis, the greater is the chance of developing diverticulitis.

Diverticulosis develops when excessive pressure from strong mass peristalsis, such as from constipation, or intestinal spasms cause the intestinal wall to bulge outward at weak spots such as haustra and blood vessels. Most cases are believed to result from inadequate amounts of fiber in the diet over a period of years.

Diverticulosis leads to diverticulitis when fecal material becomes trapped within the diverticula and produces noxious material. Irritation of diverticula by entrapped indigestible particles such as seed hulls may also cause diverticulitis.

Most people who develop diverticulitis experience significant abdominal pain, constipation, or diarrhea. Diverticulitis causes intestinal bleeding in approximately 25 percent of cases. Though the bleeding is usually slow and is not life-threatening, it may lead to anemia. Individuals with anemia usually become fatigued quickly and may be lethargic because inadequate amounts of oxygen are delivered to body cells.

Other serious complications of diverticulitis include infection and perforation of the large intestine. If the large intestine perforates, feces can pass into the surrounding body cavity. This condition can cause excruciating pain and lead to death from extremely low blood pressure or widespread infection.

Prevention of diverticulosis and diverticulitis is relatively easy if adequate dietary fiber is consumed daily and constipation is avoided. Those who have developed either abnormality can be helped by increasing their intake of dietary fiber, avoiding constipation, and avoiding foods containing seeds or other materials that increase intestinal inflammation. Antibiotics may be prescribed for those who have developed infections. Individuals with advanced cases of diverticulosis may require surgical removal of portions of the large intestine.

Hemorrhoids A fifth abnormality of the large intestine is the presence of *hemorrhoids*. These varicose veins in the rectum and anal canal were discussed in Chap. 4. Their occurrence increases and they become more of a problem as age advances because the factors that promote them increase with age.

Hemorrhoids result from conditions that cause repeated high pressure in the rectal and anal areas such as straining during bowel movements that accompany constipation, straining while lifting heavy objects, chronic coughing from bronchitis, and cirrhosis of the liver. Hemorrhoids can cause considerable pain and discomfort and, if they bleed regularly, can lead to anemia. Injured and inflamed hemorrhoids may become infected.

Preventing hemorrhoids involves avoiding circumstances that cause high pressures near the rectum and anus. Individuals with hemorrhoids can relieve discomfort by applying appropriate salves to the affected area. More advanced cases and hemorrhoids that contain clotted blood can be treated surgically.

Cancer *Cancer* is a disease characterized by uncontrolled reproduction and spreading of cells. Cancer of the large intestine is called *colorectal cancer*.

The incidence of colorectal cancer increases dramatically after approximately age 40, nearly doubling with each 5-year increase beyond that age. It occurs equally among men and women. Of all forms of cancer in the United States, only lung cancer occurs with a higher frequency.

Colorectal cancer is a common cause of death. It ranks as the third leading cause of death from cancer, accounting for about 15 percent of all deaths from cancer in adults. Only lung cancer and breast cancer cause more deaths. Since colorectal cancer increases in incidence with age, it ranks as a major cause of death among the elderly. It is the second leading cause of cancer deaths for men over age 75 and the leading cause of cancer deaths for women over age 75.

Though the causes of colorectal cancer are not known, several factors are known to increase the risk of developing it. They include diets low in fiber; diets high in meats, animal fat, or sugar; having relatives with colorectal cancer; having the identifiable gene that promotes familial colorectal cancer; having cancer of the breast or female reproductive organs; having noncancerous intestinal growths such as polyps; and conditions that cause chronic intestinal inflammation, such as ulcerative colitis.

Colorectal cancer can cause obstruction of the large intestine and destroy the intestinal wall. Obstruction can cause toxic materials from intestinal bacteria to accumulate and be absorbed into the body. Obstruction of the large intestine or destruction of the intestinal wall can lead to intestinal perforation and widespread infection. Colorectal cancer can also cause substantial bleeding. Finally, it often spreads to other parts of the body, such as the liver and lungs, and can destroy any organ it enters. The functioning of organs damaged by cancer is reduced, and their ability to help maintain homeostasis diminishes. Illness and death result.

One main way to combat colorectal cancer is to reduce the major dietary risk factors. Such changes seem to minimize the formation and accumulation of carcinogens in the large intestine. Other important preventive measures for colorectal cancer and its consequences include early detection and prompt treatment. Warning signs such as having blood in the feces and having noticeable changes in bowel functioning should be followed up by a professional examination. People over age 40 should have routine diagnostic testing for this cancer. People with a family history of colorectal cancer can be screened for the presence of the gene for familial colorectal cancer and can receive more frequent and thorough diagnostic testing if they have the gene. Finally, removal of polyps may be advisable.

Once colorectal cancer has developed, the only effective treatment is surgical removal of the affected areas. Chemotherapy or radiation therapy is sometimes used before surgical treatment of cancer in the rectum or anus.

Appendicitis Inflammation of the appendix (*appendicitis*) may be caused by entrapment of feces within the blind passageway in the appendix or infection of the lymph nodes in the appendix wall, both of which cause the appendix to become infected. The infection can be spread through the body by the circulatory system. A life-threatening crisis develops if the appendix ruptures and feces and infected material, such as pus, spread into the body cavity surrounding it. Older people face the same dangers of infection found in younger people.

The incidence of appendicitis decreases with age. However, cases in older individuals may be more severe, and the incidence of rupturing increases because age-related reductions in sensitivity to symptoms and milder signs of disease cause a delay in seeking diagnosis. In addition, an age-related reduction in blood flow allows deterioration of the appendix to occur faster and rupturing to occur sooner. Appendicitis is treated by surgical removal of the appendix. Antibiotics may be administered to combat infection.

LIVER

The *liver* is the largest gland in the body (Fig. 10.1). It is made up of microscopic units called *lobules*, which resemble each other in structure and functioning (Fig. 10.5). The liver cells making up each lobule are arranged in a radiating pattern, allowing blood from the periphery of the lobule to flow through large capillaries among the cells as it moves to the center of the lobule. These capillaries are called *liver sinusoids*.

Blood Flow

Blood enters the outer region of the lobule from arteries and veins at several points around the periphery of the lobule. The blood in the arteries comes from the heart and delivers oxygen and substances such as hormones from other organs to the liver cells. The blood in the veins comes from capillaries in the stomach, small intestine, large intestine, and pancreas. Blood from the spleen also passes through veins leading into the lobules. Since all these veins deliver blood to the

FIGURE 10.5 Structure of liver lobules.

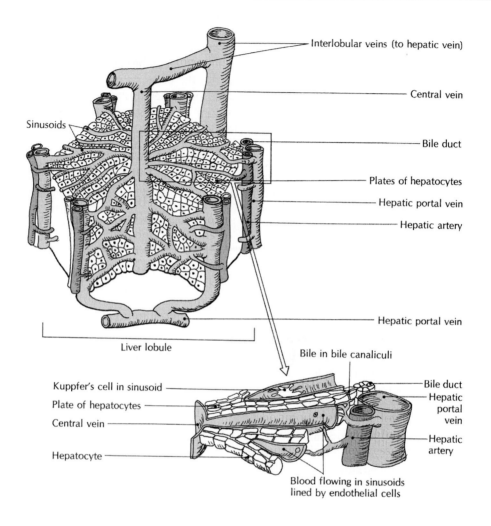

liver rather than returning it to the heart, they are called the *hepatic portal system* (Fig. 10.6).

Once blood from the arteries and the hepatic portal system has passed through the liver sinusoids, it is collected by a central vein at the center of the lobule. Blood from all the central veins moves into *hepatic veins*, which send it into a main vein going to the heart (the inferior vena cava). This arrangement of vessels permits the liver cells to adjust the contents of blood from digestive organs and the spleen before sending it to other parts of the body. The most abundant type of liver cells (*hepatocytes*) regulate the chemical makeup of blood. Other cells (*Kupffer's cells*) re-

move unwanted particles such as bacteria and damaged red blood cells from the blood (Fig. 10.5).

Bile Flow

In addition to blood vessels, each lobule contains other small passageways called *bile canaliculi* (Fig. 10.5). Bile, which is produced by hepatocytes, moves through the canaliculi to the periphery of the lobule, where it is collected into *bile ducts*. These ducts converge into one large duct, the *hepatic duct*, which carries the bile out of the liver (Fig. 10.7). Bile in the hepatic duct may flow

FIGURE 10.6 The hepatic portal system.

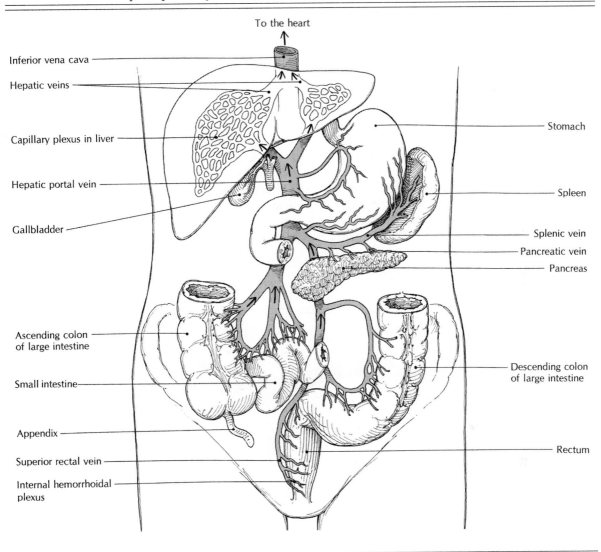

To the heart

Inferior vena cava

Hepatic veins

Capillary plexus in liver

Hepatic portal vein

Gallbladder

Ascending colon
of large intestine

Small intestine

Appendix

Superior rectal vein

Internal hemorrhoidal
plexus

Stomach

Spleen

Splenic vein

Pancreatic vein

Pancreas

Descending colon
of large intestine

Rectum

through the *cystic duct* for storage in the *gallbladder* or through the *common bile duct* into the small intestine.

Functions

Each lobule contributes to every liver function. Many of these functions were mentioned earlier in this chapter. For example, the liver helps convert foods to a usable form by secreting bile and sending it to the small intestine. *Bile* is a complex mixture of materials, including water, cholesterol, bile salts, and bile pigments. The salts and pigments are mostly waste materials removed from the blood. For example, when red blood cells are destroyed, parts of their hemoglobin molecules are converted into *bilirubin*, which is secreted into the bile. Bile also contains the breakdown products of cholesterol.

Bile helps convert foods to a usable form by breaking up droplets of fat from foods. This emulsification process allows digestive enzymes to hydrolyze the fat more easily. Bile also assists with absorption by allowing some fat to be absorbed by the small intestine without being hydrolyzed.

Since blood from the stomach and intestines flows through the liver before it is sent to other parts of the body, the liver can remove excess amounts of nutrients. The liver uses these extra nutrients to manufacture substances that are at inadequate concentrations in the blood. For example, hepatocytes remove excess sugar that is absorbed after one eats a sweet dessert. Some of the sugar may be converted into other nutrients (e.g., fat) that may be in low supply in the food, and some may be stored in the liver as glycogen. Later, when blood sugar levels drop, the liver converts the glycogen back into sugar and returns the sugar to the blood. Thus, body cells receive fat and sugar at a steady rate.

Passing blood from the stomach and intestines through the liver also allows the hepatocytes to remove harmful or toxic materials that have been ingested and absorbed, such as alcohol from alcoholic beverages. The liver also removes un-

wanted materials produced by body cells, such as ammonia and bilirubin. Ammonia produced by intestinal bacteria and absorbed by the intestine is also removed from the blood. The liver converts the toxic ammonia to a much less dangerous material called *urea*. Finally, the liver removes many medications from the blood.

The liver has several other functions. One is helping to maintain proper and fairly stable blood pressure. Because it has so many large blood vessels, the liver can hold a large volume of blood. When blood pressure begins to drop, constriction of liver vessels sends more blood to the heart and arteries, restoring blood pressure to normal levels. Alternatively, relaxation and dilation of liver vessels remove some blood from circulation and lower blood pressure when it becomes too high.

FIGURE 10.7 Pathways for bile.

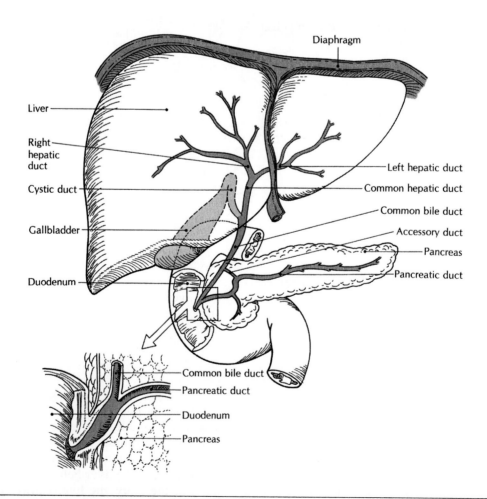

Finally, the liver regulates many substances in the blood that are not considered nutrients. For example, it manufactures several substances (e.g., fibrinogen, prothrombin) that are involved in forming blood clots. It also makes many of the blood proteins that regulate the distribution of water in the body. Without adequate amounts of these proteins, much water leaves the blood and accumulates around body cells. This condition (*edema*) can cause uncomfortable swelling; when it occurs in the lungs, respiration is seriously impaired. Finally, the liver plays a major role in removing excess hormones from the blood. Important examples include aldosterone, which increases salt and water reabsorption by the kidneys, and sex steroids.

Age Changes

Aging causes little change in the overall structure of the liver, though there seems to be a slight decrease in size, the total amount of blood flow through the liver may decline, and liver cells become somewhat altered.

These slight structural age changes seem to have little or no effect on the functional capacity of the liver. This maintenance of function probably stems from two features. First, the liver has a very large functional reserve capacity. As much as 80 percent can be removed, and the remaining portion can maintain normal body operations when conditions are favorable. Second, the liver easily regenerates new cells when older ones are damaged or destroyed. This regenerative ability is unchanged by aging. Studies of age changes in the liver suggest that both the storage of vitamin C and glycogen and the removal of a few medications (e.g., acetanilide, diazepam) decline. Elimination of particulate material by Kupffer's cells may decline with aging. It is important to note that smoking significantly reduces toxin, waste, and drug elimination by the liver.

Abnormal Changes

Cirrhosis A common and serious abnormal condition that often accompanies old age is the disease called *cirrhosis*. In this disease, the liver is converted into a lumpy scar-filled organ with greatly reduced functioning. Though many cases of cirrhosis occur among younger adults, this disease ranks among the top 10 causes of death among those over age 55.

Cirrhosis results from long-term repeated or continuous liver damage. Such damage among the elderly is most commonly caused when gallstones block large bile ducts. The resulting accumulation of bile in the liver puts pressure on liver cells and, together with chemicals in the trapped bile, damages them. Other causes include chronic alcohol consumption, hepatitis infections, and ingestion or inhalation of toxic substances such as volatile organic solvents in glue, cleaners, and paint thinners. Malnutrition, which is often associated with alcohol abuse, amplifies the effects of alcohol on the liver.

The development of cirrhosis occurs in basically the same way regardless of the cause. When liver cells are injured, the liver becomes inflamed and enlarged. Injured hepatocytes are not able to convert nutrients properly, resulting in accumulations of fat within the cells. Fibrous scar tissue then forms around the lobules. The presence of scar tissue inhibits the flow of blood and bile through the liver. With time, the flow of blood and bile is further restricted because the scar tissue shrinks, distorting and compressing blood vessels and bile passages. The liver attempts to compensate by forming new lobules. The growth of new lobules, along with compression by the scar tissue, gives the enlarged liver a lumpy appearance.

Since hepatocytes are injured, they are less able to perform their functions. Bilirubin from hemoglobin breakdown is left in a fat-soluble form called *unconjugated bilirubin* rather than being converted to the water-soluble conjugated form for excretion in bile. Inadequate amounts of bile are produced for emulsification and absorption. Since bile ducts are blocked, much of the bile cannot pass out of the liver to the hepatic duct. Therefore, digestion and absorption of fat and fat-soluble vitamins are reduced. Blood nutrient levels become unbalanced because of this and because the hepatocytes are less able to manufacture, convert, and store nutrients. All body cells become malnourished, as indicated by the onset of fatigue.

Blocked bile passages, together with the declining conversion of unconjugated bilirubin, result in accumulations of bilirubin. This gives the affected person a yellow or brown color, a condition called *jaundice*. Excessively high concentrations can eventually cause brain damage because unconjugated bilirubin accumulates in fatty myelin in the brain. In more advanced stages of cirrhosis, ammonia poses an even greater threat to the brain. Ammonia increases partly because of the dwindling conversion of ammonia to urea by hepatocytes. A second reason is that blocked blood flow causes blood from the intestines to flow through alternative routes, particularly veins in the esophagus. Thus, ammonia produced by intestinal bacteria is sent directly to the heart and from there to other organs, including the brain. Affected individuals show mental confusion, reduced muscle control, and even coma. This situation can eventually prove fatal.

The blockage of liver vessels causes other problems. Since blood cannot pass freely through the liver, it backs up into intestinal veins, causing them to swell and become varicose veins. When this happens in the rectum, hemorrhoids develop. When it happens in the esophagus, serious and even fatal bleeding can occur. As a further complication, extra fluids leak out of the stomach and intestinal capillaries into the surrounding abdominal cavity. This accumulation of fluids (*ascites*) causes abdominal swelling and imbalances in fluids in body cells in other regions.

Ascites worsens because reduced production of blood protein by injured hepatocytes allows more of the fluid to remain outside the capillaries. The reduction in blood proteins also causes edema and swelling in many parts of the body. The ascites and edema are amplified because injured hepatocytes do not remove enough steroid hormones from the blood (e.g., aldosterone, sex steroids), causing water retention by the kidneys. Edema causes a puffy appearance and discomfort, and in the lungs it significantly reduces respiratory functioning.

Many other problems result from cirrhosis. The more serious ones include bleeding because of reduced production of clotting materials, anemia because of poor hemoglobin breakdown and blood backing up into the spleen, weak bones because of reduced vitamin D activation, and reduced sexual functioning from abnormal hormone levels.

Many cases of cirrhosis are preventable. Individuals with blocked large bile ducts can usually have the blockages removed. This is especially true among the elderly, in whom the blocked ducts usually result from gallstones. Cirrhosis from chronic alcohol consumption (the other common cause of cirrhosis) can be prevented by

avoiding or reducing the consumption of alcoholic beverages. Excessive alcohol consumption is a serious problem because many elderly people suffer from loneliness, depression, boredom, and anxiety. Staying active and receiving social and emotional support can help reduce the incidence of alcohol abuse, and good nutrition reduces the effects of alcohol on the liver. Hepatitis, another cause of cirrhosis, can be prevented by (1) using good hygiene, (2) avoiding contact with affected individuals, especially their feces, blood, and body fluids, and (3) being immunized. Avoiding exposure to toxic materials can prevent other cases of cirrhosis.

Treatment of those with cirrhosis involves avoiding further liver injury by avoiding causative factors. If the cirrhosis is not very advanced, some liver regeneration and improvement in liver function can occur spontaneously. Advanced cirrhosis is essentially irreversible. Treatment at all stages includes minimizing the effects of complications from this disease.

Cancer   Most cases of *cancer* in the liver develop when cancer cells move through the hepatic portal system to the liver from other parts of the digestive system or the spleen. Movement of cancer from one location to another is called *metastasis*, and a cancer that metastasizes is called *metastatic cancer*. Metastatic cancer of the liver is often widespread and is of diverse types. It may reduce many liver functions and can cause several of the problems associated with cirrhosis.

Treatments for metastatic liver cancer, including surgery, radiation therapy, and chemotherapy, do little more than slow the progress of this fatal disease. Essentially all cases are fatal within 5 years.

GALLBLADDER

The gallbladder is a sac just under the lower edge of the liver (Fig. 10.1). Recall that the gallbladder receives bile from the liver through the cystic duct and that bile in the gallbladder may pass to the small intestine through the common bile duct (Fig. 10.7).

The gallbladder stores bile until it is needed for digestion. Recall that bile assists in the digestion and absorption of fat. The gallbladder absorbs water from bile while the bile is being stored.

Emptying of the gallbladder and passage of bile

into the small intestine is stimulated by a hormone [*cholecystokinin (CCK)*] from the small intestine and impulses in parasympathetic nerves. Operation of the CCK control mechanism is especially important when fat enters the small intestine.

Age Changes

Aging causes no significant changes in gallbladder structure. There is a decrease in the sensitivity of the gallbladder to stimulation by CCK. However, with age, the small intestine compensates by producing more CCK. Therefore, contraction of the gallbladder in response to the entrance of fat into the small intestine remains unchanged.

Two age changes associated with the gallbladder involve the bile ducts: The bile ducts widen over most of their length, and the end of the common bile duct near the small intestine becomes narrower. Normally, these changes are not important, but the former one may increase the likelihood that small gallstones formed in the gallbladder will pass down the bile ducts. The latter change inhibits the escape of such stones into the small intestine. As described below, gallstones trapped in the bile ducts can cause cirrhosis and pancreatitis.

Abnormal Changes: Gallstones

Gallstones are solid masses formed from materials in bile. They are usually formed in the gallbladder. Gallstones contain various combinations of materials, including cholesterol, bile pigments and salts, calcium, and protein. Many individuals may develop one or a few gallstones. The stones range in size, with some becoming larger than 2 cm in diameter. Some individuals may have 200 or more small stones.

The incidence of gallstones increases with age, and they are fairly common among the elderly. Approximately 25 percent of people over age 50 have gallstones. Gallstones are one of the more common reasons for surgery among older people.

An important cause of gallstone formation is having excessively concentrated bile in the gallbladder, because this situation leads to solidification of materials dissolved in the bile. These circumstances occur more frequently among the elderly because the concentration of bile produced by the liver increases with aging, particularly in obese individuals. In addition, many older per-

sons produce unusually low amounts of cholecystokinin. With less CCK, emptying of the gallbladder is delayed and may be less complete. Since more bile stays in the gallbladder for longer periods, it becomes even more concentrated and bile solidification occurs. Gallstone formation can also be initiated by infections in the gallbladder.

Individuals with gallstones often feel vague discomfort in the abdominal region and the digestive system. Many cases involve severe pain, nausea, and vomiting, especially after one has eaten foods containing fat. The painful attacks are probably caused by contraction of the gallbladder on the gallstones or by movement of a stone into the cystic duct. The gallbladder or bile duct may become injured, inflamed, and infected. In very severe cases it may perforate, spilling bile into the abdominal cavity; this can spread infection and cause a sudden drop in blood pressure.

If a gallstone moves into the bile duct and blocks it, the individual may become jaundiced because bile cannot escape and bilirubin accumulates in the body. Digestion and absorption of fat are greatly reduced, and malnutrition, including vitamin deficiencies, may develop. If the gallstone lodges below the intersection of the common bile duct and pancreatic duct, pancreatic secretions may be blocked. This can lead to inflammation of the pancreas (*pancreatitis*), which is discussed below. Prolonged blockage of the bile ducts can cause cirrhosis.

There is no effective way to prevent gallstones, though avoiding obesity reduces the risk of developing them. Gallstones can be removed by several methods, including dissolving them with solutions infused through the bile ducts or with medications, fragmenting them with ultrasound, and extracting them surgically. Surgical removal often includes removal of the gallbladder to prevent a recurrence. A person whose gallbladder has been removed can survive and digest food because the liver can store adequate amounts of bile.

PANCREAS

The pancreas is a large gland in the space below the stomach and above the first section of the small intestine (Fig. 10.1). Most of the pancreas consists of clusters of cells (*exocrine cells*) that secrete pancreatic juice into ducts within the pancreas. The ducts merge to form larger ducts, which finally converge and form one large pancreatic duct. This duct joins the common bile duct just before it enters the small intestine (Fig. 10.7). Therefore, bile and pancreatic juice enter the small intestine through the same opening.

Pancreatic juice contains several enzymes that hasten the chemical breakdown of large nutrient molecules. It also contains sodium bicarbonate, which neutralizes stomach acid and prevents it from injuring the small intestine. This process also helps provide a proper acid/base balance for the action of enzymes in the pancreatic juice and in small intestine secretions. The secretion of pancreatic juice is adjusted by the autonomic nervous system and by hormones from the small intestine.

Small clusters of cells (*endocrine cells*) that secrete hormones into the blood are widely scattered among the exocrine cells. These clusters are called *islets of Langerhans*. The endocrine cells are of two types: One type (alpha cells) secretes *glucagon*, and the other type (beta cells) secretes *insulin*. These hormones help maintain proper and fairly stable levels of glucose in the blood and affect the production and breakdown of proteins and lipids by body cells.

Age Changes

The slight age changes that occur in most components of the pancreas and its ducts do not have a significant effect on the amount of pancreatic juice produced. There may be a slight decrease in the production of certain enzymes (e.g., those for lipid digestion), sodium bicarbonate, and insulin. None of these changes are great enough to alter the ability of pancreatic secretions to digest nutrients, neutralize stomach acids, and regulate blood glucose levels. A main reason for the maintenance of pancreatic digestive function throughout life is the large reserve capacity of the pancreas. As little as 10 percent of the young adult pancreas is needed to produce enough pancreatic juice for normal digestion. Though the effectiveness of the pancreas is virtually unchanged by aging, hormone production is often significantly altered by other factors as people get older (e.g., obesity, lack of exercise) (Chap. 14).

Abnormal Changes

Pancreatitis   One of the more common abnormalities of the pancreas among the elderly is *pancreatitis*, or inflammation of the pancreas. Cases that develop quickly are called *acute pancreatitis*, and cases that develop over a prolonged period are referred to as **chronic pancreatitis**. Repeated or prolonged cases of acute pancreatitis lead to chronic pancreatitis.

Acute pancreatitis is usually caused by traumatic injury to the abdominal region (such as striking the steering wheel in an automobile accident), consumption of alcoholic beverages (especially binge drinking), or blockage of the pancreatic duct by a gallstone. Chronic pancreatitis is usually caused by chronic alcohol consumption or blockage of the pancreatic duct by a gallstone.

In addition to causing extreme pain, acute pancreatitis may quickly become life-threatening. Vomiting can severely deplete the body of fluids and minerals, leading to circulatory, nervous, and muscle malfunctions. Bleeding or leaking of pancreatic enzymes into the blood or abdominal cavity can cause blood pressure to drop dramatically, resulting in circulatory failure. Pancreatic enzymes in the abdominal cavity can begin to digest nearby organs, causing destruction and perforation of those organs. Long-lasting muscle spasms may occur as blood calcium levels drop. Blood sugar levels may also drop as injured endocrine cells release insulin. Since the endocrine cells may be permanently damaged, individuals who survive these immediate dangers may be left with diabetes mellitus. Any case of acute pancreatitis can develop into chronic pancreatitis.

Three main problems develop in patients with chronic pancreatitis: recurrent pain, poor protein and fat digestion caused by inadequate production of digestive enzymes, and inadequate regulation of blood sugar and diabetes mellitus resulting from reduced production of hormones.

Since excessive or chronic consumption of alcoholic beverages is a main cause of pancreatitis, many cases can be prevented by avoiding such drinking behaviors. Many other cases can be prevented by removing gallstones before they block the pancreatic duct.

Treatments for acute pancreatitis primarily involve preventing or reducing its effects and complications. Surgical procedures may be required to repair or remove injured organs or remove gallstones. Abstinence from alcoholic beverages is often necessary.

Pain from chronic pancreatitis is relieved with analgesics, and enzyme insufficiency is rectified by ingesting enzyme supplements with meals.

Mild cases of diabetes mellitus can be managed by regulating diet and exercise; more severe cases may require the administration of insulin. As with acute pancreatitis, surgery may be necessary, and abstinence from alcoholic beverages is often required.

Cancer Like all cancers, pancreatic cancer involves uncontrolled reproduction of cells. Almost all cases involve cancer of the exocrine cells. Cancer of the pancreas is the fifth leading cause of death from cancer, and it accounts for 3 percent of all cancers and 5 percent of all deaths from cancer. Within the digestive system, only cancer of the large intestine occurs more frequently. The incidence of pancreatic cancer rises with age, and the incidence is highest among men over age 75. Several risk factors for pancreatic cancer have been identified; smoking cigarettes, eating a diet high in animal fat, eating many foods containing high amounts of nitrates and nitrites as preservatives (e.g., bacon, cold cuts), consuming much coffee, and having diabetes mellitus.

Cancer of the pancreas is especially dangerous because it is usually detected only after it has become quite advanced. Indications include being jaundiced, losing weight, having abdominal or back pain, and having symptoms of diabetes mellitus such as unusually high thirst, hunger, and excessive urine production. This cancer causes pancreatic failure, which includes poor digestion of proteins and fat, and diabetes mellitus. Pancreatic cancer often spreads through the hepatic portal system to the liver, where it causes liver failure.

The chances of developing pancreatic cancer may be lowered by avoiding smoking and other dietary risk factors. Avoiding risk factors for diabetes mellitus, such as being obese and being sedentary, may also help.

Though surgery, chemotherapy, and radiation therapy have been employed as treatments for pancreatic cancer, no effective treatment for this disease is known. Pancreatic cancer is usually fatal within 1 year of the diagnosis.

11

DIET AND NUTRITION

The first part of this chapter introduces some principles of diet and nutrition and discusses them in general terms. Although these principles are presented primarily with the elderly in mind, they apply to all adults. The second part presents information about specific nutrients, including their uses in the body, dietary sources, reasons for having abnormal levels, and problems resulting from abnormal levels. The third part deals with abnormal conditions related to diet and nutrition.

NEED FOR NUTRITIONAL HOMEOSTASIS

A main function of the digestive system is to supply nutrients at proper and fairly steady levels to body cells and thus maintain nutritional homeostasis. Cells require nutritional homeostasis for four reasons. First, many nutrients must be constantly broken down to supply the energy needed to power the ongoing vital functions of the cells. Second, nutrients supply the raw materials needed to synthesize body substances. Some materials are used to enlarge, replace, or repair parts of the body (e.g., muscle cells, RBCs, epidermis); others are used to replace substances that are degraded (e.g., neurotransmitters, hormones) or lost (e.g., water in perspiration, digestive enzymes). Third, nutrients supply materials that assist chemical reactions (e.g., vitamins, certain minerals). Fourth, nutrients such as minerals and water maintain proper volumes and concentrations of body fluids so that blood flows properly and cells are not damaged by excessive shrinking or swelling.

RELATIONSHIPS BETWEEN DIET AND NUTRITION

Usually nutritional homeostasis can be maintained with a diet containing adequate amounts and proper types of nutrients. A healthy person's digestive system can carry out the processes involved in properly supplying the nutrients contained in such a diet. However, many adverse conditions can inhibit the proper functioning of the digestive system or unfavorably affect other body systems. This results in an inability to maintain nutritional homeostasis even though a person eats a proper diet. These adverse conditions include loss of teeth; atrophic gastritis; inadequate lactase production; atherosclerosis; diabetes mellitus; smoking; alcoholism; and certain medications. Therefore, it can be said that nutritional homeostasis cannot be achieved without a proper diet, eating a proper diet often results in nutritional homeostasis, and eating a proper diet does not guarantee nutritional homeostasis.

PROBLEMS FROM MALNUTRITION

Diversity of Problems

A person who does not maintain nutritional homeostasis is said to have *malnutrition*. This condition can cause a constellation of problems whose nature and severity depend on the specific nutrients or combinations of nutrients that are above or below the normal range; the amount of deviation from the normal range; and the frequency of the deviation or the length of time it exists. Sometimes malnutrition for one nutrient may directly affect only one body function (e.g., vitamin K and blood clot formation). In many other cases malnutrition for one nutrient (e.g., protein deficiency) may affect many or all cells in the body. However, since body cells and systems are interdependent, malnutrition that directly affects one or a few functions ultimately has widespread effects. For example, vitamin K deficiency leading to poor blood clotting can result in slow but steady blood loss, anemia, and therefore a reduced oxygen supply to all body cells.

Onset of Problems

In some cases of malnutrition problems develop quickly. For example, water deprivation on a hot day can result in overheating within a few hours. In many other cases problems develop gradually. In some of these cases, the diet may provide certain nutrients at low levels and body cells may function at only a slightly reduced or slowly diminishing capacity. For example, a slight deficiency in B vitamins may result in a barely noticeable decline in strength. In other cases of nutrient deficiency the body may be able to compensate partially by drawing on its reserves. For example, calcium deficiency may not become apparent for many years because blood calcium levels can be maintained by withdrawing calcium from the bones. Finally, in cases of nutrient excess the body may be able to convert or store large quantities of the excess nutrient and thus protect body cells from injury. For example, the liver can store much vitamin A and reduce or delay the toxic effects of an excessive dietary intake of that vitamin.

Nature of Problems

Besides developing gradually, manifestations of malnutrition often begin as vague and nonspecific abnormalities such as weakness, nausea, headache, and changes in personality. In other cases the effects of malnutrition may seem to result from a specific disease that is often not associated with malnutrition (e.g., dementia). In both situations, determining that the problem is caused by malnutrition is difficult and malnutrition is often overlooked as a cause.

Consequences

Since malnutrition is a deviation from homeostasis, body cells do not have optimum conditions and function less effectively. This decline in cell function leads to a decline in the ability of the body to adapt to other changes in internal or external conditions and therefore may result in further decrements in homeostasis. For example, calcium deficiency can lead to weaker bones that provide less support, and then fractures from traumatic injury or ordinary activities may occur. Similarly, protein deficiency can reduce the ability of white blood cells to combat bacteria, and infections may occur. Finally, in cases of excess fat and carbohydrate intakes, the obesity that may develop can inhibit the quick movements needed to avoid an accident such as a fall.

Besides reducing adaptive responses that maintain homeostasis, the malfunctions caused by malnutrition can substantially reduce the enjoyment derived from activities. As examples, obesity from an excess food intake often leads to an earlier onset of fatigue and discomfort from recreational activities, and a zinc deficiency can lead to reductions in taste sensations and decrease the pleasure derived from foods.

A third consequence of malnutrition is disease. Some types of malnutrition are known to cause specific diseases. For example, iron deficiency causes anemia, and high blood alcohol levels cause cirrhosis. Other types of malnutrition are only risk factors or contributing factors to disease; recall the dietary risk factors for osteoporosis and colorectal cancer (Chaps. 9 and 10).

A PROPER DIET

A Word of Caution

Since proper nutrition depends on consuming a proper diet, we will now explore such a diet for healthy adults. One must realize that the following recommendations may be improper for individuals who are in unusual situations or have abnormal or disease conditions. Examples include athletes undergoing intense physical training, people living in extremely hot or cold environments, individuals who are bedridden, and people with diabetes mellitus or kidney failure. Such individuals need to implement significant modifications to the recommendations to achieve nutritional homeostasis or avoid additional problems from their peculiar circumstances or diseases. This concept is especially important in regard to the elderly because people become more heterogeneous with age. Therefore, while the following recommendations can serve as guidelines in evaluating and planning diets for groups of healthy elderly people, they should be applied to individuals with caution.

Diet Based on Food Selection

Describing a diet in terms of commonly eaten foods is the most practical way to select a diet that will provide nutritional homeostasis for most healthy adults. The U.S. Department of Agriculture (USDA) has developed such a diet, the Food Guide Pyramid (Fig. 11.1). Though this plan consists of six food groups, items in the fats, oils, and sweets group provide energy but few other necessary nutrients. Foods from this group should be minimized except for individuals who are active and need to obtain more energy.

Selecting recommended servings each day from the other five food groups and varying the selections within each group will provide an adult who engages in a fairly low amount of physical activity with adequate amounts of almost all essential nutrients. Individuals who are more active should increase serving sizes or, better still, increase the number and variety of servings. Individuals who are inactive should increase their physical activity so that they can select enough servings with enough variety to achieve nutritional homeostasis without gaining weight. Trying to avoid weight gain by eliminating food groups or servings from the diet can lead to a deficiency in one or more essential nutrients.

The Food Guide Pyramid can be improved by including foods with vitamin C (e.g., citrus fruits) from the fruit group; choosing items with minimum fat content from the milk and meat groups; and selecting items made with whole grains from the bread group.

Although following the Food Guide Pyramid and these three suggestions will provide adequate dietary nutrients, this alone may not solve other problems associated with an improper diet. These problems include atherosclerosis, diabetes mellitus, tooth decay, high blood pressure, constipation, cancer, cirrhosis, and accidents. Therefore, the USDA and the Department of Health and Human Services have added several recommendations, such as (1) avoiding fat, saturated fat, cholesterol, sugar, and sodium, (2) consuming enough starch and fiber, and (3) not drinking more than a moderate amount of alcoholic beverages.

Diet Based on Chemical Composition

Recommended Dietary Allowances The recommendations for a proper diet based on food selection were developed primarily to ensure that individuals could regularly obtain adequate amounts of each nutrient. These amounts have been established by the Food and Nutrition Board of the National Academy of Science and are referred to as the *Recommended Dietary Allowances (RDAs)* of nutrients whose requirements have been well studied. The RDAs provide enough of each nutrient to maintain good health and are higher than the amounts needed just to survive. The RDAs and the dietary recommendations mentioned next can be found in tables in most textbooks on nutrition.

Several features of the RDAs warrant special attention. First, the recommendations are for individuals of certain weights and heights and should be adjusted for the individual. Second, except for infants and children, men and women have different RDAs. Third, the RDAs group individuals into categories based on age. Fourth, there are only two age categories for those over age 24. These categories are ages 25 to 50 and age

FIGURE 11.1 The Food Guide Pyramid: A Guide to Daily Food Choices. (From U.S. Department of Agriculture and U.S. Department of Health and Human Services.)

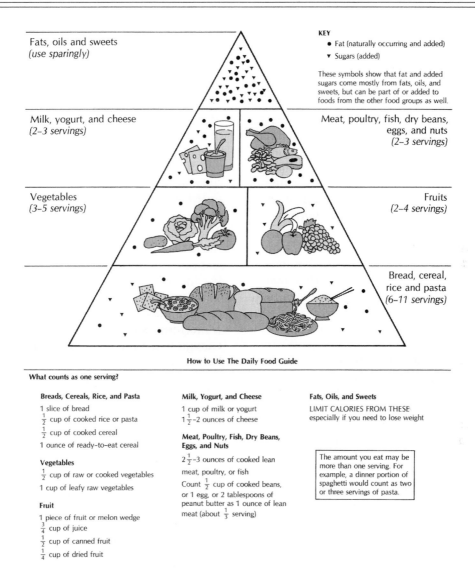

51 and over. This lumps all elderly individuals together even though it is thought that age changes after age 51 affect nutritional needs. Though little information is available about adjusting the RDAs to compensate for these age changes, a few recommendations are included in the second part of this chapter.

Besides the RDAs for specific nutrients, a table of recommended energy intake has been developed. In this table, the estimated energy requirements, indicated as *kcals*, are for persons who perform light to moderate physical activity. A *kcal*, or *kilocalorie*, also called a *Calorie* (note the uppercase C), is the unit most often used to measure the energy content in food or the energy consumed. A kilocalorie is 1,000 calories (note the lowercase c).

Persons with body dimensions that differ greatly from those listed in the table and individuals whose activity levels are very low (e.g., the bedridden) or very high (e.g., athletes) may have energy requirements as much as 1,000 kcal below or above those listed. Individuals recovering from a serious illness or accident also often need an increased energy intake to provide the energy necessary for healing. For any particular individual, the best way to estimate energy requirements is to determine the energy intake needed to maintain a desirable body weight as discussed later in this chapter.

The energy table shows that the energy requirements for people age 51 and over are lower than those for younger adults. The decline in recommended energy intake at higher ages is based on the average age-related decline in muscle mass and amount of physical activity.

Calculations using the RDAs for protein intake for people age 51 and over result in a value of approximately 0.8 gram per kilogram (0.36 gram per pound) of body weight. Other authorities suggest that the elderly can benefit from a protein intake of 1.0 gram per kilogram (0.45 gram per pound) of body weight while reducing energy intake from carbohydrates and fat. This modification can assure an adequate protein intake while preventing weight gain since elderly people tend to have lower energy requirements.

U.S. Recommended Daily Allowances   Though the RDAs are listed as daily allowances, a person need not consume each nutrient in the recommended amount each day because the body can store excess nutrients and release the stored nutrients when lesser quantities are eaten. A person need only consume enough of each nutrient over a few days or a week so that the average amounts consumed each day correspond to the RDAs.

To assist consumers and those evaluating and planning diets, the Food and Drug Administration used the values in the RDAs to develop the **U.S. Recommended Daily Allowance (U.S. RDA)** for many nutrients. The age categories for U.S. RDAs are even broader than those for RDAs, placing all individuals above age 3 into one category.

The labels on many packaged foods list the percentages of the U.S. RDA for many nutrients. This information is useful in determining the nutrient quality of foods and the contribution each food item can make to a person's daily nutrient intake.

Estimated Safe and Adequate Daily Dietary Intakes   The lists of RDAs do not include all required nutrients. Estimates of requirements for other nutrients have been made by the Food and Nutrition Board and are listed as *Estimated Safe*

and Adequate Daily Dietary Intakes (*ESADDIs*). This list places all adults into one category regardless of age.

Dietary Reference Intakes   The *Dietary Reference Intakes* (*DRIs*) make up a new comprehensive method for establishing and evaluating recommended dietary intake recommendations. This new system is under development, and some DRIs for specific nutrients have been established. The DRIs will supplant other systems on which it is based.

The DRIs are based on a combination of *Recommended Dietary Allowances* (*RDAs*), *Estimated Average Requirements* (*EARs*), *Adequate Intakes* (*AIs*), and *Tolerable Upper Intake Levels* (*ULs*). These four systems have different goals. The RDAs provide adequate intake to give 97 percent of a population adequate intake. The EARs provide adequate intake to give 50 percent of a population adequate nutrient. The AIs list the average intake for a population that will give a desired predetermined outcome (e.g., risk of a disease) based on outcomes from actual diets in that population. The ULs list maximum intakes in an actual population that provide 97 percent to 98 percent of the population with no adverse risks or effects from the high levels of intake.

Because the DRIs are based on systems with diverse standards and goals, different portions of the DRIs should be used in different circumstances so the desired outcomes are most likely to be achieved. Different portions of the DRIs satisfy different percentages of a population. Also, different levels of the DRIs are being set for different types of populations (e.g., age groups, body size, percent body fat, gender, health status, cultures, etc.). The DRIs will use 12 life stages (i.e., 0-6 months, 6-12 months, 1-3 years, 4-8 years, 9-13 years, 14-18 years, 19-30 years, 31-50 years, 51-70 years, 71+ years, and pregnant and lactating women. Different aspects of the DRIs can be used for planning diets, for assessing diets, or for assessing outcomes from programs affecting diets (e.g., institutional meal plans, school meal plans).

Comparing Proper Diets for Younger and Older Adults

A comparison of proper diets for healthy younger adults and healthy older adults reveals that with very few exceptions, these diets are basically the same. The exceptions include slight increases for the elderly in fiber, protein, and calcium intakes, while total energy and vitamin A intakes should be somewhat lower. If the diets of healthy, active younger and older adults were compared, the recommended differences in protein and total energy intakes would be eliminated, making the diets even more similar. The reason for this similarity is that the types and total amounts of cellular activities and therefore the nutritional needs of healthy active adults remain essentially the same regardless of age. However, increasing age is often accompanied by a decrease in physical activity, unusual situations, abnormal changes, and diseases, all of which may require individualized dietary adjustments.

MALNUTRITION AMONG THE ELDERLY

Malnutrition is widespread and occurs frequently among the elderly. Providing precise estimates of the extent of malnutrition and the specific nutrients involved is difficult. Interestingly, many cases occur even when adequate food and professional assistance are available, such as in nursing homes. The high incidence of malnutrition among the elderly is due to many factors, many of which become more common or severe with age. These factors may be biological, social, psychological, or economic.

Reducing and Preventing Malnutrition

Though the factors causing extensive malnutrition among the elderly are numerous and diverse, the following steps can be taken to reduce malnutrition in this segment of the population.

Evaluating Nutritional Status   To reduce and prevent malnutrition, one must determine which individuals are malnourished and what types of malnutrition they have. Several approaches are used in making such determinations. These include developing a dietary history; keeping records of food intakes; performing physical examinations; performing chemical analyses of blood samples; analyzing records of body weight; and taking body measurements such as skin fold thickness, height, and weight. The values obtained are compared with the recommended val-

ues. Combining two or more of these approaches provides determinations of nutritional status having increased accuracy and reliability.

Identifying Factors Contributing to Malnutrition   Once cases and types of malnutrition have been identified, the second step is to attempt to identify factors that have contributed to the malnutrition. Though this may be difficult, correcting malnutrition is easier if the causes can be reduced or eliminated.

Evaluating and Adjusting Diet   A third step in solving malnutrition is to compare individuals' diets with dietary recommendations based on the Food Guide Pyramid, RDAs, U.S. RDAs, and ESADDIs. When one uses these general guidelines and tables, individualized diets can be designed to provide nutritional homeostasis while minimizing problems that may be caused or amplified by specific dietary components.

Implementing dietary adjustments can be a difficult task because many factors besides hunger caused by low nutrient levels influence the amounts and types of foods people eat. Some of these factors (e.g., taste preferences) influence voluntary choices. Others (e.g., religion, culture, disease) impose dietary requirements or restrictions. Considering these factors improves the likelihood that recommended dietary adjustments will be adopted.

Many elderly people are able voluntarily, independently, and effectively to implement dietary modifications and changes in exercise to improve their nutritional status. However, many others require regular assistance. Effective programs for these people are provided by government agencies, social organizations, and volunteer groups.

Using Supplements   For some individuals, eating a proper diet or implementing dietary adjustments cannot adequately reduce or prevent malnutrition or cannot do so quickly enough. Examples include people taking certain medications (e.g., antibiotics); individuals who have or are recovering from certain diseases (e.g., atrophic gastritis, diverticulitis); people recovering from surgery or from certain types of injury (e.g., burns); and individuals who are smokers or alcoholics. In many of these cases the desired nutritional levels can be attained through other means, such as taking dietary supplements (e.g., fiber, vitamins, minerals) and drinking extra water. Other individuals are helped by taking medications that affect appetite or alter the absorption of specific nutrients such as cholesterol.

Supplements and medications should be taken only when nutritional homeostasis cannot be achieved through diet and evidence suggesting the presence of a specific nutrient deficiency is available. Inappropriate ingestion of nutrient supplements can lead to additional malnutrition, toxicity, and serious or life-threatening malfunction or failure of many body systems because of (1) age-related decreases in mechanisms, such as nutrient storage, conversion, and excretion, that permit adaptation to excess materials and (2) the likelihood that excess nutrients will aggravate existing problems or cause new ones in weakened organs or systems.

A Continuing Process The processes involved in maintaining nutritional homeostasis and reducing and preventing malnutrition should be viewed as ongoing even after malnutrition has been identified and rectified. Evaluation of nutritional status should continue since adjustments may be necessary as age changes occur and new situations and conditions arise. Examples include (1) the onset of, worsening of, or improvement in a disease, (2) changes in social situations (e.g., family structure), (3) alterations in psychological status (e.g., grief, life review process), and (4) modifications in economic conditions (e.g., declining real income).

ENERGY AND BODY WEIGHT

Energy Uses and Storage

The supply of energy from nutrients must be continuous because homeostasis and survival require that vital functions never cease. Since people eat intermittently rather than continuously, only some of the energy from food is used immediately after its nutrients have been absorbed. The remaining energy is stored in reserves until needed.

Most reserve energy is stored as fat in fat cells and as glycogen in the liver and in muscle cells. The body draws on these reserves when energy supplies from the nutrients being absorbed are lower than energy demands. The body can obtain energy from body proteins, but it usually does this only after most of the glycogen and fat reserves

have been used because proteins make up essential body components.

Dietary Sources of Energy

Most energy from foods is obtained from *carbohydrates*, *lipids*, and *proteins* (see Chapter 2). The carbohydrates in foods are either single sugar molecules such as glucose or larger molecules made of combinations of sugar molecules. Common examples include the disaccharides sucrose (table sugar) and lactose (milk sugar) and the polysaccharides starch and glycogen. Dietary lipids occur in a variety of forms, though usually more than 90 percent of them are in the form of *triglycerides*, which are also called fat on food labels. Other dietary lipids include cholesterol, monoglycerides, diglycerides, and phospholipids.

Disaccharides and polysaccharides are broken down into single sugar molecules before being absorbed into the body. Most sugars are converted into glucose by the liver before being transported by the blood to other cells. Fat is also broken down before absorption but is reconstituted before entering the circulatory system. Proteins are broken down into amino acids before they are absorbed.

Obtaining Energy from Molecules Energy in sugar, fat, and amino acid molecules is contained in the chemical bonds that hold the atoms composing each molecule together. Cells convert this energy into forms they can use by breaking the chemical bonds and transferring the released energy into motion, other chemical bonds, or heat.

Glucose molecules can yield useful energy almost immediately as they pass through a series of chemical reactions called *glycolysis*, which takes place in the cell cytoplasm. Glycolysis converts only approximately 5 percent of the energy in a glucose molecule into an immediately useful form. Some of the energy is released as heat, and the rest is in the remaining fragments of the glucose molecule. These fragments enter the mitochondria, where they are broken down further by complicated processes called the *Krebs cycle*, *electron transport*, and *oxidative phosphorylation*. More than half the energy released from glucose and other sugar molecules by these processes appears as heat. Most of the remaining energy is placed into *adenosine triphosphate* (*ATP*). The energy in ATP is used to power vital processes such as moving, manufacturing substances and body components, and transporting materials.

Fat and amino acids must also be broken down partially before they can release useful energy. Most of the fat and amino acid fragments enter the mitochondria, where chemical reactions convert most of their energy into heat and ATP molecules. Note that the initial partial breakdown of amino acids produces a toxic waste material (ammonia), which is immediately converted into a harmless substance called *urea*. The kidneys pass urea into urine for elimination.

The breakdown of molecular fragments from sugar, fat, and amino acids by mitochondrial processes requires the addition of oxygen and results in the formation of CO_2 and H_2O. If the oxygen supply is inadequate, glucose fragments are converted to *lactic acid* while most fat and amino acid fragments accumulate as *ketones* (*ketoacids*). Therefore, having an inadequate supply of oxygen limits energy extraction and leads to an excessive buildup of lactic acid or ketoacids, which disturb acid/base balance. Energy extraction can also be affected by limitations in the number of mitochondria, as occurs in unexercised muscle cells. Therefore, when oxygen supply or mitochondrial numbers are low, weakness and fatigue may result.

Kilocalories   Comparing the energy contents (kilocalories) of carbohydrates, fat, and proteins reveals that for a given weight of a nutrient (e.g., 1 gram or 1 ounce), carbohydrates and proteins contain nearly the same amount of energy. A sample of fat of the same weight contains approximately twice as much energy.

The high energy content of fat may lead to difficulties in interpreting labels on packaged foods. For example, a label indicating that a food is 95 percent fat-free may mean that only 5 percent of the weight of the food is fat. However, more than half the kilocalories may come from the fat since much of the remaining weight may come from water or other low-calorie components.

Energy Balance

If the total energy intake over a period equals the total amount of energy used by the body during that period, the body is in *energy balance*. If the energy intake is greater than the energy used, the body has a *positive energy imbalance*. The extra energy is stored as fat and glycogen, causing a gain in weight. If the energy intake is less than the energy used, the body has a *negative energy imbalance*. Additional energy needs are met by breaking down stored fat and glycogen, causing a loss of weight.

Energy Use   The energy used by the body can be placed into several categories. One category includes the energy needed to sustain body functions when a person is awake and in a state of complete rest. These conditions are called basal conditions, and the rate of energy use is called the *basal metabolic rate* (*BMR*).

About 20 percent of BMR energy use is due to muscle cells. Some of this energy is used by the few muscle cells that contract for breathing and for maintaining muscle tone, though muscle cells that are not contracting also use energy to tear down and rebuild their internal components. Although all cells constantly undergo this process of turnover, muscle cells account for a large proportion of the BMR energy used because of the relatively large amount of muscle tissue in the body. Therefore, individuals with more muscle mass have higher BMRs. Additional energy is used by muscles for respiration and heartbeat.

Other portions of BMR energy are used by turnover activities in other cells and in the normal replacement of cells in the skin, blood, GI tract lining, and in the uterine lining after menstruation. About 40 percent of BMR energy is used by ongoing brain and liver activities. Finally, some of this energy is used to keep the body warm. In children, the energy needed for growth adds to these uses. The BMR for an average young adult is approximately 1 kcal per minute, or 1,440 kcal per day.

A second portion of the energy used by the body powers the processes involved in digestion. This energy may constitute approximately 5 percent to 10 percent of the body's daily energy use. A third portion goes into muscle contractions during physical activity. Since the amount of an individual's physical activity usually varies greatly from one day to the next, so also does the amount of energy used. Days involving light physical activity (e.g., writing letters, talking with friends) may add a few hundred to more than 1,000 kcal of energy use to basal energy needs. Days involving many hours of strenuous physical activity (e.g., carpentry, caring for a household with children, hiking) may double the body's energy use at basal conditions.

A fourth portion of body energy use serves defensive and healing functions. For example, much energy is used to maintain a high fever, combat a major infection or widespread cancer, or recover from surgery, a burn, or another extensive injury. Like the energy for physical activity, this type of energy use is quite variable and can exceed the amount used at basal conditions. Pregnant women use energy in a fifth way by supporting the development of a fetus.

Age-Related Changes in Energy Use Aging is usually accompanied by a decrease in BMR energy use that is mostly caused by declining muscle mass. The BMR energy use from nonmuscle mass remains basically unchanged or declines slightly with age. A decline in nonmuscle BMR energy use may result from slower cellular turnover or replacement, slower synthesis of secretions by glands, and, in women, cessation of menstruation.

Though some of the decline in muscle mass is due to age changes in muscles, frequently a much larger proportion of the decline in muscle mass, and therefore in BMR energy use, results from decreased physical activity. Adults who remain physically active as age increases retain much of their muscle mass and have a small decline in BMR energy use. Furthermore, sedentary elderly people who increase their muscle mass through exercise have increases in BMR energy use.

There is probably no significant change in energy use from digestive processes in healthy aging adults. However, certain diseases of the digestive system may increase (e.g., GI tract spasms) or decrease (e.g., atrophic gastritis) this energy use.

On the average, energy used for physical activity decreases with age. The reasons for this decline include age changes; physical and mental disabilities; diseases; retirement children reaching adulthood and needing less care; institutionalization; changes in priorities; and following the preconceived or stereotyped sedentary lifestyles of the elderly. However, aging individuals can maintain or increase their level of physical activity and thus maintain or increase their energy use for physical activity.

Since many defense and healing processes occur more slowly with age, their rates of energy use may also decrease with age. However, since the number, frequency, and severity of problems requiring these processes seem to increase with age, the total amount of energy use for these processes may also increase.

Combining age-related changes in the five categories of energy use results in an average continuous decline in energy use with age. Estimates of energy use for men ages 23 to 50, 51 to 74, and 75 and above are 2,700 kcal a day, 2,400 kcal a day, and 2,050 kcal a day, respectively. The corresponding estimates for women are 2,000 kcal a day, 1,800 kcal a day, and 1,600 kcal a day. These values are estimated averages for healthy adults performing light physical activity. The actual values for individuals may vary from these estimates by more than 1,000 kcal a day depending on body size, amount of physical activity, and health status.

Age-Related Changes in Energy Balance On the average, changes in body weight, muscle mass, and bone material suggest that there is a positive energy imbalance until about age 50. After that, there is at most a slightly negative energy imbalance. However, many older individuals vary significantly from the average and have substantial positive or negative energy imbalances.

For many people beyond age 50, energy balance is fairly well maintained while the total amount of energy intake and use declines substantially. At the same time, the amounts of almost all nutrients needed by the body seem to remain stable or increase. This means that to obtain all required nutrients in adequate amounts while consuming a diet with fewer kilocalories, the foods being consumed must have higher concentrations of nutrients relative to the kilocalories contained in those foods. The ratio of the amount of a specific nutrient to the number of kilocalories in a portion of food is called *nutrient density*. Foods in the fats, oils, and sweets group in the Food Guide Pyramid have a very low nutrient density.

A better way to ensure an adequate intake of all required nutrients while maintaining energy balance is to increase energy use by increasing physical activity. This allows a person to eat more food while avoiding a positive energy imbalance, which can lead to excess body weight from a disproportionate amount of body fat. This strategy also increases the other benefits derived from regular exercise.

Overweight and Obesity

To understand what excess body weight and a disproportionate amount of body fat mean, one must establish values for desirable body weight and percent body fat.

Using Tables   Several tables of desirable body weights for people without serious disease have been developed. A table from the Metropolitan Life Insurance Company is based on the assumption that the weights of individuals with the longest life spans represent desirable body weights. The National Health and Nutrition Examination Surveys (NHANES) table and the Andres table include values for older people. The Andres table also considers the average age-related increase in the proportion of body fat resulting from loss of muscle tissue. Though these tables provide guidelines for desirable body weights, factors such as body frame size and amount of muscle tissue should be considered in establishing the desirable body weight for an individual.

Using Body Mass Index   Another way to determine whether a person has a desirable body weight is to find the *body mass index*. This value is calculated by dividing the person's weight in kilograms by the square of the person's height in meters. Values between 25 and 30 are considered desirable because people whose body mass indexes are within this range have the greatest longevity and the lowest risk of contracting diseases such as diabetes mellitus and high blood pressure. Those whose body mass index is above 40 have considerably reduced longevities and are at high risk for a variety of weight-related diseases.

Using Percent Body Fat   Percent body fat can be determined in several ways. These include measuring the thickness of skin folds at one or more places; measuring how much electrical resistance the body provides; and comparing a person's weight in air with body weight when that person is completely submerged in water.

Desirable percent body fat values for adults are considered to be 15 to 18 percent for men and 20 to 25 percent for women. Adults with values above 25 percent for men and 30 percent for women have shorter longevities and higher risks for developing weight-related diseases. Since the risks are somewhat higher for individuals with high concentrations of fat around the waist rather than the thighs, determining the ratio between the circumference of the waist and that of the hips gives a further indication of desirable percent body fat. Waist-to-hip ratios should be less than 0.9 for men and less than 0.8 for women.

Definitions   We can now define terms indicating deviations from accepted standards. People whose body weights are 10 percent to 20 percent greater than the desirable body weights can be considered *overweight*. People whose body weights are more than 20 percent above the desirable body weights can be considered *obese* if their percent body fat exceeds 25 percent (men) or 30 percent (women) or if their body mass index is more than 30. Very muscular individuals whose weight is more than 20 percent above desirable body weights but whose percent body fat is less than 25 percent (men) or 30 percent (women) are overweight but are not considered obese.

Consequences   Being overweight but not obese has little effect on a person's longevity or risk of developing weight-related disease. Some authorities suggest that in general elderly people may benefit from being slightly overweight because the extra energy stored in their body fat helps them maintain nutritional homeostasis and endure the adverse effects of periods of illness or other undesirable circumstances.

Though being slightly overweight may be beneficial, obesity is accompanied by increases in the incidence of a variety of problems, a greater negative impact from these and other problems, and lower longevity. The incidence and seriousness of weight-related problems are directly related to the degree of obesity. Longevity is inversely related to the degree of obesity.

Prevention and Correction   The best way to avoid the consequences of obesity is to avoid becoming obese. This is much easier than trying to lose weight once obesity has developed. Furthermore, recurring and significant fluctuations in weight decrease longevity and increase the risk of developing weight-related disease.

One of the most important steps in avoiding obesity as age increases is to maintain energy balance by decreasing total energy intake when energy use decreases. A key sign that this action is appropriate is a significant increase in weight.

Another important step is staying physically active to maintain a high level of energy use. Physical activity keeps energy use high both directly and by helping to maintain a large muscle mass, which sustains a high BMR energy use. It can also help keep energy intake down by suppressing appetite, distracting attention from eating, promoting interest in diverse activities while preventing boredom, and supporting a healthful psychological state. Finally, since people who get plenty of exercise can eat more and not gain weight, they have a higher chance of obtaining adequate amounts of all necessary nutrients without becoming obese (Chap. 8).

Solving obesity by losing weight and decreasing percent body fat can be very difficult since many factors may contribute to obesity. Key features in a successful program to lose weight include getting a physical examination; developing a long-range plan for gradual weight loss and weight maintenance; decreasing energy intake while eating foods with high nutrient densities; exercising; and receiving monitoring regularly to prevent malnutrition and other problems.

Underweight

Having a weight below the range for desirable body weight is called being *underweight*. The negative energy imbalance that causes a person to become underweight may result from inadequate energy intake, a reduced ability of the digestive system to make energy-containing nutrients available to the body, or excessively high energy use.

Consequences   Being underweight may have several undesirable consequences. These including muscle weakness; fatigue; lethargy; increased risk of low body temperature; reduced resistance to infection; and decreased ability to tolerate periods of adversity such as a prolonged disease. Since being underweight is often but not always accompanied by deficiencies in specific nutrients, body malfunctions and other problems are also commonly present, including specific diseases resulting from nutrient deficiencies. This variability helps explain why some people with very low body mass index seem to have a greater mean longevity, while others have a reduced mean longevity.

Often, being underweight results in slightly lower longevity for underweight persons or those who have a body mass index below 20. Some scientists believe that the decline in longevity is directly related to the degree to which a person is underweight, while others believe that having a low body mass index increases mean longevity. Adverse effects from being underweight may be greater for the elderly than for young adults.

Prevention and Correction   Preventing being underweight involves avoiding a negative energy imbalance. This may involve increasing energy intake when activity levels increase. When a substantial drop in weight becomes apparent, dietary energy intake may be increased, correction of or compensation for digestive system difficulties may be implemented, or activity levels may be reduced. These strategies also can be used to correct a chronic underweight condition. As with solving obesity, correcting underweight conditions in elderly persons requires special awareness of and attention to each individual's particular circumstances.

CARBOHYDRATES

Digestible Carbohydrates

Common dietary carbohydrates come in several forms. Some of them may be *monosaccharides* such as glucose and fructose, which are abundant in many packaged foods and fruits. However, a major portion of dietary carbohydrate consists of *disaccharides* such as sucrose, which is found in table sugar and in most sweet foods, and lactose, which is found in milk and milk products. Sucrose consists of glucose and fructose, and lactose consists of glucose and galactose. The other major dietary carbohydrate molecules are *polysaccharides*. The most common digestible polysaccharides are starch, which is found in plant foods, and glycogen, which is found in meats. Both are made of glucose molecules.

The specific disaccharides and polysaccharides mentioned thus far are broken down into individual monosaccharides before being absorbed. Once absorbed, galactose and much fructose are converted into glucose by the liver. The remaining fructose is broken down by a process similar to glycolysis.

Indigestible Carbohydrates: Fiber

Many polysaccharides in foods from plants cannot be broken down by digestive enzymes. These indigestible polysaccharides are called *fiber*. Fiber that does not dissolve in water is called *insoluble fiber* and includes *cellulose, hemicellulose*, and some noncarbohydrate material (lignin). Fiber that dissolves in water is called *soluble fiber* and is abundant in most fruits and vegetables, especially oats and beans.

Uses

Sugars The body uses glucose and fructose as sources of energy. Some cell types, such as brain cells and red blood cells, rely almost exclusively on glucose for energy, and active muscles and the heart consume large quantities of sugars for energy. Since glucose is an excellent energy source, the liver makes it from fragments of amino acid molecules and lactic acid. Only a small amount of glucose can be made from fat because only the glycerol portion in fat can be converted into glucose.

Some glucose helps supply energy by transferring many fragments from fat and amino acid molecules into the Krebs cycle. Without glucose, these fragments can yield no energy and are converted into ketoacids. Fragments from sugar molecules are also used to manufacture fat, other lipids, parts of amino acids, and sugar molecules in DNA and RNA.

When the blood contains more sugar than the cells require, much of the sugar is stored as glycogen or is converted into fat for storage. The level of blood sugar is controlled by several hormones.

Fiber Since fiber cannot be digested, the sugar molecules of which it is made cannot be absorbed and used in the body. However, dietary fiber is essential for good health because it stimulates intestinal motility by providing bulk. In this way, it promotes proper intestinal functioning and helps prevent several disorders of the large intestine. Furthermore, soluble fiber reduces the risk of atherosclerosis by decreasing the absorption of cholesterol. It can also help diabetics by slowing the absorption of glucose.

Recommended Dietary Intakes

Ample energy from digestible carbohydrates can be obtained by eating 50 to 100 grams of digestible carbohydrate a day. Individuals who are quite active should consume more digestible carbohydrate. Another guideline is to consume 55 to 60 percent of the total energy intake as digestible carbohydrates. Most of this intake should be in the form of starch or glycogen.

Daily fiber intake can range between 20 and 35 grams per day and probably should be close to the higher value. Individuals with a low fiber intake probably should increase it gradually to allow the GI tract to adjust to the change in diet. Individuals with special problems such as constipation, diverticulosis, atherosclerosis, and diabetes mellitus may benefit from higher fiber intake.

Carbohydrate Deficiencies

Deficiencies in digestible carbohydrates are rare because many foods have large quantities of these nutrients. However, if such a deficiency develops, body cells lack adequate energy. The consequences of a low energy supply include weakness, lethargy, and reduced mental functioning. In addition, fat and amino acid fragments that cannot be channeled into the Krebs cycle are converted into excess ketoacids, which disturb homeostasis by altering acid/base balance and causing excess sodium and potassium loss in the urine.

Inadequate fiber intake is a common problem. Several undesirable results, such as problems with the large intestine, were mentioned in Chap. 10.

Carbohydrate Excesses

Consuming excess digestible carbohydrates is common because foods high in carbohydrates are relatively inexpensive and because sweet foods are pleasant to eat. Eating excess sugars promotes tooth decay and, by suppressing the appetite, reduces the intake of other important nutrients. For example, some elderly people suffer from *protein-carbohydrate malnutrition* (*PCM*) because the carbohydrate-rich foods they eat in abundance contain little protein. A high carbohydrate consumption is also often accompanied by undesirable weight gain and hampers maintenance of blood glucose homeostasis.

A high fiber intake can lead to deficiencies in calcium, zinc, and iron because fiber inhibits the absorption of these nutrients. Excess fiber consumption also hampers maintenance of water homeostasis because fiber binds water in the large intestine and therefore reduces water absorption. For this reason, individuals on high-fiber diets should drink large quantities of water.

Certain types of indigestible carbohydrates result in gas production in the large intestine. Therefore, foods such as beans and peas, which contain high amounts of these substances, can cause discomfort and embarrassment.

LIPIDS

The body can convert the lipids contained in foods into almost all the other types of lipids it needs. It can also synthesize most types of lipids from carbohydrates and proteins. For example, the liver can manufacture cholesterol.

Though lipids are a diverse group of substances, most dietary lipids fall into only a few categories. We will examine glycerides and cholesterol here (Chap. 2).

Tri-, Di-, and Monoglycerides and Fatty Acids

The most abundant dietary lipids are *triglycerides*, which consist of three *fatty acid* molecules attached to a *glycerol* molecule. Triglycerides are also called fat on food labels and may be either solid or liquid (oils) at room temperature. Many foods contain *monoglycerides* and *diglycerides*, which contain one or two fatty acids, respectively. Monoglycerides and diglycerides are not called fat. Therefore, they are not included as part of the fat content listed on food labels even though they contain and give the body the same components as do triglycerides.

The fatty acids in fat may be *saturated* or *unsaturated*. Unsaturated fatty acids may be *monounsaturated* or *polyunsaturated*. Unsaturated fatty acids differ from one another not only in chain length and number of double bonds but also in the locations and orientations of those bonds. Foods from plants have higher proportions of monounsaturated and polyunsaturated fatty acids, while foods of animal origin have higher proportions of saturated fatty acids.

Naturally occurring unsaturated fatty acids in foods, including those in fat, may have hydrogen added during processing. Fatty acids and fat treated in this way are said to be *partially hydro-* *genated* or *hydrogenated*. Hydrogenation makes the fatty acids and fat more solid and thus improves the texture of some foods.

Two fatty acids are of special importance because they are essential parts of body molecules and must be obtained in the diet since they cannot be manufactured by the body. Therefore, these two fatty acids—linoleic acid and alpha-linoleic acid—are referred to as *essential fatty acids*. These fatty acids are also called omega-3 fatty acids because of the location of a double bond. Fat in plant oils and fish oils have high levels of these fatty acids.

Cholesterol

The other important dietary lipid is *cholesterol*. This substance, which has a molecular structure resembling chicken wire, is found only in foods from animals. Foods high in cholesterol include those containing egg yolks, cream, and fat from meats.

Uses

Dietary lipids have several functions. They make some foods more pleasant by improving flavor and texture and producing a sense of fullness and satisfaction. They also aid the absorption of vitamins A, D, E, and K. Lipids that have been absorbed are used to produce bile, supply energy, and build body components and chemicals (e.g., fat tissue, cell membranes, vitamin D, steroid hormones). The two essential fatty acids are used in building cell membranes and producing a variety of chemicals that help regulate diverse processes, including blood clotting, stomach secretion, and immune system functioning.

Recommended Dietary Intakes

Since diets in the United States typically contain more lipids than are needed, many dietary recommendations for adults provide upper rather than lower limits for dietary lipids. In general, fat should account for less than 30 percent of total daily energy intake, and saturated fat and polyunsaturated fat should each account for less than 10 percent. Cholesterol intake should be below 300 mg per day. Many steps can be taken to keep dietary lipid intakes below these limits, and the absorption of dietary cholesterol can be reduced somewhat by diets high in soluble fiber.

Lipid Deficiencies

Individuals with lipid-deficient diets may have low energy levels and deficiencies in fat-soluble vitamins. Diets lacking sufficient plant or fish oils do not provide enough essential fatty acids, resulting in varied disorders, including abnormalities in blood clotting, blood pressure, and immune system functions.

Lipid Excesses

Diets containing excess lipids promote several problems, including indigestion, obesity, colon cancer, atherosclerosis, and possibly breast cancer.

When fat and cholesterol are absorbed by the small intestine, they combine with other lipids (phospholipids) and blood proteins to form droplets of *lipoprotein*. Lipoproteins are also formed using fat and cholesterol made by the liver in cases of positive energy imbalance, especially when it is severe enough to cause obesity. Since all these lipoproteins contain much fat and are relatively very light, they are called *very low density lipoproteins* (*VLDLs*). Much of the fat in VLDLs is removed and used by body cells. The remaining particles are not as light and are called *low-density lipoproteins* (*LDLs*).

LDLs are removed from the blood by liver cells and other cells that have receptors for them. Liver cells eliminate the cholesterol from the LDLs they receive by excreting it into bile or converting it into other useful materials. Other body cells store much of the cholesterol in the LDLs they receive. When cells lining blood vessels store LDL cholesterol, it initiates the formation of atherosclerotic plaques.

Since diets low in saturated fat help liver cells remove LDLs from the blood, less LDL enters other cells and less plaque formation occurs. Conversely, diets high in saturated fat inhibit liver cells from removing LDLs from the blood, and cells in vessels remove the LDLs and form plaques. Therefore, plaque formation is kept low by low-saturated-fat diets and is raised by high-saturated-fat diets.

The number of double bonds and their locations in unsaturated fatty acids influences their amount of risk for atherosclerosis. Monounsaturated fatty acids are less likely to produce lipid peroxides (LPs). Polyunsaturated fatty acids (PUFAs) help keep blood LDLs low, but when they have double bonds at a location called the omega-6 position, they promote LP formation. PUFAs with double bonds in the omega-3 position produce less LPs and help keep blood clots and blood pressure low.

Plaque formation is also kept low by lipoproteins called *high-density lipoproteins* (*HDLs*), which have relatively little cholesterol or other lipids. HDLs carry cholesterol from body cells to the liver for elimination. This explains why a high HDL/LDL ratio partially counters the adverse effects of high blood cholesterol and high LDL levels and thus helps reduce the risk of atherosclerosis. One way exercise lowers the risk of atherosclerosis is by increasing HDLs and thus lowering blood cholesterol and LDL levels.

A total blood lipoprotein level (total cholesterol level) of approximately 200 mg per 100 milliliters of blood (200 mg/dl) is used by many people to mark the boundary between acceptable and unacceptable blood lipid levels. Individuals with total cholesterol levels substantially above 200 mg/dl have a much higher risk of developing atherosclerosis, while individuals with levels well below 200 mg/dl have a rather low risk. Having LDL levels above 165 mg/dl or HDL levels below 35 mg/dl also implies a substantial risk.

Many individuals can maintain healthful cholesterol and lipoprotein values or improve poor values by eating foods with much soluble fiber, avoiding dietary cholesterol and saturated fat, sustaining the recommended weight through energy balance, and exercising. These steps can substantially reduce the risk of developing atherosclerosis.

PROTEINS

Foods high in protein include meat and fish, egg whites, milk, beans, and peas. Proteins from animal products contain all 20 types of amino acids; plant proteins often lack or have very little of one or more types. Since different plant foods lack different amino acids, a complete set of amino acids can be obtained from plants by eating specific combinations of plant foods (e.g., rice and beans, corn and beans).

Uses

Digestive enzymes break dietary protein molecules into individual amino acids, which are then absorbed into the blood. Body cells use fragments from some amino acids for energy and to build nonprotein molecules. Other amino acids are used to build body proteins.

Body cells require large supplies of all 20 types of amino acids to build proteins. If even one type is missing or is in very short supply, synthesis of body proteins virtually stops. Although 11 of the 20 types can be manufactured in abundance by the liver and other cells, adequate amounts of the other nine types cannot be made in the body and must be supplied by dietary proteins. These amino acids are called *essential amino acids*.

Body proteins made from dietary amino acids have many uses. Many are used to build cell membranes, keratin (epidermis), collagen and elastin fibers (dermis, ligaments, cartilage), and muscle cell microfilaments. Others are used as enzymes to control chemical reactions; hormones to send messages; antibodies to fight infections; buffers to regulate acid/base balance; clotting factors to stop bleeding; blood proteins to help transport lipids (lipoproteins); and blood proteins to control the movement of water through capillary walls. Body protein molecules are broken down and their amino acids are used for energy only under starvation conditions.

In people of all ages body proteins such as keratin, muscle cell microfilaments, and digestive enzymes are continuously lost or broken down. Since the body has very limited stores of amino acids, a regular intake of dietary proteins is necessary for the continuous replacement of these structures so that their functions, and therefore homeostasis, are maintained.

Recommended Dietary Intakes

The RDA of 0.8 to 1.0 gram of protein per kilogram of body weight per day for elderly people should be increased at least to 1.0 gram per kilogram per day. Even higher protein intakes may be appropriate for individuals who must produce extra proteins, such as people recovering from a serious injury or disease and those engaged in very strenuous physical training. Diets with a reduced protein content may be recommended for individuals with certain kidney diseases because

such diets yield less urea. Diets containing mostly animal proteins rather than plant proteins also generate less urea because a higher proportion of the amino acids from animal proteins can be used to build body proteins. With reduced urea production, less urea must be eliminated by the kidneys and a urea buildup in the body is avoided. This is important because excess urea can harm body cells.

Protein Deficiencies

Since proteins contribute to many body structures and perform many functions, diets low in one or more amino acids needed for protein production cause many problems. Examples include structural and muscle weakness; slowed body reactions; increased risk of infection; loss of acid/base homeostasis; excess bleeding; edema; and poor recovery from injury.

Protein Excesses

Extra amino acids from excess dietary proteins can be used for energy or converted into other useful materials and stored fat. Therefore, diets containing more than the recommended amounts of protein usually cause no difficulty except the positive energy imbalance that may develop. However, individuals with kidney disease may accumulate harmful levels of urea.

WATER

Dietary water enters the body in beverages and foods. In addition, some is produced in the body by the complete breakdown of carbohydrates, lipids, and proteins. Water leaves the body in feces, urine, and perspiration and by evaporation from the respiratory system.

Uses

Dietary water aids the digestive system by dissolving materials, lubricating, transporting materials by bulk flow and diffusion, and reacting chemically. Once in the body, water serves the same purposes. It also helps regulate body temperature by distributing heat and eliminating heat through evaporation. Water helps cushion, maintain cell size through osmotic pressure, and regulate acid/base balance.

Body water is in three areas: the blood, the spaces between cells (intercellular space), and within cells (intracellular space). Water moves in a controlled manner between these areas. Both the total amount of water in the body and the proportionate distribution of water among these three areas are essential for healthy survival.

Recommended Dietary Intakes

One aspect of maintaining homeostasis is maintaining water balance, or obtaining as much water as is lost. On the average, an adult can maintain water balance by ingesting approximately 2 liters of water a day. The breakdown of nutrients supplies the 350 ml of additional water needed each day to achieve water balance. However, the amount of dietary water supplied may require considerable adjustment because water loss can vary greatly between individuals and from time to time. Factors that increase water loss include high environmental temperatures, low humidity, vigorous exercise, alcohol consumption, fever, vomiting, diarrhea, several kidney diseases, and various medications. In addition, individuals who spit frequently, such as those who chew tobacco, lose much water in the saliva they expectorate.

Water Deficiencies

Advancing age is accompanied by an increasing risk of water deficiency because of decreases in both the sensation of thirst and the control of water loss in urine. Many elderly people have disabilities that restrict their access to water. Others have elevated water loss from medications that reduce high blood pressure or edema. Furthermore, increasing fiber in the diet without simultaneously increasing water intake or eating a diet very high in fiber will raise the risk of water deficiency because fiber inhibits the absorption of dietary water. Since several potent factors can lead to water deficiency among the elderly, monitoring older individuals' water intake and output is advisable for care givers in clinical care and institutional settings.

Individuals who do not ingest enough water to maintain water balance may suffer from any of a variety of abnormalities, including (1) low blood pressure and the consequences of poor circulation, (2) joint stiffness, (3) sunken eyes, (4) dryness of the eyes, mouth, or skin, (5) sagging and wrinkling skin, (6) diverse brain malfunctions, (7) constipation, (8) reduced kidney functioning, and (9) urinary stones.

Water Excesses

Ingesting too much water seems to be uncommon among the elderly. Those who do so may experience high blood pressure or edema.

VITAMINS

Characteristics

Vitamins are a diverse group of substances that have five features in common.

1. They are needed in relatively small quantities.
2. They are essential because certain chemical reactions cannot occur effectively without them.
3. They must be obtained in the diet because the body cannot make them in adequate amounts.
4. They must be eaten regularly because they are stored in limited quantities and are gradually lost.
5. A deficiency in each vitamin results in at least one specific disorder.

One main group consists of the *fat-soluble vitamins* (vitamins A, D, E, and K), so called because they dissolve in fat. These vitamins are efficiently absorbed only if there is some fat in the diet. They can be stored in fairly large quantities in the liver and other areas of the body that contain high concentrations of fat.

All other vitamins are *water-soluble vitamins*. The body is less able to store these vitamins because they move freely among the water-containing spaces and are easily lost in the urine.

Sources

Vitamins are found in different amounts in various foods. Since no single food contains adequate amounts of all the vitamins, maintaining vitamin homeostasis requires one regularly to eat a variety of foods. A diet based on the Food Guide Pyramid provides a healthy adult of any age with adequate amounts of all vitamins.

Using vitamin supplements to augment dietary

intake is generally not recommended because this can reduce the dietary intake of other nutrients and because high levels of certain vitamins can injure body cells and cause other disorders (e.g., diarrhea). However, vitamin supplements may be necessary for individuals who do not eat enough varieties or quantities of food and people with abnormal conditions. Such conditions include ones that prevent adequate absorption or use of vitamins and ones that cause excess destruction or elimination. Vitamin supplements may also be used to treat certain diseases (e.g., vitamin D for osteoporosis). Supplements with vitamins A, C, and E, and β-carotene, which serve as antioxidants, seem to reduce the risk of certain diseases related to damage from free radicals (e.g., atherosclerosis, Alzheimer's disease, cancer).

Deficiencies and Excesses

Many cases of vitamin deficiency among the elderly result from eating too little food or a limited variety of food because of alcoholism or other unfavorable factors. Furthermore, some deficiencies develop because vitamins are destroyed or lost when foods are washed, cooked, or processed in other ways. Studies of vitamin status among the elderly show that deficiencies in vitamins A, D, B6, B12, and C; thiamine; riboflavin; and folate occur frequently. Examples of adverse changes from deficiencies in these vitamins include: infections and disorders of the skin and eyes (vitamin A); weak bones (vitamin D); anemia and poor functioning of the nervous and immune systems, excess blood homocysteine (vitamin B6); anemia and nervous system malfunction, excess blood homocysteine (vitamin B12); edema and poor healing (vitamin C); muscle weakness and nervous system malfunctions (thiamine); inflammation of the mouth and skin, and poor vision (riboflavin); anemia and poor functioning of the nervous and digestive systems (folate). Vitamin excesses are rare and almost always result from ingesting vitamin supplements.

Vague or nonspecific symptoms such as nausea, loss of appetite, and mental, emotional, or personality changes may occur with many cases of mild vitamin deficiency or excess. Other adverse changes that are specific for each vitamin deficiency may be barely detectable at first.

MINERALS

Characteristics

Like vitamins, minerals are a diverse group of substances. Minerals are also similar to vitamins because they must be obtained in the diet since the body cannot make them, they must be eaten regularly since gradual loss depletes the body's limited reserves, and specific disorders often accompany deficiencies or excesses of each mineral.

Minerals are different from vitamins in that some minerals (e.g., calcium, phosphorus, sodium, potassium) must be obtained in large quantities. In addition, though some minerals (e.g., zinc, magnesium, copper, selenium) assist certain chemical reactions, many others form parts of body compounds and structures. Phosphorus in DNA and calcium in bones are examples.

Sources

The previous statements about vitamins in food and supplements are also applicable to minerals.

Deficiencies and Excesses

Common among the elderly, the causes of mineral deficiencies are essentially the same as those of vitamin deficiencies. Several types of mineral deficiencies result from inadequate absorption. Many other mineral deficiencies develop because of excess mineral elimination resulting from diarrhea, kidney malfunction, or medications that increase mineral elimination by the kidneys. Deficiencies in calcium lead to weak bones; deficiencies in iron lead to anemia, weakness, and reduced immune system functioning; and deficiencies in zinc lead to slow healing, decreased immune system functioning, poor taste perception, and male impotence; deficiencies in selenium prevent certain enzymes from eliminating free radicals.

The most common mineral excess may be sodium excess from ingesting foods containing salt, which is added to improve flavor. An excess salt intake promotes water retention and therefore increases blood pressure. Other mineral excesses result from consuming diets high in specific minerals or from ingesting mineral supplements.

As with vitamins, slight to moderate deficiencies or excesses in minerals may cause only vague and nonspecific symptoms and subtle adverse changes in the body.

NUTRITION AND ALCOHOL

Alcohol consumption and alcoholism occur frequently among the elderly, just as they do among younger adults. Light or moderate alcohol consumption by healthy adults seems to have the same effects regardless of age. For example, occasional modest consumption probably does not affect nutrition, but it seems to lower the risk of atherosclerosis. However, heavy consumption, whether occasional or frequent, adversely affects nutrition and eliminates any effects alcohol has on reducing atherosclerosis. The effects may be greater as age increases because of age changes and reduced reserve capacities in some organs, more numerous diseases and more advanced stages of disease, and greater use of medications.

Alcoholism (chronic heavy alcohol consumption) often causes deficiencies in many nutrients because alcoholics are less able to plan or obtain a proper diet, eat less, and suffer damage to the digestive system that decreases its ability to provide nutritional homeostasis. Some alcoholics can be helped nutritionally by taking supplements to reduce deficiencies. Alcoholism also lowers HDL levels.

NUTRITION AND MEDICATIONS

The use of medications increases with age, and many medications used by elderly individuals have adverse effects on nutrition. However, the effects of some medications and combinations of medications are not well understood. Other factors, such as age-related modifications in the body's storage, metabolism, and elimination of medications, further complicate the effects of medications on nutrition.

Medications may alter nutrition by affecting many factors. These include mental or emotional state; physical ability and coordination; food selection (because of incompatibility with a medication or its schedule of administration); appetite; the senses of taste and smell; GI tract secretions and motility; GI tract bacteria; and nutrient absorption storage, use, or elimination. Conversely, foods can decrease the effectiveness of some medications by reducing their absorption by the GI tract (e.g., antibiotics) and can increase the effects of medications by reducing their elimination by the liver.

Many other age-related factors contribute to an age-related increase in the number and the severity of problems from medications. Examples include age-related increases in use of non-prescription drugs; number of medications taken per person; number of physicians and pharmacies used per person; decrements in sensation and cognition; erroneous compliance.

NUTRITION AND DISEASE

Proper diet and nutritional homeostasis promote survival and help prevent many diseases. Conversely, poor diets and malnutrition lead to the development of many abnormal conditions and diseases (e.g., atherosclerosis, osteoporosis, digestive system disorders) and amplify their consequences. Furthermore, age-related abnormal conditions and diseases (e.g., emphysema, dementia, digestive system diseases) contribute to poor diet and malnutrition.

NUTRITION AND MAXIMUM LONGEVITY

Animals in laboratories are often provided with food at all times. The food provided contains a balance of nutrients. The animals can eat as much as they want anytime they want. Animals able to eat freely in this way are said to be fed **ad libitum (AL)**. In other cases, the food provided and the times it is provided are restricted. These animals are undergoing **dietary restriction (DR)**.

In 1937, McCay reported that DR rats that were provided 33 percent less food than the amount eaten by AL rats had significant increases in mean longevity (ML) and maximum longevity (XL). Since then, DR has resulted in increases in ML and in XL for many types of animals including roundworms, spiders, insects, and mice.

Attempts to identify the means by which DR increases ML and XL revealed that the essential factor is the reduction in calories eaten by the animals. As long as the animals are fed balanced diets, only the amount of calories eaten affects both the ML and the XL consistently. The types of nutrients in the diet and the times of eating do not cause the increases in ML and XL. Therefore, dietary restriction is now called **caloric restriction (CR)**. CR animals have undernutrition without having deficiencies in any particular nutrient (e.g., protein, lipids, minerals, vitamins). Though extreme CR (e.g., 60 percent CR) leads to unhealthy underweight conditions, undergoing reasonable CR (e.g., 30 percent CR) and being underweight from malnutrition are different entities.

Effects from CR Though many other treatments can increase ML in animals and in humans, CR is the only means known to increase the XL of animals. Therefore, scientists believe that CR actually slows fundamental aging processes. Research with CR in monkeys and in humans is underway. CR has been shown to increase ML in monkeys.

In rats and mice (i.e., rodents), CR increases ML and XL by 25 percent to 30 percent. Animals undergoing CR also develop fewer diseases, develop diseases at later ages, and develop milder forms of disease. CR has different degrees of effect on ML, XL, and diseases depending upon when CR begins, how severe it is, and how long it lasts. CR has detrimental effects if it begins during early development, but after that, the earlier it begins, the greater the effect. Also, the longer CR lasts, the greater the effect.

CR causes many beneficial effects in the animals in which it has been tested. These effects may contribute to the increases in ML and XL and the reductions in diseases. Most of the effects relate to the causes of aging proposed in the theories of aging (see Chapter 2). However, the mechanisms by which CR increases ML and XL and reduces disease are not known.

Specific benefits from CR include slowing the age-related increases in the following: free radical production; free radical damage; protein cross-links and protein glycation; lipofuscin formation; mitochondrial damage; blood pressure; blood LDLs; blood triglycerides; blood glucose; blood insulin; obesity; diabetes mellitus; percent body fat (usually); IL-6; kidney disease; and cancers. CR slows or delays the age-related declines in

HDLS; number of muscle cells; rate of protein synthesis; DNA repair; insulin sensitivity; certain hormones (e.g., melatonin, DHEA); immune function (usually); IL-2; physical activity; and general health (e.g., skin, cardiovascular, kidneys).

Exercise produces many of these effects. Scientists have shown that CR does not increase ML and reduce diseases in the same ways that exercise produces these effects. Though both CR and exercise increase ML and reduce diseases, their effects are independent, not synergistic. The effect on ML in animals having both CR and plenty of exercise equals the sum of effects on ML seen in sedentary animals with only CR and the effects seen in AD animals with plenty of exercise. Finally, exercise does not affect XL.

How animal studies with CR can be applied to human aging is still being determined. The effects of CR on humans are still unknown. Unlike animals, humans who subsist on diets containing few kilocalories or low protein, such as people living in poor and overcrowded areas, have shorter mean longevities. No individuals like that ever exceed the maximum human longevity of 120 years. Of course, these CR people do not eat carefully compounded and analyzed laboratory feeds and do not live in the controlled environment of a laboratory, receiving inoculations, continuous monitoring, and ongoing professional care. Could humans living most of their lives in such conditions and eating limited amounts of only certain selected types and quantities of food at specified times every day have greater maximum longevity? Would an increase in maximum longevity be worth the effort? How much could longevity be increased by CR among people living more conventional lifestyles? If the mechanisms by which CR works are discovered, will it be possible to produce those mechanisms in humans without requiring them to undergo CR? Future research may eventually answer these questions.

12

URINARY SYSTEM

The Y-shaped urinary system consists of two *kidneys*, two *ureters*, the *urinary bladder*, and the *urethra* (Fig. 12.1). The bean-shaped kidneys lie behind the organs in the abdominal cavity. The kidneys perform this system's main functions for homeostasis. In doing so, they slowly produce urine, which passes through the ureters and into the urinary bladder, where it is temporarily stored. When the bladder becomes partially filled, it contracts, forcing the urine through the urethra and out of the body.

MAIN FUNCTIONS FOR HOMEOSTASIS

Several systems, including the circulatory, respiratory, skeletal, and digestive systems, play major roles in maintaining chemical homeostasis; the urinary system completes this list. Besides regulating numerous chemicals, the urinary system assists other systems in regulating blood pressure.

The urinary system makes seven contributions to homeostasis. Each activity is adjusted to compensate for changing body conditions. Each kidney function is regulated by the nervous system, the endocrine system, or the characteristics of the blood flowing through the kidneys.

Removing Wastes and Toxins

One function of the urinary system is removing wastes and toxins (e.g., heavy metals, dyes) from the blood. Major waste materials removed include urea, uric acid, and ammonia, which result primarily from the metabolism of amino acids and proteins, and *creatinine* from muscle cells. Although body concentrations of urea and creatinine can become relatively high before causing significant harm, slight elevations in uric acid cause the formation of irritating crystals (e.g., gout) and ammonia is highly toxic at very low concentrations. Many drugs which can reach toxic levels are also removed.

Regulating Osmotic Pressure

Osmotic pressure is the total concentration of dissolved materials in a liquid. Since water and many dissolved substances can pass through capillary walls, the osmotic pressures of the blood and the fluid surrounding body cells (*interstitial fluid*) are equal. The kidneys regulate the osmotic pressure

FIGURE 12.1 The urinary system.

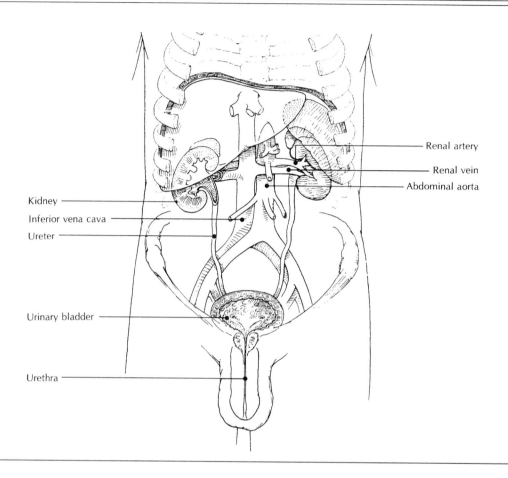

Renal artery
Renal vein
Abdominal aorta
Kidney
Inferior vena cava
Ureter
Urinary bladder
Urethra

of blood and therefore that of interstitial fluid by adjusting the amounts of water and dissolved materials that leave the body in urine.

If the osmotic pressure of the interstitial fluid is the same as the osmotic pressure inside body cells, osmotic homeostasis exists and the cells remain the same size. However, if the osmotic pressure surrounding the cells rises, water will leave the cells by the process of osmosis, causing them to shrink and their contents to become more concentrated. Conversely, if the osmotic pressure surrounding the cells falls, water will diffuse into the cells, causing them to swell and their contents to become dilute. In either situation the cells malfunction because their structure and chemical concentrations are disturbed. Swelling of brain cells is especially dangerous because the excess pressure that develops inside the skull causes neuron injury and malfunctioning.

Though all substances dissolved in the interstitial fluid contribute to its osmotic pressure, the ratio between water and sodium is the main determinant of its osmotic pressure. Therefore, the kidneys maintain osmotic homeostasis primarily by adjusting the amounts of water and sodium that remain in the blood and the amounts excreted in the urine. The kidneys must frequently alter these amounts to compensate for factors that alter osmotic pressure, including changes in intake (e.g., drinking fluids, eating salty foods) and output (e.g., perspiring, having diarrhea).

Maintaining Individual Concentrations

The urinary system maintains individual homeostatic concentrations of specific minerals such as sodium, potassium, calcium, magnesium, and phosphorus. Each mineral is important for spe-

FIGURE 12.2 Kidney structure: (a) Internal regions and blood vessels. (b) A nephron.

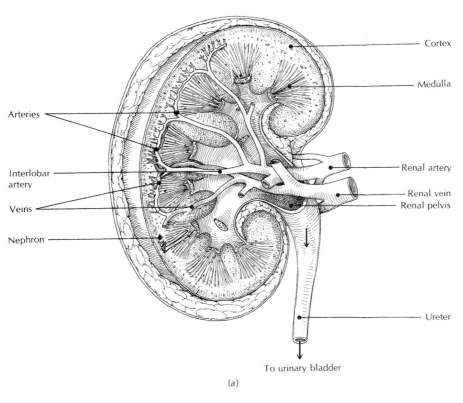

(a)

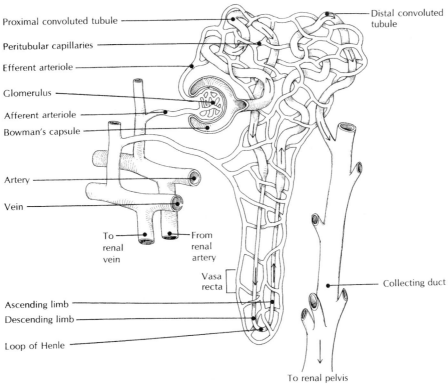

(b)

cific cell activities and must be available at the correct concentration for these activities to occur properly. The kidneys adjust the retention and excretion of each substance individually, compensating for changes in input (e.g., eating) and output (e.g., perspiring, bleeding).

Maintaining Acid/Base Balance

Maintaining acid/base balance (pH homeostasis) is important because disturbances disrupt molecular structure and functioning (e.g., enzymes). Many body activities tend to disturb acid/base balance because they produce acids (e.g., carbonic acid, lactic acid, ketoacids). Acid/base balance can also be disturbed by ingesting acidic substances such as vinegar and citrus fruits, ingesting alka-line substances such as sodium bicarbonate and other antacids, or changing CO_2 levels through altered respiratory system functioning. The kidneys help compensate for such disturbances and thus help maintain acid/base balance by adjusting acid and buffer materials (e.g., sodium bicarbonate) in the blood.

Regulating Blood Pressure

The kidneys help regulate blood pressure by adjusting the amount of water retained in the blood and thus help determine the volume of blood in the vessels. Low blood pressure can be increased by retaining more water, and high blood pressure can be reduced by allowing more water to leave in the urine.

FIGURE 12.3 Filtration, reabsorption, and secretion by a nephron.

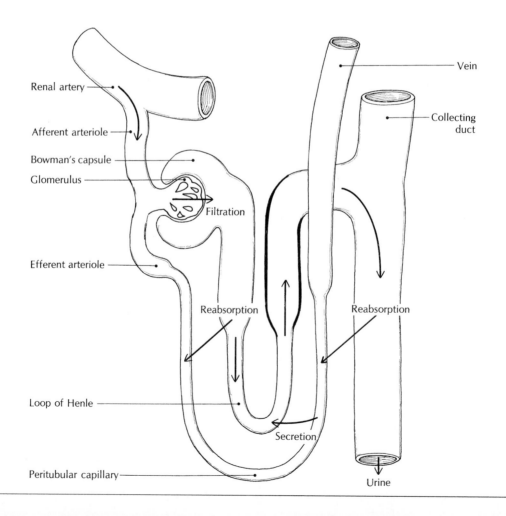

The kidneys also influence blood pressure by secreting an enzyme (renin) when blood pressure in the kidneys is low. This enzyme causes the formation of another substance in the blood (angiotensin II), which results in increased production of the hormone aldosterone. Angiotensin II and aldosterone increase blood pressure by causing small arteries to constrict and causing the kidneys to retain more water. Conversely, when blood pressure rises, less renin is produced. Then blood pressure can drop back to normal because vessels can dilate and more water can leave in the urine.

Activating Vitamin D

The urinary system helps maintain proper calcium concentrations not only by directly adjusting the retention and excretion of calcium but also by activating vitamin D. Fully activated vitamin D from the kidneys is needed for adequate absorption of calcium by the small intestine and proper calcium retention by the kidneys.

Regulating Oxygen Levels

The urinary system helps regulate oxygen levels. When oxygen levels are low, the kidneys secrete a hormone (*erythropoietin*) that stimulates red blood cell production in bone marrow. When red blood cells increase, more oxygen enters the blood in the lungs. Conversely, high oxygen levels inhibit erythropoietin production, leading to slower RBC production. As the number of RBCs declines through normal attrition, oxygen levels decrease.

KIDNEYS

Since the kidneys are virtually identical to each other in structure and functioning, we will consider only the right kidney here.

Blood Vessels

As shown in Fig. 12.1, blood vessels enter and leave the kidney where it is indented. The arteries branch into smaller vessels as they pass through the inner region (*medulla*) of the kidney (Fig. 12.2). These branches curve over the segments of the medulla and then send smaller arterial branches to the outer region (*cortex*). Within the cortex, each of the smallest branches (afferent arterioles) leads into a tuft of capillaries called a *glomerulus*. Another tiny artery (efferent arteri-

ole) leaves the glomerulus and leads into another group of capillaries (peritubular capillaries), which surround small kidney tubules. Blood from these capillaries is collected into veins, which carry it back through the medulla and out of the kidney.

The capillaries that constitute each glomerulus are much more porous than are other capillaries. Blood pressure causes much of the water and most small molecules in the blood, including both desirable and undesirable substances, to pass through the glomerular wall by the process of *filtration*. The filtrate that has passed through the glomerular wall is captured by a double layer of kidney cells called *Bowman's capsule*, which surrounds the glomerulus. The filtrate then passes into a twisted tube called the *renal tubule*, which has three sections, the proximal convoluted tubule, the loop of Henle, and the distal convoluted tubule. (Fig. 12.3). Meanwhile, blood cells, large molecules such as proteins, and some water and small molecules remain in the glomerulus and then flow through the efferent arteriole.

Tubules and Collecting Ducts

Different types of kidney cells compose each region of the renal tubule, and one region of the tubule (the loop of Henle) passes through the center of the kidney. As the filtrate passes through each region of the tubule, the tubule cells send desirable materials in the filtrate into the blood in the surrounding capillaries. These materials include essentially all the glucose and amino acids, much of the water and sodium, and smaller amounts of minerals such as calcium. This retrieval process is called *reabsorption* (Fig. 12.3). At the same time, the tubule cells cause undesirable materials remaining in the blood to move into the fluid within the tubule by the process called *secretion* (*excretion*). Finally, more water is reabsorbed as the fluid passes through the *collecting duct*. The solution of wastes, toxins, and other undesirable materials remaining in the collecting duct is *urine*. Urine passes from the kidney into the ureter, which transports it to the urinary bladder.

Nephrons

The kidney has approximately 1 million glomeruli, each of which is associated with a Bowman's capsule and a renal tubule. The combination of these three structures is called a *nephron* (Figs. 12.2 and 12.3). All nephrons function in a similar though not an identical manner. One noteworthy difference is that nephrons with glomeruli close to the medulla (juxtamedullary nephrons) seem to be especially important for reabsorbing water.

Overall Functions

Urine formation involves the three processes of filtration, reabsorption, and secretion. The rate and amount of each of these processes are carefully adjusted so that blood leaving the kidneys can compensate for any factors that tend to disturb homeostasis with respect to waste and toxin levels, osmotic pressure, the concentrations of many individual substances, acid/base balance, and blood pressure. Adjustments are made based on the quality of blood passing through the kidney and many regulatory substances including hormones, nitric oxide (NO), and sympathetic nerves. Tubule cells also add correct amounts of vitamin D and erythropoietin to the blood. Thus, the kidneys perform all the urinary system functions for homeostasis.

Under favorable living conditions, such as having comfortable temperatures, proper diet, and moderate exercise, as little as 30 percent of the working capacity of both kidneys is needed to maintain homeostasis. The additional reserve capacity becomes important when conditions are less favorable, such as when high temperatures cause profuse sweating or the diet contains excess water. However, even the most fully functional kidneys can be overburdened by extreme conditions such as complete water deprivation. Therefore, in healthy adults there is a range of living conditions within which the kidneys can maintain homeostasis. Conditions outside this range overwhelm the powers of compensation of the kidneys and lead to loss of homeostasis, cell and body malfunction, illness, and possibly death.

AGE CHANGES IN KIDNEYS

Aging causes the kidneys to gradually decrease in length, volume, and weight. The decline in size may begin as early as age 20, and the resulting changes are evident by age 50. Shrinkage of the kidneys continues thereafter.

Blood Vessels

The loss of kidney mass seems to result primarily from declining blood flow through the kidneys caused by degenerative changes in the smaller arteries and glomeruli. The smaller arteries, including arterioles attached to glomeruli, become irregular and twisted. Glomeruli can be injured by *FRs, glycation of proteins, imbalances between substances causing vasodilation and vasoconstriction, and by excess cell formation. Functional glomeruli are lost gradually, beginning before age 40. By age 80, 40 percent of the glomeruli may stop functioning. From 20 to 30 percent of glomeruli that stop functioning become solidified, and this stops all blood flow through them. Increasing numbers of other glomeruli have their capillaries replaced by one or a few arterioles that permit blood flow while preventing filtration. These shunts develop predominantly in glomeruli close to the medulla. Many remaining glomeruli become smoother and have thicker and declining surface area. These latter changes reduce their filtration rates.

Renal Blood Flow

The amount of blood flowing through kidney vessels is called *renal blood flow* (**RBF**), and age changes in kidney vessels significantly decrease RBF. The decline may begin as early as age 20 and is apparent in most individuals during the fifth decade. The average decline in RBF is 10 percent per decade, though many individuals have more rapid decreases with age. There is a greater decline in blood flow through peripheral cortical nephrons than through glomeruli close to the medulla (juxtamedullary nephrons) and the medulla itself.

This decline seems to be the main reason for most reductions in the functional capacity of the kidney, including filtration, reabsorption, and secretion. In addition, age changes seem to reduce the ability of kidney vessels to dilate and constrict and therefore to adjust kidney blood flow. This change reduces both the speed of kidney functioning and the extent to which it may increase or decrease to meet alterations in body conditions. The greater decline in blood flow in the cortical region compared with the medulla also seems to contribute to the decline in the ability of the kidneys to reduce water loss. This change reduces the ability to compensate for high osmotic pressure.

Some older individuals are at risk for even greater reductions in RBF and kidney functioning because certain abnormal or disease conditions cause less blood to pass through the kidneys. Examples include dehydration, atherosclerosis of kidney arteries, weak heart function, and edema from protein malnutrition or cirrhosis. Renal blood flow is also reduced by certain pain-relieving medications, such as nonsteroidal anti-inflammatory drugs (NSAIDs), which lead to vasoconstriction of kidney arterioles.

Glomerular Filtration Rate

One main effect of age changes in glomeruli and a declining RBF is a declining rate of filtration through the glomeruli [*glomerular filtration rate* (*GFR*)]. The GFR usually begins to drop between ages 30 and 35. However, both the age at which GFR begins to drop and the rate of decline vary greatly among individuals. In some older individuals GFR may remain steady or improve for years before declining again.

A decline in GFR is important because it reduces the elimination rate of many undesirable substances by filtration and secretion. Examples include acids, urea, uric acid, creatinine, toxins, and certain antibiotics, NSAIDs, and other drugs. Therefore, these substances may accumulate in the body and reach hazardous levels. Reductions in GFR also limit the ability of tubules to adjust the retention or elimination of materials such as water, sodium, and potassium.

Normal individual variability in the changes in GFR, together with difficulties in accurately measuring GFR, increases the possibility of making errors in establishing GFRs for older individuals. Such errors can lead to other errors in making dietary recommendations or prescribing drug doses.

Tubules and Collecting Ducts

Age changes in blood vessels are accompanied by age changes in tubules. The tubules become thicker, shorter, and more irregular as their cell numbers decrease. These changes seem to have little effect on the functioning of individual tubules. However, the total capacity for reabsorption and secretion by kidney tubules is reduced because of the decrease in GFR, which supplies filtrate to the tubules, and because whole nephrons stop functioning, shrink, and are lost. The loss of nephrons whose glomeruli are close to the medulla exceeds the loss of more peripheral nephrons.

Little information about age changes in collecting ducts is available, suggesting that these ducts undergo few age changes. There are conflicting views about whether there is an age-related decline in the responsiveness of the collecting ducts to hormones that promote water reabsorption.

Other Changes

Renin One way by which the kidneys regulate blood pressure is by adjusting the production of *renin*. The kidneys also produce renin when osmotic pressure or sodium concentrations are abnormal. Renin indirectly causes tubules to reabsorb more sodium and secrete more potassium. Therefore, adjusting renin production helps regulate blood pressure, along with osmotic pressure and concentrations of sodium and potassium.

Aging causes a gradual decrease in renin production by the kidneys, and the kidneys become less sensitive to messages initiated by renin. These changes decrease further the ability of the kidneys to maintain homeostasis of osmotic pressure, sodium and potassium concentrations, and blood pressure.

Vitamin D Activation Aging causes a decline in vitamin D activation by the kidneys, especially after age 65. Lower vitamin D activation promotes calcium deficiencies, bone fractures, and osteoporosis.

Women experience dramatic decreases in vitamin D activation before age 65 because estrogen, which stimulates vitamin D activation, drops precipitously at menopause (approximately age 50). In women, the combination of aging of the kidneys and hormonal changes results in a greatly reduced vitamin D supply and is a major reason for the higher incidence of osteoporosis among postmenopausal women.

Consequences

In summary, there is an age-related decline in the reserve capacity of the kidneys for maintaining homeostasis of osmotic pressure, concentrations of sodium and potassium, acid/base balance, and blood pressure. Elimination of wastes and toxins becomes slower, and less vitamin D is activated. As with age changes in other parts of the body, these changes begin at different times and progress at different rates. The ability to produce erythropoietin to regulate oxygen levels declines, which increases the risk of anemia. Age changes in the ability to regulate substances such as calcium and magnesium have not been well studied.

In spite of the gradual decline in many kidney functions, healthy people enter adulthood with enough kidney reserve capacity so that under favorable living conditions there is ample functioning to maintain homeostasis regardless of age. However, the declining kidney capacity results in a narrowing of the range of conditions over which the kidneys can provide compensatory adjustments. This narrowing in range, together with certain age changes and many age-related abnormal and disease changes, increases the chances that excessive demands will be placed on the kidneys. Therefore, as people get older, there is a greater likelihood that the frequency, extent, and duration of excursions outside the urinary system's adaptive capacity and beyond the bounds of homeostasis will occur. This necessitates greater conscious effort to prevent such excursions and, when they occur, to correct the conditions causing them.

The relationships between the kidneys and medications change in several ways as age increases. Age changes in the kidneys reduce their ability to destroy some drugs (e.g., morphine) and to eliminate others in the urine (e.g., aspirin, NSAIDs, antibiotics). The effects of these age changes may be enhanced or reduced by diseases (e.g., circulatory diseases, cirrhosis, urinary tract infections, kidney diseases), by some medications (e.g., NSAIDs, diuretics), and by age-related decreases in total body water and increases in percent body fat. Therefore, as age increases, types and doses of all medications should be selected in a more individualized and careful manner to provide effective therapy while minimizing the risks of complications.

ABNORMAL AND DISEASE CHANGES IN KIDNEYS

Aging is associated with an increase in the risk of developing kidney diseases such as infections.

However, the age-related rise in most kidney diseases results from an increase in factors outside the urinary system that cause adverse changes in the kidneys. Examples include age changes such as reduced white blood cell functioning, abnormal conditions such as autoimmune problems and drug toxicity, and diseases such as high blood pressure, atherosclerosis, diabetes mellitus, and prostatic hypertrophy. Abnormal and disease conditions in the kidneys can be prevented or minimized by avoiding, compensating for, or treating these nonurinary factors.

Abnormal and disease conditions of the kidneys become more important with age because aging has already reduced some kidney functions. Many conditions are serious threats since the kidneys play several essential roles in maintaining homeostasis. However, abnormal and disease conditions of the kidneys are not discussed in this book because they become neither sufficiently more frequent nor unusual in the elderly and are not among the most common disorders in older people.

URETERS

Urine passes through the *ureter* from each kidney and enters the urinary bladder (Fig. 12.1). Occasional waves of peristalsis in the muscle layer in each ureter pump urine toward the bladder. Gravity may assist the flow of urine through the ureters. The ureters seem to undergo no significant age changes.

URINARY BLADDER

The *urinary bladder* is located in the lower part of the abdominal cavity (Fig. 12.1). It has a smooth inner lining, a middle layer of smooth muscle, and an outer fibrous layer. The muscle layer, the *detrusor muscle*, is fairly thick (Fig. 12.4).

As urine enters the bladder from the ureters, the bladder wall is stretched. It can expand enough for the bladder to accommodate approximately 1 liter of urine, though the bladder usually empties before it has been filled to capacity. Emptying is accomplished by contraction of the muscular wall of the bladder and simultaneous relaxation of muscles in and around the urethra. Once emptying begins, reflexes cause it to continue until all urine has been voided. However, voluntary impulses and muscle contractions can

FIGURE 12.4 Structure of the urinary bladder, urethra, and associated structures.

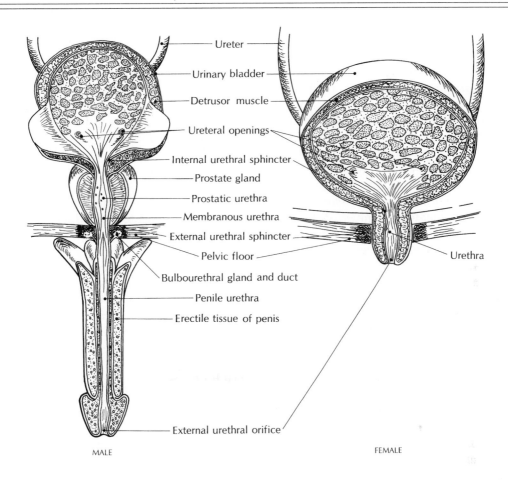

Ureter
Urinary bladder
Detrusor muscle
Ureteral openings
Internal urethral sphincter
Prostate gland
Prostatic urethra
Membranous urethra
External urethral sphincter
Pelvic floor
Bulbourethral gland and duct
Penile urethra
Erectile tissue of penis
Urethra
External urethral orifice

MALE   FEMALE

stop bladder emptying before all urine has been eliminated. Since emptying involves the coordinated actions of the bladder, the urethra, and other muscles and nerves, this function is discussed in detail after the section on age changes in the urethra.

Age Changes

Aging causes the bladder to become smaller. Bands of tissue develop within the bladder and fibrous material in the bladder wall increases. These changes reduce the bladder's ability to stretch and contract. Consequently, the bladder empties less completely and the maximum capacity of the bladder declines. Incomplete emptying of the bladder increases the risk of developing urinary tract infections from bacteria that remain in the bladder. The declining bladder capacity

results in frequent emptying, which becomes inconvenient. When the bladder must be emptied three or more times during the night, the condition is called *nocturia*. Nocturia disrupts sleep and increases the risk of falls from nighttime visits to toilet facilities. Age-related factors that increase urine production also contribute to nocturia. Finally, the age-related increase in spontaneous spastic contraction of bladder muscle (i.e., unstable bladder) increase nocturia and the risk of unintentional release of urine.

URETHRA

The urethra begins at the base of the bladder, extends through the layer of voluntary skeletal muscle at the bottom of the pelvis, and ends at an opening on the surface of the body, the external urethral meatus. The male urethra is several inches longer than the female urethra because it extends through the penis (Fig. 12.4).

The urethra has the same three layers found in the bladder, though the muscle layers in the urethra are thinner. In addition, the beginning of the urethra contains a ring of smooth muscle, the *internal urethral sphincter*. When contracted, this sphincter prevents urine from flowing from the bladder into the urethra. A second ring, composed of voluntary skeletal muscle (*external urethral sphincter*), encircles the urethra where it passes through the floor of the pelvis. Contraction of both this sphincter and the muscular floor of the pelvis can also prevent urine from passing through the urethra. Note that in men the *prostate gland*, which functions as part of the reproductive system, encircles the urethra just below the bladder.

Age Changes

The urethra as a whole becomes thinner with aging, causing increased susceptibility to injury. Thinning of the skeletal muscle seems to cause weakening of the external urethral sphincter. These changes are greater in women and seem to result largely from the decrease in estrogen after menopause.

The combination of urethral thinning and weakening of the urethral sphincter reduces the control of urination. However, significant problems such as urethral inflammation and *urinary incontinence* (inappropriate elimination of urine) develop only when other factors contribute to them.

URINATION

Elimination of urine from the bladder is called *urination*, *micturition*, or *voiding*. This process is similar in operation to the elimination of feces. When the bladder is empty, its muscular wall is relaxed and the internal urethral sphincter is contracted. As urine from the ureters enters the bladder, the bladder stretches outward. After 200 to 300 ml of urine has entered the bladder, pressure in the bladder rises and is detected by sensory neurons. Impulses are sent by nerves to the spinal cord and brain, causing the person to perceive the need to void. At this point autonomic impulses from the spinal cord reflexively stimulate contraction of the detrusor muscle and relaxation of the internal urethral sphincter, causing urine to flow out through the urethra. However, urination can be prevented by voluntary impulses from the brain that suppress the impulses from the spinal cord and cause contractions of the external urethral sphincter and muscles in the pelvic floor.

If urination is voluntarily prevented, the perception of fullness and pressure in the bladder subsides. Bladder filling, bladder stretching, and increasing pressure continue until the sensory neurons are stimulated enough to again cause the sensation of fullness. Once again reflex voiding is initiated, and bladder emptying can be voluntarily prevented. This process can be repeated until the pressure rises high enough and impulses from neurons that detect bladder pressure become powerful enough to override efforts to retain the urine.

In most circumstances maximum bladder filling and very high pressures do not develop because voluntary impulses that suppress voiding are purposely stopped. Then reflex contraction of the bladder, together with relaxation of both urethral sphincters and the pelvic floor muscles, results in forceful elimination of urine through the urethra. Once initiated, voiding usually continues reflexively until the bladder is empty, at which point the bladder relaxes and the internal urethral sphincter contracts again. Voiding can be stopped before complete emptying has been achieved by voluntarily contracting the external urethral sphincter and pelvic floor muscles.

Urination can occur voluntarily as long as the bladder contains some urine. Then voluntary contraction of abdominal muscles causes bladder pressure to rise, initiating the voiding reflex. Then voluntary relaxation of the external sphincter and pelvic floor muscles permits urine flow.

Age Changes

Age changes in the sensory nerves associated with the bladder cause a declining ability to detect bladder stretching and pressure; some individuals lose all ability to perceive bladder fullness. These sensory changes increase the risk of prolonged urine retention and therefore urinary incontinence. However, the effects of age changes in the bladder usually override the effects of changes in the sensory neurons and cause voiding to occur more frequently and at lower bladder volumes.

Urinary Incontinence

Adequate control of urination is retained regardless of age unless abnormal or disease conditions reduce it. Since the incidence and severity of many of these conditions and diseases increase with age, the incidence of abnormal and inadequate control of urination also rises with age.

One form of inadequate control that becomes more common as age increases is urinary incontinence. Estimates of its incidence vary widely depending on both the strictness applied in defining this condition and the techniques used to identify it. Among noninstitutionalized people over age 65, 5 percent to 15 percent of men and 11 percent to 50 percent of women have at least temporary urinary incontinence. However, at least 50 percent of institutionalized elderly people have urinary incontinence. The ratio of occurrence between elderly hospitalized women and men is approximately 2:1.

The very high incidence of urinary incontinence among institutionalized individuals occurs because incontinence is a main reason for institutionalizing older individuals and because many other conditions leading to institutionalization contribute to it. Examples include dementia, strokes, and severe physical disability.

Types Four distinct types of urinary incontinence can be identified. Some individuals may have two or more types simultaneously. *Overflow incontinence* is due to excess pressure in the bladder caused by excessive urine retention. This type of incontinence, which is less common than the other types, may or may not be accompanied by a strong sensation of bladder fullness. *Urge incontinence* is accompanied by a strong perception that urination is necessary even though the bladder is not filled to capacity. It is often due to excess bladder contractions. *Stress incontinence* involves urine loss from factors that weaken muscles in the sphincter and pelvic floor. Incontinent events often occur when a rise in abdominal pressure causes higher bladder pressure, such as during coughing, laughing, sneezing, and strenuous effort such as standing up and lifting a heavy object. Stress incontinence is much more common in women than in men because women have shorter urethras and postmenopausal thinning and weakening of structures used for retaining urine. *Functional incontinence* results from factors that reduce the cognitive functions needed to control urination. Factors include dementia, stroke, and strongly psychoactive medications. This type of incontinence involves no abnormalities or diseases of the urinary system. Some people have more than one type of urinary incontinence, a condition called *mixed incontinence*.

Overflow incontinence causes elimination of small volumes of urine. Stress, urge, and functional incontinence may result in loss of urine volumes ranging from a few drops to several hundred milliliters. Urge and functional urinary incontinence may cause complete bladder emptying.

Contributing Factors Urinary incontinence results from excess bladder pressure caused by excess urine production, urine retention, or stimulation of the bladder; from inadequate contraction of pelvic floor muscles due to muscle weakness or nervous system malfunction; or from a combination of these conditions. A person may have two or more factors acting simultaneously or in various sequences.

Effects and Complications Urinary incontinence has the same undesirable results that characterize fecal incontinence. These include skin inflammation, sores, and infection; social and psychological disruptions; and institutionalization. Costs for devices and supplies (e.g., absorbent undergarments) for adults with urinary incontinence reach 10 billion dollars per year.

Prevention and Treatments Some cases of urinary incontinence can be prevented by avoiding factors that substantially increase the risk of developing this condition. Examples include certain medications (e.g., diuretics, psychoactive drugs) and limited access to toilet facilities.

Many individuals with urinary incontinence can be cured or reduce their incidents of incontinence substantially. As with fecal incontinence, the nature and extent of interactions between care givers and persons with urinary incontinence can influence the degree of success achieved. Steps can also be taken to reduce the impact of incidents of urinary incontinence. The first step is to identify the factors leading to incontinence. This procedure may involve taking a patient history, performing a physical examination that includes special tests for urinary function, evaluating nervous system function, and scrutinizing the medications being taken. Once the type of incontinence and the contributing factors have been identified, an individualized care plan can be developed (Table 12.1).

TABLE 12.1 TREATMENTS FOR URINARY INCONTINENCE

Regulate intake of fluids and diuretics (e.g., alcohol, caffeine, drugs) to reduce urine formation

Regulate all medications affecting urinary or nervous system functioning

Assure accessibility to facilities such as bedpans, urinals, and care giver assistance

Urinate at scheduled times

Cure urinary tract infections to reduce bladder instability

Exercise sphincter and pelvic floor muscles to increase strength (e.g., Kegel exercises)

Use estrogen therapy in women to increase urethral strength

Take medications to modify bladder and internal sphincter function

Undergo surgery to remove obstructions (e.g., prostate surgery), enlarge the bladder, denervate the bladder, or implant an artificial sphincter

Use behavioral modification and training

Use biofeedback control to increase awareness of need to void and gain better control of muscles

Use electrical stimulators to control muscles

Use absorbent pads or male condom catheters to catch urine

Perform skin care to avoid complications

Use catheters to drain urine (can lead to complications such as infections and bladder instability)

13

REPRODUCTIVE SYSTEM

The male and female reproductive systems each consist of a pair of primary reproductive organs or *gonads* (*testes* in men, *ovaries* in women) and a variety of accessory structures, including ducts, glands, blood vessels, external reproductive structures (*genitalia*), and, in women, *breasts* (Fig. 13.1).

MAIN FUNCTIONS

Gonads

The gonads have two main functions. The first is the production of sex cells. The purpose of each male and female sex cell (*sperm cell* in men, *ovum* in women) is to unite with the other type and initiate the life of a human being. The second main function is the production of gonadal sex hormones. The main purposes of these sex hormones include stimulating the development, maintenance, and certain functions of the gonads and accessory reproductive structures. These hormones also influence structures and functions in other body systems (e.g., integumentary, circulatory, muscle, skeletal). Some of these activities contribute to homeostasis (Chap. 14).

Accessory Structures

The accessory reproductive structures may be considered to have four main functions, three of which are involved with reproduction.

Reproductive Functions By helping to create new individuals, the three reproductive functions contribute to the survival of the human species. One reproductive function is helping to unite a sperm cell with an ovum inside the female reproductive system. Male reproductive structures seem to contribute most to this function when they assist in placing sperm cells into the female reproductive system during sexual intercourse. The second and third reproductive functions of the accessory reproductive structures are supporting the development of and giving birth to new individuals. These functions are performed by the female structures. The breasts help support development after birth when they provide nourishment to children.

FIGURE 13.1 Male and female reproductive systems.

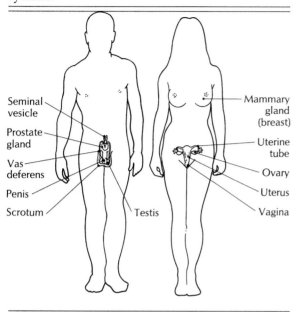

Seminal vesicle

Prostate gland

Vas deferens

Penis

Scrotum

Testis

Mammary gland (breast)

Uterine tube

Ovary

Uterus

Vagina

Sexual Activity The fourth main function of the accessory reproductive structures is performing sexual activity, which can enhance an individual's psychological well-being and social interactions. For example, sexual activity involving proper functioning of the accessory structures can provide intense physical pleasure, enhanced self-esteem, mood elevation, stress reduction, and a sense of being loved. Furthermore, it can be a means of establishing and maintaining an emotional and social bond. Finally, it can be a physical means to express affection, tenderness, and other positive feelings. However, sexual activity can be accidentally or intentionally misused. Incorrect, improper, or inappropriate sexual activity leads to many undesirable biological, psychological, social, and economic consequences.

In conclusion, a main role of the male and female reproductive systems is to act together to promote the survival of the human species. This is unlike other body systems, which generally maintain homeostasis only for the individual. The reproductive systems contribute to homeostasis indirectly, however, through certain effects of the sex hormones they produce and through the positive psychological and social effects of sexual activity.

MALE SYSTEM

The male reproductive system contains several paired tubular structures, including the *testes*, the *epididymides* (sing., epididymis), and the *ductus deferentia* (sing., ductus deferens) (Fig. 13.2). There are also two pairs of glands: the *seminal vesicles* and *bulbourethral glands*. The other important structures in this system occur singly and include the *scrotum*, the *prostate gland*, the *urethra*, and the *penis*. Finally, the scrotum and the area near the base of the penis are covered with *pubic hair*.

Testes

The two *testes* rest within the *scrotum*, a sac of skin and fibrous material suspended near the front of the body between the thighs (Fig. 13.2). Each oval-shaped testis is divided into 250 to 300 sections by fibrous sheets, and each section contains up to four long, highly coiled tubes called *seminiferous tubules*. Each tubule may be up to 100 feet long. Spaces among these tubules contain blood vessels and special cells called *interstitial cells* (*Leydig's cells*) (Fig. 13.3).

Vessels and Interstitial Cells Blood flow through the testes delivers needed materials and removes wastes and sex hormones produced by the testes. The interstitial cells produce and secrete two male sex hormones: *testosterone* and *dihydrotestosterone* (*DHT*). These hormones are essential for proper sperm production, development and maintenance of male reproductive structures, development and maintenance of other male characteristics (e.g., deep voice, beard), and interest in sexual activity (libido). They also influence several other activities (Chap. 14).

Seminiferous Tubules and Sperm Production The wall of each seminiferous tubule is many cells thick (Fig. 13.3). Many cells in the outer region of the tubule wall reproduce rapidly, and most of the newly produced cells move toward the central channel (*lumen*). As each cell moves toward the lumen, it undergoes a special type of cell division called *meiosis*, which results in the formation of four cells called *spermatids*. In men, meiosis is also called *spermatogenesis*. Each spermatid then matures into a long sperm cell through the process of spermiogenesis. Later, when fully

FIGURE 13.2 Structure of the male reproductive system: (a) Sagittal section of male pelvis.(b) Anterior view of male reproductive system.

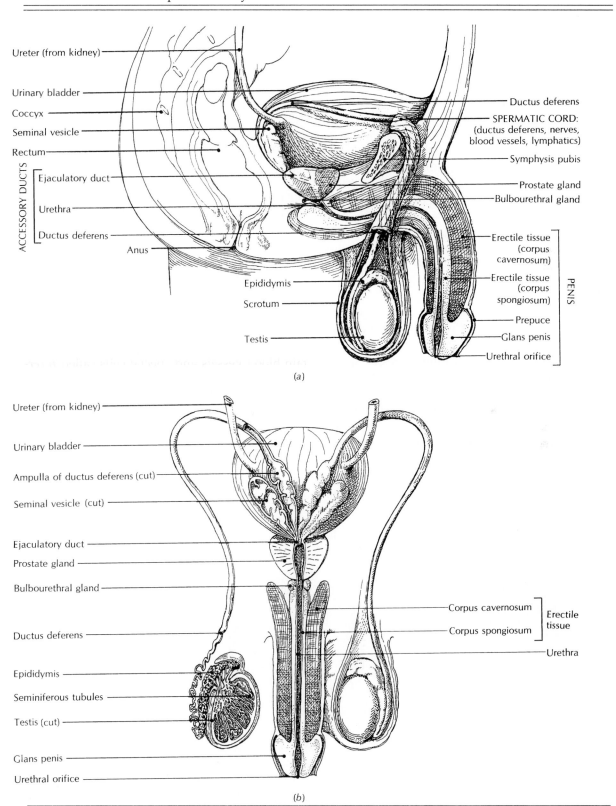

(a)

(b)

mature, each sperm cell can initiate the life of a new human being by entering an ovum during the process of *fertilization*. Since sperm production occurs continuously throughout the nearly one-eighth of a mile of seminiferous tubules in each testis, a man produces several hundred million sperm cells a day.

The seminiferous tubule wall also contains large *sustentacular cells* (*Sertoli's cells*) (Fig. 13.3), which promote sperm production in three ways. First, they produce and retain *androgen-binding protein* (*ABP*), which binds testosterone and concentrates it in the tubules, stimulating sperm production. Second, these cells protect sperm-producing cells from attack by the immune system. Third, they nourish spermatids as they mature into sperm cells. The sustentacular cells also produce a sex hormone (*inhibin*) that helps regulate testosterone levels.

Ducts

Epididymis A series of conducting tubes connect the seminiferous tubules in each testis to an *epididymis*, a coiled tube behind the testis. Sperm cells and the small amount of fluid produced by the seminiferous tubules move through the conducting tubes into the epididymis. While being stored in the epididymis for 10 days or more, sperm cells become fully mature and capable of swimming. The epididymis also secretes a small amount of fluid that seems to contain nutrients for the sperm cells.

During sexual arousal and activity, contractions of the epididymis push the sperm cells and fluid into the tubular *ductus deferens*. Mature sperm cells that are not released from the epididymis within about a month are broken down chemically.

Ductus Deferens and Ejaculatory Duct Each *ductus deferens* (also called *vas deferens*) passes up from the scrotum and into the body cavity (Fig. 13.2). There the ductus deferens loops over the urinary bladder from front to back and widens just before entering the rear of the *prostate gland*. Upon entering the prostate gland, the ductus deferens becomes the *ejaculatory duct*, which leads into the center of the prostate gland and joins the urethra. Rhythmic peristaltic contractions that occur during sexual activity propel the sperm from the ductus deferens into the urethra.

Urethra The *urethra* passes through the prostate gland, exits from the body cavity, and extends through the penis to its external opening. Rhythmic peristaltic contractions of the urethra, which occur during the peak of male sexual response (i.e., during ejaculation), propel the sperm cells and accompanying fluids through the urethra and out of the body.

FIGURE 13.3 Structure of the testes and seminiferous tubules.

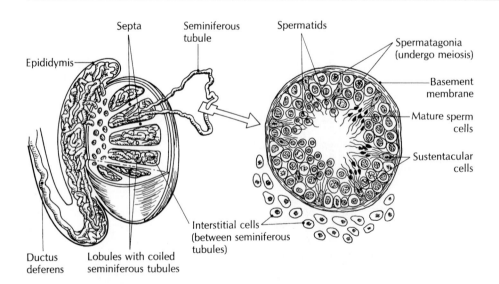

Glands

The successful delivery of functional sperm cells into the female reproductive system depends not only on the functioning of the reproductive ducts but also on secretions from reproductive glands (seminal vesicles, prostate, bulbourethral glands) (Fig. 13.2). The mixture of sperm cells and secretions released from a man's body during sexual activity usually contains 300 million to 500 million sperm cells and totals approximately 3 ml to 5 ml. Almost all of this mixture—*semen*—consists of secretions from reproductive glands.

Seminal Vesicles Approximately two-thirds of the semen comes from the *seminal vesicles*, which lie behind the urinary bladder just above the prostate gland. The wall of each seminal vesicle consists of three layers: an inner epithelial layer of secreting cells, a middle layer of smooth muscle, and an outer layer of fibrous connective tissue. Within the gland the wall is highly folded, producing many interconnected spaces that resemble a sponge.

The thick secretion from the inner layer of the seminal vesicle is stored in the internal spaces of the gland. This secretion contains a variety of substances, including water, fructose, and alkaline materials. During sexual activity contraction of the smooth muscle layer forces the secreted materials out of the spaces, through the duct in the seminal vesicle, and into the ductus deferens where it leads into the ejaculatory duct. The secretion then mixes with the sperm and fluids passing from the ductus deferens into the ejaculatory duct.

The water in this secretion dilutes the sperm cells so that they have more room to move. The fructose provides energy that allows the sperm cells to swim actively while trying to reach the ovum. The alkaline materials protect the sperm cells by neutralizing acid materials in the urethra and the female reproductive system.

Prostate Gland The *prostate gland* surrounds the upper end of the urethra. It is slightly flattened and is little more than 2.5 cm (1 inch) in diameter. It actually consists of numerous small glands that contain secreting cells and are surrounded by smooth muscle. In addition, fibrous material is found throughout the prostate gland and a distinct layer of fibrous material surrounds the entire gland.

During sexual activity, contraction by the smooth muscle forces the secretion from each small gland through its duct and into either the ejaculatory ducts or the urethra. The secretion then mixes with the sperm cells and other fluids. It contains water and alkaline materials, which serve the same functions as the corresponding substances in the secretions from seminal vesicles. Secretion from the prostate gland constitutes approximately 15 percent of the semen.

Bulbourethral Glands The *bulbourethral glands* (*Cowper's glands*) are located on opposite sides of the urethra just below the prostate gland. Each round bulbourethral gland is less than a quarter inch in diameter.

At the beginning of sexual arousal these glands secrete no more than a few drops of a clear slippery alkaline liquid into the urethra. The alkaline material helps neutralize acid materials in the urethra before contractions in the ejaculatory ducts push sperm cells into the urethra. This protects sperm cells from any acid urine in the urethra. In addition, some of the secretion often leaks out of the external urethral opening onto the end of the penis. This material (precoital fluid) lubricates the end of the penis and aids in inserting it into the vagina at the beginning of sexual intercourse.

Penis

The *penis* contains the urethra, three elongated masses of spongy *erectile tissue*, and several arteries and veins, all of which run parallel to the urethra (Figs. 13.2 and 13.4). One portion of the erectile tissue (corpus spongiosum) surrounds the portion of the urethra that passes through the penis. The other two erectile tissue masses (corpus cavernosa) are above the urethra and its erectile tissue. The spaces within the erectile tissue can be filled with blood from the penile arteries. Sheets of fibrous tissue surround all these penile components, and the surface of the penis is covered with skin.

Under resting conditions, the erectile tissues are narrow and soft because they contain little blood and the penis is *flaccid* (limp and flexible). In this state the penis is only a few inches long, though its length varies both from time to time and between individuals.

FIGURE 13.4 Penile structure and erection: (a) Longitudinal section of penis. (b) Cross section of flaccid penis. (c) Cross section of erect penis.

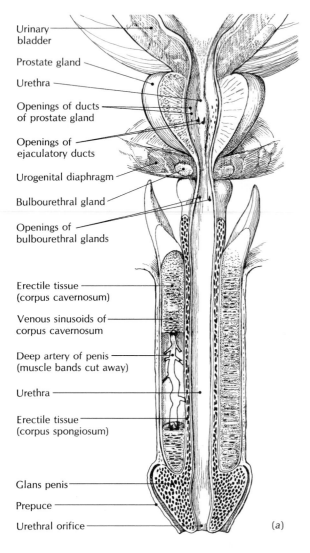

Urinary bladder

Prostate gland

Urethra

Openings of ducts of prostate gland

Openings of ejaculatory ducts

Urogenital diaphragm

Bulbourethral gland

Openings of bulbourethral glands

Erectile tissue (corpus cavernosum)

Venous sinusoids of corpus cavernosum

Deep artery of penis (muscle bands cut away)

Urethra

Erectile tissue (corpus spongiosum)

Glans penis

Prepuce

Urethral orifice

(a)

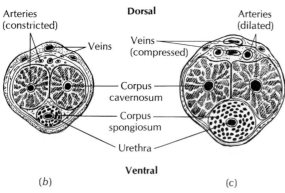

Arteries (constricted)

Dorsal

Arteries (dilated)

Veins

Veins (compressed)

Corpus cavernosum

Corpus spongiosum

Urethra

Ventral

(b)   (c)

Erection   During sexual arousal dilation of penile arteries causes blood to enter and fill the erectile tissue spaces faster than the veins carry it away (Fig. 13.4). As a result, the erectile tissues expand and become firmer, causing the penis to widen and lengthen. As the erectile tissue expands, it compresses the veins and slows the exit of blood. This further increases the extent of erectile tissue engorgement. The restricting nature of the surrounding fibrous sheets causes the pressure in the filling erectile tissue to increase, and the penis straightens and stiffens. Collectively, these changes constitute the process of *erection*. Once erection is complete, the penis can be inserted into a vagina and semen can be deposited into the female reproductive system.

An erect penis is often more than 1 inch in diameter and measures 6 inches to 7 inches in length. The final dimensions of the erect penis vary somewhat between individuals and bear only a small correspondence to its dimensions in the flaccid condition.

Erection is a reflexive action controlled primarily by autonomic nerves. It can be caused by erotic physical stimulation (stimulation of the penis or scrotum) or by erotic mental processes (sexual fantasies). Other types of physical stimulation, such as having a full urinary bladder or mild irritation of the urethra, and certain other mental processes, such as those which occur during sleep, can also result in erection. Conversely, erection can be prevented or reversed by other physical stimuli (e.g., pain) and mental processes (e.g., fear).

The results of erection are reversed when blood flow into the erectile tissue is slowed by penile artery constriction. Then blood leaves the erectile tissue faster than it enters, and the penis returns to its flaccid state.

Pubic Hair

The skin of the scrotum is covered with fairly sparse pubic hair. A thicker mat of pubic hair covers the skin near the base of the penis.

AGE CHANGES IN THE MALE SYSTEM

Testes

On the average, the testes decrease in size and weight with age. The degree of shrinkage is highly variable among individuals and is not present in all men. Therefore, it may be caused by factors other than aging, such as poor nutrition and diseases in other parts of the body. Within the testes, the interstitial cells decrease in number but show no significant changes in structure. Changes in their production of gonadal sex hormones and the effects of those changes are described in Chap. 14.

Changes in the seminiferous tubules include thinning of the wall and narrowing of the lumen. At first, the changes occur in small and widely scattered patches. These modified patches increase gradually in size and number. In some places the lumen becomes so narrow that the tubule is completely blocked. Most of the age-related changes in the seminiferous tubules seem to be due to factors other than aging. Factors include declining blood flow caused by blood vessel changes; injury to the tubules by the immune system; and alterations in sex hormone production. Age-related compensatory increases in stimulatory hormones from the pituitary gland [luteinizing hormone (LH) and follicle-stimulating hormone (FSH)] may help minimize tubule alterations caused by other changes.

Because of these changes, there is an average decline in the rate of sperm production and a higher percentage of irregular nonfunctional sperm cells are produced. However, the decrease in sperm production is highly variable among men, and some men show no decline. Furthermore, sperm production never ceases in a healthy man because only some tubules stop production completely. In other tubules production slows only in the modified patches. Therefore, the reserve capacity for sperm production assures that the number of normal sperm cells produced by a healthy man always remains adequate for fertilization.

Ducts

The effects of aging on the structure and sperm-carrying abilities of the ducts have not been well studied. There is no evidence that aging results in significant changes in their contributions to reproduction.

Glands

Seminal Vesicles   Age changes in the seminal vesicles have no known effect on their contribu-

tions to reproduction.

Prostate   Age changes in the lining and muscle layer of the prostate gland, which become thinner, are apparent by about age 40. At first the changes occur in widespread patches, but by age 60 essentially all of the prostate has changed. The thickening basement membrane and declining blood flow may reduce nutrient supply and waste removal and therefore may account for some of the shrinkage in the lining and muscle layer. None of the changes in the prostate seem to adversely affect its contributions to reproduction. However, increased binding of testosterone may promote the development of an abnormal condition called benign prostatic hypertrophy. In contrast to the prostate, most body structures show an age-related decline in the binding of testosterone.

Bulbourethral Glands   The decrease in fluid production by the bulbourethral glands is not sufficient to reduce their contributions to reproduction significantly.

Penis

Age changes in the penis appear between ages 30 and 40. The accumulation of fibrous material occurs first in the erectile tissue surrounding the urethra. Some years later fibrous buildup in the other two masses of erectile tissue begins. By age 55 to 60 all areas of erectile tissue show gradually increasing fibrosis.

Age changes in the penis seem to contribute to the gradual decline in the speed with which erection occurs. However, in healthy men under ordinary circumstances the penis retains the ability to become erect and enter a vagina and therefore to assist in placing sperm cells into the female reproductive system.

MALE SEXUAL ACTIVITY

Although the testes function continuously, many activities in other parts of the male system are largely restricted to periods of sexual activity. During these active periods the chronological sequence in which these parts operate is different from their anatomic sequence. In keeping with Masters and Johnson's thorough description of human sexual response, which forms the basis for the following discussion of sexual activity, this chronological sequence can be divided into five phases. These phases are listed in Table 13.1 along with bodily changes that occur in almost all instances (universal changes) and changes that occur less frequently (common or occasional changes).

Different men are affected by diverse intensities and combinations of factors. Therefore, there is great heterogeneity in age-related changes in sexual activity. Additional heterogeneity occurs because for each man, each incident of sexual activity may vary as a result of conditions such as physical environment, partner's involvement, mood, and length of time since the last sexual experience.

Excitement Phase

The *excitement phase* involves the development and persistence of several changes. Sometimes this phase may last for only a few seconds, while at other times it may be extended for many minutes.

The first action during this phase is erection of the penis. This is accompanied by reflexive contraction of the muscular sphincters below the urinary bladder so that urine cannot leave the bladder and injure sperm cells. These developments may be completed within several seconds and may be sustained for many minutes. The onset of erection is accompanied by or shortly followed by secretion from the bulbourethral glands. These three changes are often accompanied by other bodily changes, some of which occur more commonly than others.

Plateau Phase

As sexual arousal increases during the excitement phase, peristaltic contractions of the epididymis and ductus deferens move sperm cells and fluids into the ejaculatory ducts; as the penis reaches peak erection, the *plateau phase* begins. This phase is often accompanied by increases in the bodily changes seen during the excitement phase. Other changes may also occur. Up to this point distractions or a reduction in sexual stimulation may cause reduced or complete loss of erection and reversal of all the other concomitant sexual response changes. Renewed sexual stimulation can restore the conditions of the plateau phase.

At the end of this phase muscle contractions in the seminal vesicles and prostate gland force their secretions into the ejaculatory ducts and urethra. Simultaneously, the external urinary sphincter relaxes, allowing the semen to flow into the urethra. The internal urinary sphincter contracts forcefully, preventing semen from entering the bladder and preventing urine from leaving the bladder and entering the urethra.

Orgasmic Phase

The expansion of the urethra by semen often gives the impression that loss of voluntary control of sexual activity is imminent, and within a few seconds it triggers the *orgasmic phase*. During this phase completely reflexive rhythmic peristaltic contractions of the urethra, the other ducts, and muscles at the base of the penis force the semen out of the urethra (*ejaculation*). The first few contractions are the strongest and occur at intervals of slightly less than 1 second. After the first few contractions, the rhythm slows and the force of contraction diminishes. These activities are often accompanied by extreme levels of pleasurable sensations, which are usually centered in the penis, and by increases in the other changes that occur during the excitement and plateau phases. They may be followed by widespread perspiration. The orgasmic phase may last for a number of seconds.

Resolution Phase

As the orgasmic phase subsides, the penis begins to become flaccid. From seconds to minutes may be required to achieve a completely flaccid state. All the other bodily changes that occur during orgasm also quickly diminish and gradually disappear. This return to resting conditions constitutes the *resolution phase*.

Refractory Period

The resolution phase is followed by the *refractory period*, during which erection of the penis and the accompanying changes cannot occur. This period may last a few minutes to many hours, after which erection and the other activities in the cycle of male sexual activity may occur again.

TABLE 13.1 PHASES IN MALE SEXUAL ACTIVITY

Excitement phase
  Universal changes
    Erection of penis
    Reflexive contraction of urethral sphincters
    Secretion from bulbourethral glands
    Peristaltic contractions of epididymis and ductus deferens
  Common changes
    Pleasurable sensations
    Tensing of scrotum
    Swelling of testes
    Moving upward of testes
  Occasional changes
    Erection of breast nipples
    Increases in breathing rate, heart rate, blood pressure
    Reflexive and voluntary tensing of various muscles
  Age changes
    Increasing stimulation required to initiate excitement phase
    Slowing in development of erection
    Declining firmness of erection
    Declining intensity and occurrence of most excitement phase changes

Plateau phase
  Universal changes
    Increases in excitement phase changes
    Contractions in seminal vesicles and prostate gland
    Relaxation of external urinary sphincter
    Contraction of internal urinary sphincter
    Perception of imminent orgasm
  Occasional changes
    Reflexive and voluntary movements of body parts (e.g., hands, feet)
    Flushing of skin
  Age changes
    Declining firmness of erection
    Increasing occurrence of inability to reestablish erection

Increasing voluntary control of onset of orgasmic phase
Weakening of contractions of seminal vesicles and prostate gland
Changing perceptions of imminent orgasm
Increasing occurrence of failure to achieve orgasm
Declining intensity of pleasurable sensations
Declining intensity and occurrence of most excitement phase and plateau phase changes

Orgasmic phase
  Universal changes
    Reflexive rhythmic peristaltic contractions of ducts and penile muscles
    Ejaculation
    Pleasurable sensations
    Increases in excitement phase and plateau phase changes
  Occasional changes
    Perspiration
  Age changes
    Declining number and strength of ejaculatory contractions
    Declining intensity and occurrence of other muscle contractions
    Declining intensity of pleasurable sensations

Resolution phase
  Universal changes
    Loss of erection
    Reversal and disappearance of previous changes
  Age changes
    Accelerating loss of erection
    Accelerating descent of testes
    Retarding of loss of scrotal tensing and nipple erection

Refractory period
  Universal changes
    Inability to undergo changes in previous phases
  Age changes
    Lengthening of refractory period

AGE CHANGES IN MALE SEXUAL ACTIVITY

Significant age-related alterations occur in the functioning of the male reproductive system during sexual activity. Some of these changes seem to be due to age changes within the reproductive structures. For example, alterations in erection seem to be due in part to an increase in fibrous material in the penis. However, many age-related changes seem to be due to age changes in other systems, abnormal and disease conditions, or other age-related factors. Let us consider five examples.

First, since much of the functioning during sexual activity depends on conscious sensations and reflexive actions, some age-related changes probably derive from age-related changes in the nervous system.

Second, since several events in male sexual activity (e.g., erection, scrotal tensing, testes enlargement) depend on substantial increases in blood flow to reproductive structures, some age-related changes seem to be due largely to age changes in arteries, atherosclerosis, and other age-related conditions that reduce the ability of the circulatory system to quickly increase blood flow to structures.

Third, advancing age is accompanied by increased use of medications, many of which adversely affect sexual functioning. Some medications for high blood pressure reduce erectile ability and the intensity of erection, many medications that regulate nervous system functioning (e.g., sedatives, tranquilizers, antidepressants) suppress erection and ejaculation, and certain hormones (e.g., ACTH) and drugs decrease blood levels of testosterone. In this case, deterioration of sexual activity occurs because the maintenance and function of reproductive structures require sustaining adequately high levels of testosterone.

Fourth, declining sexual performance by the reproductive system seems to be directly correlated with declining frequency of use, which occurs in many aging men.

Fifth, since sexual activity is highly influenced by psychological factors, some changes in sexual activity may result from age-related psychological changes such as a deteriorating self-image. These changes may result from age-related social or economic changes such as institutionalization, retirement, and loss of income.

These and other factors that seem to affect male sexual activity are discussed in the section on impotence, below.

Excitement Phase

Alterations in erection of the penis are perhaps the most noticeable age-related changes in the excitement phase. Major reasons for these changes include diminishing penile sensitivity to touch, increasing amounts of fibrous material in the penis, alterations (e.g., stiffening, narrowing) of arteries leading to and through the penis, and increased leakage of blood through veins that drain the penis.

Plateau Phase

Since the plateau phase consists largely of a heightening of the excitement phase, age-related changes in these phases are similar. Though most age-related changes in the plateau phase may be of little consequence, three of them seem to be significant. One is a more frequent inability to reestablish erection if it is lost because of decreased stimulation or distraction. The effect is entry into a refractory period without having achieved ejaculation. Such occurrences may be considered unsatisfactory attempts at sexual activity.

A second change, which is often considered desirable, is an increasing ability to delay ejaculation and prolong the plateau phase. By so doing, a man can extend his sexually aroused state and the duration of sexual intercourse and therefore may provide more stimulation for his female partner. As a result, the woman is more likely to achieve sexual satisfaction by attaining orgasm before the man ejaculates and loses his erection.

Third, the strength of contractions in the seminal vesicles and prostate and the amount of fluid forced into the ejaculatory ducts and urethra often become inadequate to initiate the sensation of imminent ejaculation. Therefore, ejaculation may begin with no warning sensations. By contrast, occasionally the perception of imminent ejaculation develops and lasts several seconds but is not followed by ejaculation. Both situations may diminish physical pleasure during the transition from the plateau phase to the orgasmic phase.

Orgasmic Phase

As the number and strength of ejaculatory contractions diminish, the duration of ejaculation becomes shorter. These changes may result from altered reflex functioning and the decline in the quantity of semen released. They are accompanied by a diminishing intensity of the pleasant sensations associated with ejaculation.

Resolution Phase

Most aspects of the resolution phase occur faster.

Refractory Period

The refractory period may last only a few minutes in young men. With advancing age, it may last from many minutes to several days.

In summary, as age increases, the activities and alterations in the male reproductive system that occur during sexual activity generally take longer to develop, reach lower peak levels, and return to resting conditions more rapidly. However, the male reproductive system largely retains the ability to provide satisfactory sexual experiences.

FEMALE SYSTEM

The reproductive system in women contains certain paired structures, including the *ovaries*, the tubular *oviducts*, external genital structures (*labia minora* and *labia majora*), and breasts (Figs. 13.1, 13.5, and 13.7). Important female structures that occur singly include the *uterus, vagina,* and *clitoris* (Fig. 13.5).

This section is abbreviated because reproductive functioning in women ends at *menopause.* Menopause usually occurs some time between ages 45 and 55 and is evidenced by the absence of menstrual periods for at least 1 year.

Ovaries

The *ovaries* are held in place within the lower region of the abdominal cavity by several ligaments (Fig. 13.5b). Each ovary is shaped like a slightly flattened oval and is approximately the size of a large almond (2.5 to 5.0 cm long, 1.0 to 2.5 cm wide, 0.5 to 1.0 cm thick). The bulk of the ovary consists of the *stroma*, which contains fibrous material with many blood vessels. Embedded within the stroma and near its surface are many small clusters of cells called *follicles*. Each follicle contains an immature ovum. The entire ovary is surrounded by a thin layer of cells (germinal epithelium) (Fig. 13.6).

Ovarian Cycles Unlike the testes, which produce hormones and sperm cells at a fairly steady rate, ovarian functioning consists of a sequence of events during which hormones and ova are produced periodically. Since this sequence is repeated over and over, it is called the *ovarian cycle.*

An ovarian cycle begins when two hormones (LH and FSH) from the pituitary gland stimulate the cells in a few follicles to make more follicle cells and secrete the hormones *estrogen* and *progesterone.* The blood levels of these hormones rise as the follicles grow and increase their hormone production. These hormones also cause the immature ovum in each stimulated follicle to begin to mature. For unknown reasons, one of the follicles develops faster than do the others, and after several days the other stimulated follicles begin to degenerate and become masses of scar tissue.

Approximately 14 days after the cycle has begun, elevated blood estrogen levels cause the pituitary gland to increase production of LH and FSH. The surge in these hormones causes the fully mature follicle to rupture and release its ovum in a process called *ovulation.* The freed ovum is then transported down the oviducts, where it degenerates unless it is fertilized within 3 days.

After ovulation, LH and FSH cause the ruptured follicle, which remains in the ovary, to grow into a mass called a *corpus luteum.* The corpus luteum produces estrogen and great quantities of progesterone for about 10 days after ovulation. Then, as high levels of progesterone cause blood levels of the pituitary hormones to decline, the corpus luteum degenerates. As it does so, the production and blood levels of estrogen and progesterone fall sharply. The degenerated corpus luteum remains in the ovary as a pale mass of scar tissue. Once estrogen and progesterone levels have become very low, the pituitary gland initiates the beginning of the next ovarian cycle. Though a typical ovarian cycle spans 28 days, each cycle may vary by several days.

The estrogen and progesterone produced by the ovaries are required for the complete development and maintenance of female reproductive system structures and other female characteristics (e.g., body contour). The ovary also produces a very small amount of testosterone, which seems to stimulate interest in sexual activity (libido), just as it does in men. These three hormones also influence several other activities (Chap. 14).

Approximately 200,000 immature follicles are present in each ovary when ovarian cycles begin and sexual maturation occurs during adolescence. Once begun, ovarian cycles are repeated until menopause. Since usually only one follicle matures fully and releases its ovum during each ovarian cycle, not more than approximately 500 follicles release ova before ovarian cycles cease. Most of the other follicles degenerate into atretic follicles.

FIGURE 13.5 Structure of the female reproductive system: (a) Mid-sagittal section of female pelvis. (b) Anterior view of female reproductive system.

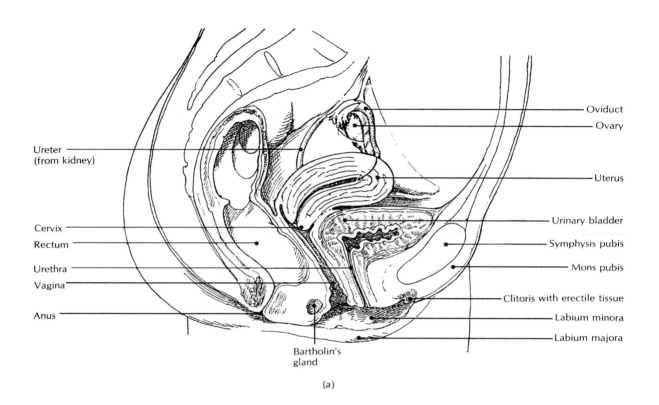

(a)

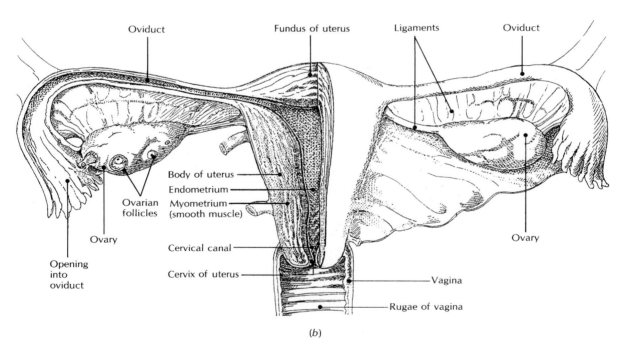

(b)

FIGURE 13.6 Ovarian structure, follicle development, and ovulation.

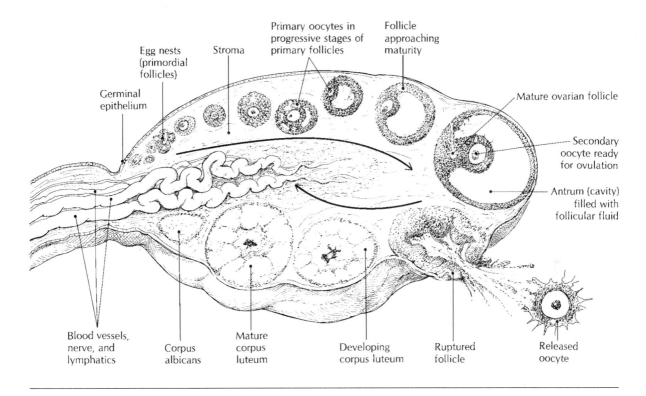

Labels in figure:
- Germinal epithelium
- Egg nests (primordial follicles)
- Stroma
- Primary oocytes in progressive stages of primary follicles
- Follicle approaching maturity
- Mature ovarian follicle
- Secondary oocyte ready for ovulation
- Antrum (cavity) filled with follicular fluid
- Blood vessels, nerve, and lymphatics
- Corpus albicans
- Mature corpus luteum
- Developing corpus luteum
- Ruptured follicle
- Released oocyte

Oviducts

An ovum that has been ovulated enters the funnel-shaped opening of the nearby *oviduct* (*uterine tube, fallopian tube*) (Fig. 13.5b). Each oviduct is approximately 10 cm (4 inches) long and extends from the region near its corresponding ovary to its point of entry into the upper part of the uterus. The wall of each oviduct consists of an inner lining of cells, a middle layer containing smooth muscle, and an outer layer containing fibrous material that helps hold the oviduct in place.

The cells covering the open end of the oviduct and lining its interior have motile projections called cilia. The cilia beat in an organized pattern that sweeps fluids from the body cavity into the funnel-shaped opening of the oviduct. The current produced carries each ovulated ovum into the oviduct. Movement of the *cilia* and peristaltic contractions of the smooth muscle keep the ovum moving toward the uterus, a journey that takes almost 9 days. The cells in the lining of the oviduct also secrete fluid that seems to nourish the ovum.

The oviducts also serve as an upward passageway for sperm cells deposited into the female system during sexual intercourse. This function is important because an ovum is viable for not more than 3 days after ovulation. Therefore, for fertilization to occur, the sperm must reach the ovum while it is in the upper third of the oviduct. The mechanism by which sperm cells move quickly up the oviduct while an ovum is carried downward toward the uterus is not clearly understood.

Ova that are not fertilized degenerate. However, once an ovum is fertilized, the embryo begins to develop immediately. It divides into many cells and begins to form a hollow sphere of cells before reaching the uterus. Secretions from the oviduct help support the development of the new individual by providing it with nutrients.

Uterus

The *uterus* is suspended near the bottom of the abdominal cavity and has a broad upper region into which the oviducts enter. It is held in place by several ligaments and receives support from the muscular floor of the pelvic cavity, the urinary bladder, and the end of the large intestine. The uterus tapers to a narrow lower portion, the cervix, which protrudes into the vagina (Fig. 13.5). An average uterus in a young adult woman is 7.5 cm (3 inches) long and 5.0 cm (2 inches) wide at its broadest point.

Like the oviducts, the uterine wall is composed of three layers, but the two inner layers are much thicker. The innermost layer (*endometrium*) becomes especially thick when its growth is stimulated by estrogen and progesterone. The middle layer of smooth muscle (*myometrium*) is the thickest layer. The outermost layer contains much fibrous material that attaches to ligaments that hold the uterus in place. The uterine wall surrounds a narrow space called the *uterine cavity*, which connects the passageways in the oviducts with the central channel in the vagina.

Since the uterine cavity extends from the vagina to the oviducts, it serves as a passageway for sperm cells in the vagina to reach the ovum. Several days after fertilization, the embryo reaches and enters the uterus. The embryo remains adrift in the uterine cavity for a few days, after which it embeds itself into the endometrium.

The endometrium contributes to the formation of the placenta and thus nourishes the developing child until birth. The placenta also produces estrogen and progesterone, which further stimulate breast development and help maintain pregnancy. As the developing child and the placenta grow, the myometrium stretches to accommodate them. When prenatal development is complete, contractions of the myometrium (labor contractions) push the infant through the vagina and out of the mother's body.

Menstrual (Uterine) Cycles Because of hormonal changes during an ovarian cycle, the uterus also undergoes cyclic changes. These changes constitute a *menstrual cycle* or *uterine cycle*.

A menstrual cycle begins within 3 days to 4 days after blood levels of estrogen and progesterone start to fall, near the end of the previous ovarian cycle. Since the endometrium is no longer strongly stimulated by these hormones, the arteries serving it constrict, resulting in inadequate blood flow to this thickened layer. Then, except for a thin layer of endometrial cells close to the myometrium, the endometrium dies and is shed

along with some blood from the damaged vessels. This material passes through the central passageways in the cervix and the vagina and leaves the woman's body as the *menstrual flow*. The period of 3 days to 5 days required for endometrial shedding is often called the *menstrual period*, and the woman is said to be *menstruating* or "having a period."

By the time *menstruation* is completed, the next ovarian cycle has begun and blood levels of estrogen and progesterone rise. These rising hormone levels stimulate the remaining endometrial cells to proliferate, and the endometrium thickens considerably for the next 20 days. This prepares the endometrium to receive and nourish an embryo if fertilization and the embedding of an embryo occur. If embedding does not occur, estrogen and progesterone levels fall and the menstrual cycle ends approximately 3 days thereafter, when the next menstrual period begins. Thus, the end of one menstrual cycle is marked by the beginning of the next. Since these cycles are controlled by and parallel ovarian cycles, both cycles take approximately 28 days.

If an embryo is embedded in the endometrium, the developing placenta produces hormones that stimulate the corpus luteum to continue hormone production so that menstruation does not occur. Therefore, the developing child is retained and pregnancy continues. Hormones from the placenta and corpus luteum also inhibit the production of LH and FSH by the pituitary gland and prevent additional ovarian cycles until birth has occurred.

Vagina

The vagina is a tube approximately 7.5 to 10.0 cm (3 to 4 inches) long that extends downward behind the urinary bladder and urethra. It leads from the cervix to the outside of the body (Fig. 13.5).

The wall of the vagina is thin and is composed of an inner lining of cells covering a layer containing smooth muscle, blood vessels, and much fibrous elastic material. Under resting conditions the wall of the vagina is wrinkled and collapsed inward so that the inner surfaces touch and close its central channel. However, the wrinkles (rugae) and the elasticity of the wall allow the vagina to be stretched considerably in both length and width. This allows the entrance of a penis during sexual intercourse and the exit of an infant during birth.

The vagina makes five contributions to the reproductive functioning of the female system. It permits the menstrual flow to leave the woman's body; serves as part of the passageway for sperm cells to reach an ovum; helps sperm cells reach an ovum by accommodating the entrance of a penis and permitting the cells to be deposited close to the opening to the uterus; provides a warm moist environment for sperm cell survival; and provides a birth canal through which an infant can leave the mother's body during birth.

Since the vagina undergoes physical trauma during sexual intercourse and provides a relatively wide entry into a woman's body, its lining has three adaptations to resist abrasion and the entry of microbes. First, the lining cells form many layers which resemble the epidermis except that no keratin is present. The surface cells of the lining steadily peel away and are replaced by underlying cells. Second, lubricating fluids, which seem to seep through the lining from underlying blood vessels, help reduce friction during sexual intercourse. Third, lining cells contain glycogen, which is released from these cells after they peel away. Healthful bacteria in the vagina use the glycogen as a nutrient to produce acidic waste products that prevent the growth of harmful microbes. Sperm cells survive the acids because alkaline materials in the vaginal lubricating fluid and in semen neutralize the acids.

External Structures (Genitalia)

Externally, the vaginal opening is flanked by a pair of thin fleshy folds called the *labia minora* (Figs. 13.5 and 13.7). These folds also flank the urethral opening, which lies in front of the vaginal orifice. The labia minora meet a short distance in front of the urethral opening.

Under resting conditions, the free edges of the labia minora tend to meet at the midline and cover the vaginal orifice, inhibiting the entrance of microbes and foreign materials. The labia are very sensitive to touch because they have many sensory nerve endings. Their surface also contains many sebaceous glands. Between the rear of the labia minora and the vaginal orifice lie a pair of *Bartholin's glands*. During sexual arousal, these glands secrete a small amount of lubricating fluid.

To the side of the labia minora lie the labia majora, two thick fleshy folds that contain fat and have hair (pubic hair) on their exposed surfaces. These labia meet in front of the junction of the labia minora and blend with the *mons pubis*, a fatty hair-covered pad overlying the front of the pelvis at the center. As in the labia minora, the free edges of the labia majora meet under resting conditions and help block the entrance to the vagina.

The junction of the labia minora marks the location of the *clitoris*. Most of the clitoris is embedded between the front limits of the labia minora and the junction of the labia majora. However, the tip of the clitoris (glans) protrudes slightly just behind the junction of the labia minora. The clitoris is approximately 2.5 cm long and less than 1.3 cm wide.

The clitoris consists primarily of two masses of erectile tissue. Though much smaller, these masses correspond to the two masses of erectile tissue along the top of the penis (corpus cavernosa). Also like the penis, the clitoris is very sensitive to touch and its erectile tissue becomes engorged with blood during sexual arousal.

Except for preventing the entrance of foreign materials and providing a small amount of lubricant, the external genitalia contribute little to the reproductive role of the female system. However, they make major contributions to the pleasurable sensations derived from sexual activity.

FIGURE 13.7 Female external genitalia.

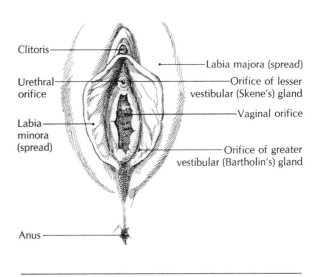

Clitoris
Labia majora (spread)
Urethral orifice
Orifice of lesser vestibular (Skene's) gland
Labia minora (spread)
Vaginal orifice
Orifice of greater vestibular (Bartholin's) gland
Anus

Breasts

The *breasts* are attached to the layer of fibrous material that overlies the large chest muscles (pectoralis major) (Fig. 13.8). Except when a woman is pregnant, each breast consists mostly of fat tissue. The breast is divided by sheets of fibrous material into approximately 20 segments, each of which contains some glandular material (mammary glands). The glands remain small unless the woman becomes pregnant. During pregnancy the very high blood levels of estrogen and progesterone stimulate them to enlarge. When a woman is not pregnant, the fat tissue makes the breast firm and the sheets of fibrous material support the breast, causing the breast to protrude from the chest wall.

The circular pigmented patch of skin on the front of each breast is called an *areola*. The *nipple* is the protrusion at the center of the areola. Both structures contain many sensory nerve endings which make them especially sensitive to touch. When the areola or nipple is stimulated by touch or other factors (e.g., cold) or when a woman becomes sexually aroused, smooth muscle cells contract and cause the nipple to become firmer and protrude farther, a process called erection of the nipple.

The reproductive role of the breasts is to help support the development of a child by providing nourishment after birth. This is accomplished when a hormone (prolactin) produced by the mother's pituitary gland after giving birth causes the enlarged mammary glands to produce milk. Another pituitary hormone (oxytocin) causes the breasts to eject the milk through ducts leading out of the nipples.

AGE CHANGES IN THE FEMALE SYSTEM

Aging of the female reproductive system can be divided into two phases, which are separated from each other by menopause. *Menopause* is the time when age changes in the ovaries cause menstrual cycles to cease for at least 1 year. The average age at which menopause occurs is 51, though it can occur any time between ages 45 and 55. Women who have not experienced menopause are called *premenopausal*, and those who have passed through it are called *postmenopausal*.

Women lose reproductive ability quickly during menopause. Female animals do not have menopause. Animals lose their ability to reproduce very gradually as they age, though there may be a few animals that have menopause (e.g.,

FIGURE 13.8 Structure of a breast: (a) Sagittal section. (b) Internal anatomy.

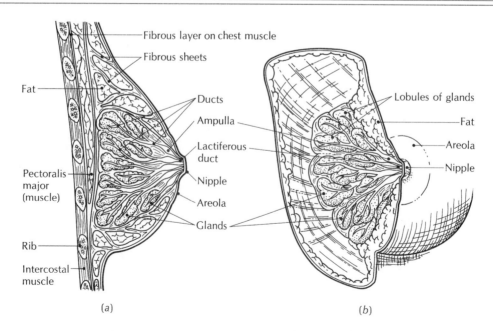

Fibrous layer on chest muscle

Fibrous sheets

Fat

Ducts

Ampulla

Lactiferous duct

Pectoralis major (muscle)

Nipple

Areola

Glands

Rib

Intercostal muscle

Lobules of glands

Fat

Areola

Nipple

(a)

(b)

whales). Scientists speculate about why menopause occurs in humans and in essentially no other animals. One explanation rests on the great amount and length of care human infants require after they are born. Evolution by natural selection promotes menopause because it allows women to nurture children more effectively. Bearing more children at advanced ages would prevent women from devoting enough time and energy to nurture the children they already had. Women past menopause could also help raise children from other women.

Unlike humans, who lose reproductive ability quickly during menopause, female animals lose their ability to reproduce gradually as they age. Female animals do not have menopause.

Menopause is important for two reasons. First, pregnancy, and therefore reproduction, is no longer possible because ovulation has stopped. Second, bodily structures and functions that are influenced by estrogen and progesterone undergo significant changes after menopause because blood levels of these hormones drop and stay low when follicles no longer mature. Some of these hormone-related age changes were described in Chaps. 4 and 9 (e.g., changes in blood lipoproteins, skeletal changes). These and other postmenopausal changes are also described in Chap. 14. This chapter will discuss premenopausal, menopausal, and postmenopausal age changes in the reproductive system.

Ovaries

Before menopause, as more follicles ovulate or become atretic, the number of follicles capable of ovulation decreases, and by age 50 almost no viable follicles remain. While the number of viable follicles is decreasing, the follicles that develop during each ovarian cycle develop less completely, fewer of them ovulate and form a corpus luteum, and therefore less estrogen and progesterone are produced. Eventually no follicles mature fully, ovulation ceases, ovarian cycles disappear, and reproduction becomes impossible. Since the occurrence of menstrual cycles depends on the hormones produced by ovarian cycles, menstrual cycles also cease and menopause occurs.

After menopause, the ovaries produce very small quantities of estrogen and progesterone, and this production gradually diminishes. These hormonal changes are responsible for many other changes, including most age changes in other parts of the reproductive system.

Oviducts

In spite of age changes, the oviducts do not entirely lose the ability to carry ova and sperm cells. Of course, this point is moot once ovulation ceases.

Uterus

The most noticeable change in the uterus before menopause is a decrease in the degree of endometrial thickening during menstrual cycles. Since this decline is due to reduced ovarian hormone production, this change is observed most frequently during menstrual cycles that accompany ovarian cycles with no ovulation.

The menstrual flow resulting from cycles that have reduced endometrial thickening may be so slight that it is not viewed as constituting a true menstrual period. Menstrual cycles may seem to become very long because two or more cycles may occur before enough menstrual flow is produced to definitively mark the end of a cycle.

As the proportion of ovarian cycles involving no ovulation increases, the time between menstrual periods also increases. When no periods happen for 1 year, menopause has occurred. Note that occasional periods may occur after menopause. However, pregnancy becomes impossible when ovulation ends or when the endometrium fails to thicken enough to form a placenta capable of supporting the development of an embryo.

Both before and after menopause, the uterus often tips backward and settles lower in the abdominal cavity as its supporting ligaments weaken and surrounding structures shift. If the lowering of the uterus is excessive, it may descend into or through the vagina. This abnormal condition is called *uterine prolapse*. Uterine prolapse may also develop as a complication from childbirth or from surgery in the pelvic area.

After menopause, the entire uterus shrinks. It may decrease in size by 50 percent within 15 years after menopause and may eventually shrink to less than 2.0 cm in width.

Vagina

Age changes have no significant effect on the reproductive functions of the vagina. However, they significantly reduce its role as a barrier against abrasion and microbes because as the vagina becomes smaller, thinner, and less elastic, it is more easily damaged by even mild physical trauma, such as friction during sexual intercourse. The increased risk of injury and the decline in lubrication are also more likely to result in pain from sexual intercourse. As the risk of vaginal injury rises, so does the risk of developing sores and vaginal infections. The risk of vaginal infections rises also because as age increases, the lining cells contain and release less glycogen. As glycogen declines, less acidic material is produced by healthful microbes and injurious microbes can flourish.

Finally, since the urethra lies immediately in front of the vagina, thinning of the vaginal wall increases the frequency of urethral and bladder irritation and inflammation (cystitis) from agitation of the vagina during intercourse. Painful urination and temporary urinary incontinence may result.

The rate of age changes in the vagina can be slowed by continued frequent sexual intercourse and by administered estrogen. Therefore, the undesirable consequences of these changes can be minimized. Topical application of estrogen-containing creams directly to the vagina is especially effective. In addition, because the postmenopausal vagina is relatively thin, much of the estrogen enters the blood and can have beneficial effects in other areas.

External Structures (Genitalia)

Though the changes in the vagina are the main reasons for the increased risk of developing vaginal infections, age-related shrinkage of the labia majora may also play a contributing role. This shrinkage causes the labia to remain separated more of the time, allowing microbes to enter the vagina more easily. Other age changes in the external genitalia are discussed in connection with age changes in female sexual activity, below.

Breasts

Age changes in the breasts result primarily from the decline in ovarian hormones after menopause. Shrinkage of glandular material, increases in fat, and weakening of fibrous materials in the breasts reduce their firmness and support, causing them to sag and droop rather than protrude from the chest. Breasts that have stretched fibers from years with little support from undergarments may show more sagging. These changes have significant cosmetic effects. Other age changes in the breasts are discussed in connection with female sexual activity.

FEMALE SEXUAL ACTIVITY

The sequence of changes during female sexual activity can be divided into phases that resemble the phases seen in men. However, female sexual activity involves only the first four of these phases because women can pass from the resolution phase into another excitement phase without an intervening refractory period. Many women can even cycle between the plateau and orgasmic phases several times before entering a resolution phase. In addition and in contrast to the male system, the female system can perform its reproductive role without undergoing the changes involved in female sexual activity. If ovulation and endometrial development have occurred, all that is required for reproduction to begin is the placement of sperm cells into the female reproductive system.

Several changes occur during female sexual activity in young adult women (Table 13.2). These physical changes are accompanied by pleasurable sensations that vary in nature and degree from person to person and from one sexual experience to another. The intensity of these pleasurable sensations usually increases from the beginning of the sexual experience through much of the orgasmic phase and then subsides during the resolution phase.

This section is based largely on the work of Masters and Johnson. The previous comments on heterogeneity in sexual activity in men apply to sexual activity in women. Therefore, the age-related changes in female sexual activity described below represent typical alterations.

Excitement Phase

Erection of the nipples and lubrication of the vaginal lining are the first signs of sexual arousal and mark the beginning of the excitement phase. The lubricating fluid in the vagina eases entry of the penis and neutralizes some of the vaginal acids to prevent damage to sperm cells. Substantial nipple erection and vaginal lubrication may be achieved within seconds after sexual stimulation begins. Both conditions persist and may increase throughout the excitement and plateau phases.

Nipple erection and vaginal lubrication are quickly followed by changes in the clitoris and labia minora. Swelling of the labia minora, which results from an increased input of blood, causes the labia minora to double or triple in thickness and push outward. In so doing, they separate the labia majora and protrude from between them. Simultaneously, changes in the labia majora cause further separation, making the vaginal opening more accessible. Swelling of the breasts results from an increase in blood flow.

As the excitement phase progresses, the inner portion of the vagina opens. This change and others in the vagina, uterus, and glands seem to further prepare the vagina for the entry of the penis. Other changes occur toward the end of this phase.

Plateau Phase

Movement of the clitoris marks the transition from the excitement phase to the plateau phase. If flushing of the skin occurs, it may spread over more of the body. The changes in the vagina and labia minora result from increases in blood flow into these structures. Once these changes have occurred, the orgasmic phase is imminent.

Orgasmic Phase

The orgasmic phase begins with rhythmic contractions of the outer portion of the vagina. As happens during ejaculation in men, the first few

---

TABLE 13.2 PHASES IN FEMALE SEXUAL ACTIVITY

Excitement phase
  Universal changes
    Erection of nipples
    Lubrication of vaginal lining
    Swelling of clitoris and labia minora
    Swelling of breasts
    Lengthening and widening of vagina
    Upward movement of uterus
    Secretion by Bartholin's glands
  Common changes
    Swelling of areola
    Tensing of body muscles
  Occasional changes
    Flushing of skin below breasts
  Age changes
    Declining speed, intensity, and occurrence of excitement phase changes (except erection of nipples)

Plateau phase
  Universal changes
    Upward and inward movement of clitoris
    Increase in respiration, heart rate, and blood pressure
    Increase in excitement phase changes
    Thickening of vaginal lining near external opening
    Darkening color of labia minora
  Age changes
    Little change in movement of clitoris

    Declining intensity and occurrence of excitement phase and plateau phase changes

Orgasmic phase
  Universal changes
    Rhythmic contractions of outer portion of vagina

Pleasurable sensations
  Common changes
    Contractions of uterus
    Increase in several excitement phase and plateau phase changes (e.g., swelling of breasts, muscle contractions, flushing of skin, increases in breathing, heart rate, and blood pressure)
  Age changes
    Decreasing number of vaginal contractions
    Increasing incidence of painful uterine contractions

Resolution phase
  Universal changes
    Gradual cessation of vaginal contractions
    Reversal and diminishing of all changes in previous phases
  Common changes
    Return to excitement phase or plateau phase
  Occasional changes
    Widespread perspiration
  Age changes
    Shortening of resolution phase

contractions are the strongest and occur at intervals of slightly less than 1 second. After the first few contractions, the rhythm slows and the force of contraction diminishes. Vaginal contractions, which may number a dozen or more, are often accompanied by extreme levels of pleasant sensations.

Resolution Phase

As the vaginal contractions of the orgasmic phase subside, the resolution phase begins. The changes that occurred during the previous three phases are reversed, starting with the ones that began last. Thinning of the outer region of the vagina and fading of the color of the labia minora occur within a few seconds. Many seconds to several minutes may be required to completely reverse other changes and fully reestablish resting conditions. However, if sexual stimulation continues or is repeated any time after the orgasmic phase, restoration of resting conditions may cease. The changes of the excitement and plateau phases may then be repeated, and another orgasmic phase may occur.

AGE CHANGES IN FEMALE SEXUAL ACTIVITY

Age-related changes in sexual activity in women result from diverse combinations of many of the same factors that produce these changes in men (e.g., age changes, less frequent sexual activity, nonbiological factors). However even more frequently than in aging males, sexual activity in aging females is subject to the effects of reproductive system diseases (e.g., cervical cancer) and hormone therapies (e.g., estrogen replacement therapy).

Excitement Phase

The only aspect of the excitement phase that remains essentially unaffected by aging is erection of the nipples. Because less vaginal lubrication is present, insertion and movement of the penis is more difficult and sexual intercourse may become painful. The decline in lubricant production seems to result primarily from a declining frequency of sexual activity. Older women who frequently participate in sexual activity maintain high levels of vaginal fluid production.

Difficulties from low lubricant production can be overcome by applying a lubricant either near the vaginal opening or to the penis before sexual intercourse. Lubricated condoms and slippery contraceptive materials can also be helpful.

Plateau Phase

During the plateau phase upward and inward movement of the clitoris changes little with age. Though there is less thickening of its outer region, the vagina narrows with age, and the final size of the passage though this region of the vagina remains the same.

Orgasmic Phase

As with the decline in vaginal fluid production, there is less of a decrease in the number of vaginal contractions in women who frequently engage in sexual activity. Painful uterine contractions may result from shrinkage of the uterus; they may be relieved by hormone therapy.

Resolution Phase

The pattern of changes during the resolution phase remains the same, though virtually all changes occur more rapidly and resting conditions are achieved more quickly.

In summary, as age increases, the activities and alterations that occur in the female reproductive system during sexual activity generally take longer to develop, reach lower peak levels, and return to resting conditions more rapidly. In spite of this, the female reproductive system largely retains its ability to provide satisfactory sexual experiences.

FREQUENCY AND ENJOYMENT OF SEXUAL ACTIVITY

Trends

A brief examination of age-related alterations in the frequency and enjoyment (subjective quality) of sexual activity follows. Except where specific differences are noted, this discussion describes both men and women.

On the average, the frequency of sexual activity decreases with age. This decrease accompanies average declines in desire for, interest in, and enjoyment of sexual activity. However, the degree of change in these three parameters is highly variable among individuals, and some people experience increases rather than decreases in one or more parameters.

Contributing Factors

Among the most important biological factors that reduce the frequency and enjoyment of sexual activity is declining health of one of the partners, especially the man. Furthermore, the influence of declining health is often amplified by treatments that affect sexual functioning (e.g., medications, radiation therapy, surgery on reproductive or other organs). Other biological factors that tend to reduce the frequency and enjoyment of sexual activity include age changes in the reproductive, nervous, and circulatory systems; faster onset of fatigue; overeating; and excessive consumption of alcohol.

Menopause leads to an average reduction in sexual activity. However, it results in increased sexual activity for some women because they no longer fear pregnancy or because sexual activity helps prevent or reverse negative self-images resulting from menopause.

Male sexual activity is not affected by ordinary changes in testosterone levels until after approximately age 80. However, abnormally severe decreases in testosterone reduce the desire for sexual activity and adversely affect the functioning of the reproductive system.

The frequency and enjoyment of sexual activity are usually reduced by age-related social changes (e.g., loss of spouse, change in household) and psychological factors. Relevant psychological factors include fear of aggravating a heart condition, boredom from lack of variation in sexual expression, diminished self-image, perceptions of altered physical appearance, depression, stereotyping, and fear of failure. Conversely, especially for men, novelty in sexual activity (e.g., new spouse, modified types of sexual expression) may result in temporary increases in the frequency of sexual activity.

The consequences of economic changes may also diminish the frequency and enjoyment of sexual activity. For example, a reduced income may cause altered living arrangements (e.g., living with relatives) and an accompanying loss of privacy. Finally, institutionalization can create or amplify adverse effects in all these categories (biological, social, psychological, economic).

Making Adjustments

The physical, social, psychological, and emotional benefits of sexual activity continue to be important for many older people. Therefore, actions that prevent or ameliorate factors that adversely affect sexual activity can help maintain a high quality of life. Such actions include maintaining good health; accepting age changes (e.g., cosmetic changes, slowed responsiveness); and using compensatory strategies (e.g., lubricants, modified sexual techniques). When the frequency or enjoyment of sexual activity becomes unsatisfactory, medical and psychological evaluation and therapy can help identify and resolve problems.

ABNORMAL AND DISEASE CONDITIONS

Several abnormal and disease conditions affect the reproductive systems and become especially common or serious with advancing age.

Benign Prostatic Hypertrophy

In men, one of these conditions is *benign prostatic hypertrophy (BPH)*, which is a noncancerous enlargement of the prostate gland.

Recall that the prostate gland surrounds the urethra immediately below the urinary bladder (Fig. 13.2). Though the prostate of most 40-year-old men has begun to shrink, in some men of this age it enlarges because of increases in fibrous material and muscle cells.

By age 40 few prostate glands have grown enough to cause problems. However, the percentage of men with substantially enlarged prostates increases with age so that approximately 90 percent of all men who reach age 80 have a prostate large enough to cause significant problems.

Causes   The causes of this abnormal growth are unknown, though age-related changes in sex hormones and increased binding of testosterone by the prostate are suspected to be contributing factors.

Consequences   Benign prostatic hypertrophy is a serious disorder primarily because the enlarged gland compresses the urethra. The resulting partial or complete blockage of urine flow is especially harmful to the urinary system because it promotes difficult and painful urination; enlargement and weakening of the bladder; bladder spasms; urinary incontinence; urinary tract infections; urinary stone formation; kidney malfunction; kidney damage; and impotence (see below). All these effects can reduce the quality of life, and most of them diminish the ability of the urinary system to maintain homeostasis.

Prevention and Treatment   Nothing can be done to prevent the initial abnormal enlargement of the prostate gland in men with BPH. However, BPH develops gradually and can be detected in the early stages. Once it has been discovered, treatments to prevent the adverse effects can be initiated.

The simplest method for early detection of BPH is to include evaluation of the prostate in an annual physical examination. Other conditions suggesting the development of BPH include (1) slow urine flow, (2) difficulty starting, continuing, or stopping urine flow, (3) discomfort or pain during urination, (4) frequent need to urinate, and (5) urinary incontinence.

Some cases of BPH can be treated by regulating dietary fluid intake and with medications. Many cases are treated surgically. One of the simpler surgical procedures, transurethral resection of the prostate (TURP), involves using surgical instruments to remove the inner region of the prostate piecemeal through the urethra. More advanced cases require more involved surgical procedures. Though these procedures rarely affect sexual functioning directly, their negative psychological consequences may adversely affect sexual activity.

Impotence

The essential feature of *impotence* is an inability to engage in sexual intercourse because the penis is not sufficiently erect (not stiff enough) to be inserted into the vagina. In some cases (primary impotence) adequate erection is not achieved in spite of significant amounts of sexual stimulation; in other cases (secondary impotence) an adequate erection is achieved but subsides before insertion of the penis.

Occasional incidents of impotence occur in many men at every age and are not considered abnormal. Impotence is considered abnormal only when it occurs in a high percentage of attempts at sexual intercourse. Opinions vary widely regarding what rate of impotence constitutes an abnormal frequency. Identifying abnormal impotence in older men is complicated because the refractory period may last for several days.

Abnormal impotence seems to be present in far less than 10 percent of men under age 40. Its incidence increases slowly between ages 40 and 50, though it seems to remain below 10 percent. The incidence rises slightly more rapidly after age 50, reaching perhaps 15 percent by age 60. Thereafter the incidence rises more rapidly, and impotence may be present in more than 50 percent of men over age 80.

Causes   The age-related increase in abnormal impotence occurs because of the age-related increase in both the incidence and severity of many factors that contribute to this condition. Since proper functioning of the nervous and circulatory systems is essential for erection, factors that adversely affect these systems contribute substantially to impotence. The highest ranking among these factors are medications, especially neuroactive drugs and drugs that reduce blood pressure; diabetes mellitus; and atherosclerosis. Other common contributing factors are nervous system diseases (e.g., strokes, dementia), surgery of reproductive or adjacent structures (e.g., prostate, rectum), and alcoholism. Less common contributing factors include hormone imbalances (e.g., inadequate testosterone production), malnutrition (e.g., inadequate zinc), and other diseases (e.g., emphysema, kidney disease). Many older men have more than one contributing factor. Aging of arteries in the penis may augment the effects of these contributing factors.

Probably more than 50 percent of all cases of abnormal impotence result primarily from one or more of these physical factors, though psychological factors may also contribute. Psychological conditions are the primary cause of all other cases of abnormal impotence, though some degree of physical impairment may also be present. Relevant psychological conditions include anxiety, depression, fear of aggravating a physical problem such as heart disease, boredom, lack of confidence (e.g., fear of repeated impotence), and poor self-image.

Note that virtually none of these factors are age changes. Therefore, contrary to a common stereotype of older men, becoming impotent is not an inevitable part of becoming old but an abnormal condition. Recall that unless abnormal or disease conditions develop, the male reproductive system retains the ability to perform its reproductive functions and its operations in sexual activity throughout life.

Consequences   The onset of abnormal impotence is of concern to many aging adults because it can cause extensive adverse psychological and social effects, including the breakdown of a relationship. Therefore, preventing, ameliorating, or eliminating abnormal impotence can provide a much higher quality of life.

Prevention and Treatment   Obviously, preventing impotence means avoiding or minimizing factors that contribute to its development. Chaps. 4, 6, 7, and 14 describe methods for avoiding or minimizing many relevant physical factors (e.g., diabetes mellitus, atherosclerosis, strokes). Good physical health, positive social interactions, and economic security also help prevent or minimize some contributing psychological factors.

When abnormal impotence occurs, identifying the specific contributing factors is the first step in establishing treatment strategies. Once the principal factors have been identified, appropriate treatments can be applied. This may involve reducing or removing the cause, which may include modifying medications, repairing blood vessels surgically, administering hormones, or instituting counseling or psychotherapy.

When reducing or removing the cause is impossible or unsuccessful, other techniques can be used. One involves injecting a vasodilating drug (e.g., papaverine) into vessels in or near the penis when erection is desired. External pumps that draw blood into the penis can be used to achieve erection. Various prostheses, pumps, and other devices that provide temporary or permanent erection can be surgically implanted into the penis or surrounding areas. Both the number and extent of treatments attest to the seriousness with which this disorder is regarded.

The drug *sildenafil* is a recent addition to treatments for impotence. Sildenafil is produced by Pfizer Labs and is sold under the brand name *Viagra.* The drug helps produce and sustain va-sodilation of vessels in the penis by assisting the actions of nitric oxide (*NO).

During sexual stimulation and arousal, neurons stimulate the production of *NO in the penis and elsewhere in the body. The *NO causes smooth muscle cells in penile arteries to produce a special form of nucleotide - *cyclic GMP (cGMP)*. The cGMP causes the smooth muscle cells to relax, allowing blood pressure to expand the arteries and produce erection. Eventually the cGMP is broken down by an enzyme (i.e., cGMP phosphodiesterase), the smooth muscle contracts, and the penis returns to the flaccid condition.

Sildenafil helps develop and sustain erection by inhibiting the enzyme that breaks down cGMP. By inhibiting the enzyme, more cGMP can accumulate and it can last longer, so erection occurs easier and lasts longer. When sildenafil is taken orally as the drug Viagra, it is absorbed within minutes. The sildenafil is slowly removed from the blood by the liver, so it becomes ineffective within a few hours.

The body contains at least six forms of the enzyme that breaks down cGMP. Different cells have different proportions of these enzyme forms. Sildenafil has a much greater effect on enzyme form 5, the form that predominates in penile vessels. Therefore, sildenafil has little effect in other parts of the body. However, since sildenafil has some effect on other forms of the enzyme, it may cause extra vasodilation in other vessels. For example, sildenafil affects vessels in the retina, leading to side effects in vision such as altered perception of blue and green colors. If sildenafil affects many body vessels, it can cause widespread vasodilation and low blood pressure. Blood pressure can become abnormally low if the effects from sildenafil are amplified by other medications. Examples include medications that promote *NO formation and medications for vasodilation that contain nitrates (e.g., nitroglycerine).

Prostate Cancer

Like all cancers, *prostate cancer* consists of cells whose relentless reproduction and spreading are not stopped by the body's normal regulatory mechanisms.

Prostate cancer occurs rarely before age 50; its incidence rises steadily afterward. In men, the incidence of prostate cancer is second only to that of lung cancer. Prostate cancer ranks second to melanoma as a cause of death in men from cancer. For men over age 55, it is the third leading cause of death from cancer, exceeded only by lung cancer and colorectal cancer. The fact that cancer ranks second only to heart disease in causing deaths among older men highlights the importance of these statistics.

Causes   Since the causes of prostate cancer are not known, the specific reasons for the age-related increase are also unclear. The development of prostate cancer is not related to having BPH. Since prostate cancer is 50 percent more common in black males than in white males, a genetic factor may be involved.

Consequences   At first the cancerous cells remain within the prostate gland. As the mass of cells enlarges, the prostate compresses the urethra. Since this obstructs urine flow, the consequences are similar to those of BPH.

The cancer eventually spreads out of the prostate and usually invades the pelvic region first. Because cancer cells can be carried by blood and lymph, they also spread to other regions. Common sites include the vertebrae and other bones, the lungs, and the liver. Several organs may be invaded simultaneously.

The cancer destroys the normal structure and functioning of every part of the body it enters. The ability to sustain homeostasis deteriorates, illness develops, and death ensues. Three examples will be presented. First, prostate cancer weakens bones, causing pain and leading to fractures and their complications. Second, prostate cancer in the lungs may block airways, thicken membranes, fill air spaces, and cause hemorrhaging, significantly reducing respiratory functioning. Third, prostate cancer can severely impair many of the numerous functions of the liver and may cause problems similar to those caused by cirrhosis of the liver.

Prevention and Treatment   Since the causes of prostate cancer are not known, virtually nothing can be done to prevent its onset. However, as with BPH, early detection can lead to early treatment, which may prevent, delay, or minimize the effects. Unfortunately, prostate cancer produces few signs and symptoms until it is well developed. Since some cases can be detected by feeling the prostate during a rectal exam, such an examination

should be part of an annual physical exam, especially for men over age 40. A newer and more convenient method involves evaluating blood samples for the presence of *prostate-specific antigen* (*PSA*). The PSA test is more accurate than other diagnostic procedures and is used to test many men. Ultrasound imaging (sonograms) is used to test for prostate cancer, and small pieces of the prostate can be removed and tested for the presence of cancer cells.

Sometimes the best treatment is to retest periodically to see how the disease is progressing. Prostate cancers that grow very slowly may require no further treatment. Sometimes prostate cancer is treated with radiation therapy, surgery, or medications that suppress testosterone production.

Vaginal Infections

In aging women changes in the vagina increase the risk of developing vaginal infections. Perhaps the most common type of vaginal infection which results from the age-related decrease in vaginal acidity is yeast infection. This type of infection often causes intense itching and is usually accompanied by excessive vaginal discharge.

Wearing underwear made of cotton and avoiding clothing that fits tightly in the genital area reduce the risk of developing vaginal infections. Yeast infections can be treated effectively with antibiotic creams or suppositories.

Breast Cancer

In aging women as in aging men, reproductive system cancers are common and serious disorders. The most common cancer of the female reproductive system is breast cancer.

In women, the incidence of breast cancer is exceeded only by that of lung cancer. Breast cancer occurs in 10 percent of all women at some time. The rate of new cases increases with age throughout life, with the most rapid increase occurring between ages 45 and 65. For women over age 55, breast cancer is second only to heart disease as a cause of death.

Risk Factors   A woman's chances of developing breast cancer are increased by many risk factors besides age. One of the strongest factors is having a mother or sister with breast cancer, especially if it occurred during early adulthood or in both breasts. Other risk factors include using oral contraceptives containing estrogen, undergoing estrogen replacement therapy, drinking alcoholic beverages, being exposed to high doses of radiation, having no children, and, possibly, consuming a high-fat diet.

Consequences   Though the dozen or more types of breast cancer have various effects on the breasts, the most dangerous ones are those which tend to spread easily to other parts of the body. Spreading usually occurs through lymph and blood vessels. The structures more frequently invaded include certain bones (skull, vertebrae, ribs, pelvis), the lungs, the liver, and the kidneys. The loss of homeostasis resulting from fractures or inadequate functioning of other vital organs leads to illness and death.

Prevention and Treatment   In spite of the serious threat posed by breast cancer, most of the complications, illnesses, and deaths it causes can be prevented by early detection and treatment. Knowing and routinely checking for signs of breast cancer can be helpful. These signs include (1) a thickening or lump in the breast, (2) changes on the breast skin, areola, or nipples (e.g., wrinkling, puckering, sores), (3) enlarged lymph nodes near the armpits, and (4) irregularly shaped or asymmetrical breasts. These signs can be detected by monthly breast self-examination and by having a breast examination as part of a routine physical examination. However, the most effective way to detect breast cancer in the early stages is to receive mammograms (x-rays of the breast). An annual mammogram is especially recommended for women over age 45 because the risk of developing breast cancer increases markedly after that age.

Another important aspect of prevention is minimizing risks from estrogen intake by limiting the amount of estrogen used for oral contraception or estrogen replacement therapy, administering estrogen on a cyclic basis rather than continuous one, and including progesterone along with the estrogen.

When breast cancer is suspected, the preliminary diagnosis can be confirmed or negated by examining a sample of the tissue (biopsy). If cancer is present, the specific type and its extent are determined. Depending on the results of these investigations, treatment plans designed to cure the cancer or reduce the effects by slowing its progress are developed. Such treatment plans may involve surgery, radiation therapy, chemotherapy, and hormone therapy.

Endometrial Cancer

Endometrial cancer is cancer of the uterine lining. It has the highest incidence among cancers of the female reproductive structures, occurring in slightly more than 2 percent of all women. New cases develop most frequently between ages 50 and 64. Risk factors include eating excess calories in the diet, having a lowered glucose tolerance, having no children, having relatives with endometrial cancer, and receiving estrogen therapy.

The risk of developing endometrial cancer can be reduced by avoiding overeating and adjusting estrogen therapies in ways similar to those recommended for preventing breast cancer. Once initiated, endometrial cancer reveals its presence by causing bleeding from the vagina between menstrual periods or after menopause. This type of cancer is not as dangerous as others because it is usually detected early. The most common indicator is abnormal bleeding from the reproductive system. Because endometrial cancer is usually detected early, it is easily and effectively treated by surgery and hormone therapy.

Ovarian Cancer

Ovarian cancer ranks fifth in occurrence among cancers in women and in older women, it ranks second among cancers of the female reproductive structures. Ovarian cancer causes more deaths than any other female reproductive system cancer and is the fifth leading cause of death for women.

Risk factors include never being pregnant, inhaling cigarette smoke, and estrogen replacement therapy. Avoiding or minimizing these factors can help reduce the incidence of this cancer. However, little can be done to prevent it from spreading and destroying other organs because it is difficult to detect before it is well established in many areas. Surgery, radiation, and chemotherapy usually can only slow its destructive progress somewhat.

Cervical Cancer

Cervical cancer ranks third in older women among cancers of the female reproductive structures.

Recall that the cervix is the lower part of the uterus and protrudes into the upper part of the vagina (Fig. 13.5). Cervical cancer occurs in 2 to 3 percent of all women before age 80. Most new cases develop between ages 40 and 60.

Risk Factors   The most important risk factor for cervical cancer is having sexual intercourse soon after sexual maturation. Other risk factors include having many male sex partners; having male sex partners who have had sexual intercourse with other women with cervical cancer; inhaling cigarette smoke; using oral contraceptives; and having sexually transmitted diseases such as human papillomavirus. This virus sometimes causes genital warts and often occurs together with genital herpes and chlamydia. Hence, the incidence of cervical cancer is higher in women with these diseases.

Consequences   Once present, cervical cancer usually spreads by infiltrating nearby organs. Later, it is carried to more distant organs by the lymphatics. Like other reproductive system cancers, it causes illness and death by destroying the structure and functioning of any organ it invades.

Prevention and Treatment   A primary strategy in the prevention of cervical cancer is avoiding or minimizing behaviors that increase its risk factors. Once cervical cancer begins, it provides few indications of its presence, though it may cause slight bleeding or a watery discharge from the vagina between menstrual periods or after menopause. However, cervical cancer can be easily detected in the early stages by a Pap smear, which involves examining a sample of cells scraped off the cervix. It is recommended that younger women and women with abnormal cervical cells have a Pap smear as part of an annual physical examination. Older women who repeatedly have normal Pap smears may require smears at 2- to 3-year intervals rather than annually. As with most cancers, early detection and treatment can prevent the development of complications, illness, and death.

If a Pap smear reveals the presence of cervical cancer, one or a combination of treatments (e.g., surgery, radiation therapy, chemotherapy) may be used to cure it or slow its progress.

Uterine Fibroids

One type of growth in the female reproductive system that becomes less of a problem as age increases after menopause is *uterine fibroids*, or leiomyomas. A uterine fibroid consists of a spherical mass of smooth muscle within the muscular wall of the uterus.

Uterine fibroids may begin to develop during or after puberty and may continue to grow until menopause. They occur in various sizes and sometimes become larger than a grapefruit. They usually develop in the upper part of the uterus, though they may occur anywhere in its wall. Uterine fibroids occur in 20 percent to 25 percent of women beyond age 35, and affected women often have more than one. Since uterine fibroids do not invade neighboring regions or metastasize, they are not cancerous.

Most women with uterine fibroids suffer no adverse effects. However, some fibroids cause excessive bleeding during menstrual periods, and unusually large ones may cause problems such as constipation, frequent urination, and kidney disease by putting pressure on adjacent structures. Finally, fibroids occasionally become painful. Treatment may consist of removing the fibroids and the affected part of the uterus or removing the entire uterus (hysterectomy).

Women with uterine fibroids who do not experience significant problems before menopause rarely develop fibroid-related problems afterward because fibroids shrink when blood levels of estrogen and progesterone decline. However, postmenopausal women on estrogen replacement therapy may develop fibroid-related problems because this therapy can cause fibroids to enlarge.

Sexually Transmitted Diseases

Both male and female reproductive systems are affected by *sexually transmitted diseases* (STDs). Commonly encountered examples include bacterial types (e.g., gonorrhea, syphilis, chlamydia) and viral types (e.g., herpes type II, human papillomavirus, AIDS). The incidence of STDs is much lower among older people than among younger adults, perhaps because older people have sexual encounters with fewer partners. However, increasing age seems to have little impact on the causes, effects, methods of prevention, and treatments for STDs. Therefore, these diseases are not discussed in this book.

14

ENDOCRINE SYSTEM

The *endocrine system* consists of all body structures that secrete *hormones* (Fig. 14.1). Hormones are substances that result from manufacturing processes in cells, are secreted into the blood, and alter the activities of cells in other parts of the body. Body materials with only one or two of these characteristics are not hormones (e.g., CO_2, lactic acid, oil from sebaceous glands, blood-clotting materials). Many scientists now consider vitamin D a hormone rather than a dietary nutrient because it can be manufactured in the body, enters the blood, and increases intestinal calcium absorption.

Many endocrine system structures (e.g., heart, stomach) have additional important functions. Usually, only endocrine structures that seem to have hormone secretion as their primary function are called *endocrine glands* (e.g., pituitary gland, thyroid gland, adrenal glands).

MAIN FUNCTIONS FOR HOMEOSTASIS

Like the nervous system, the overall job of the endocrine system is to regulate parts of the body and ensure that they contribute to maintaining homeostasis. The main functions for homeostasis of this system are similar to four of those performed by the nervous system: monitoring, communicating, stimulating, and coordinating (Chap. 6). However, the endocrine system performs these functions somewhat differently.

When an endocrine structure detects that a body condition is changing or has strayed from homeostasis, it secretes a hormone, which communicates information about the errant condition to other body cells. The hormone also causes alterations in cell activities to stop the change from occurring and bring the straying condition back into the homeostatic range. Often the hormone alters functions in many types of cells in several ways and therefore coordinates adaptive responses in various organs. For example, if the concentration of calcium in the blood declines, the parathyroid gland detects this change and secretes parathormone. The parathormone causes the blood calcium level to rise back to normal through its effects on bones, the small intestine, and the kidneys. The parathyroid gland then detects the rise in blood calcium and diminishes parathormone secretion, preventing the calcium concentration from rising too high. If calcium levels rise

FIGURE 14.1 The endocrine system: (a) Hormone-producing structures. (b) Hypothalamus and pituitary gland.

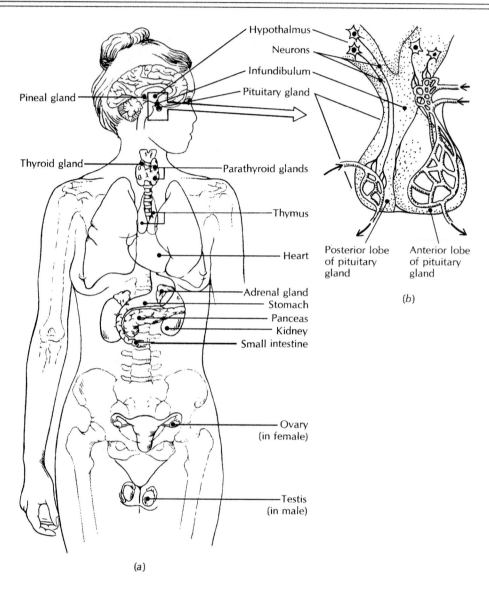

Pineal gland

Thyroid gland

Hypothalmus

Neurons

Infundibulum

Pituitary gland

Parathyroid glands

Thymus

Heart

Adrenal gland
Stomach
Panceas
Kidney
Small intestine

Ovary
(in female)

Testis
(in male)

Posterior lobe
of pituitary
gland

Anterior lobe
of pituitary
gland

(b)

(a)

above a satisfactory level, the thyroid gland secretes thyrocalcitonin, which lowers blood calcium through its effects on bone matrix. Therefore, like the nervous system, the endocrine system performs the first two steps in negative feedback systems and contributes to the third by causing adaptive responses by parts of the body.

Negative feedback responses are not governed simply by turning hormone secretions and the functions being controlled on and off as thermostats do in systems operating heaters and air conditioners. Negative feedback control by the endocrine system can produce a continuous gradation and modulation in the rates of hormone secretion and functional alterations. Therefore, negative feedback responses in the endocrine system operate more like a driver who maintains a proper and fairly steady speed on a hilly highway by adjusting foot pressure on the accelerator and brake pedals of a vehicle. Furthermore, like slightly excessive foot pressure on an accelerator or a brake pedal, the amount of circulating hormone can become slightly too high or low. Though a small error may cause no harm over a short period, sustaining such a condition can result in substantial deviations in the condition being regulated. Normally, negative feedback responses prevent such occurrences. Finally, like an alert driver, the endocrine system is a highly sensitive and rapidly responsive negative feedback system that can reverse shifts in hormone levels quickly so that minimal fluctuations occur.

There are a few situations in which the endocrine system provides positive feedback responses, which increase rather than decrease change. These situations involve functions in the female reproductive system that cause ovulation by the ovary and milk production by the breasts. These functions end during menopause.

COMPARING ENDOCRINE AND NERVOUS SYSTEMS

Since the overall job and main functions of the endocrine system are very similar to several functions performed by the nervous system, why does the body have both systems? A comparison reveals that these systems complement rather than duplicate each other.

The nervous system can cause adaptive responses such as the blink of an eye with pinpoint accuracy and within a fraction of a second because it uses nerve impulses and highly specialized and localized connections. It is also able to provide remembering and thinking. However, the nervous system can directly control only neurons, muscle contractions, and secretions from a few glands (e.g., sweat glands, salivary glands). Furthermore, this system has difficulty sustaining long-term control of activities because neurotransmitters become depleted.

While minutes or hours may be required for an endocrine structure to secrete enough hormone to cause an adaptive response, the hormone may remain in the blood and cause the adaptive response to continue for many hours. Additional hormone can be secreted gradually to continue long-term responses for days, weeks, or months (e.g., growth, maturation). Furthermore, hormones can directly control many body cells and functions that are not influenced by neurotransmitters, including epidermis, bone, cartilage, and blood cells. Representative functions include growth and gene activity. In fact, every type of

body cell has at least some of its functions regulated by hormones.

Though the nervous and endocrine systems have exclusive control of certain body activities, they share responsibility for regulating others, such as GI tract functioning and blood pressure.

Coordinated Operation

Coordination of the nervous and endocrine systems is provided by three communication links. The hypothalamus and infundibulum at the base of the brain provide one main link by which the brain can influence hormone secretion by the pituitary gland (Fig. 14.1).

The hypothalamus uses neurons to send hormones through the infundibulum and into the blood in the posterior pituitary gland. In contrast, the hypothalamus uses blood vessels in the infundibulum to send hormones to the anterior pituitary gland. These hormones regulate the production of other hormones by the anterior pituitary gland. Since the pituitary secretes many hormones, some of which regulate the secretion of other hormones, influencing the pituitary gland produces widespread effects on the endocrine system.

Production of most hormones by the hypothalamus is controlled by negative feedback mechanisms. In addition, the production of hormones destined for the anterior pituitary gland can be influenced by brain activities involved in psychological states and emotional reactions.

The nervous system also uses certain sympathetic nerves to stimulate epinephrine and norepinephrine secretion by the adrenal medulla (inner part of the adrenal glands). Epinephrine and norepinephrine have similar effects. Recall that many sympathetic nerves in other parts of the body release norepinephrine as a neurotransmitter.

The third link between these systems is the circulatory system, which delivers hormones to the brain. Some hormones significantly alter brain function. Of course, altered brain function in turn can result in modified hormone secretion.

HORMONES

Control of Secretion

The body has three ways to control hormone secretion. First, it can be controlled by nervous impulses that result from internal or external stimuli. For example, the secretion of norepinephrine can be controlled by sympathetic nerve impulses initiated by fear. Second, secretion can be controlled by other hormones. For example, the secretion of anterior pituitary hormones is controlled by hormones from the hypothalamus.

Third, secretion can be controlled by the substance or condition being regulated by a hormone. For example, it was noted earlier in this chapter that the rate of parathormone secretion is controlled by calcium in the blood, which in turn is regulated by parathormone. Control of hormone secretion by a substance or condition acted on by the hormone is called *substrate control*.

Hormone Elimination

The concentration of a hormone in the blood is determined by the balance between the rate at which the hormone is secreted and the rate at which it is eliminated. Hormones are removed from the blood by being chemically broken down, converted to other materials, or excreted. The liver and kidneys are very active in these processes. Elimination of substantial amounts of some hormones requires only minutes, while elimination of significant amounts of others may require several hours.

Receptors and Responses

Since hormones are secreted into the blood, which transports them to virtually all parts of the body, each hormone contacts many cell types, yet most hormones affect only certain cells and organs. The affected structures, called the hormone's *targets*, respond because their cells contain receptor molecules to which the hormone molecules bind. Receptors for some hormones are on the target cell membranes, while receptors for others are in the target cell cytoplasm.

Different targets exposed to the same amount of a particular hormone respond to different degrees because they have fewer or more receptors or because their receptors bind the hormone more weakly or strongly. In addition, the strength of each target's response can be changed by modulating the number or binding strength of its receptors. Finally, the effectiveness of a hormone can be influenced by conditions in the target that affect its response mechanism.

Hormone Effectiveness

We have seen that hormone effectiveness can be influenced by the rate of hormone secretion, the rate of hormone elimination, target receptors, and conditions within the target cells. However, many other factors, such as substances that bind hormones, changes in rhythms of secretion, and interference by nerve impulses or other hormones, can influence the effectiveness of a hormone.

With so many factors influencing hormonal effectiveness, determining the rates of hormone secretion or measuring the concentrations of hormones in the blood at any one time provides only a small portion of the information needed to evaluate endocrine system performance and the ability of the aging endocrine system to continue contributing effectively to homeostasis.

SPECIFIC HORMONES

The endocrine system produces a prodigious number and assortment of hormones that have myriad influences on various cell types and parts of the body. There is significant information about the presence or absence of age changes and the nature and effects of such changes for only a few hormones. The following section contains additional information about these few hormones.

Abnormal and disease conditions will be discussed only with regard to insulin and glucagon because abnormalities and diseases of other hormones are not common among the elderly. When such conditions arise, they usually involve having an inadequate amount or an excess of a hormone. Inadequacies are usually treated simply by administering more of the hormone or stimulating its secretion. Excesses are often treated by destroying or removing part or all of the structure secreting the hormone or administering other hormones or medications that reduce the secretion or effects of the excess hormone.

Finally, little mention will be made of changes in endocrine structures. Usually aging results in decreases in size, increases in fibrous material or lipofuscin, and certain changes in the cells. Except for the thymus gland and ovaries, these age changes do not seem to have a significant effect on the ability of the endocrine structure to perform its functions. This situation may exist because of the large reserve capacity in many endocrine structures.

GROWTH HORMONE

Source and Control of Secretion

Growth hormone (*GH*) secretion by the anterior pituitary gland is regulated primarily by hormones from the hypothalamus using negative feedback mechanisms (Fig. 14.2). A decline in blood levels of GH results in increased GH secretion, which causes an elevation in GH blood levels. GH secretion is also increased by low blood levels of *insulin-like growth factor* (*IGF-1*), also known as *somatomedin C*, and by exercise. Conversely, elevations in GH result in decreased GH secretion and a decline in GH blood levels. GH secretion is also slowed by high blood levels of IGF-1. Other factors that influence GH secretion include brain neurons and blood levels of glucose, fatty acids, and amino acids.

Growth hormone is secreted in short bursts (i.e., pulses). The accumulation of GH from many large frequent pulses causes blood levels of GH to rise. In young adults, GH secretion and blood levels rise during the night. As secretion diminishes later during the night and GH is removed from the blood, blood levels begin to decline. The blood level of GH reaches a minimum during the following day. Since this cycle is repeated with each succeeding night and day, it is called a *circadian rhythm* (*diurnal rhythm*). Though GH blood levels follow a diurnal rhythm, IGF-1 levels remain steady.

Effects

The pattern of GH pulses plus the total blood level of GH cause its effects. Growth hormone causes the liver and other target cells to secrete IGF-1. The IGF-1 affects target cells and diffuse to neighboring cells, producing the effects from GH. IGF-1 in the blood also promotes the effects from GH. The local effects from IGF-1 may be more important than the effects from IGF-1 in the blood.

The IGF-1 increases passage of amino acids into cells and increased synthesis of proteins from those acids. These chemical changes result in growth, especially of bone and muscle. Growth hormone also causes an increased breakdown of fat to supply energy. The combination of these changes increases the proportion of lean body mass, which consists mainly of the skeletal and muscle systems and the skin, spleen, liver, kidneys and immune cells. Finally, GH increases blood glucose levels and therefore is considered to antagonize insulin.

Age Changes

Age changes in growth hormone have been studied mostly in men. On the average, GH secretion and IGF-1 levels during the day remain unchanged. During the night, there is less rise in GH pulses and blood levels and IGF-1 secretion. The declines in secretions begin for many men at age 30. These age-related decreases have been called *somatopause*.

When the nighttime rise in GH secretion finally disappears, the diurnal rhythm in GH blood levels vanishes and these levels become steady at all times. As a result, both the total amount of GH produced in each 24-hour period and the blood level of IGF-1 decrease. IGF-1 levels in women are also known to decrease with age.

Though many men show the age changes just

FIGURE 14.2 Control of growth hormone secretion: a negative feedback mechanism.

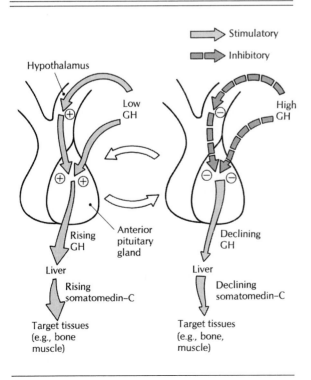

Stimulatory

Inhibitory

Hypothalamus

Low GH

High GH

Anterior pituitary gland

Rising GH

Declining GH

Liver

Liver

Rising somatomedin–C

Declining somatomedin–C

Target tissues (e.g., bone muscle)

Target tissues (e.g., bone, muscle)

described, decline in the nighttime surge in GH blood levels shows considerable heterogeneity among individuals. Some elderly men have nighttime surges that are approximately equal to those found in young men. However, virtually no individuals show substantial increases in GH or IGF-1 levels with advancing age.

Decreasing GH and IGF-1 seem to contribute to a gradual decrease in lean body mass. For example, decreased stimulation of bone and muscle may contribute to the age-related decline in the thickness and strength of bone matrix and muscles plus the age-related increase in body fat. The increase in body fat may then reduce GH secretion, and a downward spiral of GH secretion begins as body fat increases. Age changes in the skin and kidneys may also be due in part to decreased GH secretion. The smoothing of the diurnal peaks in GH blood levels may contribute to changes in other diurnal rhythms, such as sleep patterns. Finally, the effects of lowered GH levels may be amplified because there is an age-related decrease in the responsiveness of cells to IGF-1.

Growth Hormone Supplementation

Though GH production declines, target structures seem to retain their ability to respond to it by producing IGF-1. For example, when older men are injected with GH or artificial substances that stimulate GH secretion, the levels of IGF-1 and lean body mass increase and body fat decreases. Also, loss of matrix from some bones occurs more slowly or is reversed; blood LDLs decline and HDLs increase; skin thickens; immune function; and mental functions improve. Recall that an increase in exercise produces almost all these results, perhaps by stimulating GH secretion. However, unlike exercise, injections of GH or GH stimulants cause undesirable increases in blood pressure and blood glucose levels, and they may prevent normal circadian rhythms. Questions remain about the desirability of GH-stimulated enlargement of parts of the body such as the spleen, liver, and kidneys. Other potential problems include heart disease; high blood pressure; arthritis; high blood glucose levels and diabetes mellitus; and faster growth of cancers. The lack of adequate information about the short-term and long-term effects of GH administration argues against the routine use of GH supplementation to stop or reverse GH-related age changes.

ANTIDIURETIC HORMONE

Source and Control of Secretion

Antidiuretic hormone (*ADH*) is released by neurons that originate in the hypothalamus and extend through the infundibulum to the posterior pituitary (Fig. 14.1). ADH secretion is stimulated by an increase in osmotic pressure or a decrease in blood pressure. Conversely, it is inhibited by decreased osmotic pressure, increased blood pressure, and alcohol.

Effects

Antidiuretic hormone provides a communication link in the negative feedback mechanisms that maintain homeostatic levels of body osmotic pressure and blood pressure. When more ADH is secreted, more water is allowed to pass from the filtrate in the kidney collecting ducts back into the blood. The reabsorbed water helps prevent increases in body osmotic pressure and, by helping maintain a substantial blood volume, prevents lowering of blood pressure. The constriction of blood vessels caused by ADH also helps sustain adequate blood pressure.

Conversely, a decrease in ADH secretion causes more water to escape in the urine, increasing body osmotic pressure or lowering blood volume and blood pressure.

Age Changes

Though there is wide variation among individuals, aging results in an average increase in blood levels of ADH, particularly when body osmotic pressures are high. Since the extra ADH stimulates more water reabsorption by the kidneys, this age-related increase seems to compensate partially for the corresponding decline in the ability of the kidneys to retain water when needed. The ability to increase ADH secretion in response to low blood pressure seems to remain intact or decrease slightly. There is no apparent age change in the ability to decrease ADH secretion to eliminate more water in the urine. Therefore, the contributions to homeostasis made by ADH do not decline significantly, and one of them seems to improve in many individuals.

MELATONIN

Source and Control of Secretion

Melatonin is secreted by the *pineal gland* (Fig. 14.1). Its secretion seems to be controlled by alterations in brain impulses caused by light detected by the eyes. Since an increase in light exposure inhibits melatonin secretion, blood levels follow a diurnal rhythm, with low levels during the day and high levels at night. Secretion is also influenced by the qualities of light, including time of exposure, light intensity, and wavelengths (colors). Melatonin is also produced by the retina, where it seems to have an antioxidant effect.

Effects

Melatonin seems to inhibit sexual maturation until the teenage years. Its diurnal rhythm regulates circadian rhythms, including sleep and body temperature. Also, oscillations in melatonin seem to affect psychological parameters such as mood and depression. Therefore, adverse effects from alterations in diurnal secretion may occur in people who receive little exposure to natural light, who are exposed to different wavelengths of light (e.g., fluorescent light), or who receive exposure to artificial light very different from the natural schedule of daylight.

Melatonin travels easily to all body parts. It is a powerful antioxidant. Also, it stimulates production of enzymes that remove *FRs, and it promotes the formation and effectiveness of other antioxidant substances. It seems to have a major role in protecting the brain and other body parts (e.g., lungs) from *FR damage.

Age Changes

Aging is accompanied by a decrease in the amplitude of the circadian rhythm of melatonin, which results from a decrease in the maximum blood levels attained at night. This leveling in rhythm may influence age-related changes in circadian rhythms such as sleep patterns and hormone secretion (e.g., GH, testosterone).

Melatonin Supplementation

Using melatonin supplements has become popular. Melatonin is inexpensive, easy to obtain and can be taken orally. Though it is a hormone, its sale and use are not regulated like most other hormones. Melatonin is used commonly to reduce the effects from "jet lag" and to help establish or reestablish circadian rhythms. Some people take melatonin to slow, stop, or reverse age changes because it is an antioxidant, caloric restriction helps sustain melatonin levels as it increases ML and XL, and melatonin supplements increase ML in animals. Hazards from melatonin supplementation include disruption of circadian rhythms, unbalancing other hormones, and overdosing or toxicity from the unregulated production and testing of melatonin supplements.

THYROID HORMONES (T_3 AND T_4)

Source and Control of Secretion

Most of the hormone-producing cells of the thyroid gland secrete two related *thyroid hormones: thyroxine* (T_4) and *triiodothyronine* (T_3). Target cells convert some T_4 to T_3 and then release the T_3 back into the blood.

Thyroid hormone secretion is controlled primarily by a negative feedback mechanism that acts through the hypothalamus and anterior pituitary gland. Low blood levels of thyroid hormone result in increased thyroid production, and vice versa. However, thyroid hormone secretion can be influenced by other factors. For example, a low metabolic rate stimulates thyroid hormone secretion. By contrast, high blood levels of somatostatin, glucocorticoids, and sex steroids (e.g., testosterone, estrogen) inhibit secretion.

Effects

Though T_3 is more powerful than T_4, both cause a general increase in the metabolic rate by increasing the rates of many chemical reactions in most cells. Only certain cells in the brain, spleen, testes, uterus, and thyroid gland are unaffected by thyroid hormones. The general increase in metabolic rate increases heat production, which assists in maintaining normal body temperature. Regulating the metabolic rate also assures that each organ will grow, repair itself, and perform its functions at a proper rate.

Age Changes

As age increases, average thyroid hormone blood levels decline slightly, though they remain in the normal range. In addition, there is a decrease in the T_3/T_4 ratio. These changes may be compensatory because they prevent the development of excessively high metabolic rates in certain cells (e.g., muscle) as lean body mass declines. Therefore, aging does not alter the ability of thyroid hormones to provide proper regulation of their target tissues.

There is an age-related increase in damage to the thyroid gland by the immune system, which occurs usually in women. The result is inadequate thyroid hormone production, which affects as many as 10 percent of elderly women. Some diagnostic procedures and therapies that use iodine also damage the thyroid gland. The treatment is thyroid hormone supplementation.

Excess thyroid production is rare in the elderly. It can result from Grave's disease, excess iodine intake, or thyroid nodules. These conditions are treated easily by removing part or all or the thyroid gland or reducing iodine intake.

CALCITONIN (THYROCALCITONIN)

Source and Control of Secretion

Though most thyroid gland cells secrete thyroid hormones, others secrete *calcitonin (thyrocalcitonin)*. Calcitonin secretion is controlled by blood calcium levels, using a simple negative feedback mechanism that does not involve the hypothalamus or pituitary gland but instead uses substrate control. High blood levels of calcium stimulate calcitonin secretion, which causes blood calcium levels to decline by stimulating removal of calcium from the blood. The resulting lower blood calcium levels then reduce calcitonin secretion and allow parathormone to raise blood calcium levels. When blood calcium rises, calcitonin secretion increases again. As a result, homeostasis of blood calcium is maintained.

Effects

Calcitonin decreases blood calcium by stimulating osteoblasts to incorporate blood calcium into bone matrix, inhibiting osteoclasts from removing calcium from bone matrix, and allowing the kidneys to release more calcium in the urine.

Age Changes

Calcitonin levels may decrease with age, increasing the risk of osteoporosis in some individuals. Furthermore, the possible decrease in the effectiveness of calcitonin which may be caused by decreases in estrogen, may play an additional role in the development of osteoporosis in women (see Sex Hormones in Women, below).

PARATHORMONE

Source and Control of Secretion

Parathormone is produced by the parathyroid gland, which consists of several clusters of cells on the back of the thyroid gland (Fig. 14.1). The secretion of parathormone is controlled by a negative feedback system similar to the one regulating calcitonin secretion. However, unlike calcitonin, parathormone secretion is stimulated by low blood calcium and inhibited by high blood calcium.

Effects

Parathormone raises blood calcium levels by causing several alterations in target cell activities. It stimulates the removal of calcium from bone matrix; reduces the release of calcium by the kidneys into the urine; stimulates directly the absorption of calcium by the small intestine; and stimulates the activation of vitamin D by the kidneys. The increase in active vitamin D aids the absorption of calcium by the small intestine.

Note that parathormone directly antagonizes the effects of calcitonin. Maintenance of blood calcium homeostasis depends on providing a proper balance between these two hormones, proper functioning of their target structures, and adequate supplies of active vitamin D and dietary calcium. Besides building and maintaining bone matrix, homeostasis of blood calcium is important for cell and muscle movements, nervous impulse transmission, blood clotting, and regulation of cell activities.

Age Changes

During aging, the parathyroid gland retains the ability to increase or decrease the secretion of parathormone in response to changes in blood calcium levels. Blood levels of active parathormone do not change or increase slightly. Paradoxically, blood calcium levels decline, though they remain satisfactory. This decline may result from decreases in dietary calcium intake and vitamin D levels.

Individuals with increased parathormone levels may be at greater risk for both rapid loss of bone matrix and osteoporosis. In addition, as noted for calcitonin, the drop in estrogen levels in postmenopausal women may have adverse effects on the ability of the parathyroid gland and parathormone to regulate blood calcium and to help maintain bone matrix.

THYMOSIN

Source and Control of Secretion

Thymosin is a group of several related hormones produced by the *thymus gland* (Fig. 14.1). The mechanisms controlling thymosin secretion and blood levels are not understood, though secretion and blood levels of thymosin seem to be positively correlated with the size of the thymus. As the thymus increases in size during childhood, thymosin secretion and blood levels rise. Later, the increase in sex steroid hormones that accompanies puberty causes the thymus to stop growing. At about age 20 the thymus begins to shrink, and thymosin production and blood levels begin to decline slowly. After age 30, thymosin secretion and blood levels decrease more quickly as shrinkage of the thymus continues. By age 60, secretion stops and thymosin levels reach zero.

Effects

Thymosin is necessary for the maturation of lymphocytes, which are specialized white blood cells. Lymphocytes constitute a major part of the immune system, whose overall function is defense. This system identifies, destroys, and eliminates many types of undesirable materials, microbes, viruses, and cancer cells that may either enter the body or be produced within it.

Age Changes

As thymosin decreases with age, fewer immature lymphocytes are able to mature and become functional defense cells. Therefore, the ability of the immune system to protect the body declines. The consequences include an increased susceptibility to infection and an increased risk of cancer.

CHOLECYSTOKININ

Source and Control of Secretion

Cholecystokinin (*CCK*) is secreted by the inner lining of the first section of the small intestine (duodenum). Its secretion is stimulated by dietary fat that enters the small intestine from the stomach.

Effects

Cholecystokinin stimulates contraction of the gallbladder, causes relaxation of the muscular valve that controls the passage of bile and pancreatic digestive juices from the common bile duct into the duodenum, and stimulates the pancreas to secrete digestive juices. Recall that both bile and enzymes in pancreatic juice are essential for the proper digestion and absorption of fat.

Age Changes

Aging is accompanied by an increase in the length of time during which rapid CCK secretion occurs after fat enters the duodenum. Therefore, the total amount of CCK produced increases and blood levels of CCK remain elevated longer. These changes seem to compensate for an age-related decrease in the sensitivity of the gallbladder to CCK. The overall result is no age change in the rate of gallbladder emptying or the total amount of bile ejected from the gallbladder. This compensatory mechanism helps sustain normal digestion and absorption of fat.

GLUCOCORTICOIDS

Source and Control of Secretion

Glucocorticoids are a mixture of more than 12 steroid hormones secreted by selected cells in the *adrenal cortex*, the outer region of the *adrenal glands* (Fig. 14.1). The main glucocorticoid is *cortisol*.

Glucocorticoid secretion is controlled by a negative feedback system that resembles the mechanism controlling thyroid hormone secretion. Activities in the hypothalamus and other brain areas cause a diurnal rhythm in glucocorticoid secretion and blood levels. Peak secretion is at about the time of awakening in the morning, and slowest secretion occurs at approximately midnight.

Effects

Glucocorticoids have many types of target cells and produce a variety of responses, including increases in glucose production and release by the liver, leading to an increase in the blood glucose level; fat breakdown to supply energy for cells; breakdown of proteins to supply free amino acids; and vasoconstriction to increase blood pressure. Glucocorticoid secretion increases in times of stress because these four responses help the body overcome threats to homeostasis. Glucocorticoids also inhibit the inflammatory response and therefore reduce the accompanying pain, itching, and/or swelling. This effect has led to the widespread use of several forms of glucocorticoidlike medications (e.g., cortisone, hydrocortisone, prednisone) to treat many physical injuries, skin disorders, and chronic inflammatory diseases (e.g., arthritis).

Besides these responses, glucocorticoids cause five undesirable effects. These effects are suppression of cartilage and bone formation; stimulation of bone demineralization; promotion of GI tract bleeding and ulcer formation; damage to memory centers in the brain (i.e., hippocampus); and inhibition of portions of the immune response. The last effect increases susceptibility to infection when a person experiences severe stress.

When glucocorticoidlike steroids (corticosteroids) are administered therapeutically to reduce inflammation, their blood levels often rise above those established by internal negative feedback controls. This situation increases the number of unwanted effects. Older persons are especially susceptible to higher risks of glucose intolerance and diabetes mellitus, high blood pressure, osteoporosis, and infections. These risks can be reduced by using minimal doses, administering alternative forms of corticosteroids, or simultaneously instituting other medications or diet modifications to counteract the side effects. Alternatively, using nonsteroidal anti-inflammatory drugs (NSAIDs) can reduce these risks. How-

ever, NSAIDs promote certain problems, including ulcer formation in the GI tract and kidney malfunction.

Age Changes

In healthy adults, aging causes no change in blood levels of glucocorticoids or the diurnal rhythm of those levels. Therefore, aging seems to have no adverse effect on the contribution of glucocorticoids to homeostasis. However, there is a small age-related decrease in the sensitivity of the negative feedback mechanisms that control glucocorticoid levels. This change may lead to abnormally high levels in elderly individuals with disorders (e.g., Alzheimer's disease, depression) that reduce the effectiveness of the glucocorticoid control mechanism even further.

MINERALOCORTICOIDS (ALDOSTERONE)

Source and Control of Secretion

Like glucocorticoids, *mineralocorticoids* are a mixture of steroid hormones from the adrenal cortex. In humans, *aldosterone* is essentially the only mineralocorticoid that affects body functions.

Aldosterone secretion is controlled by four negative feedback mechanisms that operate through the kidney. These mechanisms help maintain homeostasis by regulating blood pressure, osmotic pressure, and blood levels of sodium and potassium.

In the most influential of these mechanisms, aldosterone secretion increases when the kidney secretes *renin* in response to low blood pressure, high osmotic pressure, or adverse changes in sodium concentrations. Aldosterone increases sodium and water reabsorption and retention by the kidneys, causing an increase in blood pressure and adjustments to osmotic pressure and sodium concentrations. Conversely, high blood pressure, low osmotic pressure, and the opposite changes in sodium concentration can suppress renin secretion and aldosterone production, allowing more sodium and water to leave in the urine. This lowers blood pressure, raises osmotic pressure, and corrects sodium concentrations.

Aldosterone secretion is regulated secondarily by the effects of blood levels of sodium and potassium on the adrenal cortex, by a hormone (atrial natriuretic factor) secreted by the heart when blood volume is high, and by a hormone (ACTH) secreted by the anterior pituitary gland during stress. In each case, the adjustment in aldosterone secretion helps maintain proper blood pressure, osmotic pressure, and blood levels of sodium and potassium.

Effects

Aldosterone and other mineralocorticoids cause these adaptive responses by stimulating the kidney tubules to reabsorb sodium and water and secrete potassium and/or acids (hydrogen ions).

Age Changes

Though aldosterone secretion decreases with aging, blood levels remain steady under ideal body conditions because the decline in secretion is accompanied by a compensatory decrease in elimination. However, aging is accompanied by a decrease in the ability to raise aldosterone secretion and blood levels when needed, leading to a decrease in aldosterone reserve capacity.

These changes are not due to age changes in the adrenal cortex, which largely retains the ability to increase aldosterone levels when needed. The age-related decrease in aldosterone reserve capacity is due primarily to the declining ability of the kidneys to secrete renin when needed. Aging is also accompanied by a declining ability to increase aldosterone secretion during stress. There is an age-related decrease in kidney sensitivity to aldosterone.

Because of the interrelationships between aldosterone secretion and kidney functioning, there is age-related decrease in the ability to maintain normal conditions when faced with adverse conditions such as low blood pressure, dehydration, and disease. Body conditions that are likely to become abnormal include blood pressure; osmotic pressure; concentrations of sodium and potassium; and acid/base balance.

DHEA

Source and Control of Secretion

DHEA (dehydroepiandrosterone) is a steroid hormone produced by the adrenal cortex.

DHEA is converted to DHEAS, testosterone, estrogen, and other steroids plus unknown substances.

Effects

The functions of DHEA and DHEAS are unknown. Levels of DHEA and DHEAS in rats, mice, and most other research animals are very low and do not show age-related changes like those found in humans. Therefore, research on DHEA and its possible effects in humans is limited.

Age Changes

Between conception and birth of a child, the child's blood levels of DHEA rise to almost adult levels, dropping to almost zero as birth approaches. During childhood, DHEA levels rise until age 20, after which they gradually decline throughout life. DHEAS levels show a similar pattern.

DHEA Supplementation

Like melatonin, using DHEA supplements has become popular. DHEA is inexpensive, easy to obtain and can be taken orally. Though it is a hormone, its sale and use are not regulated like most other hormones. People take melatonin to slow, stop, or reverse age changes. Some reports suggest it can slow, stop, or reverse aging and age-related diseases in the circulatory system, nervous system, muscle system, skeletal system, and immune system and cancer growth. However, research reports reveal contradictory results depending upon many variables (e.g., age, sex, other hormone levels, dosages, animals used).

Hazards from DHEA and DHEAS supplementation include increasing certain age changes and age-related diseases, unbalancing other hormones, promoting cancers, and unpredictable effects from the unregulated production and testing of DHEA and DHEAS supplements.

SEX HORMONES IN MEN

Sources and Control of Secretion

Testosterone is the main sex steroid in men. Nearly all testosterone is secreted by the *interstitial cells*

(*Leydig's cells*), which lie between the seminiferous tubules in the testes (Fig. 14.1). The small amount of testosterone secreted by the adrenal cortex does not play a significant role in men unless the testes are removed. Another sex hormone, *inhibin*, is secreted by the sustentacular cells in the seminiferous tubules.

Secretion of testosterone and inhibin is stimulated by hormones from the anterior pituitary gland. *Luteinizing hormone* (*LH*) stimulates the interstitial cells to secrete testosterone. Because of this action, LH is also called *interstitial cell-stimulating hormone* (*ICSH*). *Follicle-stimulating hormone* (*FSH*) stimulates inhibin secretion. The secretion of all these hormones is controlled primarily by negative feedback mechanisms similar to those which regulate the secretion of thyroid hormones and glucocorticoids. Testosterone secretion can also be influenced by brain activities such as those involved in emotional reactions.

Besides stimulating inhibin secretion, FSH stimulates the sustentacular cells to manufacture a protein called *androgen-binding protein* (*ABP*), which helps testosterone stimulate sperm production by binding testosterone and concentrating it in the seminiferous tubules.

Testosterone secretion in young men occurs in a diurnal rhythm. The blood level reaches its peak value during the night or early morning. Testosterone secretion and blood levels then decline during the day, reaching a minimum value by evening.

Forms of Testosterone

Much of the testosterone that passes out of the testes binds to molecules in the blood called *sex hormone-binding globulin* (*SHBG*). Testosterone that is bound to SHBG is inactive, and only free testosterone molecules significantly alter target cell activities.

Testosterone can be converted to other sex steroids. The main alternative form is *dihydrotestosterone* (*DHT*). Though the testes produce some DHT, approximately 80 percent of DHT in the blood results from the conversion of testosterone to DHT by target tissues. For example, the prostate gland releases DHT back into the blood. Some target cells respond to testosterone (e.g., skeletal muscle), while others respond to DHT (e.g., most reproductive system structures). A small amount of testosterone is converted to the hormone *estrogen* by certain brain regions and fat tissue.

Effects

Testosterone and DHT stimulate numerous responses in men, including (1) sperm production, (2) development and maintenance of all reproductive structures, (3) development and maintenance of male secondary sex characteristics such as deep voice, beard, thick body hair, and little fat on the hips and thighs, (4) interest in sexual activity (libido), (5) involuntary nocturnal erections during sleep, (6) thickening and strengthening of bones and muscles, and (7) a high basal metabolic rate (BMR).

Age Changes

On the average, aging is accompanied by a gradual decrease in blood testosterone levels that becomes evident after age 40 in many men. However, there is great variability in age changes in testosterone, and some older men have levels equal to or greater than the normal values for young adult men. There is also an average decrease in the proportion of free (active) testosterone. Furthermore, there is a gradual decline in the early morning peak levels, which tends to flatten the diurnal rhythm, and the peaks and valleys in daily testosterone levels occur up to 2 hours later. The effects of aging on DHT levels remain controversial.

Causes of Age Changes   Age-related changes in testosterone seem to result from several age changes. These include decreasing effects of LH on interstitial cells; decreasing numbers of interstitial cells; decreasing reserve capacity for LH and FSH secretion; and changing rhythms of LH secretion. However, the age-related increase in blood levels of LH that occurs in many aging men may help compensate for these changes. The increase may also explain why less than 10 percent of older men have blood testosterone levels low enough to be considered clinically abnormal.

Other age-related factors that may reduce testosterone levels include aging of the brain; adverse changes in the circulatory system; poor nutrition; obesity; alcohol consumption; medications; institutionalization; other specific diseases; and poor general health status.

Other Factors Affecting Testosterone and DHT   Besides age-related changes in testosterone levels, men are subject to age-related changes in the effectiveness of testosterone and DHT. First, the effectiveness of testosterone is reduced by the decrease in the proportion of free testosterone. Second, the effectiveness is reduced by an age-related decrease in the number of testosterone receptors in most target cells. In contrast, the prostate gland has an age-related increase in testosterone binding, which may contribute to benign prostatic hypertrophy. Third, the effectiveness of both testosterone and DHT is reduced by the age-related increase in estrogen, which results from increased conversion of testosterone and other hormones to estrogen by fat tissue. The increase in estrogen also may be partially responsible for the age-related increase in the incidence of benign prostatic hypertrophy.

Effects of Changes in Sex Hormones The age-related changes that affect testosterone and DHT in healthy men result in most age changes in the male reproductive system. However, testosterone levels remain adequate to sustain enough reproductive system functioning to achieve reproductive success and sexual satisfaction throughout life. Except in very old men, lower testosterone levels are not correlated with a decreased frequency of sexual activity. Furthermore, the age-related increase in blood levels of FSH and the resulting increase in stimulation of the sustentacular cells may be a compensatory factor. It may contribute to the lifelong ability to produce adequate numbers of functional sperm cells.

Using testosterone supplementation can benefit men who have a severe testosterone deficiency. Such cases are unusual. Men who have normal levels of testosterone and who take testosterone supplements receive little benefit while increasing their risks from atherosclerosis, benign prostatic hypertrophy, and possibly from prostate cancer.

Other alterations associated with age-related changes in testosterone and DHT and their effectiveness include reductions in body hair and in secretion by apocrine sweat glands and sebaceous glands. Finally, declining testosterone activity seems to be a main factor in the more rapid loss of bone matrix and the increased incidence of type II osteoporosis in older men, particularly men over age 65.

SEX HORMONES IN WOMEN

Sources and Control of Secretion

Young adult women produce two main sex steroids: *estrogen* and *progesterone*. There are two main forms of estrogen: *estriol* and *estrone*. Estriol, which constitutes approximately 60 percent of total estrogen, is more powerful than estrone.

Before menopause, almost all estrogen and progesterone come from *follicle cells*, which surround the developing egg cells in the ovaries (Fig. 14.1). Follicle cells also secrete inhibin. *Stroma cells*, which surround each group of follicle cells, secrete some testosteronelike hormone. Almost all of this hormone is converted to estrogen by the follicle cells. In addition, small amounts of estrogen, testosterone, and androstenedione are produced by the adrenal cortex, but these secretions do not play a significant role in women unless the ovaries are removed or *menopause* occurs.

Secretion of estrogen, progesterone, and inhibin by the ovaries is controlled by mechanisms that result in dramatic and rhythmic increases and decreases in hormone levels. These hormone cycles are accompanied by cycles of development and degeneration of follicles, egg cells, and the uterine lining.

The negative feedback mechanisms that control the secretion of estrogen, progesterone, and inhibin are similar to those which regulate testosterone and inhibin in men. However, in women a positive feedback mechanism becomes operative for a few days at about the middle of each cycle. This mechanism results in high LH levels, which cause ovulation. The high blood levels of estrogen and progesterone that also occur then produce a negative feedback effect again, resulting in decreasing LH and FSH levels; degeneration of the follicle; diminishing estrogen and progesterone levels; destruction of the uterine lining; and finally, menstruation. These changes lead to the next cycle.

Sex hormone secretion can be influenced by various brain activities; this may result in irregular cycles or the cessation of cycles.

Effects

The main effects of estrogen include (1) developing and maintaining reproductive system structures, including the breasts but not the ovaries,

(2) developing and maintaining female secondary sex characteristics (e.g., fat deposits on the hips and thighs, female pattern of hair distribution), (3) maintaining low blood levels of LDLs and high levels of HDLs, and (4) increasing and maintaining bone matrix. Estrogen seems to affect bone matrix in several ways. These ways include directly stimulating osteoblasts; increasing the secretion or effectiveness of calcitonin; inhibiting the effects of parathormone on bone cells; and inhibiting the production and effects of IL-6. IL-6 stimulates osteoclast activity and bone removal.

Progesterone stimulates the development of glandular tissues in the uterine lining and breasts. These functions become important only if a woman becomes pregnant or nurses her child. Finally, as in men, testosterone stimulates interest in sexual activity.

Age Changes

Before Menopause   At about age 45 the length of each female cycle begins to shorten because the time between the end of one cycle and ovulation in the next cycle decreases. Since this phase of the cycle produces much estrogen, its shortening results in a decline in estrogen secretion, and estrogen levels fall. These levels become so low that an increasing number of cycles do not produce a positive feedback effect, and ovulation does not occur. These changes seem to be caused initially by a decrease in the responsiveness of the ovaries to FSH and LH.

Because of these changes, there is less progesterone production, less development of the uterine lining, and an increasing number of cycles with little or no menstrual flow. Since menstrual flow is the most noticeable indicator of female cycles, the occasional absence of menstrual flow when it is expected indicates that the cycles are becoming irregular. By age 50 to 51 progesterone is essentially absent, menstrual flow happens less than once each year, ovarian and menstrual cycles (uterine cycles) have ended, and menopause has occurred.

After Menopause   When menopause occurs, estrogen secretion by the ovaries and blood estrogen levels decline quickly. During the four years after menopause estrogen secretion by the ovaries dwindles to zero. Blood estrogen levels do not reach zero, however, because small amounts of estrogen are produced by conversion of testosterone and androstenedione to estrogen and by the adrenal cortex. In spite of this estrogen production, blood estrogen levels usually drop to slightly below the lowest levels that were present during premenopausal hormone cycles. Furthermore, postmenopausal estrogen levels may be less than 5 percent of those present during midcycle estrogen peaks before menopause. The low estrogen levels reached within a few years after menopause do not fluctuate cyclically.

As with testosterone levels in older men, estrogen levels among postmenopausal women show considerable variation. Obese women generally have higher levels because fat tissue converts much androstenedione to estrogen. In extreme cases these elevated estrogen levels may equal or exceed average levels in premenopausal women. However, the estrogen in postmenopausal women is not as potent as that in premenopausal women because most postmenopausal estrogen is estrone rather than estriol.

Testosterone secretion from both the ovaries and the adrenal glands in postmenopausal women declines slightly, resulting in a small decrease in the already low blood levels. This testosterone has a greater impact, however, because the ratio of testosterone to estrogen increases.

Effects of Age Changes

Up to the time of menopause slow age changes in sex hormones result in gradually shorter and increasingly irregular menstrual cycles.

Temporary Effects   As menopause occurs, ovarian sex hormone levels plummet, causing many menopausal women to experience temporary signs and symptoms that are often considered part of menopause. Among these common phenomena, *hot flashes* involve sudden dilation of skin blood vessels in the head and neck, which often spreads downward over other regions. Affected women may feel a sense of pressure in the head, followed by sensations of heat or burning in areas where vessel dilation is occurring. The affected areas appear flushed, and profuse sweating may occur. Hot flashes seem to be caused by low estrogen levels.

These flashes usually last approximately four minutes but may last from a few seconds to over

30 minutes. It is difficult to estimate the incidence among menopausal women because of the extreme individual variations in the intensity of hot flashes. Hot flashes in some women are barely noticeable, while other women may be briefly disabled or may be awakened by very intense flashes.

In approximately 85 percent of menopausal women who experience hot flashes, the flashes occur for more than a year after menopause. They continue to occur for up to five years in 25 to 50 percent of the women who experience them after menopause.

Other consequences of altered sex hormone levels during and shortly after menopause may include depression, anxiety, irritability, nervousness, fatigue, and impaired memory and ability to concentrate. Some of these psychological changes may result from sleep disturbances caused by low estrogen levels or nocturnal hot flashes. All these undesirable features usually subside. These and other effects from menopause vary greatly among different cultures.

Permanent Effects Since hot flashes and the psychological alterations accompanying menopause are almost always temporary, they do not seem to fit the definition of age changes. Other changes caused primarily by the paucity of estrogen after menopause are permanent unless estrogen levels are raised, and many of these changes intensify into very old age. They include shrinkage and decreased functioning of all reproductive system structures; alterations in secondary sex characteristics (e.g., shrinkage of the breasts, increase in visible facial hair, decrease in axillary and pubic hair); increases in LDLs and decreases in HDLs; and more rapid loss of bone matrix.

The changes resulting from plunging estrogen levels and menopause are diverse. As for the reproductive system, the loss of childbearing ability is considered by some people to be a negative effect which can lead menopausal women into depression or other psychological disturbances. Others view the loss of childbearing ability as a positive outcome because it eliminates concerns about unwanted pregnancies. As a result, some women have an increase in sexual activity. Other consequences of reproductive system changes that follow menopause are discussed in Chap. 13.

Alterations in secondary sex characteristics after menopause are often considered cosmetically undesirable. Other changes in the skin include decreased secretion by apocrine sweat glands and sebaceous glands and an increased incidence of skin abnormalities. The changes in blood lipoproteins raise the risk of developing atherosclerosis and its complications (e.g., heart attacks, strokes). Shrinkage of the urethra promotes urinary stress incontinence. Finally, the rapid loss of bone matrix is a main risk factor for type I osteoporosis.

Estrogen Replacement Therapy

Since many target structures retain much of their responsiveness to estrogen, many postmenopausal changes can be slowed, stopped, or reversed by administering estrogenlike substances. This type of treatment is often called *estrogen replacement therapy* (ERT).

Benefits In many women ERT reduces or eliminates hot flashes. It also restores low LDL levels and high HDL levels, lowering the risk of atherosclerosis, and it seems to maintain cognitive functions and reduce the risk of getting Alzheimer's disease. To be most effective against osteoporosis, ERT should begin within six months after menopause, before significant bone loss has occurred. Furthermore, prevention of osteoporosis may require ERT for 10 or more years after menopause because stopping ERT allows the rate of bone resorption to increase to pretreatment levels. Postmenopausal prevention of osteoporosis may also require measures such as exercise, calcium supplements, and vitamin D supplements.

Risks Estrogen replacement therapy is especially recommended for women who have an early menopause and those at high risk for developing osteoporosis. Because ERT increases the risks of certain disorders (e.g., thrombus formation, breast cancer, gallbladder disease, endometrial cancer), it is not recommended for women with risk factors for certain conditions. These conditions include breast cancer or other reproductive system cancers; circulatory abnormalities such as high blood pressure, thrombus formation, and varicose veins; liver or gallbladder disease; or endometriosis. Women who are heavy smokers or are obese are also poor candidates for ERT.

The risks from ERT can be greatly reduced in several ways. These include using small doses of estrogen; administering estrogen by injection rather than orally; administering estrogen in cycles that mimic natural cycles; and administering low levels of certain progesteronelike substances (progestins). Cyclic administration of estrogen, especially when supplemented with progestins, often results in continued uterine cycles and periodic menstrual flow, which is usually less than premenopausal menstrual flow. However, no egg cells are produced and pregnancy is impossible.

In conclusion, ERT can relieve several serious consequences of low estrogen levels in postmenopausal women. Though ERT increases certain risks somewhat, it greatly reduces others. Therefore, the net effect in many women is an increase in the quality of life and life expectancy.

Alternatives   For some women, the risks from estrogen supplementation are too high when compared with the possible benefits. Scientists have found artificial compounds that promote some beneficial effects from estrogen supplements (e.g., reduce risk of atherosclerosis, reduce bone thinning) while not increasing certain risks from estrogen supplements (e.g., blood clots, cancers). Examples include tamoxifen and raloxifene. Other alternatives to traditional estrogen supplement therapy are being developed and evaluated. These alternatives include using different methods of estrogen administration (e.g., skin patches, vaginal creams); minimizing other risk factors (e.g., high fat diet); increasing alternative beneficial practices (e.g., improved diet; exercise; vitamin and calcium supplements; exercise; foods containing natural estrogens; herbal remedies).

INSULIN AND GLUCAGON

Sources and Control of Secretion

Cells in the pancreas that secrete hormones are located in pinhead-sized clusters called *islets of Langerhans*, which are scattered throughout the pancreas (Fig. 14.1). Some islet cells secrete *insulin*, and others secrete *glucagon*.

The secretion of both insulin and glucagon is regulated primarily by negative feedback mechanisms involving substrate control by blood glucose. High blood glucose levels stimulate insulin secretion and inhibit glucagon secretion, leading to a decline in blood glucose. Conversely, low blood glucose levels inhibit insulin secretion and stimulate glucagon secretion, leading to a rise in blood glucose. These mechanisms help maintain homeostasis of blood glucose.

Effects

Insulin   The principal target structures of insulin are muscle, liver, and fat cells. Insulin helps provide energy for these cells while simultaneously reducing blood glucose levels by stimulating entry of glucose into the cells; the breakdown of glucose for energy; storage of glucose as glycogen; and storage of glucose as fat. The first three processes occur primarily in muscle and liver cells, while liver and fat cells carry out most of the fourth. These three target cell types obtain and use blood glucose only if adequate insulin is supplied. Body cells other than muscle, liver, and fat cells can use glucose without the presence of insulin. Of all the glucose removed from the blood because of insulin, approximately 75 percent enters muscle cells. Insulin also stimulates both the passage of amino acids into cells and the synthesis of proteins from amino acids.

Glucagon   Unlike insulin, glucagon causes blood glucose levels to rise by stimulating liver cells to produce and release glucose. The glucose is produced from glycogen in the liver, amino acids in the blood and liver, and fats in the liver and fat cells.

Combined Effects   The antagonistic effects of insulin and glucagon make up the body's main mechanism for providing proper and fairly stable blood glucose levels. Insulin is essentially the only control signal that causes a decrease in blood glucose. By contrast, though glucagon is the main control signal that causes an increase in blood glucose, sympathetic nerves and other hormones (e.g., growth hormone, epinephrine, glucocorticoids) can also cause such an increase.

Maintaining blood glucose homeostasis is important for two reasons. First, preventing low glucose levels assures that all body cells receive enough glucose to obtain energy and building materials. Second, avoiding high levels helps prevent many problems associated with the disease *diabetes mellitus* (see Diabetes Mellitus, below).

The effects of insulin on amino acids and protein synthesis are also helpful to the body because they assist in the formation and replacement of parts of the body and secretions that contain protein.

Blood Glucose Homeostasis

The blood glucose level is often expressed as glucose per 100 ml (per deciliter) of blood plasma in a sample taken at least 2 hours after eating. The result, called the *fasting plasma glucose* (*FPG*) value, is normally 80 to 115 mg/dl. Glucose levels may fluctuate irregularly within this range because of changes in body activity, sympathetic nerve impulses, and hormone levels.

The ability of the body to reverse a dramatic rise in blood glucose and restore glucose homeostasis is called its *glucose tolerance*. For example, soon after a person has ingested a large quantity of sugar, the blood glucose level may exceed 200 mg/dl. If that person has good glucose tolerance, blood glucose is brought down below 140 mg/dl within 2 hours of ingesting the sugars, and glucose levels stabilize at 80 to 115 mg/dl soon afterward.

Glucose tolerance can be tested by an *oral glucose tolerance test* (*OGTT*). In this procedure, blood glucose levels are measured during the 2-hour period after ingesting a large amount (75 grams) of glucose.

Age Changes

As age increases, the pancreas retains the ability to quickly increase blood insulin levels and glucagon levels and maintain them within the normal range for young adults. Furthermore, aging causes no significant changes in the ability of insulin and glucagon to regulate blood glucose levels. Because of the continued effectiveness of insulin and glucagon in regulating glucose levels, aging causes no important change in FPG values or glucose tolerance.

Abnormal Changes

Although aging has essentially no effect on the ability of the pancreas to regulate insulin levels, there is an average age-related increase in blood insulin levels. This is not an age change but is associated with reductions in physical activity and ˙Vo2max and increases in body fat. Elevated insulin levels are closely associated with increased abdominal body fat near the waist (increased waist/hip ratio) which is common in aging men.

The age-related increase in blood insulin seems to result from a decrease in target cell responsiveness to insulin, which is called *insulin resistance*. Because of insulin resistance, the target cells (muscle, liver, and fat cells) remove little blood glucose even when blood glucose and insulin levels are high. Since blood glucose remains high, the pancreas secretes additional insulin, further elevating insulin levels. Insulin levels are increased until they are high enough to stimulate the somewhat unresponsive target cells to remove some blood glucose and lower blood glucose levels.

In individuals with insulin resistance, once insulin levels become high, they stay high because more insulin is needed to counter the effects of glucagon. (Recall that the blood levels and the effectiveness of glucagon do not change with advancing age.) By themselves, small increases in insulin resistance and insulin levels do not cause problems, but they can be warning signs that more substantial changes in insulin may occur. Therefore, it is advisable to monitor these changes and take steps to prevent or reverse them.

Restoring Insulin Levels and Target Sensitivity
Elderly people with an age-related increase in insulin resistance can regain much insulin sensitivity and reduce high blood insulin levels through a program that combines vigorous exercise and weight loss. Although improvements in insulin sensitivity occur within a few days of starting such a program, regular exercise must be continued because exercise-induced gains in insulin sensitivity begin to decline within three days of ending the program. If the program is not reinstated, the original low levels of insulin sensitivity are reached within several days to 2 weeks.

Blood Glucose Levels   Individuals who have normal FPG values and take almost 2 hours during an OGTT to reduce blood glucose to less than 140 mg/dl are considered to have normal glucose regulation but *decreased glucose tolerance*. Individuals who have FPG values below 140 mg/dl and whose blood glucose levels at the end of an OGTT are 140 to 200 mg/dl have an abnormal condition called *impaired glucose tolerance* (*IGT*). Individuals with FPG values greater than 115 mg/dl have an abnormal condition called *hyperglycemia*. Finally, individuals with FPG values below 80 mg/dl have an abnormal condition called *hypoglycemia*.

The proportion of people with decreased glucose tolerance rises rapidly after age 45. Since glucose tolerance depends on the action of insulin, the decrease in glucose tolerance usually occurs in those who also have insulin resistance and increased insulin levels. Like age-related changes in insulin, decreased glucose tolerance is probably not an age change.

Small decreases in glucose tolerance do not constitute a problem, though they may warn of impending difficulties with glucose regulation. Monitoring, preventing, and reversing decreases in glucose tolerance may be appropriate. Decreased glucose tolerance in older people can be improved by the same techniques that improve high insulin levels and low insulin sensitivity (see above).

Decreased glucose tolerance often worsens and becomes impaired glucose tolerance (IGT). Other than abnormally high OGTT values, individuals with IGT have no signs or symptoms of the disorder. This allows problems from IGT to develop insidiously.

Impaired glucose tolerance develops in up to 40 percent of those over age 60. Among these individuals, approximately 30 percent will improve with no medical intervention and their glucose tolerance will enter the normal range. Another 50 percent will continue to have only IGT. The remaining 20 percent will develop diabetes mellitus.

Impaired glucose tolerance can be caused by the same factors that cause insulin resistance and decreased glucose tolerance and by other factors that contribute to diabetes mellitus (see below). However, IGT can be prevented or reversed by methods used in connection with decreased glucose tolerance and diabetes mellitus. Such actions are highly advisable because of the complications that result from continued IGT and the subsequent development of diabetes mellitus. For example, elderly individuals who continue to have IGT are at high risk for developing atherosclerosis and its related disorders.

DIABETES MELLITUS

Definition and Types

Diabetes mellitus (DM) is a group of diseases characterized by chronic hyperglycemia, poor glucose tolerance, and usually other abnormalities in metabolism. Diabetics have FPG values above 140 mg/dl and final OGTT values above 200 mg/dl. High glucose levels, which may exceed 800 mg/dl, persist because of inadequate insulin production or high insulin resistance. As a result, muscle, liver, and fat cells cannot lower blood glucose adequately and the effect of glucagon, which raises blood glucose, remains unchecked.

There are four types of DM. Individuals with *type I*, or *insulin-dependent*, *diabetes mellitus* (*IDDM*) cannot survive unless they receive insulin therapy. This type of DM is also frequently called *juvenile-onset diabetes mellitus* because most cases develop before age 20 and very few develop after age 30. A third name is *ketosis-prone diabetes mellitus* because these patients usually develop high blood levels of ketoacids. Many individuals who develop IDDM as children or young adults receive adequate treatment and survive long enough to enter the elderly population.

Victims of *type II*, or *non-insulin-dependent*, *diabetes mellitus* (*NIDDM*) have low insulin levels or high insulin resistance but do not require insulin therapy to survive. However, insulin treatment may be of significant help in more severe cases. NIDDM is frequently called *maturity-onset diabetes mellitus* because most cases develop after age 40. It is also called *non-ketosis-prone diabetes mellitus* because few cases involve high ketoacid levels. NIDDM may advance to become IDDM.

Another type of DM is called *secondary diabetes mellitus* because it develops as a complication from another abnormality or disease, such as alcoholism, pancreatitis, excess growth hormone, or excess glucocorticoids. The fourth type is *gestational diabetes*, which occurs in some women during pregnancy because high sex hormone levels reduce the effectiveness of insulin.

We will consider only IDDM and NIDDM in detail because no cases of gestational diabetes occur among older people and because most cases of secondary diabetes, though differing in cause, have the same effects as NIDDM.

Incidence

Diabetes mellitus ranks as the eighth leading chronic condition among both people over age 45 and those over age 65. Its incidence nearly doubles between ages 45 and 65. Because of its many complications, DM is among the most common factors causing older people to visit a physician, and it is the sixth or seventh leading cause of death among people over age 65.

NIDDM is the most common type of DM among the elderly. It is estimated that 10 percent of all people over age 56 have NIDDM. Furthermore, the incidence increases with age, and NIDDM is present in approximately 20 percent of those age 65 to 74 years and approximately 40 percent of those over age 85.

Causes

IDDM seems to be caused by autoimmune reactions against insulin-producing cells in the pancreas. Virtually no insulin can be produced because essentially all the insulin-producing cells are destroyed.

NIDDM has a strong tendency to run in families, certain ethnic groups, and blacks. Therefore, it seems to rely heavily on genetic predispositions that interact with other factors. One important factor is obesity; another is eating large quantities of carbohydrates. These factors are important because they are associated with high insulin resistance. The significance of high insulin resistance becomes evident when one realizes that though some people with NIDDM produce little insulin, many have normal or high insulin levels but high insulin resistance. Finally, contrary to previous beliefs, aging makes little or no contribution to the development of NIDDM.

Main Effects and Complications

The most important effect of diabetes mellitus is the maintenance of abnormally high blood glucose levels. The glucose causes several alterations, each of which contributes to one or more of the complications of DM. The development and consequences of most of the complications mentioned below are described in greater detail in Chaps. 3-10, 12, 13, and 15 on the body systems in which these complications appear.

Excess Glucose One outcome of high blood glucose levels is excess conversion of glucose to a sugar called *sorbitol*, which accumulates in certain areas. Within the eye, sorbitol causes cataracts in the lens and initiates diabetic retinopathy. These conditions are, respectively, the leading eye disease and the leading cause of blindness among the elderly. Within nerves, sorbitol causes degeneration of neurons and the Schwann cells that surround them, resulting in deterioration of sensory and motor neuron functioning and leading to diminished reflex response, conscious sensation, and voluntary muscle control. Effects include reduced sweating, increased traumatic injury, gangrene of the feet, abnormal GI tract peristalsis, and fecal and urinary incontinence.

A second outcome of high blood glucose levels is the formation of glucose cross-links between protein molecules both inside and outside cells. No enzymes are needed to initiate this process, which is called *nonenzymatic glycosylation*. Furthermore, once started, the process continues even if glucose levels return to normal. Since the cross-links formed by glucose are strong and permanent, the movement of protein molecules and the passage of materials between proteins, such as collagen fibers, are restricted. Harmful consequences of nonenzymatic glycosylation in the circulatory system adversely affect all parts of the body. The most frequent serious complications of circulatory system changes are heart attacks, strokes, and gangrene of the feet and legs. Nonenzymatic glycosylation promotes free radical formation, and it can cause degeneration and failure of the kidneys.

A third effect of high blood glucose levels is high osmotic pressure. A high concentration of glucose in the blood causes some dehydration of cells because water moves from the cells into the blood, causing the cells to malfunction. The situation is compounded because the extra glucose prevents the collecting ducts in the kidneys from reabsorbing enough water, leading to excess water loss in urine.

The consequences of altered kidney functioning caused by high blood glucose levels are revealed by three classic signs and symptoms of diabetes mellitus: an increase in urine production (*polyuria*), the presence of glucose in the urine (*glycosuria*), and an increase in appetite and eating (*polyphagia*). An abnormally high loss of water in the urine tends to lower blood pressure while further increasing blood osmotic pressure and cell dehydration. This is indicated by the fourth classical sign and symptom of DM: increased thirst and drinking (*polydipsia*). Finally, as more water leaves the body and dehydration worsens, additional minerals (e.g., sodium, potassium) are lost and mineral deficiencies develop. Outcomes may include circulatory failure, brain malfunction, coma, and death.

The final effect of high blood glucose levels is an increased risk of infection. This occurs because the glucose provides abundant nutrients for microbes, encouraging their rapid proliferation.

Ketoacidosis   In all cases of IDDM and some cases of NIDDM, insulin levels are so low that very little glucose enters liver, muscle, and fat cells. These cells then attempt to obtain energy from the breakdown of fats and amino acids. However, this requires the use of some glucose because glucose provides the substance that channels fatty acid fragments and amino acid fragments into the Krebs cycle. With inadequate insulin to assist the entry and use of glucose, the fragments from the fat and amino acids cannot be broken down but are converted into substances called *ketoacids* (*ketones*). Ketoacids enter and accumulate in the blood, producing a condition called *ketoacidosis*. The resulting disturbance in acid/base balance causes cells to malfunction, especially in the brain.

Excess ketones leave the body in exhaled breath and in the urine (ketonuria). The ketones are noticed as a sweet fruity aroma, which indicates the presence of ketoacidosis. When ketoacids are eliminated in the urine, they carry minerals (e.g., sodium, potassium, zinc, magnesium) out of the body, causing increased mineral deficiencies.

Prevention

There are no preventive measures for IDDM. However, the development over a period of weeks of polyuria, polydipsia, and polyphagia, which are usually accompanied by rapid weight loss, is a clear warning sign that this disease has developed. The intensity of these conditions forces affected individuals to seek medical attention. Failure to obtain treatment results in serious acute illness and death.

NIDDM develops subtly and insidiously over a period of years. The classic warning signs and symptoms may not be apparent or may develop gradually and be ignored or attributed to other conditions. Therefore, the first indication of NIDDM to be noticed is often a complication it causes, such as eye diseases, heart attack, stroke, and foot infections. At this stage many other com-

plications are also well established. Therefore, early prevention of NIDDM by avoiding modifiable risk factors is very important, especially for blacks and anyone with close relatives who have NIDDM.

Perhaps the most important modifiable risk factor for NIDDM is having an abundance of body fat, particularly when it reaches the level of obesity. Obesity accompanied by a diet high in carbohydrates creates an especially high risk of developing NIDDM. Furthermore, the greater the amount of body fat, obesity, and carbohydrate intake, the greater the risk. Therefore, one of the best ways to prevent NIDDM is to maintain a desirable body weight and avoid excess body fat.

The other important modifiable risk factor for NIDDM is having a low level of physical activity. Exercise reduces the risk of NIDDM by helping to increase insulin sensitivity, improve glucose tolerance, and prevent obesity by increasing energy use and influencing appetite (Chap. 8).

Treatment

The goals for treating DM include maintaining homeostasis of blood glucose levels and avoiding ketoacidosis. One key to achieving these goals is dietary regulation. Careful planning includes controlling the times food is consumed; the amounts of sugar, fiber, and other carbohydrates consumed; and the total intake of kilocalories.

A second aspect of these treatment plans is regulating the amount of physical activity. Exercise tends to lower glucose levels by causing cells to consume glucose and glycogen and to have increased insulin sensitivity. Sedentary periods have the opposite effects. Exercise that is accompanied by reductions in body fat can cure some cases of NIDDM. However, exercise programs must be tailored to individual needs to maximize the benefits while minimizing the adverse effects.

Treatment plans may include the administration of insulin or medications that stimulate insulin production. Various types, doses, and schedules of insulin administration are used in different individuals. Furthermore, the administration of insulin and other medications is adjusted frequently to compensate for fluctuations in diet, exercise, and other aspects of daily living.

In spite of all efforts at planning, numerous factors militate against maintaining glucose homeostasis and preventing ketoacidosis at all times. These factors include an inability to adequately regulate diet and exercise; age changes; the presence of abnormal conditions or diseases; other medications; physical and mental limitations and disabilities; and a host of social, psychological, and economic factors. Furthermore, the number and severity of these factors and the complexity of interactions among them increase with age. Therefore, it is likely that an elderly person with diabetes mellitus will develop at least some complications. However, the adverse effects can be minimized by regularly checking for complications and promptly treating any that appear.

15

THE
IMMUNE
SYSTEM

The *immune system* contains a diverse group of structures and cells that are spread throughout the body (Fig. 15.1). These structures include the *thymus gland*, *red bone marrow*, *spleen*, *lymph nodes*, *lymph vessels*, other structures containing *lymphatic tissues* (tonsils, lining of the respiratory system and GI tract), and *skin*. These structures promote the development of the cells of this system and, together with blood and lymph, serve as the main repositories for them. Immune system cells include *macrophages*, *Langerhans cells*, and several types of *lymphocytes*. Special protein molecules called *antibodies* (see below) are also considered part of the immune system.

MAIN FUNCTIONS FOR HOMEOSTASIS

The immune system is a main defense mechanism for the body against harmful agents, including many foreign substances, bacteria, parasites, viruses, and its own cancer cells. This system uses different defense strategies against different agents, including blocking their entrance, abolishing those found in the body, and helping neutralize or eliminate undesirable substances produced by invading organisms.

Therefore, a person's immune system serves like a combination of a nation's agencies for customs and immigration, drug enforcement, and counterinsurgency. Unfortunately, as with such agencies, immune system activities occasionally injure innocent constituents through misidentification or overzealous actions such as autoimmunity and allergy.

UNIQUE CHARACTERISTICS

Three characteristics of the immune system make it unique among the body's defense mechanisms. First, it shows *self-recognition*, which means that the immune system attempts to distinguish between substances that are normal constituents of a person's body and substances that are foreign to it. Upon identifying a substance as foreign, the immune system mounts an *immune response* against it. It can perform an immune response against some types of cancer cells because they display molecules identified as foreign. Any substance that causes an immune response is called an *antigen*.

FIGURE 15.1 Immune system structures.

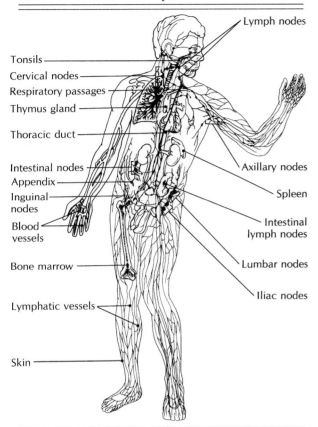

The second unique characteristic is *specificity*, which means that an immune response will operate only against one antigen. Therefore, a different immune response must be produced each time a different antigen is encountered. For example, specificity explains why an immune response against the virus that causes measles provides no protection against the virus that causes chickenpox. In contrast to the immune system, other bodily defense mechanisms against harmful chemicals, microbes, viruses, and cancer cells are called *nonspecific* because each mechanism helps protect the body against a variety of these agents. Some nonspecific defense mechanisms, such as the skin, mucous membranes, and mucus, prevent harmful agents from entering the body, while others, such as movements of cilia, defecation, and urination, help expel them. Other mechanisms, such as fever, perspiration, sebum, and acidic conditions in the stomach and vagina, inhibit the growth of harmful microbes. Finally, phagocytic cells (e.g., WBCs) and natural killer (NK) cells act nonspecifically in killing microbes and cancer cells.

The third unique characteristic is *memory*.

When the immune system responds against an antigen, it develops a residual set of lymphocytes called *memory cells* and usually develops a group of long-lasting antibodies. *Antibodies* are protein molecules that adhere to antigens and help combat them. Each time a particular antigen is encountered, the memory cells and antibodies developed for that antigen cause a quicker and more intense attack and thus eliminate it faster. In contrast, nonspecific defense mechanisms function with the same speed and intensity each time an injurious agent presents itself, allowing for the same risk of injury from the agent before it is eliminated.

DEVELOPMENT OF THE IMMUNE SYSTEM

A burst of development in the immune system occurs over several weeks before and after birth. At first the prenatal liver and spleen produce *monocytes*, which are phagocytic white blood cells, and *lymphocytes*. As the time of birth approaches, production of monocytes and lymphocytes shifts to the red bone marrow, which continues to produce these cells thereafter (Fig. 15.2).

Macrophages and Langerhans Cells

Many monocytes pass through capillary walls and enter the spaces among body cells and within lymph nodes and other lymphatic tissues. These migrating monocytes are then called *macrophages*. Similar cells called *Langerhans cells* develop in the epidermis. Lifelong monocyte production by red bone marrow helps sustain the population of macrophages, and the epidermis attempts to maintain adequate numbers of active Langerhans cells.

Thymus and T Cells

As macrophage formation begins, the blood transports a portion of the new lymphocytes into the *thymus* which lies above the heart and behind the sternum (breastbone) (Figs. 15.1 and 15.2). The thymus converts these lymphocytes into a special type of cell called *T lymphocytes* (*T cells*).

HLA Receptor Formation   One process that occurs during T-cell development involves varying T cells so that they produce cell surface receptor

FIGURE 15.2 Development of macrophages, T cells, and B cells. (1) Bone marrow cells produce monocytes and lymphocytes. (2) Monocytes enter blood vessels and are transported to capillaries throughout the body. (3a) Some lymphocytes enter blood vessels and are transported to the thymus. (3b) Some lymphocytes enter blood vessels and are transported to other areas, such as bone marrow. (4) Some monocytes leave capillaries and become macrophages (M) among body cells. (5) Lymphocytes in the thymus reproduce and develop HLA receptors and antigen-specific receptors and become T cells (T). (6) T cells with antigen-specific receptors for self-antigens are destroyed (clonal selection). (7) Remaining T cells are transported to lymphatic tissues such as the lymph nodes and spleen. (8) T cells in lymph tissues reproduce and mature to form T cell clones. (9) Lymphocytes in bone marrow reproduce and develop HLA proteins and antigen-specific receptors to become B cells (B). (10) B cells with antigen-specific receptors for self-antigens are destroyed (clonal selection). (11) Remaining B cells enter blood vessels and are transported throughout the body.

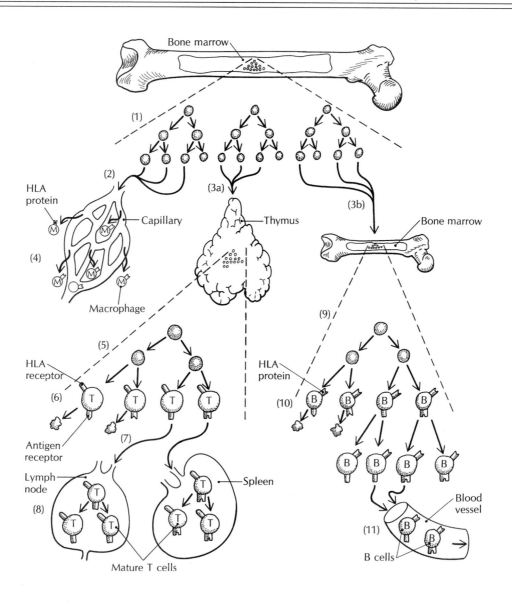

molecules called *human leukocyte-associated* ecules of *HLA protein*, which are found on virtually every cell in the body.

Each person has certain types of HLA proteins on his or her cells, and these proteins differ from the proteins in every other person. Therefore, each person's HLA protein identifies each cell as belonging only to that person's body. Exceptions occur with genetically identical people (e.g., identical twins), whose cells have identical HLA proteins.

Antigen-Specific Receptor Formation A second process during T-cell development results in each T cell producing a second type of surface receptor, an *antigen-specific receptor*. All the antigen-specific receptors on each T cell can bind to only one substance, and each T cell develops a different type of antigen-specific receptor. There may be 100 million types of antigen-specific receptors and therefore an equal number of different types of T cells.

Clonal Selection and Suppression Many scientists believe that during the formation of antigen-specific receptors samples of all surface materials on body cells are carried into the thymus. Once these materials enter the thymus, that gland selectively destroys T cells with antigen-specific receptors that bind to any of those materials. Surface materials that bind to antigen-specific receptors are called *self-antigens* because they are native body materials that could start an immune response. T cells that are incapable of binding to self-antigens that enter the thymus during this period survive and begin to reproduce. Therefore, in each person, each surviving T cell forms a clone of identical cells. Each cell has HLA receptors for that person's HLA protein plus one type of antigen-specific receptor for one substance that is not a self-antigen.

The entire process of T-cell development is called *clonal selection*. Members from each clone are carried throughout the body by the circulatory system, with many of them deposited in the spleen, lymph nodes, and other lymphatic tissues. Thymic hormones continue to cause the dispersed T cells to reproduce and mature. Once mature, the T cells can use their HLA receptors to distinguish the individual's cells from any other cells. The T cells will be activated to participate in an immune response whenever both their HLA receptors and their antigen-specific receptors are bound to the surface of a cell. They are therefore said to be *immunocompetent* and have also developed both self-recognition and specificity.

Almost all clonal selection is believed to occur in the thymus within 1 month after birth. Furthermore, at least some members of each clone probably survive outside the thymus for many years. Therefore, each clone represents a widespread reserve of T cells that can attack one antigen each. However, as long as the thymus secretes ample thymic hormones, these hormones may be able to stimulate additional conversion of lymphocytes, clonal selection, and maturation of T cells outside the thymus. These processes could form T-cell clones for additional antigens and bolster or reestablish some older clones that had dwindled or vanished through gradual T-cell death.

Many scientists believe that during clonal selection some T cells actually form T-cell clones against self-antigens. These clones lack self-recognition and therefore could begin immune responses against the body's own cells. They are prevented from doing this by mechanisms that suppress their participation in an immune response. At least part of the suppression may be performed by special T cells called *suppressor T cells (sT cells)*.

B Cells

Recall that only some lymphocytes produced by the red bone marrow are converted into T cells by the thymus and thymic hormones. Other new lymphocytes are converted into *B lymphocytes (B cells)*. B-cell formation does not depend on the thymus. Though its site is unknown, this process seems to be very similar to the clonal selection that produces T-cells (Fig. 15.2). However, there are two important differences. First, B cells do not develop HLA receptors and therefore need have only their antigen-specific receptors bound to an antigen to begin participation in an immune response. Second, B cells develop HLA protein, which allows them to bind to HLA receptors on T cells.

IMMUNE RESPONSES

Once macrophages, Langerhans cells, T cells, and B cells have developed, the immune system is ready to initiate immune responses. The system begins to monitor substances in the body in an attempt to detect foreign materials.

Immune responses and other immune system activities are regulated by signaling substances from the nervous system and endocrine system, and from the immune system cells themselves. Some regulating substances act at great distances from their sites of production, and others affect cells close to their sources. Like hormones, these signaling substances contact many cell types, yet they affect only certain cells and organs. Usually the affected cells respond because they contain receptor molecules to which the signaling molecules bind.

Different cells exposed to the same amount of a signal respond to different degrees because they have fewer or more receptors or because their receptors bind weakly or strongly to the signal molecules. The strength of each target's response can be changed by modifying the number or binding strength of its receptors. The effectiveness of a signal can be influenced by conditions in the target cells. Finally, as with hormones, signal effectiveness can be influenced by rates of formation and elimination. Therefore, determining secretion rates or measuring concentrations of signaling substances provides only a small portion of the information needed to evaluate immune system performance. In this section, only a few of the signaling substances will be mentioned and only some of their main effects will be described.

Processing and Presentation

The macrophages perform this surveillance by phagocytizing microbes, viruses, and unusual molecules (Fig. 15.3). As a macrophage or Langerhans cell digests and destroys these items, it trans-

FIGURE 15.3 Processing and presentation of antigens and the formation of specialized T cells. (1) Macrophages (M) ingest antigen. (2) Macrophages digest antigen and present antigen fragments. (3) T cells (T) with antigen-specific receptors for the antigen join to the presenting macrophage, using HLA receptors and antigen-specific receptors. IL-1 (dashed arrow) from macrophages stimulates the joined T cells. (4) T cells reproduce and form specialized T cells (hT, cT, dT, sT). (5) Specialized T cells reproduce.

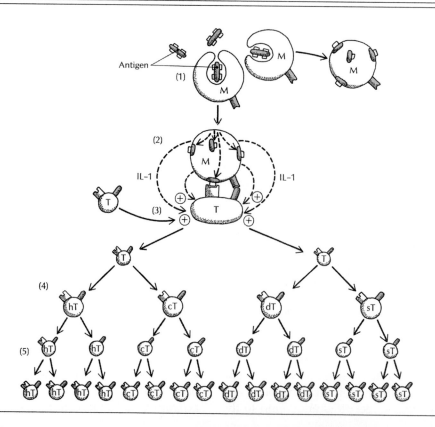

ports fragments of each one through its cell membrane. Then the cell presents the fragments, together with its own HLA protein, to neighboring T cells. If a fragment and the HLA protein bind to a T cell, the fragment and the item from which it came (microbe, virus, or molecule) are considered antigens. Both the macrophage and the T cell are activated to initiate an immune response against these antigens.

T-Cell Participation

T-Cell Specialization The activated macrophage secretes *interleukin-1* (*IL-1*), which stimulates the T cell to produce more identical T cells. The new T cells specialize into any combination of four different types, depending on the source of the antigen fragment and the type of HLA used: *helper T cells* (*hT cells*), *cytotoxic T cells* (*cT cells*), *delayed-hypersensitivity T cells* (*dT cells*), and *suppressor T cells* (*sT cells*). Some of these cells have other names: hT cells are CD4+ cells; cT cells are CD8+ cells.

Langerhans cells act like macrophages except that they do not produce IL-1. However, neighboring keratinocytes produce IL-1, and so similar immune activities occur in the skin. Besides stimulating T cells, IL-1 evokes inflammation and fever, two nonspecific defense mechanisms.

In the rest of this section, note that IL-1 and several specialized T-cell secretions (IL-2, lymphokines) cause positive feedback effects that amplify the immune response until sT cells come into play.

Helper T-Cell (hT-Cell) Activities There are two main types of hT cells. TH-1 hT cells produce signaling substances that stimulate cT cells and that promote inflammation. These substances from TH-1 hT cells include IL-2, interferon- (IFN-), and tumor necrosis factor (TNF). Macrophages also produce TNF and stimulate inflammation. TH-2 hT cells produce signaling substances that stimulate B cells. Examples include IL-4, IL-5, IL-6, and IL-10. Many cell types including monocytes, macrophages, endothelial cells, mast cells, keratinocytes, and osteoblasts produce IL-6. IL-6 promotes inflammation, bone matrix removal, and other body activities. The signaling substances from hT cells and from other cells regulate the hT cells and other immune response cells so immune system activities remain balanced.

The hT cells that are produced bind to the original presenting macrophage or to any other macrophage that presents the same antigen. Then the hT cells secrete *interleukin-2* (*IL-2*). IL-2 initially enhances the developing immune response in several ways (Fig. 15.4). First, it stimulates macrophages to phagocytize more antigen, leading to the digestion and presentation of more antigen and the activation of more T cells specific for that antigen. Second, it stimulates the production of more hT cells and cT cells. Third, it stimulates the proliferation and activity of any B cells that have bound to the original undigested antigen.

While the hT cells are producing IL-2, they also secrete other helpful defense substances called *lymphokines*. These substances increase macrophage phagocytosis in several ways and protect normal body cells from viruses.

Cytotoxic T-Cell (cT-Cell) Activities Unlike hT cells, which bind to antigen and HLA protein on macrophages and Langerhans cells, cT cells bind to antigen and HLA protein on other body cells (Fig. 15.5). Such combinations occur on cells infected with viruses, fungi, or bacteria; certain types of cancer cells; and cells transplanted into the body from a person with different HLA protein or from an animal. When a cT cell binds to an antigen-bearing cell, it is activated and proliferates, producing a clone of cT cells that bind to other cells with the same antigen. With the assistance of IL-1 and IL-2, each cT cell destroys the cell to which it binds, using secretions that damage the cell membrane. This type of immune response is called a *cell-mediated response* because the cT cells make direct contact with each antigen-bearing cell they attack. It contrasts with the *humoral response* by B cells, which secrete antibodies that attack antigens at a distance (see below).

Every cT cell can move from cell to cell, selectively destroying each antigen-bearing cell that binds to both of its types of receptors. The cT cells also release lymphokines, which activate macrophages and other types of T cells while protecting normal cells from viruses. Certain cT-cell lymphokines also stimulate *natural killer cells* (*NK cells*), nonspecific lymphocytes that destroy cancer cells.

Delayed-hypersensitivity T-cell (dT-cell) Activities Delayed-hypersensitivity T cells are similar to cT cells in the way in which they identify abnormal cells. However, these cells do not kill cells

FIGURE 15.4 Activation and activities of hT cells. (1) Macrophages (M) ingest antigen. (2) Macrophages digest and present the antigen. (3) hT cells with specific receptors for the antigen join the presenting macrophage, using HLA receptors and antigen-specific receptors. (4) hT cells produce IL-2 (dashed arrow), which stimulates macrophages, hT cells, cT cells, and B cells that are joined to the antigen. (5) hT cells produce lymphokines (jagged arrow).

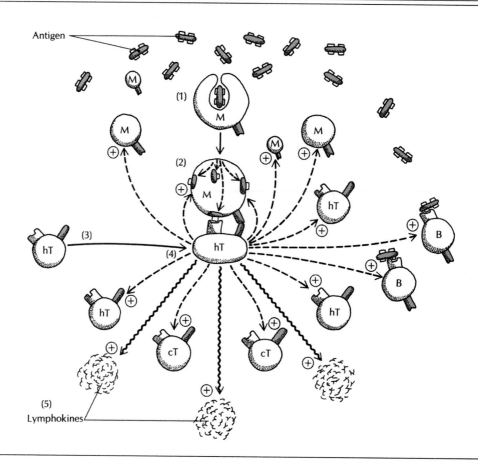

directly. Rather, the lymphokines produced by these cells stimulate other immune system cells (e.g., macrophages) to destroy cells with surface antigens. These lymphokines also cause inflammation, which increases the defense of the affected area (Chap. 3). The inflammation is evident during excessive dT-cell reactions, such as those resulting from poison ivy or positive skin tests for tuberculosis (e.g., tine tests). Delayed-hypersensitivity cells received their name because at least 1 day is required for them to cause significant inflammation. By contrast, allergic reactions caused by B cells are called *immediate-hypersensitivity reactions* because they produce significant effects within minutes or hours. Examples of immediate-hypersensitivity reactions include forms of asthma and allergic reactions to penicillin, bee stings, and foods.

Suppressor T-Cell (sT-Cell) Activities   We have seen that IL-2 stimulates immune activity by acting on hT cells, cT cells, and dT cells. However, IL-2 also stimulates the proliferation of sT cells. Since this occurs slowly, its takes approximately 1 week to develop a large number of sT cells specific for the antigen being attacked. When enough sT cells have developed, their secretions overpower and quell the immune activities of the attacking immune system cells and the immune response to that antigen subsides. By this time the antigen usually is being reduced or has been eliminated.

Suppression of the immune response helps prevent the adverse effects that may accompany ex-

FIGURE 15.5 Activation and activities of cT cells. (1a) Antigen invades body cells (bc). (2a) cT cells (cT) with specific receptors for the antigen join to the infected body cell, using HLA receptors and antigen-specific receptors. (3a) cT cells produce lymphokines (jagged arrow) against infected cells. (4a) Infected cells and enclosed antigen are destroyed. (5a) cT cells produce more identical cT cells to attack other body cells that are infected with the antigen. (1b) Body cells become cancer cells (cc) and produce antigens. (2b) cT cells with specific receptors for the antigen join to the cancer cells, using HLA receptors and antigen-specific receptors. (3b) cT cells produce lymphokines (jagged arrows) against the cancer cells. (4b) Cancer cells are destroyed. (5b) cT cells produce more identical cT cells to attack other identical cancer cells.

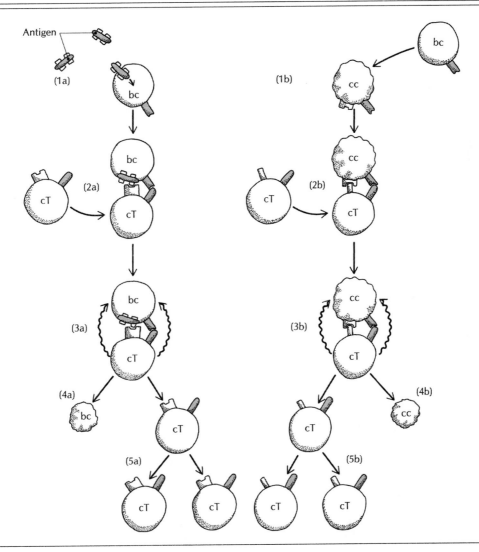

cessive or prolonged immune activity. Examples include discomfort and damage from accidental immune injury to normal body components and from inflammation. Therefore, sT cells help maintain homeostasis by providing timely negative feedback that reverses the positive feedback effects of other immune system cells.

Though some sT cells are antigen-specific and therefore suppress specific immune responses, others suppress immune responses to many different antigens simultaneously. This provides ongoing regulation of the entire immune system. One benefit is a reduction in *autoimmune reactions*, which are immune responses against normal parts of the body, such as rheumatoid arthritis and insulin-dependent diabetes mellitus.

Another benefit is a reduction in *allergic responses*, which are excessive immune responses against foreign antigens, such as hay fever, asthma, and food and drug allergies.

B-Cell Participation

Most aspects of the immune response mentioned up to this point result in nonspecific defense reactions against an antigen (e.g., phagocytosis, inflammation, fever). Only the portion of the antigen bound to cells bearing HLA protein is specifically attacked. To understand how unbound antigen, such as antigen suspended in body fluids, is attacked, we must examine the operations of B cells and the antibodies they produce.

B-Cell Activation   Since antigen-specific receptors on B cells are more complete than antigen-specific receptors on T cells, B cells can bind to an antigen even when HLA protein is not present (Fig. 15.6). No macrophages or other cells are needed to process or present the antigen to B cells, and B cells specific for the antigen bind to it wherever they meet it. This stimulates the attached B cells to proliferate and produce two special types of B cells-*memory B-cells* (*mB cells*) and *plasma cells*-both of which continue to reproduce. All the mB cells and plasma cells have the same antigen specificity as the B cells from which they were derived. Memory B cells are described below in connection with memory. The plasma cells manufacture and secrete *antibodies (immunoglobulins)*, which combat the antigen in several ways (see below). However, the original antigen-bound B cells and their progeny usually function inadequately unless they are stimulated by IL-2. The hT cells provide a concentrated application of IL-2 to the antigen-bound B cells, plasma cells, and mB cells by binding to them. This cell-to-cell binding is similar to other types employed by various T cells. That is, the two types of hT-cell receptors bind simultaneously to their corresponding HLA proteins and antigens on the surfaces of the B cells and their progeny.

Several days after the antigen is detected by the T cells and B cells, many fully activated plasma cells are produced and secrete antibodies profusely. Each plasma cell may continue antibody production for up to 1 week, after which it dies. Exhausted plasma cells may be replaced by new ones.

FIGURE 15.6 Activation and activities of B cells. (1) Antigen binds to B cells (B) that have specific receptors for the antigen. (2) B cells with bound antigen join hT cells (hT) that have specific receptors for the antigen, using HLA receptors and antigen-specific receptors. (3) hT cells release IL-2 (dashed arrow), which stimulates B cells to reproduce. (4) Stimulated B cells produce mB cells (mB) and plasma cells (p). (5) Plasma cells produce antibodies that have specific bonding sites for the antigen. (6) Antibodies bind to the antigen.

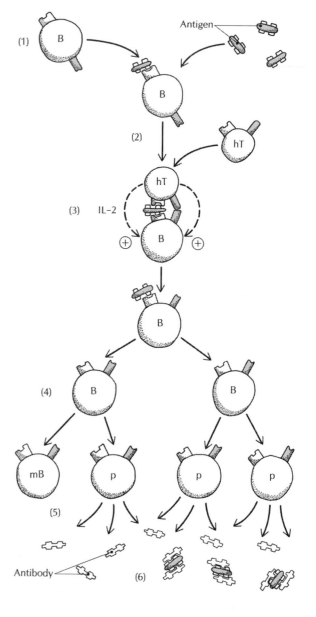

Antibodies   All antibodies from a plasma cell have the same antigen specificity as did the B cell that first bound to the antigen. Therefore, these antibodies also bind to the antigen wherever the two meet. However, different classes of antibodies are produced and concentrated in different places in the body.

The different antibody classes are designated by different letters. IgA antibodies are concentrated in the secretions from mucous membranes lining body systems (e.g., respiratory system, digestive system). When they bind to antigens, they help block the entry of the antigens into the body. Most antibodies in the IgM and IgG classes are found in blood and lymph. IgM and IgG prevent injury from antigens in several ways. For example, they cause some antigens to become more easily phagocytized by clumping them together and coating them with phagocyte-stimulating substances. IgM and IgG chemically neutralize other antigens. They also lead to the destruction of other antigens by activating a group of substances in the blood called the *complement system*. Complement substances can kill antigenic cells such as bacteria directly and intensify defense activities by promoting phagocytosis and inflammation. IgE antibodies bind to mast cells. When antigen later binds to this IgE, the mast cells release histamine and cause inflammation. Antibodies assist only in fighting antigens; they do not destroy antigens.

Memory

We have seen that an antigen causes the production of a large number of specialized T cells, plasma cells, and antibodies that have specificity. It takes several days to produce enough of these cells and antibodies to combat a large dose of antigen the first time it is encountered. This is called a *primary immune response* (Fig. 15.7).

Secondary Immune Response   Many specialized T cells and plasma cells produced during the primary immune response are eliminated once the antigen has been reduced. The remaining specialized T cells constitute a cadre of *memory T cells* (*mT cells*). The mB cells and much of the antibody produced during the primary immune response also remain, though the amount of antibody declines over a period of weeks. Since memory cells are abundant and specialized, they swiftly produce many specialized T cells and plasma cells if the antigen is detected again. Furthermore, the antibody level rises precipitously, reaching a value far above the peak level attained during the primary immune response. Therefore, the old antibody, along with the new specialized T cells and the antibody from new plasma cells, produces a more rapid and intense attack on the antigen if it appears in the body again. An immune response to an antigen encountered a second or subsequent time is called a *secondary immune response*. This

FIGURE 15.7 Primary immune response.

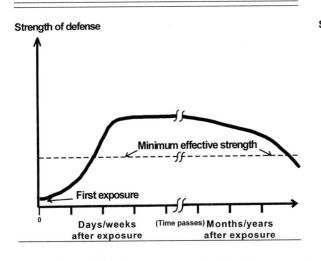

FIGURE 15.8 Secondary immune responses.

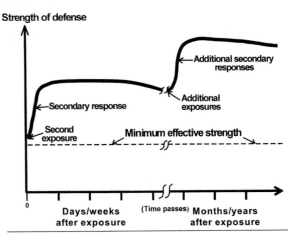

response may effectively eliminate an antigen within 1 day of its detection by the immune system.

With each secondary immune response to an antigen, additional memory cells and antibody for that antigen may accumulate and the antibody level may decline more slowly. When this happens, each subsequent *secondary immune response* is faster and more effective than the previous one. This further decreases the risk of sustaining injury from the antigen each time it is present. It can also cause allergic responses to become worse with repeated exposure to certain antigens (e.g., penicillin, bee stings) (Fig. 15.8).

Acquired Active Immunity   sometimes the formation of memory is so effective that only one exposure to an antigen prompts complete and permanent resistance. For example, most people have measles or chickenpox only once. Once the secondary immune response is strong enough to prevent significant adverse effects from the next encounter with an antigen, the person has an *acquired active immunity* against that antigen. Such immunity may result when the first dose of antigen is sufficient to cause serious illness, as occurs with full-blown measles. However, this immunity is sometimes acquired after a person receives an antigen in one or more doses that are not strong enough to cause significant illness. Acquired active immunity against antigens such as polio, tetanus, diphtheria, and influenza can be intentionally induced in this way, using *vaccines* that contain the antigen.

Though acquired active immunity against antigens such as polio usually lasts indefinitely, the memory cells and antibodies for other antigens may diminish substantially if the cells do not encounter the antigen (e.g., tetanus) again for months or years . Then the antigen may cause approximately the same degree of injury that it did when it was first encountered because several days may be required to produce enough of an immune response to eliminate it. This sometimes can be prevented by receiving a vaccine again (a booster dose) at appropriate intervals.

AGE CHANGES

Distinguishing age changes from other changes in the immune system is difficult for various reasons. Reasons include limited understanding of this complex system; confounding influences from changes in nonspecific defense mechanisms; diverse and rapidly changing methods of research; controversy over the interpretation of results; and diversity in the factors affecting it. Examples of such factors include chronic exposure to sunlight; cirrhosis; malnutrition; diabetes mellitus; cancer; chemotherapy; radiation therapy; anesthesia; surgery; and stress. Common age-related increases in most of these factors reduce immune system effectiveness. Finally, comparing immune systems between individuals reveals an age-related increase in heterogeneity. Therefore, unless otherwise noted, the immune system changes described below represent average age-related changes that may be due in part to aging of the system.

Trends

As the age of a population increases, the proportion of people with a declining immune function increases and the level of immune function within the individual decreases. Decreases in immune function include reductions in the speed, strength, and duration of immune responses and in the regulation of the immune system.

Developmental Changes

Most developmental changes in the immune system occur before the end of puberty and result in the formation of a mature immune system. The few developmental changes after this period may be considered age changes, most of which result in declining immune system functioning.

Macrophages and Langerhans Cells   Age changes and age-related changes in macrophage production and numbers have not been well studied, suggesting that these changes are small. However, the rate of production of Langerhans cells falls below the rate of destruction, leading to significant decreases in cell numbers. Areas of skin chronically exposed to sunlight have much greater reductions than do unexposed areas. These reductions lead to decreases in the processing and presentation of antigens in the epidermis, causing increasing risks of skin infection and cancer and decreasing delayed-hypersensitivity reactions, including allergic reactions. The latter change causes a reduction in signs and symptoms (e.g., local swelling, itching, rash) that warn of the presence of potentially harmful substances (e.g., topical medications) and are used in skin tests to detect previous exposure to tuberculosis (e.g., tine test).

Thymus and T Cells The thymus begins to shrink during or shortly after puberty and continues to do so until about age 50, when it may be only 5 percent of the original size. Thymic hormone production declines in a parallel fashion, and circulating levels of these hormones reach zero by age 60. These changes cause a precipitous decline soon after puberty in the conversion of unspecialized lymphocytes to T cells and the clonal selection of T cells. Lymphocyte conversion and clonal selection for T cells finally cease, ending the development of immune response capabilities against additional antigens.

Though changes in T cells in human lymph tissues have not been well studied, healthy people seem to retain a steady rate of production of antigen-specific T cells in lymphoid tissue since the total number of T cells in the blood remains stable. However, declining thymic hormone causes reduced maturation of T cells, resulting in a decrease in the ratio of mature to immature T cells in the blood. This change decreases the number of circulating T cells that can respond to and combat each antigen. The degree of change varies considerably between individuals because of differences in aging, abnormal conditions, diseases, and other factors. Up to 25 percent of older people may show no decrease in T-cell functioning, while approximately 50 percent have moderate declines. The remaining 25 percent experience major decreases in T cell responses to an antigen.

B Cells As with T cells, B-cell production from B-cell clones that were established early in life seems to remain steady in many people since the total number of these cells in the blood usually remains stable. However, the number of circulating B cells decreases in some individuals. Ordinary changes in B-cell numbers are not important since limitations in B-cell functioning derive from decreases in T-cell stimulation of B cells rather than from changes in B cells themselves.

Immune Responses

Decreases in immune system function result not only from developmental changes in the immune system but also from age-related changes in immune responses.

Processing and Presentation Aging seems to have little or no effect on the functioning of macrophages and Langerhans cells during the initial processing and presentation stage of an immune response. However, the effectiveness of macrophages and Langerhans cells during an immune response decreases with aging because they receive less IL-2 stimulation from hT cells that bind to them. The effectiveness of Langerhans cells also declines because their total number decreases with age.

T-Cell Participation Age-related changes in the ability of macrophages and Langerhans cells to convert T cells into specialized T cells (hT cells, etc.) are unclear. However, when T cells from older people are experimentally stimulated to reproduce, fewer T cells can reproduce and those which reproduce do so fewer times. These decreases seem to be caused by a combination of declining production of IL-2 and declining T-cell responsiveness to IL-2. Since the progeny from T cells control all subsequent aspects of the immune response, age-related reductions in T-cell proliferation lead to reductions in all the subsequent parts of an immune response. Part of the decline in cell reproduction may result from loss of telomeres.

Regardless of reductions in the proliferation of T cells or specialized T cells, there is an age-related decline in both the production and effectiveness of IL-2 from hT cells. This decline stems from a decrease in IL-2 receptors on T cells and specialized T cells. By reducing the positive feedback effects of IL-2, these changes diminish the intensity of both the cell-mediated and humoral parts of an immune response, reducing all its defensive capabilities. Delayed-hypersensitivity responses also decline, reducing their roles in defense and as warning mechanisms.

Since aging decreases the ratio between TH-1 hT cells and TH-2 hT cells, immune responses become unbalanced. This loss of balance causes reductions in certain aspects of the response while producing excesses in other aspects including IL-6 production and autoimmune responses. The extra IL-6 contributes to loss of bone matrix and unwanted inflammation. The excess inflammation contributes to increased damage from free radicals and age-related diseases (e.g., atherosclerosis, arthritis, Alzheimer's disease, kidney disease). These and other adverse effects from weaker, unbalanced, and poorly regulated immune responses form the basis for the immune theory of aging (see Chapter 2).

The average activity level of NK cells, which receive stimulation from cT cells, does not change. However, total NK-cell activity becomes more heterogeneous between individuals. Additional NK-cell heterogeneity develops because of individualized increases and decreases in NK cells for different types of cancer. People with a reduction in NK-cell activity may have an increased risk of developing cancer.

Finally, age-related decreases in IL-2 may contribute to a decline in sT-cell numbers or effectiveness, which may be a main factor in the age-related reduction in the regulation of the immune system. This reduction is evident as an increase in the production of *autoantibodies*, which are antibodies against self-antigens. No significant consequences from autoantibodies resulting from aging have been discovered. However, while they are suspected of contributing to age-related detrimental changes such as seminiferous tubule degeneration, they may be beneficial by helping rid the body of abnormally altered proteins. Autoantibodies resulting from processes other than aging may contribute to abnormal or disease conditions such as rheumatoid arthritis.

B-cell Participation and Antibodies   Aging seems to have little or no direct effect on the ability of B cells to bind to antigen, be activated, or perform their other functions during a primary immune response. As mentioned previously, however, all aspects of B-cell participation decline with aging because B cells receive less IL-2 stimulation from hT cells. Consequently, more antigen is needed to prompt antibody production, antibody production is slower, antibody production ends sooner, a lower peak antibody concentration is achieved, and the antibody level declines faster. Furthermore, the effectiveness of antibodies against certain antigens declines because some antibodies bind less well to their antigens and because of the increased variability in the proportions of the different classes of antibodies. Of note is a decline in IgE. Finally, there is an increase in autoantibody production.

The first six changes contribute to the age-related decline in the effectiveness of primary immune responses. The decline in IgE contributes to the age-related decline in allergic reactions. All changes in B cells and antibodies develop slowly until approximately age 60, after which they occur more rapidly.

Memory   As primary immune responses diminish with aging, they leave the body with fewer memory cells and less residual antibody for memory. Furthermore, residual antibody dissipates faster. As immune memory declines with aging, the speed and strength of the initial secondary immune response against an antigen also decline. Therefore, the higher the age at which an antigen is first encountered, the greater injury caused by a second encounter with that antigen. The antigen also may cause injury many times because additional secondary responses may be needed before acquired active immunity develops. Because of these changes, aging is accompanied by a decline in the effectiveness of initial vaccinations.

Though memory produced from primary immune responses declines substantially with aging, there is much less of a decline in the ability to maintain memory produced during youth or young adulthood. Therefore, secondary immune responses resulting from such early memory remain effective, especially if the antigen is encountered occasionally, as occurs with booster doses of vaccines.

Furthermore, since the decline in establishing memory becomes particularly evident after age 60 and advances more rapidly afterward, vaccinations should be received well before age 60. However, vaccines can be beneficial at any age, especially for those who are weakened by other factors and are at high risk for exposure to certain bacterial pneumonias (e.g., pneumococcal pneumonia) or strains of influenza virus.

Consequences

In conclusion, age changes in the immune system contribute to a decline in the ability to maintain homeostasis because they decrease resistance to harmful foreign materials and lead to an increase in the incidence and severity of infections and cancer. The risks increase with the age at which antigens or carcinogenic factors are first encountered. The risks also increase, though less so, with the number of years between encounters with an antigen. The effects on the immune system of many other age-related factors magnify these consequences, as do many age-related changes in nonspecific defense mechanisms. By contrast, the undesirable effects of allergic reactions decrease with aging.

There is an increased incidence of renewed injury from the bacteria causing tuberculosis (TB) and the virus causing chickenpox. In both cases the disease-causing agent may reside within body cells indefinitely, where it is hidden from immune cells after the disease seems to have disappeared. As immune memory against these diseases fades and age-related changes and factors such as stress weaken the immune system, the bacteria or virus is no longer held in check. TB bacteria, which reside in lung cells, may then cause a reactivated infection and additional lung damage. The chickenpox virus, which resides close to the spinal cord in sensory neurons, will be transported down sensory neurons to the areas of the skin they serve. Once there, the virus can cause excruciating pain and severe skin eruptions known as *herpes zoster* or *shingles*.

Finally, age changes in the immune system increase the progress and the severity of effects from HIV infection.

Minimizing Consequences

Since the consequences of declining immune system effectiveness often lead to a reduced quality of life and lower life expectancy, researchers are seeking ways to prevent, reduce, or delay the deterioration of the immune system caused by aging. Though some success has been achieved in animals (e.g., diet regulation), no practical and effective methods for humans are available. Other research is aimed at restoring immune system effectiveness lost because of age changes. Studies with animals involving supplements (e.g., thymic hormones, sex hormones) and other drugs have been somewhat successful, but safe and effective methods for aging humans have not been developed. A potential hazard from stimulating immune functioning is the activation of harmful immune activities (e.g., autoimmunity, allergic reactions) along with beneficial ones.

Though there are no practical methods for controlling aging of the human immune system, steps can be taken to help minimize other undesirable changes in this system, including avoiding or reducing factors known to suppress immune system functioning. Other actions may reduce the risks of developing the adverse consequences of decreases in immune functioning. These actions include receiving vaccinations in a timely fashion and minimizing exposure to potentially harmful agents such as bacteria, viruses, and carcinogens. Finally, risks can be reduced by preventing or treating abnormal conditions and diseases that promote infections and cancer.

ABNORMAL AND DISEASE CONDITIONS

Recall that aging is accompanied by a decrease in the balanced regulation of the immune system and an increase in the production of new autoantibodies. The autoantibodies produced as part of aging are of little known importance. In addition, there is only a small increase in the incidence of new-onset autoimmune diseases among the elderly. However, the damage that seems to result from excess inflammation and autoimmune response activities that are part of abnormal or disease conditions often become more serious with age. This occurs largely because chronic inflammation and many autoimmune responses initiated during childhood or young adulthood continue to injure or destroy body components for many years. Some age-related diseases associated with chronic inflammation and autoimmune responses are atherosclerosis, valvular heart disease, chronic obstructive pulmonary diseases, Alzheimer's disease, and chronic renal diseases.

Some abnormal autoimmune responses follow a steady course and cause unremitting progressive damage (e.g., atrophic gastritis). Other autoimmune diseases occur as periodic or occasional flare-ups separated by periods of remission (e.g., rheumatoid arthritis). In these disorders the affected individual's condition worsens in a stepwise manner. However, in all autoimmune disorders there is great variability among individuals regarding the rate at which deleterious effects develop.

A few abnormal autoimmune conditions important in the elderly have been described in Chaps. 9 and 10 (e.g., atrophic gastritis, rheumatoid arthritis). Other abnormal and disease conditions important to older people that seem to involve autoimmune responses and are serious and relatively common will be mentioned briefly here.

Bullous pemphigoid causes blistering of the skin and accompanying itching and discomfort. *Rheumatic heart disease* results when rheumatic fever leads to autoimmune damage to valves in the heart. Common outcomes include the failure of one or more chambers of the heart and respiratory problems from pulmonary edema. *Multiple sclerosis* involves patchy deterioration of myelin in the CNS and can result in diverse deficits depending on the portions of the CNS affected. This disorder is characterized by flare-ups and remissions of varying duration. *Myasthenia gravis* involves autoimmune damage to receptors for acetylcholine at neuromuscular junctions and leads to progressive muscle weakness and paralysis, including the muscles for respiration. *Regional enteritis (Crohn's disease)*, which often involves flare-ups and remissions, causes inflammation of the small intestine and large intestine and often results in decreased absorption of nutrients, pain, and diarrhea. *Ulcerative colitis* is similar to regional enteritis; though it affects only the large intestine, it often causes intestinal bleeding and increases the risk of developing colorectal cancer. *Graves' disease* results in abnormally high blood levels of thyroid hormone which increase the metabolic rate and cause bulging and deterioration of the eyes.

16

INTO THE FUTURE

People want to live long lives. People want to live healthy lives. People want to live quality lives. Many people want to bear children. These facts lead to higher mean longevities (MLs) and ongoing attempts to increase maximum longevity. (For a few people, increasing their MLs and XLs even means preserving their dead bodies by freezing so they can be revitalized in a future better time.) Chapter 1 described results in terms of past and projected births, birthrates, death rates, mean longevities, and population trends. The trends focused on changes in the population of elders.

Results from these facts can also be viewed as changes in *survival curves*. A survival curve shows the percentages of an initial population that are still alive as the age of the population increases. Survival curves often begin with a population at birth and follow that population through time. Survival curves can also be developed mathematically for people at different ages (Fig. 16.1).

In the year 1900, the birthrate and the number of births in the U.S. were high, but many children died during the first year of life. Many others died before reaching adulthood. For many who reached adulthood, life still contained many risks. Risks were high because of inadequate public health, crowded urban living, low economic status, limited education, dangerous working conditions, and limited knowledge of techniques for preventing and curing diseases. Death rates remained substantial at all ages. Therefore, the percentage of the population born in 1900 continued to decline significantly as the age of that population increased (Fig. 16.1).

During the century following 1900, conditions that contributed to high death rates at all ages improved. Infant mortality rates plummeted, and death rates at all ages declined. By 1990, almost all children survived their first year of life and childhood. Those that reached adulthood continued to have a much lower death rate. Therefore, the survival curve for a population born in 1990 is more rectangular in shape than is the survival curve for the population born in 1900. Mean longevity at all ages increased (Fig. 16.1).

People making projections believe that many conditions affecting ML will continue to improve, and ML will continue to increase. This book contains many suggestions for increasing ML. If conditions do improve, the survival curves for future populations in the U.S. will be even more rectangular (Fig. 16.1).

If ML increases as projected, the total population will gain many more elders, and they will make up a much larger proportion of the total population. If birthrates do not decline much, the size of the total population will also increase because many more people would be living longer. Using moderate estimates, the U.S. population will increase from 249 million in 1990 to 298 million in 2010, 347 million in 2030, and 394 million in 2050. Using the 1990 population as the reference value, these increases are 20 percent, 40 percent, and 58 percent respectively. If conditions do not improve from those in 1990, mean longevities will not increase, but the numbers of elders and the total population will still increase because of continued births.

What would happen if conditions were not changed but only fundamental aging processes and age changes could be slowed? Caloric restriction seems to slow age changes, resulting in increases in maximum longevity (XL) for animals. If caloric restriction or some other technique slowed human aging and people adopted that technique, the human XL would increase. Most people would still die at the usual ages from the usual causes of death. These causes are not age changes (e.g., atherosclerosis, Alzheimer's disease, diabetes mellitus, cancer). However, some people would survive longer than ever before. The survival curve would become less rectangular at high ages (Fig. 16.2). This change would add a small number of very old people to the total population. Also, the percentage of elders in the total population would increase slightly, and the total population would increase slightly.

What would happen if conditions improved AND aging processes were slowed? Both ML and XL would increase. The survival curve would have little change in shape, but it would become extended (Fig. 16.3). The number of elders would increase; there would be many elders who are very old; and elders would make up a larger proportion of the total population. The total population would increase substantially. The total population would become much larger than if only ML were increased.

If any of these three scenarios occurred globally, the same trends in the total population on Earth would occur. In fact, the first scenario has occurred. Historically, survival curves for humans have become much more rectangular as MLs have increased. The human ML in prehistoric times was probably under 20 years. The ML in the Roman empire at its peak was 23 years. By 1800, the ML in England was approaching 40 years. By 1900, the ML in the U.S. was 47 years. Of course, MLs in less developed areas were probably much lower and continue to be so. However, human MLs are increasing across the globe. Elders are increasing in numbers and as a percentage of the total population. The total human population is growing so rapidly that many people speak of it as a population bomb that is exploding (Fig. 16.4).

These trends create complex and difficult challenges. Present examples include providing adequate living accommodations, health care, income, and quality of life for elders. These challenges will likely grow in size and in complexity because aging and age-related changes in diverse realms interact (e.g., biological, social, psychological, economic, spiritual). Moreover, the challenges related to elders amplify challenges for society as a whole. As more elders live longer with less disability and disease, they not only increase problems and challenge, they also can have a larger role in solving and meeting them.

Problems and challenges create needs and opportunities. Here are a few problems and challenges that are creating needs and opportunities related to aging.

To deal with the growing population of elders, people must be better educated about all aspects of aging (e.g., biological, social, psychological, economic, spiritual). Gerontology research and gerontology education must expand and become more multidisciplinary and interdisciplinary. Gerontology must become more multicultural, cross cultural, and international.

To accommodate present and future changes in demographics, people who help formulate economic and public policy must be more creative and truly effective. The elders of the future are already here.

People seeking careers should consider the burgeoning opportunities brought about by the enlarging elder population. Of course, a background in gerontology can be invaluable. Almost any career path can be modified to take advantage of and serve the special concerns, needs, and desires of elders. Beyond health care, diverse ex-

FIGURE 16.1 Survival curves: 1900, 1990, 1990 plus increase in ML.

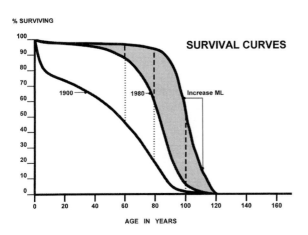

FIGURE 16.3 Survival curves: 1900, 1990, 1990 plus increases in ML and LS.

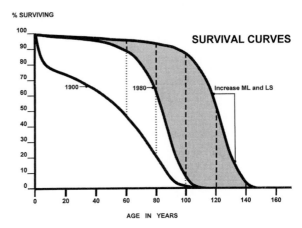

FIGURE 16.2 Survival curves: 1900, 1990, 1990 plus increase in LS.

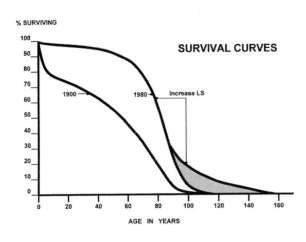

FIGURE 16.4 World population growth, past and projected.

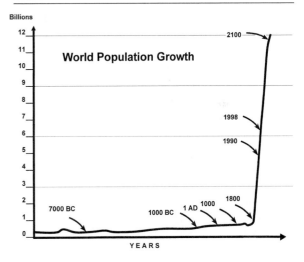

amples include architecture, art, community and regional planning, education, law, marketing, management, recreation and leisure activities, and creating ways by which elders can be contributors to society.

Finally, people must take more responsibility for protecting the environment as the global population expands rapidly. A desirable human civilization and quality of life cannot exist without a healthy ecosystem.

Now you know. What will you do about it?

GLOSSARY

*FR - free radical

*LP - lipid peroxide

*NO - nitric oxide

*O2⁻ - super oxide free radical

*OH - hydroxyl free radical

β-A - beta- amyloid

Absorption - the passage of materials contained within the GI tract into the circulatory system

Accommodation - the process of adjusting the curvature of the lens to focus light from objects located at different distances from the eye

Acetylcholine - a neurotransmitter used in numerous activities (e.g., memory; controlling skeletal muscle, cardiac muscle, smooth muscle, and certain glands)

Acquired active immunity - the ability to produce a secondary immune response that is strong enough to prevent significant adverse effects from any subsequent encounter with the antigen

Action potential - the process by which an impulse travels along a neuron or a muscle cell

Acute gastritis - an abnormal condition consisting of short-term inflammation of the stomach

Ad libitum (AL) - with reference to diet, being able to eat as much as desired at any time

AD - Alzheimer's disease

ADH - antidiuretic hormone

Advanced glycation end-product (AGE) - a protein chain containing glucose cross-links

Afterimage - a lingering perception that an image is present after the actual image in the eye has changed or disappeared

Age changes - developmental changes that occur in the later years of life

AGE - advanced glycation end product

Age-related macular degeneration (AMD) - a disease of the eye that reduces vision by causing degeneration of the cones in the macular region of the retina

Aging - developmental changes that occur in the later years of life

AL - ad libitum

Allergic response - an excessive undesirable immune response against a foreign antigen

Alveolus (pl., **alveoli**) - a small cup-shaped outpocketing in the lung where gas exchange occurs

Alzheimer's disease (AD) - a form of dementia characterized by memory loss and by the presence of excessive senile plaques and neurofibrillar tangles in the brain

Amino acid - a molecular unit that makes up protein molecules

Amyloid - a specific type of protein found within or between cells

Amyloid precursor protein (APP) - a protein in cell membranes whose breakdown yields beta-amyloid and other products

Anemia - an abnormal condition consisting of a low concentration of red blood cells in the blood

Aneurysm - an abnormal outpocketing of either a blood vessel or the wall of the heart

Antagonistic pleiotropy - having several effects that oppose one another

Antibody - a protein molecule that is produced by immune system plasma cells and that adheres to antigens and assists in combating antigens

Antigen - a substance that causes an immune response

Antigen-specific receptor - a type of surface receptor on a T-cell that allows that T-cell to bind to one and only one substance capable of initiating an immune response

Antioxidants - substances that convert free radicals into substances that are not free radicals by donating one or more electrons to other molecules

APOE - apolipoprotein E

Apolipoprotein E (APOE) - a lipoprotein that helps move cholesterol and other lipoproteins from cell to cell and through cell membranes, and also seems to help in neuron development and repair

Apoptosis - programmed cell death

APP - amyloid precursor protein

Appendicitis - an abnormal condition involving inflammation of the appendix

Aqueous humor - the liquid that fills the region in the eye behind the cornea and in front of the lens

Arthritis - an abnormal condition that involves inflammation of joints between bones

Ascites - the presence of extra fluid in the abdominal cavity

Atherosclerosis - a disease involving the formation and enlargement of a weak scar-like material in the walls of arteries

ATP (adenosine triphosphate) - a nucleotide used for transferring energy within a cell

Atrophic gastritis - an abnormal condition resulting in excessive thinning of the lining of the stomach

Auerbach's plexus - a network of nerve cells in the wall of the GI tract that helps to control functions of the GI tract

Autoantibody - an antibody that can act against a native body material

Autoimmune reaction - an immune response against a normal body part

Autonomic motor neuron - a motor neuron involved in controlling either cardiac muscle, smooth muscle, or a gland

Axon - a neuron projection that conducts impulses away from the nerve cell body of that neuron

B-cell - a type of specialized lymphocyte that can produce antibody-secreting cells as part of an immune response; B-lymphocyte

Basement membrane - a non-cellular layer between a layer of cells and the material under the layer of cells

Basilar membrane - a flexible membrane within the cochlea of the ear that supports the organ of corti

Bed sore - a patch of skin that has died because it received insufficient blood flow; a decubitus ulcer

Benign neoplasm - a neoplasm that does not spread to other areas

Benign prostatic hypertrophy (BPH) - an abnormal non-cancerous condition in which the prostate gland grows and becomes excessively enlarged

Beta-amyloid (β-A) - a protein of unknown function that is abundant in senile plaques

Bile - a liquid secreted by the liver and stored in the gall bladder and that contains bilirubin, cholesterol and other waste materials

Bilirubin - a substance consisting of remnants of hemoglobin molecules from red blood cells

Biological aging - aging in the physical structures and functioning of the body that affects either the body's ability to survive or its appearance

Biorhythm - a cyclic fluctuation or oscillation of a biological process

BMR - the rate of energy use needed to sustain body functions when a person is awake and in a state of complete rest; basal metabolic rate

Bone matrix - the material between bone cells in bone tissue

Bone marrow - the specialized tissue that is housed within bones and that produces blood platelets and some types of blood cells

BPH - an abnormal non-cancerous condition in which the prostate gland grows and becomes excessively enlarged; benign prostatic hypertrophy

Bruch's membrane - a non-cellular membrane that lies between the pigmented epithelium of the retina and the choroid layer

Caloric restriction (CR) - receiving a diet that is limited in the number of calories it contains

Cancer - a disease consisting of the uncontrolled reproduction and spreading of cells; a neoplasm that spreads to other areas; a malignant neoplasm

Carbohydrate - a molecule consisting of one or more simple sugar molecules

Cardiac output (CO)- the amount of blood pumped by each ventricle of the heart in one minute

Cardiac muscle - the type of muscle that is found only in the heart

Cataracts - the condition of having so many opacities close to the center of the lens of the eye that vision is greatly impaired

Cavity - a spot of decay on a tooth

CCK - a hormone from the small intestine that stimulates emptying of the gall bladder; cholecystokinin

Cells - the living structural and functional units of the body

Central nervous system (CNS) - the combination of the brain and the spinal cord

Cerumen - a semi-solid waxy material that is secreted into the ear canal by ceruminous glands; ear wax

Cervix - the narrow lower portion of the uterus

Chain reaction - with free radicals, a sequence of chemical reactions where free radicals react with substances that are not free radicals and form new free radicals

Chronological aging - the passage of time since birth

Cilia - microscopic hair-like structures that project from cells and that swing back and forth

Circadian rhythm - a cyclic fluctuation or oscillation that repeats itself approximately every 24 hours

Cirrhosis - an abnormal condition characterized by the presence of much scar tissue within the liver because of chronic liver damage

Clonal selection - the process of destroying lymphocytes capable of binding to native body materials and initiating an immune response against native body materials

CNS - central nervous system

Collagen - a tough non-elastic fibrous structural protein that is found outside of cells

Colorectal cancer - cancer of the large intestine

Compact bone - the type of bone tissue that is dense and that forms the outer layer of a bone; cortical bone

Compliance - the ability to increase or decrease in volume as a result of a change in pressure

Conduction - the process by which an impulse travels along a neuron

Cones - photoreceptors in the eye that have different responses to different colors of light

Constipation - the abnormal condition wherein periods between bowel movements become abnormally long or defecation requires excessive straining

Coronary artery - a blood vessel that transports blood to capillaries within heart muscle

Cortical bone - the type of bone tissue that is dense and that forms the outer layer of a bone; compact bone

Cosmetic aging - changes in outward appearance with advancing age

Cough reflex - a reflex that causes bursts of air to be expired rapidly to force materials out of airways located below the pharynx

CR - caloric restriction

Creatinine - a waste material from muscle cells

Cross-link - a chemical connection between molecules

Cross-sectional study - a study that places individuals with similar ages into groups and evaluates them one time

Cross-sequential study - a study method that combines the cross-sectional method and the longitudinal method. Cross-sectional studies on age categories are repeated on the same population as time passes.

Crush fracture - a type of fracture caused by compression forces acting on a bone that result in collapse of the bone

Crystallized intelligence - using cognitive skills with familiar learned activities

Crystallized intelligence - using cognitive skills with familiar learned activities

cT-cells - cytotoxic T-cell

Cytoskeleton - a network of microfilaments and microtubules that provides support within a cell

Dead space - the volume of air in the lower airways that does not reach the alveoli during ventilation but that remains in the lower airways during ventilation

Declarative memory - remembering specific facts that a person tried to learn so they could be remembered

Defecation - the elimination of feces through the anus; a bowel movement

Dementia - an abnormal condition that involves a serious decline in memory functioning accompanied by a major decline in at least one other mental function

Dementia with Lewy bodies - a type of age-related dementia characterized by the presence of excess round masses of clumped microfilaments in neurons

Dendrite - a neuron projection that conducts impulses toward the nerve cell body of that neuron

Dentin - a firm layer located internal to the enamel of a tooth

Depth perception - the perception of the distance from the eyes to an object being viewed; binocular vision

Dermis - the layer of the skin underlying the epidermis

Development - developmental changes that occur before birth or during childhood

Developmental changes - irreversible normal changes in a living organism that occur as time passes

DHEA - dehydroepiandrosterone; a steroid hormone of unknown function

DHT - a principal sex hormone in men, much of which is produced by the testes and the prostate gland; 5-alpha-dihydrotestosterone

Diabetes mellitus - a disease that involves inadequate insulin production or inadequate sensitivity of cells to insulin and that results in inhibition of the regulation of blood glucose levels

Diabetic retinopathy (DR) - a disease of the eye that is associated with diabetes mellitus and that reduces vision because of sorbitol accumulation in the retina

Diarrhea - the condition of having more than three relatively liquid and voluminous bowel movements in one day

Diastolic pressure - the lowest blood pressure attained between contractions of the ventricles of the heart

Dietary restriction (DR) - receiving a diet that is limited in the amount of food

Dietary Reference Intakes (DRIs) - a comprehensive method for establishing and evaluating recommended dietary intake recommendation

Diffusion - the movement of materials from an area where they are in high concentration to an area where they are in lower concentration

Diglyceride - a lipid molecule composed of one glycerol and two fatty acid molecules

Disaccharide - a molecule consisting of two simple sugar molecules

Diurnal rhythm - a repetitive cycle of events that occurs once in each 24-hour period; circadian rhythm

Diverticula - deep outpocketings in the wall of the large intestine

Diverticulitis - the abnormal condition of having inflamed diverticula

Diverticulosis - a disease condition consisting of the presence of deep outpocketings in the wall of the large intestine

Dizziness - the sensation that the body is unstable

DM - diabetes mellitus

DNA (deoxyribonucleic acid) - the nucleic acid that makes up genes

DR - dietary restriction

dT-cells - delayed hypersensitivity T-cell

Duodenal peptic ulcer - a peptic ulcer in the duodenum

Early onset AD - a form of Alzheimer's disease with onset before age 65, usually during the sixth decade of life

Economic aging - age changes in financial status

Edema - the presence of extra fluid between cells

Ejaculation - the expulsion of semen from the penis caused by reflexive rhythmic peristaltic contractions of the urethra, other male reproductive ducts, and muscles at the base of the penis

Elastase - an enzyme that breaks down elastic fibers into elastin peptides

Elastin peptides - short pieces of the protein chain that composed elastin

Elastin - a resilient structural protein that is found outside of cells

Electron - small negatively charged particle that moves about at a distance from the nucleus of an atom

Electron transport - a series of steps where electrons lose energy

Embolus (pl., **emboli**) - a mass (usually a blood clot) that is transported within a blood vessel

Enamel - a hard substance that covers the exposed surface a tooth

Endocrine gland - a structure that has hormone secretion as its primary function

Endolymph - the fluid that fills the membranous chambers of the inner ear

Endometrium - the innermost layer of the uterus

Endothelium - the innermost layer of a blood vessel

Endurance - the length of time that vigorous activities can be performed without stopping

Energy - the power that makes things happen; the ability to do work

Energy balance - the condition of having the amount of energy taken in over a period of time equal the amount of energy used during the same period

Enzyme - a protein substance that increases the rate of chemical reactions in a regulated manner

Epidermis - the outer layer of the skin

Episodic memory - remembering the times and places at which events happened

Erectile tissue - the spongy tissue in the penis and the clitoris that fills with blood and becomes firm during sexual arousal

Erection - the swelling, elongating, and stiffening of the penis due to engorgement of the erectile tissue with blood

ERT - the administration of estrogen-like substances to bolster the effects of estrogen in the body when levels of endogenous estrogen become low; estrogen replacement therapy

Estrogen - a principal sex hormone, most of which is produced by the ovaries

Evolutionary theory - a theory that attempts to explain how aging developed in concert with natural selection

Expiration - the movement of air out of the respiratory system; exhalation

Expiratory reserve volume (**ERV**) - the amount of air that can be forcefully expired after expiring the tidal volume

Explicit memory - remembering specific facts that a person tried to learn so they could be remembered

External ear - the region of the ear external to the eardrum

Familial AD (**FAD**) - a form of Alzheimer's disease with onset before age 65, usually during the sixth decade of life

Fat - lipid molecules composed of one glycerol and three fatty acid molecules; triglycerides

Fecal incontinence - the abnormal condition of eliminating feces at inappropriate times at least once per month

Feces - a mixture consisting largely of undigested materials, waste products, and bacteria that is located in the large intestine

Fiber - dietary polysaccharides that cannot be broken down by digestive enzymes

Fibroblast - a cell that can produce and secrete proteins that form collagen and elastin fibers

Fibrocartilage - the type of cartilage that consists of a smooth, slippery, and resilient substance that contains many thick collagen fibers and that has the consistency of hard rubber

Filtration - the separation of small substances from large ones when fluid pressure forces the small substances through the pores in a membrane

Fluid intelligence - using cognitive skills in new situations to learning novel problem solving, motor activities, or reasoning

Foam cell - a fat-filled macrophage in a vessel wall

Follicle - in the ovary, a cluster of cells that surrounds a developing egg cell

Fovea (centralis) - the central region of the macula lutea that contains the highest concentration of cones

Free radical (*FR) - an atom or molecule with an unpaired electron (* = unpaired electron)

FSH - follicle stimulating hormone

Gag reflex - a reflex that causes closure of the openings into the larynx and the esophagus when irritating materials enter the pharynx

Gall stone - a solid mass formed from materials in bile

Gas exchange - the movement of gases between the atmosphere and the blood

Gastric peptic ulcer - a peptic ulcer in the stomach

Gastric refluxing - the movement of stomach contents upward and into the esophagus

Gene - a length of DNA used to direct the formation of an amino acid chain

Genitalia - external reproductive structures

Gerontology - the study of aging

GFR - the rate of filtration through the glomeruli; glomerular filtration rate

GH - growth hormone

GI tract - the portion of the digestive system that consists of a tube extending from the mouth to the anus; gastrointestinal tract

Glare - seeing bright areas in the wrong places in the field of view because of scattered light striking the retina in the wrong places and in a disorganized way

Glaucoma - a disease of the eye that reduces vision by causing an increase in pressure within the eye due to an accumulation of aqueous humor

Glomerulus - a tuft of capillaries in each nephron in the kidney

Glucagon - a hormone from the pancreas that tends to increase blood sugar levels

Glucose - a simple sugar that is found in abundance in the body and that provides a major source of energy for cells

Glycation - using glucose to form cross-links between protein chains

Glycoprotein - a molecule consisting of a combination of carbohydrate and protein

Glycosylation - using glucose to form cross-links between protein chains

Glycoxydation - using glucose to form cross-links between protein chains

Gonads - the primary reproductive organs (i.e., testes in males, ovaries in females), which produce hormones and sperm cells or egg cells

H_2O_2 - hydrogen peroxide

Hair cells - neurons in the inner ear that respond to bending caused by different stimuli (e.g., vibrations, gravity, rotation of the head)

Hayflick limit - the maximum number of divisions a cell can undergo

HCl - an acid secreted by the stomach; hydrochloric acid

HDL - high density lipoprotein

Hearing - the conscious perception of sound vibrations

Heart rate (HR)- the number of times the heart beats in one minute

Heat shock proteins - proteins produced by cells when they are stressed. Heat shock proteins have a protective effect.

Hemorrhoid - a varicose vein in the rectum or the anal canal

Heterochromatin - masses of tightly wound DNA and protein

Histamine - a substance from cells that initiates inflammation

HLA receptor - a surface receptor molecule on a T-cell that is able to bind to specific identification molecules (i.e., HLA protein) found on other body cells

HLA protein - a type of protein molecule on a cell surface that serves to identify that cell as belonging to a specific person

Homeostasis - the state of having proper and fairly steady conditions

Hormone - a substance that result from manufacturing processes in cells, that is secreted into the blood, and that alters the activities of cells in other parts of the body

hT-cell - helper T-cell

Hyaline cartilage - the type of cartilage that consists of a smooth, slippery, white substance having the consistency of hard rubber

Hydrogen peroxide (H_2O_2) - a reactive oxygen species containing two hydrogen atoms and two oxygen atoms

Hydroxyl radical (*OH) - a free radical containing hydrogen atom and one oxygen atom and having 17 electrons

IDDM - the type of diabetes mellitus wherein death ensues unless insulin therapy is provided; insulin-dependent diabetes mellitus

IL-1 - a secretion that stimulates T-cells to produce more identical T-cells during an immune response; interleukin-1

IL-2 - a substance secreted by hT-cells that enhances an immune response by stimulating macrophages to phagocytize more antigen, stimulating the production of more of the hT-cells and cT-cells, and stimulating the proliferation and activity of B-cells; interleukin-2

Il-6 - an interleukin that stimulates inflammation, macrophages, osteoclasts, and other activities

Immune response - the activities of the immune system whereby the system combats an antigen

Implicit memory - remembering specific facts that a person did not try intentionally to learn so they could be remembered

Impotence - the inability of a man to engage in sexual intercourse because the penis is not sufficiently erect to be inserted into the vagina

Incidental memory - remembering information or skills that were self-taught

Inhibin - a sex hormone from gonads that helps regulate sex hormone production by the anterior pituitary gland

Initiation - with free radicals, a chemical that converts a molecule to a free radical

Inner ear - the region of the ear internal to the oval window

Inspiration - the movement of air into the respiratory system; inhalation

Inspiratory reserve volume (IRV) - the amount of air that can be forcefully inspired after inspiring the tidal volume

Insulin - a hormone from the pancreas that tends to decrease blood sugar levels

Insulin-like growth factor (IGF) - signaling substances from cells that stimulate growth and regulate other cell activities

Intercellular material - substances located between cells

Interleukin (IL) - a signaling substance that helps regulate immune responses and other body functions

Interneuron - a neuron that carries impulses from one neuron to another neuron

Interstitial cells (**of Leydig**) - the cells among the seminiferous tubules in the testes that produce sex hormones

Intervertebral disc - a pad of fibrocartilage located between the vertebrae in the spinal column

Intrinsic factor - a substance secreted by the stomach that promotes the absorption of vitamin B12 by the small intestine

Ion - an atom that has lost or gained one or more electrons

Ischemia - the condition of having inadequate blood flow

Jaundice - the condition wherein parts of the body have a yellow or brown color because of the accumulation of bilirubin

Joint capsule - the thick layer of fibrous material that encases a synovial membrane and that helps to bind bones together

Kcal - the unit of energy most often used in measuring the energy content in food; kilocalorie

Keratin - a structural protein found in the outermost layer of the epidermis, in hair, and in the nails of the fingers and toes

Keratinocyte - a cell that can make keratin

Ketoacid - a small molecule resulting from the breakdown of a fatty acid or an amino acid; ketone

Ketoacidosis - the abnormal condition of having excess ketoacids in the body

Ketone - a small molecule resulting from the breakdown of a fatty acid or an amino acid; ketoacid

Lacrimal fluid - a liquid secreted by the lacrimal glands that bathes the front surface of the eye

Lactase - an enzyme that assists in the breakdown of lactose

Lactic acid - a substance produced by the partial breakdown of glucose in cells that have a low amount of oxygen

Lactose intolerance - the condition of producing an insufficient quantity of lactase to properly digest dietary lactose

Lactose - a type of sugar that consists of one glucose molecule joined to one galactose molecule; milk sugar

Langerhans cell - a cell in the epidermis of the skin that carries out phagocytosis and assists in initiating an immune response

Late onset AD - a form of Alzheimer's disease with onset usually after age 60

LDL - low density lipoprotein

Lewy body - a round mass of clumped microfilaments in neurons found in excess numbers in types of dementia including Parkinson's disease

LH - luteinizing hormone

Life expectancy of a population - the average age at which death occurs for the members of a population

Ligament - a cable-like structure composed largely of collagen fibers that binds one bone to another bone

Lipid peroxide (*LP) - a lipid molecule that is a free radical

Lipofuscin - a mixture of certain chemical waste products from normal cell activities

Lipoprotein - a molecule consisting of a combination of lipid and protein

Localization of sound - the perception of the direction from which a sound originates

Longitudinal study - a study that evaluates individuals over a period of time

Lower airways - air passages within the lungs that conduct air to and from alveoli

Lymphocyte - a type of white blood cell that functions as part of the immune system

Lymphokines - a group of substances from lymphocytes that augments the defensive effects of an immune response in several ways including by activating macrophages and certain lymphocytes, by causing inflammation, and by protecting normal body cells from viruses

Macrophage - a cell outside of the blood that performs phagocytosis and that assists in immune responses

Macula (lutea) - the region of the retina that is in line with the center of the cornea and the lens and that has a high concentration of cones

Maillard reaction - a reaction that forms glucose cross-links between protein chains

Malignant neoplasm - a neoplasm that spreads to other areas, cancer

Malnutrition - the condition of lacking nutritional homeostasis

Maltose - a molecule that consists of two glucose molecules linked together

Maturation - developmental changes that result in the transformation of a child into an adult

Maximum longevity (XL)- the greatest age ever achieved by any member of a species

mB-cell - memory B-cell

Mean longevity (ML) - the average age at which death occurs for the members of a population

Melanin - a brown pigment found in the epidermis, in hair, and in other places

Melanocyte - a cell that can make melanin

Memory - with regard to the immune system, the ability of the system to respond to an antigen more quickly and more intensely when the antigen is encountered a second or subsequent time

Menopause - the time when age changes in the ovaries cause menstrual cycles to cease for at least one year

Menstrual cycle - a sequence of events in the uterus that repeats itself and that involves thickening, deterioration, and shedding of the endometrium; uterine cycle

Menstruation - the shedding and passage of the endometrium and blood through the vagina

Metastasis - the spreading of cancer from one location to another

Metastatic cancer - cancer that is transported to other parts of the body through blood, lymph, or other body fluids

Middle ear - the region of the ear between the eardrum and the oval window

Minute volume - the volume of air that is inspired or expired per minute

Mitochondria - organelles that release useful energy for cells

Mitochondrial DNA (mtDNA) - DNA in mitochondria

ML - mean longevity

Molecule - a group of atoms or ions bound together in a specific ratio

Monoglyceride - a lipid molecule composed of one glycerol and one fatty acid molecule

Monosaccharide - a single simple sugar molecule

Motor end plate - the modified area of a muscle cell membrane that contains receptor molecules to receive and respond to neurotransmitter molecules from a motor neuron

Motor unit - the combination of one motor neuron and all the muscle cell it controls

Motor neuron - a nerve cell that conducts impulses that control the actions of a muscle or a gland

mRNA (messenger RNA) - a nucleic acid that carries information to determine the sequence of amino acids in a protein molecule

mT-cell - memory T-cell

mtDNA - mitochondrial DNA

Myelin - a substance that coats axons and dendrites and that allows impulses to travel faster along these structures

Myocardium - the layer of muscle in the heart, which is the middle layer of the heart

Myofilament - a protein filament in a muscle cell that is involved in producing contraction of the cell

Myoglobin - a type of protein in muscle cells that attracts oxygen from the blood into muscle cells and stores oxygen

Myometrium - the muscular middle layer of the uterus

Near point (of accommodation) - the smallest distance from the eye that an object can be located and still be seen clearly

Negative energy imbalance - the condition wherein the amount of energy taken in over a period of time is less than the amount of energy used during the same period

Negative feedback - a process that prevents or reverses changes

Neoplasm - a group of cells that continue cell reproduction in an uncontrolled fashion

Nephron - a unit of structure and function in the kidney consisting of a glomerulus, a Bowman's capsule, and a renal tubule

Neurofibrillar tangle (NTs) - a network of fibers composed of one or two protein fibers in a neuron twisted into a helix

Neuromuscular junction - the site of transmission of an impulse from a neuron to a muscle cell

Neuron - a nerve cell

Neurotransmitter - a substance that transmits an impulse from a neuron to another structure

NIDDM - the type of diabetes mellitus wherein survival is not dependent upon the administration of insulin therapy; non-insulin dependent diabetes mellitus

Nitric oxide (*NO) - a free radical that serves as a signaling substance in the body

NK cell - a non-specific lymphocyte that destroys cancer cells; natural killer cell

Non-enzymatic glycation - reactions forming glucose cross-links between protein chains without the use of enzymes

Norepinephrine - a substance that is used as a neurotransmitter and as a hormone and the causes effects of the sympathetic nervous system other than those in the skin

NT - neurofibrillar tangle

Obesity - having a body weight that is greater than 20 percent more than the desirable body weight and a percent body fat that exceeds 25 percent (men) or 30 percent (women), or having a body mass index of more than 30

Opacities - opaque spots in the lens of the eye

Optic disc - the region of the retina to which the optic nerve is attached

Oral mucosa - the lining of the oral cavity

Organ of Corti - the rows of neurons on the basilar membrane of the inner ear that respond to vibrations

Organ - an organized group of different tissues that performs certain functions

Organelle - a structural component of a cell that is composed of several or many molecules

Osmotic pressure - a measure of the total concentration of dissolved materials in a liquid

Osteoarthritis - a type of arthritis that is characterized by the deterioration of cartilage and the formation of extra bone

Osteoblast - the type of cell in bone tissue that produces bone matrix

Osteoclast - the type of bone cell that removes bone matrix from a bone

Osteocyte - the type of bone cell that remains quiescent unless some severe condition develops in the bone

Osteon - a long thick tube of bone matrix which, when fused with other osteons, composes compact bone tissue

Osteoporosis - a disease of the skeletal system that causes substantial reductions in the quantity and strength of bone matrix

Otolith - a heavy crystal that is embedded in the gel attached to the ends of the protruding hair cells in the utricle and the saccule of the inner ear

Ovarian cycle - a sequence of events in the ovaries that repeats itself and that results in the production of hormones and ova

Overweight - having a body weight that is 10-20 percent greater than the desirable body weight

Ovulation - the release of an ovum from an ovary

Oxidation - removing one or more electrons from an atom or molecule

Oxidative phosphorylation - chemical reactions where phosphate groups and the energy from electrons are transferred to molecules (e.g., ATP)

Pancreatitis - inflammation of the pancreas

Pentosidine - glucose cross-links between certain specific amino acids in protein chains

Peptic ulcer - an abnormal condition consisting of a pit created by the death and pealing away of cells because of the presence of secretions from the stomach

Perfusion - the passage of blood through the vessels in a body part

Perilymph - the fluid that fills the bony chambers of the inner ear

Periodontal disease - a disease at the base of teeth

Peripheral nervous system (PNS) - all parts of the nervous system outside of the brain and the spinal cord, all nerves

Peristalsis - a wave of contraction that travels along a tube-shaped structure and that causes material contained within the tube to move along progressively

Peroxyl radical (*ROO) - a free radical containing an organic molecule joined to two oxygen atoms

Peroxynitrite anion (ONOO-) - a reactive oxygen species that is toxic to cells

Phagocytosis - the process by which a cell engulfs a particle and takes it into the cell

Photoreceptor - a neuron in the sensory retina that responds to light by starting impulses in the form of action potentials

Physiological theory - a theory that attempts to explain how aging occurs in a living animal

Pigmented epithelium - the outer layer of the retina

Plasma - the liquid portion of blood

Plasma cells - a B-cell that produces antibodies

PMN - polymorphonuclear leukocyte (a type of white blood cell)

PNS - peripheral nervous system

Polysaccharide - a molecule consisting of many simple sugar molecules

Polyunsaturated fatty acid (PUFA) - a fatty acid molecule with more than one double bond in the carbon chain

Positive energy imbalance - the condition wherein the amount of energy taken in over a period of time is greater than the amount of energy used during the same period

Post-menopausal osteoporosis - the type of osteoporosis that occurs most commonly in women during the years following menopause; Type I osteoporosis

Presbycusis - the ear disorder consisting of a significant decrease in the ability to hear due to aging

Presbyopia - farsightedness caused by age-related stiffening of the lens

Presenilin - a protein which when mutated is correlated with increased risk of Alzheimer's disease

Primary immune response - the immune response initiated by the first encounter with an antigen

Procedural memory - remembering how to perform a process or series of steps

Progeroid syndrome - a series of manifestations of aging that occur prematurely

Progesterone - a principal sex hormone, most of which is produced by the ovaries

Programmed theory - a theory based on the premise that aging occurs in a directed manner

Propagation - with free radicals, a sequence of chemical reactions where free radical convert other molecules to a free radicals, leaving the original free radicals in an abnormal shape

Prostate gland - a donut-shaped gland in men that surrounds the urethra below its point of connection to the bottom of the urinary bladder

Protein - a molecule composed of one or more chains of amino acids

Psychological aging - age changes that affect the way a person thinks and behaves

PUFA - polyunsaturated fatty acid

Pulmonary embolism - the disease condition of having one or more emboli move into the lungs

Pulmonary edema - an accumulation of fluid in the lungs from blood vessels in the lungs

Pulmonary congestion - an accumulation of blood in pulmonary vessels

Pulmonary vessel - a blood vessel involved in perfusion of the lungs (e.g., pulmonary artery, pulmonary capillary, pulmonary vein)

Pulp - the nerves, blood vessels, and other soft materials in the innermost region of a tooth

RBC - red blood cell

RBF - the amount of blood flow through the kidneys per unit time; renal blood flow

RDA - the recommended dietary intake of a nutrient as established by the Food and Nutrition Board of the National Academy of Science; Recommended Dietary Allowance

Reaction time - the time needed to begin a voluntary motion in response to a stimulus

Reactive oxygen species (**ROS**) - highly reactive substances that contain oxygen

Reception - the process in a neuron of having an impulse generated in response to environmental conditions or to messages from other neurons

Receptor molecule - a molecule in a cell or on a cell membrane that binds to a substance that is carrying a message to the cell

Receptor for advanced glycation end-product (**RAGE**) - a receptor molecule on a cell that bonds to an advanced glycation end product (AGE)

Reflex - an involuntary response to a stimulus

Reinitiation - chemical reactions where molecules that were formerly free radical interact and form one or more new free radicals

Renin - a substance produced by the kidneys that helps to regulate blood pressure, osmotic pressure, and sodium concentration in the body

Replicative senescence (**RS**) - the cessation of cell division by a cell as it ages

Residual volume (**RV**) - the amount of air left in the lungs after expiring as much as possible

Respiratory rate - the number of breaths per minute

Rheumatoid arthritis - a type of arthritis characterized by the destruction of cartilage and bone in a joint

RNA (**ribonucleic acid**) - the nucleic acid used to guide the synthesis of protein

Rods - photoreceptors in the eye that can respond to dim light but that cannot respond differently to different colors of light

ROS - reactive oxygen species

Sarcolemma - the cell membrane of a muscle cell

Sarcomere - a cluster of myofilaments in a muscle cell

Sarcopenia - gradual loss of muscle mass by shrinkage of muscle

Sarcoplasmic reticulum - the membranes within a muscle cell that constitutes its endoplasmic reticulum

Saturated fat - a molecule of fat having fatty acids containing the maximum number of hydrogen atoms

Schwann cells - cells in the peripheral nervous system that form myelin

SDAT - Senile dementia of the Alzheimer's type

Sebaceous gland - a skin gland that secretes sebum

Sebum - an oily substance secreted by sebaceous glands

Secondary immune response - an immune response produced by the second or subsequent encounter with an antigen

Self-antigen - a native body material that can bind to an antigen-specific receptor and therefore can initiate an immune response against a body component

Semen - the mixture of sperm cells and secretions that is released from a man's body during sexual activity

Seminiferous tubule - a long and highly coiled tube that produces sperm cells in the testes

Senescence - age changes that have detrimental effects

Senile plaque (SPs) - a round microscopic masses found near brain synapses and having various mixtures and densities of materials

Senile osteoporosis - the type of osteoporosis that occurs most commonly in people of more advanced age, especially those over age 60; Type II osteoporosis

Senile dementia of the Alzheimer's type (SDAT) - a form of Alzheimer's disease with onset usually after age 601

Sensory neuron - a nerve cell that monitors conditions and conducts impulses toward the brain or spinal cord

Sensory retina - the layer of the retina that is closest to the vitreous humor and that contains several layers of neurons including rods and cones

Skeletal muscle - the type of muscle that is almost always attached to bones of the skeletal system

Sleep apnea (SA) - a condition consisting of having least five temporary cessation of ventilation per hour or exhibiting at least 10 occasions of depressed ventilation and cessation of ventilation per hour when asleep

Sliding hiatal hernia - a structural abnormality wherein a portion of the stomach is situated above the diaphragm and is between the diaphragm and the lower end of the esophagus

Smooth muscle - a type of muscle that lacks striations and that is under involuntary control by the nervous system and the endocrine system; visceral muscle

Social aging - age changes in the interactions people have with others

Somatic motor neuron - a motor neuron that controls skeletal muscle

Sorbitol - a type of sugar that is derived from glucose and that tends to accumulate in certain body parts

SP - senile plaque

Speed of movement - the time from the beginning of a motion to the end of that motion

Spongy bone - the type of bone tissue that consists of an open network of trabeculae that are fused together; trabecular bone

Sporadic Alzheimer's disease - a form of Alzheimer's disease with onset usually after age 60

sT-cells - suppressor T-cell

Stamina - the ability to perform vigorous activity continuously for more than a few seconds

Stochastic theory - a theory based on the premise that aging occurs by random chance events that occur with no direction or control

Stricture - a ring of scar tissue that inhibits the movement of materials through a tube-shaped structure by causing a narrowing of the structure

Stroke - a disease condition involving an injury to or death of brain cells caused by low blood flow or bleeding in an area of the brain

Stroke volume (SV) - the amount of blood pumped by one contraction of a ventricle of the heart

Subcutaneous layer - the layer of the integumentary system that underlies the skin

Superoxide radical (*O2-) - an oxygen molecule containing two oxygen atoms and having 17 electrons

Suprachiasmatic nucleus (SCN) - a region in the hypothalamus of the brain that regulates circadian rhythms

Surface tension - the force of attraction between substances on a surface

Surfactant - a substance that reduces surface tension

Survival curve - a graph depicting the percentage of a population that is still alive as time passes from a reference point

Suspensory ligaments - fibers within the eye that radiate outward from the lens and attach to the ciliary body

Sustentacular cells - cells in the wall of seminiferous tubules that promote sperm production; Sertoli cells

Swallowing reflex - a reflex that clears the pharynx by pushing materials down into the esophagus

Synapse - the site of transmission of an impulse from one neuron to another neuron

Synovial fluid - the fluid that lubricates and cushions the ends of the bones joined by a freely movable joint

Synovial cavity - the space between bones joined by a freely movable joint

Synovial membrane - the membrane that surrounds a synovial cavity and that produces and removes synovial fluid

Systemic vessel - a blood vessel other than one involved in perfusion of the lungs (i.e., other than a pulmonary vessel)

Systolic pressure - the peak blood pressure attained by contraction of the ventricles of the heart

T-cell - a type of immune system cell formed from an unspecialized lymphocyte because of the influence of the thymus on the lymphocyte; T-lymphocyte

Target - with regard to a hormone, a structure that responds to the hormone

τ **-protein** - tau protein

Tau protein (τ–**protein**) - a protein that seems to promote microtubule formation and that is abundant in neurofibrillar tangles

Telomerase - an enzyme that repairs and replaces telomeres on chromosomes during DNA replication

Telomere - a strand of DNA on the end of a chromosome

Termination - a chemical reaction where a free radical is converted to a substance that is not a free radical without forming another free radical

Testosterone - a principal sex hormone, most of which is produced by the testes

Thrombus (pl., **thrombi**) - a stationary blood clot inside a vessel

Tidal volume (**TV**)- the volume of air that is inspired or expired per breath when a person is at rest and breathing quietly

Time-lag study - a study method that uses a cross-sectional study procedure more than one time. The cross-sectional procedure is repeated on different groups that have the same age at different times in the study.

Tinnitus - the perception of sound by a person when there is no sound external to the person

Tissue - an organized group of similar cells and associated intercellular material that performs certain functions

TMJ - the joint that attaches the bone of the lower jaw to the skull; temporomandibular joint

Total lung capacity (TLC) - the total amount of air that the lungs can hold

Trabeculae - small pieces of bone tissue which, when fused together, compose trabecular bone

Trabecular bone - the type of bone tissue that consists of an open network of trabeculae that are fused together; spongy bone

Transmission - the process by which an impulse is passed from a neuron to another structure

Triglyceride - a lipid molecule composed of one glycerol and three fatty acid molecules; a molecule of fat

Type I fiber - a muscle cell that contracts slowly and can work longer than other types of muscle cells before becoming fatigued

Type IIB fiber - a type of muscle cell that contracts quickly and becomes fatigued quickly

Type IIC fiber - a type of muscle cell that contracts at a moderate speed and that becomes fatigued at a moderate period of contraction

Type IIA fiber - a type of muscle cell that contracts quickly and resists becoming fatigued

U.S. RDA - the recommended daily intake of a nutrient as established by the Food and Drug Administration; U.S. Recommended Daily Allowance

Underweight - having a body weight that is below the range for desirable body weight

Unsaturated fat - a molecule of fat having one or more fatty acids that can contain additional hydrogen atoms linked to their carbon atoms

Upper airways - air passages that conduct air into and out of the lungs

Urea - a waste product resulting from the detoxification of ammonia by the liver or from the breakdown of proteins

Urinary incontinence - the elimination of urine at inappropriate times

Uterine fibroid - an abnormal growth consisting of a spherical mass of smooth muscle within the muscular wall of the uterus

Vaccine - a substance that contains an antigen and that is administered to produce an acquired active immunity to that antigen or to antigens that are very similar to it

Varicose vein - a vein that has developed and retains an abnormally large diameter

Ventilation - the movement of air into and out of the respiratory system; breathing

Vertigo - the sensation that either the body or the surrounding environment is spinning when no spinning is actually occurring

Vision - the conscious perception of images formed on the retina of the eye

Visual acuity - the amount of detail that can be seen

Vital capacity (VC) - the maximum volume of air that a person can expire after taking the deepest possible inspiration

Vitamin D - a vitamin that assists in the absorption of calcium by the small intestine

Vitreous humor - the transparent soft gel that fills the region of the eye behind the lens and in front of the retina

˙Vo2max - the amount of oxygen used per kilogram of body weight per minute while engaging in exercise at the fastest rate attainable by the person performing the exercise; aerobic capacity

Voiding - releasing urine from the body

WBC - white blood cell

Work of breathing - the amount of energy used or the amount of oxygen consumed to perform ventilation

Working memory - remembering information at or close to the level of consciousness so it can be used in cognitive processing

XL - maximum longevity

BIBLIOGRAPHY
Second Edition

BOOKS

(no author): "Historical Statistics of the United States: Colonial Times to 1970: Part 1." U.S. Department of Commerce, September (1975).

(no author): "Statistical Abstract of the United States 1997." The National Data Book, 117th Edition, U.S. Department of Commerce, October (1997).

Abrams WB, Beers MH, Berkow R: "The Merck Manual of Geriatrics" Second Edition Merck Research Laboratories, Whitehouse Station, NJ (1995).

Arking R: "Biology of Aging: Observations and Principles." Prentice Hall, Englewood Cliffs, NJ (1991).

Austad SN: "Why We Age: What Science Is Discovering about the Body's Journey Through Life." John Wiley & Sons, Inc., NY (1997).

Hayflick L: "How and Why We Age." Ballantine Books, NY (1994).

Kurian GT: "Datapedia of the United States 1790-2000 America Year by Year" Bernan Press, Lanham, MD (1994).

Papalia DE, Camp CJ, Feldman RD: "Adult Development and Aging." The McGraw-Hill Companies, Inc., NY (1996).

Rowe JW, Kahn RL: "Successful Aging." Pantheon Books, NY (1998).

Schulz R, Salthouse T: "Adult Development and Aging: Myths and Emerging Realities." Third edition, Prentice Hall, Upper Saddle River, NJ (1999).

Weindruch R, Walford, RL: "The retardation of aging and disease by dietary restriction." Charles C. Thomas, Springfield, IL, (1998).

Yu BR ed.: "Modulation of aging processes by dietary restriction." CRC Press, Boca Raton, FL,57-87, (1994).

ARTICLES

(no author): "Births, Marriages, Divorces, and Deaths for November 1997." Monthly Vital Statistics Report, U.S. Department of Health and Human Services, 4(12) July 28 (1998).

(no author): "Births, Marriages, Divorces, and Deaths for November 1996." Monthly Vital Statistics Report, U.S. Department of Health and Human Services 45(11) May 15 (1997).

(no author): "Consensus Recommendations for the Postmortem Diagnosis of Alzheimer's Disease." Neurobiology of Aging (1997) 18(S4):S1-S2.

(no author): "Report of the Working Group on: 'Molecular and Biochemical Markers of Alzheimer's Disease.'" Neurobiology of Aging (1998) 19(2):109-116.

(no author): "Resident Population of the United States: Estimates, by Age and Sex." U.S. Bureau of the Census, August 28 (1998).

(no author): "Sixty-five Plus in the United States." U.S. Census Bureau: the Official Statistics Economics and Statistics Administration, U.S. Department of Commerce, May (1995).

(no author): "Trends in the Health of Older Americans: United States, 1994." Vital and Health Statistics, U.S. Department of Health and Human Services, Series 3: Analytic and Epidemiological Studies No. 30, DHHS Publication No. (PHS) 95-1414 April (1995).

(no author): "Dietary reference intakes." Nutrition Reviews (1997) Sep;55(9):319-326.

(no author): "Uses of dietary reference intakes." Nutrition Reviews (1997) Sep;55(9):327-331.

Ajmani RS, Rifkind JM: "Hemorheological changes during human aging." Gerontology (1998) 44(2):111-120.

Akisaka M, Suzuki M, Inoko H: "Molecular genetic studies on DNA polymorphism of the HLA class II genes associated with human longevity." Tissue Antigens (1997) Nov;50(5):489-493.

Allsopp RC: "Models of initiation of replicative senescence by loss of telomeric DNA." Experimental Gerontology (1996) Jan;31(1-2):235-243.

Amar K, Wilcock GK, Scot M, Lewis T : "The presence of leuko-araiosis in patients with Alzheimer's disease predicts poor tolerance to tacrine, but does not discriminate responders from non-responders." Age Ageing. (1997) Jan;26(1) 25-9.

Aminoff T, Smolander J, Korhonen O, Louhevaara V: "Physical work capacity in dynamic exercise with differing muscle masses in healthy young and older men." European Journal of Applied Physiology and Occupational Physiology (1996) 73(1-2):180-185.

Arvat E, Camanni F, Ghigo E: "Age-related growth hormone-releasing activity of growth hormone secretagogues in humans." Acta Paediatrica. Supplement (1997) Nov;423:92-96.

Arvin AM:" Varicella-zoster virus: overview and clinical manifestations." Seminars in Dermatology (1996) Jun;15(2 Suppl 1):4-7.

Atchison DA: "Accommodation and presbyopia." Ophthalmic and Physiological Optics (1995) Jul;15(4):255-272.

Aviv A, Aviv H: "Reflections on telomeres, growth, aging, and essential hypertension." Hypertension (1997) May;29(5):1067-1072.

Aviv JE: "Effects of aging on sensitivity of the pharyngeal and supraglottic areas." American Journal of Medicine (1997) Nov 24;103(5A):74S-76S.

Bains JS, Shaw CA: "Neurodegenerative disorders in humans: the role of glutathione in oxidative stress-mediated neuronal death." Brain Research. Brain Research Reviews (1997) Dec;25(3):335-358.

Balagopal P, Proctor D, Nair KS: "Sarcopenia and hormonal changes." Endocrine (1997) Aug;7(1):57-60.

Balagopal P, Rooyackers OE, Adey DB, Ades PA, Nair KS : "Effects of aging on in vivo synthesis of skeletal muscle myosin heavy-chain and sarcoplasmic protein in humans." Am J Physiol. (1997) Oct;273(4 Pt 1) E790-800.

Ballou SP, Kushner I: "Chronic inflammation in older people: recognition, consequences, and potential intervention." Clinics in Geriatric Medicine (1997) Nov;13(4):653-669.

Barber CE: "Olfactory acuity as a function of age and gender: a comparison of African and American samples." International Journal of Aging and Human Development (1997) 44(4):317-334.

Barzel US, Massey LK: "Excess dietary protein can adversely affect bone." Journal of Nutrition (1998) Jun;128(6):1051-1053.

Battmann A, Battmann A, Jundt G, Schulz A: "Endosteal human bone cells (EBC) show age-related activity in vitro." Experimental and Clinical Endocrinology and Diabetes (1997) 105(2):98-102.

Baylis C, Corman B: "The aging kidney: insights from experimental studies." Journal of the American Society of Nephrology (1998) Apr;9(4):699-709.

Beck J, Garcia R, Heiss G, Vokonas PS, Offenbacher S: "Periodontal disease and cardiovascular disease." Journal of Periodontology (1996) Oct;67(10 Suppl):1123-1137.

Beckman KB, Ames BN: "The free radical theory of aging matures." Physiol Rev (1998) Apr;78(2):547-581.

Beffert U, Poirier J : "Apolipoprotein E, plaques, tangles and cholinergic dysfunction in Alzheimer's disease." Ann N Y Acad Sci. (1996) Jan 17;777 166-74.

Bemben MG: "Age-related alterations in muscular endurance." Sports Medicine (1998) Apr;25(4):259-269.

Benbassat CA, Maki KC, Unterman TG: "Circulating levels of insulin-like growth factor (IGF) binding protein-1 and -3 in aging men: relationships to insulin, glucose, IGF, and dehydroepiandrosterone sulfate levels and anthropometric measures." Journal of Clinical Endocrinology and Metabolism (1997) May;82(5):1484-1491.

Benedetto AV: "The environment and skin aging." Clinics in Dermatology (1998) Jan;16(1):129-139.

Bernstein EF, Uitto J: "The effect of photodamage on dermal extracellular matrix." Clinics in Dermatology (1996) Mar;14(2):143-151.

Benitez del Castillo JM, del Rio T, Garcia-Sanchez J: "Effects of estrogen use on lens transmittance in postmenopausal women." Ophthalmology (1997) Jun;104(6):970-973.

Benjamin BJ: "Speech production of normally aging adults.".: "Seminars in Speech and Language (1997) May;18(2):135-141.".

Bernarducci MP, Owens NJ: "Is there a fountain of youth? A review of current life extension strategies." Pharmacotherapy (1996) Mar;16(2):183-200.

Bernier L, Wang E: "A prospective view on phosphatases and replicative senescence." Experimental Gerontology (1996) Jan;31(1-2):13-19.

Bikle DD: "Biochemical markers in the assessment of bone disease." American Journal of Medicine (1997) Nov;103(5):427-436.

Birge SJ, Mortel KF : "Estrogen and the treatment of Alzheimer's disease." Am J Med. (1997) Sep 22;103(3A) 36S-45S.

Birren JE, Fisher LM: "Aging and speed of behavior: possible consequences for psychological functioning." Annual Review of Psychology (1995) 46:329-353.

Bitsios P, Prettyman R, Szabadi E: "Changes in autonomic function with age: a study of pupillary kinetics in healthy young and old people." Age and Ageing (1996) Nov;25(6):432-438.

Bizbiz L, Alperovitch A, Robert L: "Aging of the vascular wall: serum concentration of elastin peptides and elastase inhibitors in relation to cardiovascular risk factors. The EVA study." Atherosclerosis (1997) May;131(1):73-78.

Bohlhalter S, Murck H, Holsboer F, Steiger A: "Cortisol enhances non-REM sleep and growth hormone secretion in elderly subjects." Neurobiology of Aging (1997) Jul;18(4):423-429.

Bonithon-Kopp C, Coudray C, Berr C, Touboul PJ, Feve JM, Favier A, Ducimetiere P: "Combined effects of lipid peroxidation and antioxidant status on carotid atherosclerosis in a population aged 59-71: The EVA Study. Etude sur le Vieillisement Arteriel." American Journal of Clinical Nutrition (1997)

Jan;65(1):121-127.

Bonjour JP, Schurch MA, Rizzoli R: "Proteins and bone health." Pathologie Biologie (1997) Jan;45(1):57-59.

Bonnefoy M, Kostka T, Arsac LM, Berthouze SE, Lacour JR: "Peak anaerobic power in elderly men." European Journal of Applied Physiology and Occupational Physiology (1998) 77(1-2):182-188.

Boonen S, Aerssens J, Dequeker J: "Age-related endocrine deficiencies and fractures of the proximal femur. I implications of growth hormone deficiency in the elderly." Journal of Endocrinology (1996) Apr;149(1):7-12.

Boonen S, Aerssens J, Dequeker J, Nicholson P, Cheng X, Lowet G, Verbeke G, Bouillon R: "Age-associated decline in human femoral neck cortical and trabecular content of insulin-like growth factor I: potential implications for age-related (type II)osteoporotic fracture occurrence." Calcified Tissue International (1997) Sep;61(3):173-178.

Boonen S, Broos P, Dequeker J: "Age-related factors in the pathogenesis of senile (Type II) femoral neck fractures." American Journal of Orthopedics (1996) Mar;25(3):198-204.

Booth FW, Weeden SH, Tseng BS: "Effect of aging on human skeletal muscle and motor function." Medicine and Science in Sports and Exercise (1994);May;26(5):556-560.

Borst SE, Lowenthal DT: "Role of IGF-I in muscular atrophy of aging." Endocrine (1997) Aug;7(1):61-63.

Bowman MA, Spangler JG: "Osteoporosis in women." Primary Care; Clinics in Office Practice (1997) Mar;24(1):27-36.

Brant LJ, Fozard JL: "Age changes in pure-tone hearing thresholds in a longitudinal study of normal human aging." Journal of the Acoustical Society of America (1990) Aug;88(2):813-820.

Brawer MK: "Screening and early detection of prostate cancer will decrease morbidity and mortality from prostate cancer: the argument for." European Urology (1996) 29 Suppl 2:19-23.

Broderick GA: "Intracavernous pharmacotherapy: treatment for the aging erectile response." Urologic Clinics of North America (1996) Feb;23(1):111-126.

Brown M, Sinacore DR, Host HH: "The relationship of strength to function in the older adult." J Gerontol A Biol Sci Med Sci (1995) Nov;50 Spec No:55-9.

Brownlee M: "Advanced protein glycosylation in diabetes and aging." Annual Review of Medicine (1995) 46:223-234.

Bryan TM, Reddel RR: "Telomere dynamics and telomerase activity in *in vitro* immortalised human cells." Eur J Cancer (1997) Apr;33(5):767-773.

Brzezinski A: "Melatonin in humans." New England Journal of Medicine (1997) Jan 16;336(3):186-195.

Buckwalter JA: "Aging and degeneration of the human intervertebral disc." Spine (1995) Jun 1;20(11):1307-1314.

Buckwalter JA, Mankin HJ: "Articular cartilage: degeneration and osteoarthritis, repair, regeneration, and transplantation." Instructional Course Lectures (1998) 47:487-504.

Buee L, Hof PR, Delacourte A: "Brain microvascular changes in Alzheimer's disease and other dementias."Annals of the New York Academy of Sciences (1997) Sep 26;826:7-24.

Burr DB: "Muscle strength, bone mass, and age-related bone loss." Journal of Bone and Mineral Research (1997) Oct;12(10):1547-1551.

Cagnacci A, Krauchi K, Wirz-Justice A, Volpe A: "Homeostatic versus circadian effects of melatonin on core body temperature in humans." Journal of Biological Rhythms (1997) Dec;12(6):509-517.

Campbell WW, Evans WJ: "Protein requirements of elderly people." European Journal of Clinical Nutrition (1996) Feb;50 Suppl 1:S180-S183.

Campisi J, Dimri GP, Nehlin JO, Testori A, Yoshimoto K: "Coming of age in culture." Experimental Gerontology (1996) Jan;31(1-2):7-12.

Caprio-Prevette MD, Fry PS: "Memory enhancement program for community-based older adults: development and evaluation." Experimental Aging Research (1996) Jul;22(3):281-303.

Carlson JC, Riley JC: "A consideration of some notable aging theories." Experimental Gerontology (1998) Jan;33(1-2):127-134.

Carr DB, Goate A, Phil D, Morris JC: "Current concepts in the pathogenesis of Alzheimer's disease." American Journal of Medicine (1997) Sep 22;103(3A):3S-10S.

Caruso AJ, McClowry MT, Max L: "Age-related effects on speech fluency." Seminars in Speech and Language (1997) May;18(2):171-179.

Caruso C, Candore G, Cigna D, DiLorenzo G, Sireci G, Dieli F, Salerno A: "Cytokine production pathway in the elderly." Immunologic Research (1996) 15(1):84-90.

Casper RC: "Nutrition and its relationship to aging." Experimental Gerontology (1995) May;30(3-4):299-314.

Castanet J, Ortonne JP: "Pigmentary changes in aged and photoaged skin." Archives of Dermatology (1997) Oct;133(10):1296-1299.

Castleden CM : "The Marjory Warren Lecture. Incontinence—still a geriatric giant?." Age Ageing. (1997) Dec;26 Suppl 4 47-52.

Chan K, Spencer EM: "General aspects of insulin-like growth factor binding proteins." Endocrine (1997) Aug;7(1):95-97.

Chapleau MW, Cunningham JT, Sullivan MJ, Wachtel RE, Abboud FM: "Structural versus functional modulation of the arterial baroreflex." Hypertension (1995) Aug;26(2):341-347.

Charlesworth B: "Evolution of senescence: Alzheimer's disease and evolution." Current Biology (1996) Jan 1;6(1):20-22.

Charlesworth B, Partridge L: "Ageing: levelling of the grim reaper." Current Biology (1997) Jul 1;7(7):R440-R442.

Cheraskin E: "Antioxidants in health and disease." Journal of the American Optometric Association (1996) Jan;67(1):50-57.

Chigira M: "Mechanical optimization of bone." Medical Hypotheses (1996) Apr;46(4):327-330.

Chodzko-Zajko WJ: "Normal aging and human physiology." Seminars in Speech and Language (1997) May;18(2):95-104.

Christopher-Hennings J, Kurzman ID, Haffa AL, Kemnitz JM, Macewen EG: "The effects of high fat diet and dehydriepiandrosterone (DHEA) administration in the rhesus monkey." In Vivo (1995) 9:415-420.

Chumlea WC, Guo SS, Vellas B, Guigoz Y: "Techniques of assessing muscle mass and function (sarcopenia) for epidemiological studies of the elderly." J Gerontol A Biol Sci Med Sci (1995) Nov;50 Spec No:45-51.

Chung WS, Park YY, Kwon SW: "The impact of aging on penile hemodynamics in normal responders to pharmacological injection: a Doppler sonographic study." Journal of Urology (1997) Jun;157(6):2129-2131.

Chutka DS, Fleming KC, Evans MP, Evans JM, Andrews KL: "Urinary incontinence in the elderly population." Mayo Clinic Proceedings (1996) Jan;71(1):93-101.

Coggan AR: "Muscle biopsy as a tool in the study of aging." J Gerontol A Biol Sci Med Sci (1995) Nov;50 Spec No:30-4.

Cohen MP, Ziyadeh FN: "Role of Amadori-modified nonenzymatically glycated serum proteins in the pathogenesis of diabetic nephropathy." Journal of the American Society of Nephrology (1996) Feb;7(2):183-190.

Collins KJ, Abdel-Rahman TA, Goodwin J, McTiffin L: "Circadian body temperatures and the effects of a cold stress in elderly and young subjects." Age and Ageing (1995) Nov;24(6):485-489.

Conley KE, Cress ME, Jubrias SA, Esselman PC, Odderson IR: "From muscle properties to human performance, using magnetic resonance." J Gerontol A Biol Sci Med Sci (1995) Nov;50 Spec No:35-40.

Conover CA: "IGFs and cellular aging." Endocrine (1997) Aug;7(1):93-94.

Cook JL, Dzubow LM: "Aging of the skin: implications for cutaneous surgery." Archives of Dermatology (1997) Oct;133(10):1273-1277.

Copinschi G, Van Cauter E: "Effects of ageing on modulation of hormonal secretions by sleep and circadian rhythmicity." Hormone Research (1995) 43(1-3):20-24.

Cossarizza A, Ortolani C, Monti D, Franceschi C: "Cytometric analysis of immunosenescence." Cytometry (1997) Apr 1;27(4):297-313.

Couillard DR, Webster GD: "Detrusor instability." Urologic Clinics of North America (1995) Aug;22(3):593-612.

Cowen T, Gavazzi I: "Plasticity in adult and ageing sympathetic neurons." Progress in Neurobiology (1998) Feb;54(3):249-288.

Cristofalo VJ, Pignolo RJ: "Molecular markers of senescence in fibroblast-like cultures." Experimental Gerontology (1996) Jan;31(1-2):111-123.

Da Silva FC: "Benign prostatic hyperplasia: natural evolution versus medical treatment." European Urology (1997) 32 Suppl 2:34-37.

Dahse R, Fiedler W, Ernst G: "Telomeres and telomerase: biological and clinical importance." Clinical Chemistry (1997) May;43(5):708-714.

de Grey AD: "A proposed refinement of the mitochondrial free radical theory of aging." Bioessays (1997) Feb;19(2):161-166.

de Leon MJ, Convit A, George AE, Golomb J, de Santi S, Tarshish C, Rusinek H, Bobinski M, Ince C, Miller D, Wisniewski H : "In vivo structural studies of the hippocampus in normal aging and in incipient Alzheimer's disease." Ann N Y Acad Sci. (1996) Jan 17;777 1-13.

De Jong GI, De Vos RA, Steur EN, Luiten PG: "Cerebrovascular hypoperfusion: a risk factor for Alzheimer's disease? Animal model and postmortem human studies." Annals of the New York Academy of Sciences (1997) Sep 26;826:56-74.

Dean RT, Fu S, Stocker R, Davies MJ: "Biochemistry and pathology of radical-mediated protein oxidation." Biochem J (1997) May 15;324(Pt 1):1-18.

Delagrange P, Guardiola-Lemaitre B: "Melatonin, its recep-

tors, and relationships with biological rhythm disorders." Clinical Neuropharmacology (1997) Dec;20(6):482-510.

Diamond P, Cusan L, Gomez JL, Belanger A, Labrie F: "Metabolic effects of 12-month percutaneous dehydroepiandrosterone replacement therapy in postmenopausal women." Journal of Endocrinology (1996) Sep;150 Suppl:S43-S50.

Dickson DW: "The pathogenesis of senile plaques." Journal of Neuropathology and Experimental Neurology (1997) Apr;56(4):321-339.

Dillon J: "UV-B as a pro-aging and pro-cataract factor." Documenta Ophthalmologica (1994);88(3-4):339-344.

Diplock AT: "Will the 'good fairies' please prove to us that vitamin E lessens human degenerative disease?." Free Radic Res (1997) Nov;27(5):511-532.

Donahue JL, Lowenthal DT: "Nocturnal polyuria in the elderly person." American Journal of the Medical Sciences (1997) Oct;314(4):232-238.

Doria G, Frasca D: "Genes, immunity, and senescence: looking for a link." Immunological Reviews (1997) Dec;160:159-170.

Dossey BM: "Complementary and alternative therapies for our aging society." Journal of Gerontological Nursing (1997) Sep;23(9):45-51.

Dubbels R, Reiter RJ, Klenke E, Goebel A, Schnakenberg E, Ehlers C, Schiwara HW, Schloot W: "Melatonin in edible plants identified by radioimmunoassay and by high performance liquid chromatography-mass spectrometry." J. Pineal Res (1995) 18:28-31.

Duchek JM, Hunt L, Ball K, Buckles V, Morris JC: "The role of selective attention in driving and dementia of the Alzheimer type." Alzheimer Disease and Associated Disorders (1997) Jun;11 Suppl 1:48-56.

Dutta C, Hadley EC, Lexell J: "Sarcopenia and physical performance in old age: overview." Muscle and Nerve. Supplement (1997) 5:S5-S9.

Dutta C, Hadley EC: "The significance of sarcopenia in old age." J Gerontol A Biol Sci Med Sci (1995) Nov;50 Spec No:1-4.

Edelstein DR: "Aging of the normal nose in adults." Laryngoscope (1996) Sep;106(9 Pt 2):1-25.

Effros RB: "Loss of CD28 expression on T lymphocytes: a marker of replicative senescence." Developmental and Comparative Immunology (1997) Nov;21(6):471-478.

Effros RB: "Replicative senescence in the immune system: impact of the Hayflick limit on T-cell function in the elderly." American Journal of Human Genetics (1998) May;62(5):1003-1007.

Ershler WB, Harman SM, Keller ET: "Immunologic aspects of osteoporosis." Developmental and Comparative Immunology (1997) Nov;21(6):487-499.

Ershler WB, Longo DL: "The biology of aging: the current research agenda." Cancer (1997) Oct 1;80(7):1284-1293.

Estivariz CF, Ziegler TR: "Nutrition and the insulin-like growth factor system." Endocrine (1997) Aug;7(1):65-71.

Evans JG, Bond J: "The challenges of age research." Age and Ageing (1997) Dec;26 Suppl 4:43-46.

Evans WJ: "What is sarcopenia?" J Gerontol A Biol Sci Med Sci (1995) Nov;50 Spec No:5-8.

Feuerstein TJ, Seeger W: "Modulation of acetylcholine release in human cortical slices: possible implications for Alzheimer's disease." Pharmacology and Therapeutics (1997) 74(3):333-347.

Finch CE, Tanzi RE: "Genetics of aging." Science (1997) Oct 17;278(5337):407-411.

Forbes WF, Gentleman JF, Park E : "Dementia and age at death." Exp Gerontol. (1995) Sep-Oct;30(5) 445-53.

Frost HM: "On our age-related bone loss: insights from a new paradigm." Journal of Bone and Mineral Research (1997) Oct;12(10):1539-1546.

Fultz NH, Herzog AR: "Epidemiology of urinary symptoms in the geriatric population." Urologic Clinics of North America (1996) Feb;23(1):1-10.

Gambert SR, Schultz BM, Hamdy RC: "Osteoporosis. Clinical features, prevention, and treatment." Endocrinology and Metabolism Clinics of North America (1995) Jun;24(2):317-371.

Gao X, Porter AT, Grignon DJ, Pontes JE, Honn KV: "Diagnostic and prognostic markers for human prostate cancer." Prostate (1997) Jun 1;31(4):264-281.

Gardner P, Rosenberg HM: "Leading Causes of Death by Age, Sex, Race, and Hispanic Origin: United States, 1992." Vital and Health Statistics Series 20: Data on Mortality, No. 29 U.S. Department of Health and Human Services DHHS Publication No. (PHS) 96-1857 June (1996).

Garnero P, Delmas PD: "Osteoporosis." Endocrinology and Metabolism Clinics of North America (1997) Dec;26(4):913-936.

Gasiorowski K, Leszek J: "A proposed new strategy of immunotherapy for Alzheimer's disease." Medical Hypotheses (1997) Oct;49(4):319-326.

Gates GA, Rees TS: "Hear ye? Hear ye! Successful auditory aging." Western Journal of Medicine (1997) Oct;167(4):247-252.

Geirsson G, Fall M, Lindstrom S: "Subtypes of overactive bladder in old age." Age and Ageing (1993);Mar;22(2):125-131.

Gelato MC, Frost RA: "IGFBP-3. Functional and structural implications in aging and wasting syndromes." Endocrine (1997) Aug;7(1):81-85.

Geller J: "Benign prostatic hyperplasia: pathogenesis and medical therapy." Journal of the American Geriatrics Society (1991) Dec;39(12):1208-1216.

Gerbitz KD, Gempel K, Brdiczka D: "Mitochondria and diabetes. Genetic, biochemical, and clinical implications of the cellular energy circuit." Diabetes (1996) Feb;45(2):113-126.

Giambra LM: "Sustained attention and aging: overcoming the decrement?." Experimental Aging Research (1997) Apr;23(2):145-161.

Giannakopoulos P, Hof PR, Michel JP, Guimon J, Bouras C: "Cerebral cortex pathology in aging and Alzheimer's disease: a quantitative survey of large hospital-based geriatric and psychiatric cohorts." Brain Research. Brain Research Reviews (1997) Oct;25(2):217-245.

Gilchrest BA: "A review of skin ageing and its medical therapy." British Journal of Dermatology (1996) Dec;135(6):867-875.

Gilchrest BA : "Treatment of photodamage with topical tretinoin: an overview." J Am Acad Dermatol. (1997) Mar;36(3 Pt 2) S27-36.

Gilmartin B: "The aetiology of presbyopia: a summary of the role of lenticular and extralenticular structures." Ophthalmic and Physiological Optics (1995) Sep;15(5):431-437.

Glasser SP, Selwyn AP, Ganz P: "Atherosclerosis: risk factors and the vascular endothelium." American Heart Journal (1996) Feb;131(2):379-384.

Gloth FM 3rd, Tobin JD: "Vitamin D deficiency in older people." Journal of the American Geriatrics Society (1995) Jul;43(7):822-828.

Goedert M : "Tau protein and the neurofibrillary pathology of Alzheimer's disease." Ann N Y Acad Sci. (1996) Jan 17;777 121-31.

Going S, Williams D, Lohman T: "Aging and body composition: biological changes and methodological issues." Exercise and Sport Sciences Reviews (1995) 23:411-458.

Goldsmith LA: "Genetic skin diseases with altered aging." Archives of Dermatology (1997) Oct;133(10):1293-1295.

Grammas P, Botchlet TR, Moore P, Weigel PH : "Production of neurotoxic factors by brain endothelium in Alzheimer's disease." Ann N Y Acad Sci. (1997) Sep 26;826 47-55.

Gray R, Stern G, Malone-Lee J: "Lower urinary tract dysfunction in Parkinson's disease: changes relate to age and not disease." Age and Ageing (1995) Nov;24(6):499-504.

Green K: "Free radicals and aging of anterior segment tissues of the eye: a hypothesis." Ophthalmic Research (1995) 27 Suppl 1:143-149.

Griebling TL, Nygaard IE: "The role of estrogen replacement therapy in the management of urinary incontinence and urinary tract infection in postmenopausal women." Endocrinology and Metabolism Clinics of North America (1997) Jun;26(2):347-360.

Grimby G: "Muscle performance and structure in the elderly as studied cross-sectionally and longitudinally." J Gerontol A Biol Sci Med Sci (1995) Nov;50 Spec No:17-22.

Grose JH: "Binaural performance and aging." Journal of the American Academy of Audiology (1996) Jun;7(3):168-174.

Grossman A, Rabinovitch PS, Lane MA, Jinneman JC, Ingram DK, Wolf NS, Cutler RG, Roth GS: "Influence of Age, Sex, and Dietary Restriction on Intracellular Free Calcium Responses of CD4+ Lymphocytes in Rhesus Monkeys (*Macaca mulatta*),." J. Cell Physiol (1995) 162(2):298- 303.

Guldner J, Schier T, Friess E, Colla M, Holsboer F, Steiger A: "Reduced efficacy of growth hormone-releasing hormone in modulating sleep endocrine activity in the elderly." Neurobiology of Aging (1997) Sep;18(5):491-495.

Guyuron B: "The aging nose." Dermatologic Clinics (1997) Oct;15(4):659-664.

Hachinski V, Munoz DG : "Cerebrovascular pathology in Alzheimer's disease: cause, effect or epiphenomenon?." Ann N Y Acad Sci. (1997) Sep 26;826 1-6.

Hakim AA, Petrovitch H, Burchfiel CM, Ross GW, Rodriguez BL, White LR, Yano K, Curb JD, Abbott RD: "Effects of walking on mortality among nonsmoking retired men." New England Journal of Medicine (1998) Jan 8;338(2):94-99.

Hamerman D: "Aging and the musculoskeletal system." Annals of the Rheumatic Diseases (1997) Oct;56(10):578-585.

Hammond CB: "Menopause and hormone replacement therapy: an overview." Obstetrics and Gynecology (1996) Feb;87(2 Suppl):2S-15S.

Harman D: "Extending functional life span." Experimental Gerontology (1998) Jan;33(1-2):95-112.

Hart RW, Turturro A: "Evolution and dietary restriction." Experimental Gerontology (1998) Jan;33(1-2):53-60.

Harvell JD, Maibach HI: "Percutaneous absorption and inflammation in aged skin: a review." Journal of the American Academy of Dermatology (1994);Dec;31(6):1015-1021.

Hattori A, Migitaka H, Iigo M, Itoh M, Yamamoto K, Ohtani-Kaneko R, Hara M, Suzuki T, Reiter RJ: "Identification of melatonin in plants and its effect on plasma melatonin levels and binding to melatonin receptors in vertebrates." Biochem. Mol. Biol. Int. (1995) 35:627-634.

Henderson VW : "Estrogen, cognition, and a woman's risk of Alzheimer's disease." Am J Med. (1997) Sep 22;103(3A) 11S-18S.

Hepple RT, Mackinnon SLM, Thomas SG, Goodman JM, Plyley MJ: "Quantitating the capillary supply and the response to resistance training in older men." Pflugers Archiv. European Journal of Physiology (1997) Jan;433(3):238-244.

Heymsfield SB, Gallagher D, Visser M, Nunez C, Wang ZM: "Measurement of skeletal muscle: laboratory and epidemiological methods." J Gerontol A Biol Sci Med Sci (1995) Nov;50 Spec No:23-9.

Hindmarsh PC, Fall CH, Pringle PJ, Osmond C, Brook CG: "Peak and trough growth hormone concentrations have different associations with the insulin-like growth factor axis, body composition, and metabolic parameters." Journal of Clinical Endocrinology and Metabolism (1997) Jul;82(7):2172-2176.

Hobbs MV, Ernst DN: "T cell differentiation and cytokine expression in late life." Developmental and Comparative Immunology (1997) Nov;21(6):461-470.

Hock C, Villringer K, Muller-Spahn F, Hofmann M, Schuh-Hofer S, Heekeren H, Wenzel R, Dirnagl U, Villringer A: "Near infrared spectroscopy in the diagnosis of Alzheimer's disease." Annals of the New York Academy of Sciences (1996) Jan 17;777:22-29.

Hodge WG, Whitcher JP, Satariano W: "Risk factors for age-related cataracts." Epidemiologic Reviews (1995) 17(2):336-346.

Hof PR: "Morphology and neurochemical characteristics of the vulnerable neurons in brain aging and Alzheimer's disease." European Neurology (1997) 37(2):71-81.

Hofman MA: "Lifespan changes in the human hypothalamus." Experimental Gerontology (1997) Jul;32(4-5):559-575.

Holt SE, Wright WE, Shay JW: "Multiple pathways for the regulation of telomerase activity." Eur J Cancer (1997) Apr;33(5):761-766.

Horan MA, Ashcroft GS: "Ageing, defence mechanisms and the immune system." Age and Ageing (1997) Dec;26 Suppl 4:15-19.

Hornick TR, Kowal J: "Clinical epidemiology of endocrine disorders in the elderly." Endocrinology and Metabolism Clinics of North America (1997) Mar;26(1):145-163.

Hornsby PJ: "DHEA: a biologist's perspective." Journal of the American Geriatrics Society (1997) Nov;45(11):1395-1401.

Hotta K, Gustafson TA, Ortmeyer HK, Bodkin NL, Nicolson MA, Hansen BC: "Regulation of Obese (ob) mRNA and plasma leptin levels in rhesus monkeys: effects of insulin, body weight, and non-insulin-dependent diabetes mellitus." J. Biol. Chem. (1996) 271:2527-2533.

Hoyer S: "Oxidative metabolism deficiencies in brains of patients with Alzheimer's disease." Acta Neurolologica Scandinavica. Supplementum (1996) 165:18-24.

Huang Z, Bodkin NL, Ortmeyer HK, Hansen BC, Shuldiner AR: "Hyperinsulinemia is associated with altered insulin receptor mRNA splicing in muscle of the spontaneously obese diabetic Rhesus monkey." J. Clin. Invest. (1994); 94:1289-1296.

Huppert F, Wilcock G: "Ageing, cognition and dementia." Age and Ageing (1997) Dec;26 Suppl 4:20-23.

Hurley BF: "Age, gender, and muscular strength." J Gerontol A Biol Sci Med Sci (1995) Nov;50 Spec No:41-4.

Hurley DL, Khosla S: "Update on primary osteoporosis." Mayo Clinic Proceedings (1997) Oct;72(10):943-949.

Hyman BT, Gomez-Isla T, West H, Briggs M, Chung H, Growdon JH, Rebeck GW : "Clinical and neuropathological correlates of apolipoprotein E genotype in Alzheimer's disease. Window on molecular epidemiology." Ann N Y Acad Sci. (1996) Jan 17;777 158-65.

Ingram DK, Cutler RG, Weindruch R, Renquist DM, Knapka JJ, AprilM, Belcher CT, Clark MA, Hatcherson CD, Marriott, BM, Roth GS: "Dietary restriction and aging: the initiation of a primate study." Journal of Gerontology Biological Sciences (1990) 45(5):B148-B163.

Ingram DK, Lane MA, Cutler RG, Roth GS: "Longitudinal study of aging in monkeys: effects of diet restriction." Neurobiol. Aging. (1993);14:687-688.

Ivy JL: "Role of exercise training in the prevention and treatment of insulin resistance and non-insulin-dependent diabetes mellitus." Sports Medicine (1997) Nov;24(5):321-336.

Janssen JA, Stolk RP, Pols HA, Grobbee DE, Lamberts SW: "Serum total IGF-I, free IGF-I, and IGFB-1 levels in an elderly population: relation to cardiovascular risk factors and disease." Arteriosclerosis, Thrombosis, and Vascular Biology (1998) Feb;18(2):277-282.

Jelicic M: "Aging and performance on implicit memory tasks: a brief review." International Journal of Neuroscience (1995) Jun;82(3-4):155-161.

Jensen-Urstad K, Storck N, Bouvier F, Ericson M, Lindblad LE, Jensen-Urstad M: "Heart rate variability in healthy subjects is related to age and gender." Acta Physiologica Scandinavica (1997) Jul;160(3):235-241.

Johnell O: "The socioeconomic burden of fractures: today and in the 21st century." American Journal of Medicine (1997) Aug 18;103(2A):20S-25S.

Johnson TE: "Genetic influences on aging." Experimental Gerontology (1997) Jan;32(1-2):11-22.

Jorgensen JO, Vahl N, Fisker S, Norrelund H, Nielsen S, Dall R, Christiansen JS: "Somatopause and adiposity." Hormone Research (1997) 48 Suppl 5:101-104.

Kalaria RN: "Cerebral vessels in ageing and Alzheimer's disease." Pharmacology and Therapeutics (1996) 72(3):193-214.

Kalaria RN, Cohen DL, Premkumar DR, "Apolipoprotein E alleles and brain vascular pathology in Alzheimer's disease." Ann N Y Acad Sci. (1996) Jan 17;777 266-70.

Kang S, Fisher GJ, Voorhees JJ: "Photoaging and topical tretinoin: therapy, pathogenesis, and prevention." Archives of Dermatology (1997) Oct;133(10):1280-1284.

Kanungo MS, Gupta S, Upadhyay R: "Molecular biology of ageing." Indian Journal of Medical Research (1997) Oct;106:413-422.

Karlsson MK, Johnell O, Obrant KJ: "Is bone mineral density advantage maintained long-term in previous weight lifters?." Calcified Tissue International (1995) Nov;57(5):325-328.

Kenney WL, Buskirk ER: "Functional consequences of sarcopenia: effects on thermoregulation." J Gerontol A Biol Sci Med Sci (1995) Nov;50 Spec No:78-85.

Kenney WL: "Thermoregulation at rest and during exercise in healthy older adults." Exercise and Sport Sciences Reviews (1997) 25:41-76.

Khorram O: "DHEA: a hormone with multiple effects." Current Opinion in Obstetrics and Gynecology (1996) Oct;8(5):351-354.

Khorram O, Laughlin GA, Yen SS: "Endocrine and metabolic effects of long-term administration of [Nle27] growth hormone-releasing hormone-(1-29)-NH2 in age-advanced men and women." Journal of Clinical Endocrinology and Metabolism (1997) May;82(5):1472-1479.

Khorram O, Yeung M, Vu L, Yen SS: "Effects of [norleucine27] growth hormone-releasing hormone (GHRH) (1-29)-NH2 administration on the immune system of aging men and women." Journal of Clinical Endocrinology and Metabolism (1997) Nov;82(11):3590-3596.

Kirkwood TB: "Human senescence." Bioessays (1996) Dec;18(12):1009-1016.

Kirkwood TB : "The origins of human ageing." Philos Trans R Soc Lond B Biol Sci. (1997) Dec 29;352(1363) 1765-72.

Kirkwood TB, Kowald A: "Network theory of aging." Experimental Gerontology (1997) Jul;32(4-5):395-399.

Kirkwood TB, Ritter MA: "The interface between ageing and health in man." Age and Ageing (1997) Dec;26 Suppl 4:9-14.

Kirollos MM: "Statistical review and analysis of the relationship between serum prostate specific antigen and age." Journal of Urology (1997) Jul;158(1):143-145.

Klein R, Klein BE, Moss SE: "Relation of smoking to the incidence of age-related maculopathy. The Beaver Dam Eye Study." American Journal of Epidemiology (1998) Jan 15;147(2):103-110.

Klemera P, Doubal S: "Human mortality at very advanced age might be constant." Mechanisms of Ageing and Development (1997) Nov;98(2):167-176.

Kline DW: "Optimizing the visibility of displays for older observers." Experimental Aging Research (1994);Jan;20(1):11-23.

Knight JA: "Reactive oxygen species and the neurodegenerative disorders." Annals of Clinical Laboratory Science (1997) Jan;27(1):11-25.

Koehler KM, Pareo-Tubbeh SL, Romero LJ, Baumgartner RN, Garry PJ: "Folate nutrition and older adults: challenges and opportunities." Journal of the American Dietetic Association (1997) Feb;97(2):167-173.

Kohrt WM, Holloszy JO: "Loss of skeletal muscle mass with aging: effect on glucose tolerance." J Gerontol A Biol Sci Med Sci (1995) Nov;50 Spec No:68-72.

Kondo H, Lane MA, Yonezawa Y, Ingram DK, Culter RG, Roth GS: "Effects of aging and dietary restriction on activity of monkey serum in promoting fibroblast migration." Mech. Ageing Dev. (LMJ) (1995) 79:141-150.

Kostka T, Bonnefoy M, Arsac LM, Berthouze SE, Belli A, Lacour JR: "Habitual physical activity and peak anaerobic power in elderly women." European Journal of Applied Physiology and Occupational Physiology (1997) 76(1):81-87.

Krentz AJ, Nattrass M: "Insulin resistance: a multifaceted metabolic syndrome. Insights gained using a low-dose insulin infusion technique." Diabetic Medicine (1996) Jan;13(1):30-39.

Kubeck JE, Delp ND, Haslett TK, McDaniel MA: "Does job-related training performance decline with age?." Psychology and Aging (1996) Mar;11(1):92-107.

Kumazaki T, Wadhwa R, Kaul SC, Mitsui Y: "Expression of endothelin, fibronectin, and mortalin as aging and mortality markers." Experimental Gerontology (1997) Jan;32(1-2):95-103.

Lakatta EG: "Cardiovascular regulatory mechanisms in advanced age." Physiol Rev (1993); 73:413-467.

Lamberts SW, van den Beld AW, van der Lely AJ: "The endocrinology of aging." Science (1997) Oct 17;278(5337):419-424.

Lane MA, Baer DJ, Rumpler WV, Weindruch R, Ingram DK, Tilmont EM, Cutler RG, Roth GS: "Calorie restriction lowers body temperature in rhesus monkeys, consistent with a postulated anti- aging mechanism in rodents." Proc. Natl Acad. Sci USA (1996) 93:4159-4164.

Lane MA, Ball SS, Ingram DK, Cutler RG, Engel J, Read V, Roth GS: "Diet restriction in rhesus monkeys lowers fasting and glucose-stimulated glucoregulatory end points." Amer. J. Physiol. (1995) 268(Endocrinol. Metab. 31):E941-E948.

Lane MA, Ingram DJ, Roth GS: "Beyond the rodent model: calorie restriction in rhesus monkeys." Age (1997) 20:39-50.

Lane MA, Ingram DK, Culter RG, Knapka JJ, Barnard DE, Roth GS: "Dietary restriction in nonhuman primates progress report on the NIA study." Ann. N.Y. Acad. Sci. (1992) 673:36-45.

Lane MA, Reznick, AZ, Tilmont, EM, Lanir A, Ball SS, Read V, Ingram DK, Cutler RG, Roth GS: "Aging and food restriction alter some indices of bone metabolism in male rhesus monkeys (Macaca mulatta)." J. Nutr. (1995) 125:1600-1610.

Lane MA, Reznik A: "Aging and food restriction alter some aspects of bone metabolism in male rhesus monkeys (Macca mulatta)." J. Nutr., (1995). 125:1600-1610.

Langton CM, Langton DK: "Male and female normative data for ultrasound measurement of the calcaneus within the UK adult population." British Journal of Radiology (1997) Jun;70(834):580-585.

Larsson L, Ansved T: "Effects of ageing on the motor unit.". Progress in Neurobiology (1995) Apr;45(5):397-458.

Laurent S: "Arterial wall hypertrophy and stiffness in essential hypertensive patients." Hypertension (1995) Aug;26(2):355-362.

Laurent S, Lacolley P, Girerd X, Boutouyrie P, Bezie Y, Safar M: "Arterial stiffening: opposing effects of age- and hypertension-associated structural changes." Canadian Journal of Physiology and Pharmacology (1996) Jul;74(7):842-849.

Laval J: "Role of DNA repair enzymes in the cellular resistance to oxidative stress." Pathologie Biologie (1996) Jan;44(1):14-24.

Lazarus R, Sparrow D, Weiss ST: "Effects of obesity and fat distribution on ventilatory function: the normative aging study." Chest (1997) Apr;111(4):891-898.

Lee C, Kozlowski JM, Grayhack JT: "Intrinsic and extrinsic factors controlling benign prostatic growth." Prostate (1997) May 1;31(2):131-138.

Lee C, Kozlowski JM, Grayhack JT: "Etiology of benign prostatic hyperplasia." Urologic Clinics of North America (1995) May;22(2):237-246.

Lee CM, Weindruch R, Aiken JM: "Age-associated alterations of the mitochondrial genome." Free Radic Biol Med (1997) 22(7) 1259-1269.

Lee YL, Yip KM: "The osteoporotic spine." Clinical Orthopaedics and Related Research (1996) Feb;323:91-97.

LeMaoult J, Szabo P, Weksler ME: "Effect of age on humoral immunity, selection of the B-cell repertoire and B-cell development." Immunological Reviews (1997) Dec;160:115-126.

Lesourd BM, Mazari L, Ferry M: "The role of nutrition in immunity in the aged." Nutrition Reviews (1998) Jan;56(1 Pt 2):S113-S125.

Levy P, Pepin JL, Malauzat D, Emeriau JP, Leger JM: "Is sleep apnea syndrome in the elderly a specific entity?." Sleep (1996) Apr;19(3 Suppl):S29-S38.

Lexell J: "Evidence for nervous system degeneration with advancing age." Journal of Nutrition (1997) May;127(5 Suppl):1011S-1013S.

Lexell J: "Human aging, muscle mass, and fiber type composition." J Gerontol A Biol Sci Med Sci (1995) Nov;50 Spec No:11-6

Lichtman R: "Perimenopausal and postmenopausal hormone replacement therapy. Part 2. Hormonal regimens and complementary and alternative therapies." Journal of Nurse-Midwifery (1996) May;41(3):195-210.

Lieberman SA, Hoffman AR: "The somatopause: should growth hormone deficiency in older people Be treated?." Clinics in Geriatric Medicine (1997) Nov;13(4):671-684.

Lindle RS, Metter EJ, Lynch NA, Fleg JL, Fozard JL, Tobin J, Roy TA, Hurley BF: "Age and gender comparisons of muscle strength in 654 women and men aged 20-93yr." Journal of Applied Physiology (1997) Nov;83(5):1581-1587.

Lips P: "Epidemiology and predictors of fractures associated with osteoporosis." American Journal of Medicine (1997) Aug 18;103(2A):3S-8S.

Liu AY, Lee YK, Manalo D, Huang LE: "Attenuated heat shock transcriptional response in aging: molecular mechanism and implication in the biology of aging." EXS (1996) 77:393-408.

Loehr J, Verma S, Seguin R: "Issues of sexuality in older women." Journal of Womens Health (1997) Aug;6(4):451-457.

Looker AC, Orwoll ES, Johnston CC Jr, Lindsay RL, Wahner HW, Dunn WL, Calvo MS, Harris TB, Heyse SP: "Prevalence of low femoral bone density in older U.S. adults from NHANES III." Journal of Bone and Mineral Research (1997) Nov;12(11):1761-1768.

Lord A, Eastwood H: "Detrusor hyperreflexia—are there two types?." Age and Ageing (1994);Jan;23(1):32-33.

Low PA, Denq JC, Opfer-Gehrking TL, Dyck PJ, O'Brien PC, Slezak JM: "Effect of age and gender on sudomotor and cardiovagal function and blood pressure response to tilt in normal subjects." Muscle and Nerve (1997) Dec;20(12):1561-1568.

Lubinski R, Welland RJ : "Normal aging and environmental effects on communication." Semin Speech Lang. (1997) May;18(2) 107-25.

Lubran MM: "Renal function in the elderly." Annals of Clinical and Laboratory Science (1995) Mar;25(2):122-133.

Lundh U, Nolan M: "Aging and quality of life. 1: Towards a better understanding." British Journal of Nursing (1996) Nov 14;5(20):1248-1251.

Lyons AR : "Clinical outcomes and treatment of hip fractures." Am J Med. (1997) Aug 18;103(2A) 51S-63S.

Macieira-Coelho A: "Genetics of ageing." Pathologie Biologie (1997) Jun;45(6):500-505.

Macieira-Coelho A: "The last mitoses of the human fibroblast proliferative life span, physiopathologic implications." Mechanisms of Ageing and Development (1995) Aug 8;82(2-3):91-104.

Macieira-Coelho A: "The implications of the 'Hayflick limit' for aging of the organism have been misunderstood by many gerontologists." Gerontology (1995) 41(2):94-97.

Mackall CL, Gress RE: "Thymic aging and T-cell regeneration." Immunological Reviews (1997) Dec;160:91-102.

Madden KS, Rajan S, Bellinger DL, Felten SY, Felten DL: "Age-associated alterations in sympathetic neural interactions with the immune system." Developmental and Comparative Immunology (1997) Nov;21(6):479-486.

Madersbacher S, Klingler HC, Schatzl G, Stulnig T, Schmidbauer CP, Marberger M: "Age related urodynamic changes in patients with benign prostatic hyperplasia." Journal of Urology (1996) Nov;156(5):1662-1667.

Madersbacher S, Pycha A, Schatzl G, Mian C, Klingler CH, Marberger M: "The aging lower urinary tract: a comparative urodynamic study of men and women." Urology (1998) Feb;51(2):206-212.

Mahley RW, Nathan BP, Pitas RE : "Apolipoprotein E. Structure, function, and possible roles in Alzheimer's disease." Ann N Y Acad Sci. (1996) Jan 17;777 139-45.

Maki BE, McIlroy WE: "The role of limb movements in maintaining upright stance: the"change-in-support." strategy." Physical Therapy (1997) May;77(5):488-507.

Malone-Lee J, Wahedna I: "Characterisation of detrusor contractile function in relation to old age." British Journal of Urology (1993);Dec;72(6):873-880.

Marcus R: "Relationship of age-related decreases in muscle mass and strength to skeletal status." J Gerontol A Biol Sci Med Sci (1995) Nov;50 Spec No:86-7.

Marcus R, Hoffman AR: "Growth hormone as therapy for older men and women." Annual Review of Pharmacology and Toxicology (1998) 38:45-61.

Marcus R: "Skeletal effects of growth hormone and IGF-I in adults." Endocrine (1997) Aug;7(1):53-55.

Mariotti S, Franceschi C, Cossarizza A, Pinchera A: "The aging thyroid." Endocrine Reviews (1995) Dec;16(6):686-715.

Maroulis GB: "Ovarian aging." Annals of the New York Academy of Sciences (1997) Jun 17;816:22-26.

Martin GM: "Genetic modulation of the senescent phenotype of Homo sapiens." Experimental Gerontology (1996) Jan;31(1-2):49-59.

Martin GM: "The genetics of aging." Hosp Pract (Off Ed) (1997) Feb 15;32(2):47-50.

Martin GM, Austad SN, Johnson TE: "Genetic analysis of ageing: role of oxidative damage and environmental stresses." Genet (1996) May;13(1):25-34.

Martin JT: "Development of an adjuvant to enhance the immune response to influenza vaccine in the elderly." Biologicals (1997) Jun;25(2):209-213.

Martin SE, Mathur R, Marshall I, Douglas NJ: "The effect of

age, sex, obesity and posture on upper airway size." European Respiratory Journal (1997) Sep;10(9):2087-2090.

Martyn CN, Greenwald SE: "Impaired synthesis of elastin in walls of aorta and large conduit arteries during early development as an initiating event in pathogenesis of systemic hypertension." Lancet (1997) Sep 27;350(9082):953-955.

Masliah E, Mallory M, Veinbergs I, Miller A, Samuel W: "Alterations in apolipoprotein E expression during aging and neurodegeneration." Progress in Neurobiology (1996) Dec;50(5-6):493-503.

Masoro EJ: "Dietary restriction." Exp Gerontol. (1995) May-Aug;30(3-4) 291-8.

Masoro EJ: "Possible mechanisms underlying the antiaging actions of caloric restriction." Toxicologic Pathology (1996) Nov;24(6):738-741.

Masoro EJ, Austad SN: "The Evolution of the Antiaging Action of Dietary Restriction: A Hypothesis." J. Gerontol Biol. Sci. Med. Sci. (1996) 51(6):B387-B391.

May H, Murphy S, Khaw KT: "Age-associated bone loss in men and women and its relationship to weight." Age and Ageing (1994);May;23(3):235-240.

McCann SM: "The nitric oxide hypothesis of brain aging." Experimental Gerontology (1997) Jul;32(4-5):431-440.

McCarter RJ: "Exercise and Aging." In: Annual Review of Gerontology and Geriatrics Vol. 15 Chapter 8 (187-228) Springer Publishing Company, New York (1995).

McCarty MF: "Promotion of interleukin-2 activity as a strategy for 'rejuvenating' geriatric immune function." Medical Hypotheses (1997) Jan;48(1):47-54.

McClearn GE: "Biogerontologic theories." Experimental Gerontology (1997) Jan;32(1-2):3-10.

McClearn GE: "Biomarkers of age and aging." Experimental Gerontology (1997) Jan;32(1-2):87-94.

McDonald RB: "Influence of dietary sucrose on biological aging." American Journal of Clinical Nutrition (1995) Jul;62(1 Suppl):284S-292S.

McEwen BS, Magarinos AM: "Stress effects on morphology and function of the hippocampus." Annals of the New York Academy of Sciences (1997) Jun 21;821:271-284.

Meek PD, McKeithan K, Schumock GT: "Economic considerations in Alzheimer's disease." Pharmacotherapy (1998) Mar;18(2 Pt 2):68-73.

Meikle AW, Stephenson RA, Lewis CM, Middleton RG: "Effects of age and sex hormones on transition and peripheral zone volumes of prostate and benign prostatic hyperplasia in twins." Journal of Clinical Endocrinology and Metabolism (1997) Feb;82(2):571-575.

Melton LJ 3rd, Thamer M, Ray NF, Chan JK, Chesnut CH 3rd, Einhorn TA, Johnston CC, Raisz LG, Silverman SL, Siris ES: "Fractures attributable to osteoporosis: report from the National Osteoporosis Foundation." Journal of Bone and Mineral Research (1997) Jan;12(1):16-23.

Merriam GR, Buchner DM, Prinz PN, Schwartz RS, Vitiello MV: "Potential applications of GH secretagogs in the evaluation and treatment of the age-related decline in growth hormone secretion." Endocrine (1997) Aug;7(1):49-52.

Meston CM: "Aging and sexuality." Western Journal of Medicine (1997) Oct;167(4):285-290.

Miquel J: "An update on the oxygen stress-mitochondrial mutation theory of aging: genetic and evolutionary implications." Experimental Gerontology (1998) Jan;33(1-2):113-126.

Miquel J, Ramirez-Bosca A, Soler A, Diez A, Carrion-Gutierrez MA, Diaz-Alperi J, Quintanilla-Ripoll E, Bernd A, Quintanilla-Almagro E: "Increase with age of serum lipid peroxides: implications for the prevention of atherosclerosis." Mechanisms of Ageing and Development (1998) Jan 12;100(1):17-24.

Mohan S, Baylink DJ: "Serum insulin-like growth factor binding protein (IGFBP)-4 and IGFBP-5 levels in aging and age-associated diseases." Endocrine (1997) Aug;7(1):87-91.

Monk TH, Buysse DJ, Reynolds CF 3rd, Kupfer DJ, Houck PR: "Circadian temperature rhythms of older people." Experimental Gerontology (1995) Sep;30(5):455-474.

Moody DM, Brown WR, Challa VR, Ghazi-Birry HS, Reboussin DM: "Cerebral microvascular alterations in aging, leukoaraiosis, and Alzheimer's disease." Annals of the New York Academy of Science (1997) Sep 26;826:103-116.

Mooradian AD: "Normal age-related changes in thyroid hormone economy." Clinics in Geriatric Medicine (1995) May;11(2):159-169.

Morais JA, Gougeon R, Pencharz PB, Jones PJ, Ross R, Marliss EB: "Whole-body protein turnover in the healthy elderly." American Journal of Clinical Nutrition (1997) Oct;66(4):880-889.

Moriguchi S: "The role of vitamin E in T-cell differentiation and the decrease of cellular immunity with aging." Biofactors (1998) 7(1-2):77-86.

Morin GB: "Telomere control of replicative lifespan." Experimental Gerontology (1997) Jul;32(4-5):375-382.

Morley JE: "Anorexia of aging: physiologic and pathologic." American Journal of Clinical Nutrition (1997) Oct;66(4):760-773.

Morley JE, Miller KD, Johnson LE: "Vitamins and aging." Annual Review of Gerontology and Geriatrics, Springer Publishing Company, NY Chapter 7 Volume 15 (143-186) (1995).

Morris RJ, Brown WS Jr: "Age-related differences in speech variability among women." Journal of Communication Disorders (1994);Mar;27(1):49-64.

Morrison JH, Hof PR: "Life and death of neurons in the aging brain." Science (1997) Oct 17;278(5337):412-419.

Mosinger BJ: "Human low-density lipoproteins: oxidative modification and its relation to age, gender, menopausal status and cholesterol concentrations." Europena Journal of Clinical Chemistry and Clinical Biochemistry (1997) Mar;35(3):207-214.

Mountz JD, Wu J, Zhou T, Hsu HC: "Cell death and longevity: implications of Fas-mediated apoptosis in T-cell senescence." Immunological Reviews (1997) Dec;160:19-30.

Mueller PB: "The aging voice." Seminars in Speech and Language (1997) May;18(2):159-168.

Muir JL: "Acetylcholine, aging, and Alzheimer's disease." Pharmacology, Biochemistry and Behavior (1997) Apr;56(4):687-696.

Muller DC, Elahi D, Tobin JD, Andres R: "The effect of age on insulin resistance and secretion: a review." Seminars in Nephrology (1996) Jul;16(4):289-298.

Multhaup G, Ruppert T, Schlicksupp A, Hesse L, Beher D, Masters CL, Beyreuther K: "Reactive oxygen species and Alzheimer's disease." Biochemical Pharmacology (1997) Sep

1;54(5):533-539.

Munch G, Thome J, Foley P, Schinzel R, Riederer P: "Advanced glycation end products in ageing and Alzheimer's disease." Brain Research. Brain Research Reviews (1997) Feb;23(1-2):134-143.

Mundal R, Kjeldsen SE, Sandvik L, Erikssen G, Thaulow E, Erikssen J: "Predictors of 7-year changes in exercise blood pressure: effects of smoking, physical fitness and pulmonary function." Journal of Hypertension (1997) Mar;15(3):245-249.

Murray TM: "Mechanisms of bone loss." Journal of Rheumatology. Supplement (1996) Aug;45:6-10.

Myers BL, Badia P: "Changes in circadian rhythms and sleep quality with aging: mechanisms and interventions." Neuroscience Biobehavior Reviews (1995) 19(4):553-571.

Nachbar F, Korting HC: "The role of vitamin E in normal and damaged skin." Journal of Molecular Medicine (1995) Jan;73(1):7-17.

Nakamura E, Lane MA, Roth GS, Cutler RG, Ingram DK: "Evaluation Measures of Hematology and Blood Chemistry in Male Rhesus Monkeys as Biomarkers of Aging." Exp. Gerontol. (1994);29:151-177.

Nakamura S, Kobayashi Y, Tozuka K, Tokue A, Kimura A, Hamada C: "Circadian changes in urine volume and frequency in elderly men." Journal of Urology (1996) Oct;156(4):1275-1279.

Nam SY, Lee EJ, Kim KR, Cha BS, Song YD, Lim SK, Lee HC, Huh KB: "Effect of obesity on total and free insulin-like growth factor (IGF)-1, and their relationship to IGF-binding protein (BP)-1, IGFBP-2, IGFBP-3, insulin, and growth hormone." International Journal of Obesity and Related Metabolic Disorders (1997) May;21(5):355-359.

Narayanan S: "Laboratory markers as an index of aging." Annals of Clinical and Laboratory Science (1996) Jan;26(1):50-59.

Nicoll JA, Roberts GW, Graham DI: "Amyloid beta-protein, APOE genotype and head injury." Annals of the New York Academy of Sciences (1996) Jan 17;777:271-275.

Nikitin AG, Shmookler Reis RJ: "Role of transposable elements in age-related genomic instability." Genet Res (1997) Jun;69(3):183-195.

Nohl H, Staniek K, Gille L: "Imbalance of oxygen activation and energy metabolism as a consequence or mediator of aging." Experimental Gerontology (1997) Jul;32(4-5):485-500.

Norman DC, Yoshikawa TT: "Fever in the elderly." Infectious Disease Clinics of North America (1996) Mar;10(1):93-99.

Nyberg F: "Aging effects on growth hormone receptor binding in the brain." Experimental Gerontology (1997) Jul;32(4-5):521-528.

O'Boyle CA: "Measuring the quality of later life." Philos Trans R Soc Lond B Biol Sci. (1997) Dec 29;352(1363) 1871-9.

Oddis CV: "New perspectives on osteoarthritis." American Journal of Medicine (1996) Feb 26;100(2A):10S-15S.

Oliveira RJ: "The active ear canal." Journal of the American Academy of Audiology (1997) Dec;8(6):401-410.

Olshansky SJ, Carnes BA: "Ever since Gompertz." Demography (1997) Feb;34(1):1-15.

Ooyama T, Sakamato H: "Elastase in the prevention of arterial aging and the treatment of atherosclerosis." Ciba Foundation Symposium (1995) 192:307-317.

Ortmeyer HK, Bodkin NL, Hansen BC: "Insulin-mediated glycogen synthetase activity in muscle of spontaneously insulin-resistant and diabetic rhesus monkeys." Am. J. Physiol. (1993); 265(3PT2):R552-R558.

Ortmeyer HK, Bodkin NL, Hansen BC: "Adipose tissue glycogen synthetase activation by *in vivo* insulin in spontaneously insulin-resistant and type 2 (non-insulin-dependent) diabetic rhesus monkeys." Diabetologia (1993);36:200-206.

Oshimura M, Barrett JC: "Multiple pathways to cellular senescence: role of telomerase repressors." Eur J Cancer (1997) Apr;33(5):710-715.

Ouslander JG: "Aging and the lower urinary tract." American Journal of the Medical Sciences (1997) Oct;314(4):214-218.

Owsley: "Clinical and research issues on older drivers: future directions." Alzheimer Disease and Associated Disorders (1997) Jun;11 Suppl 1:3-7.

Ozawa T: "Genetic and functional changes in mitochondria associated with aging." Physiol Rev (1997) Apr;77(2):425-464.

Pandit L, Ouslander JG: "Postmenopausal vaginal atrophy and atrophic vaginitis." American Journal of the Medical Sciences (1997) Oct;314(4):228-231.

Pantoni L, Garcia JH: "Cognitive impairment and cellular/vascular changes in the cerebral white matter." Annals of the New York Academy of Sciences (1997) Sep 26;826:92-102.

Papa S, Skulachev VP: "Reactive oxygen species, mitochondria, apoptosis and aging." Mol Cell Biochem (1997) Sep;174(1-2):305-319.

Papanicolaou DA, Wilder RL, Manolagas SC, Chrousos GP: "The pathophysiologic roles of interleukin-6 in human disease." Annals of Internal Medicine (1998) Jan 15;128(2):127-137.

Parker CR Jr, Mixon RL, Brissie RM, Grizzle WE: "Aging alters zonation in the adrenal cortex of men." Journal of Clinical Endocrinology and Metabolism (1997) Nov;82(11):3898-3901.

Parr T: "Insulin exposure controls the rate of mammalian aging." Mechanisms of Ageing and Development (1996) Jul 5;88(1-2):75-82.

Parr T: "Insulin exposure and aging theory." Gerontology (1997) 43(3):182-200.

Parsons PA: "The limit to human longevity: an approach through a stress theory of ageing." Mechanisms of Ageing and Development (1996) Jun 25;87(3):211-218.

Paul RG, Bailey AJ: "Glycation of collagen: the basis of its central role in the late complications of ageing and diabetes." Int J Biochem Cell Biol (1996) Dec;28(12):1297-1310.

Pawelec G, Adibzadeh M, Solana R, Beckman I: "The T cell in the ageing individual." Mechanisms of Ageing and Development (1997) Feb;93(1-3):35-45.

Perazella MA, Mahnensmith RL: "Hyperkalemia in the elderly: drugs exacerbate impaired potassium homeostasis." Journal of General Internal Medicine (1997) Oct;12(10):646-656.

Perez-Campo R, Lopez-Torres M, Cadenas S, Rojas C, Barja G: "The rate of free radical production as a determinant of the rate of aging: evidence from the comparative approach." Journal of Comparative Physiology [B] (1998) Apr;168(3):149-158.

Perry E, Kay DW: "Some developments in brain ageing and dementia." British Journal of Biomedical Science (1997)

Sep;54(3):201-215.

Personelle J, De Campos S, Ruiz R de O, Ribeiro GQ: "Injection of all-trans retinoic acid for treatment of thin wrinkles." Aesthetic Plastic Surgery (1997) May;21(3):196-204.

Peterlik M: "Aging, neuroendocrine function, and osteoporosis." Experimental Gerontology (1997) Jul;32(4-5):577-586.

Petrella RJ, Cunningham DA, Paterson DH: "Effects of 5-day exercise training in elderly subjects on resting left ventricular diastolic function and VO2max." Canadian Journal of Applied Physiology (1997) Feb;22(1):37-47.

Pharoah PD, Day NE, Duffy S, Easton DF, Ponder BA: "Family history and the risk of breast cancer: a systematic review and meta-analysis." International Journal of Cancer (1997) May 29;71(5):800-809.

Pickle LW, Mungiole M, Jones GK, White AA: "Atlas of United States Mortality." U.S. Department of Health and Human Services, DHHS Publication No. (PHS) 97-1015 December (1996).

Pierscionek BK, Weale RA: "The optics of the eye-lens and lenticular senescence. A review." Documenta Ophthalmologica (1995) 89(4):321-335.

Pincus SM, Mulligan T, Iranmanesh A, Gheorghiu S, Godschalk M, Veldhuis JD: "Older males secrete luteinizing hormone and testosterone more irregularly, and jointly more asynchronously, than younger males." Proceedings of the National Academy of Sciences of the United States of Ameri(1996) Nov 26;93(24):14100-14105.

Playfer JR: "Parkinson's disease." Postgraduate Medical Journal (1997) May;73(859):257-284.

Poehlman ET, Toth MJ, Fishman PS, Vaitkevicius P, Gottlieb SS, Fisher ML, Fonong T: "Sarcopenia in aging humans: the impact of menopause and disease." J Gerontol A Biol Sci Med Sci (1995) Nov;50 Spec No:73-7.

Poole DC, Sexton WL, Farkas GA, Powers SK, Reid MB: "Diaphragm structure and function in health and disease." Medicine and Science in Sports and Exercise (1997) Jun;29(6):738-754.

Poulin MJ, Cunningham DA, Paterson DH: "Dynamics of the ventilatory response to step changes in end-tidal PCO2 in older humans." Canadian Journal of Applied Physiology (1997) Aug;22(4):368-383.

Prinz PN: "Sleep and sleep disorders in older adults." Journal of Clinical Neurophysiology (1995) Mar;12(2):139-146.

Proctor DN, Joyner MJ: "Skeletal muscle mass and the reduction of VO2max in trained older subjects." Journal of Applied Physiology (1997) May;82(5):1411-1415.

Rantanen T, Era P, Heikkinen E: "Physical activity and the changes in maximal isometric strength in men and women from the age of 75 to 80 years." Journal of the American Geriatrics Society (1997) Dec;45(12):1439-1445.

Rattan SI: "Synthesis, modifications, and turnover of proteins during aging." Experimental Gerontology (1996) Jan;31(1-2):33-47.

Raz N, Dupuis JH, Briggs SD, McGavran C, Acker JD: "Differential effects of age and sex on the cerebellar hemispheres and the vermis: a prospective MR study." AJNR American Journal of Neuroradiology (1998) Jan;19(1):65-71.

Reed, M.J., P.E. Penn, et al.: "Enhanced cell proliferation and biosynthesis mediate improved wound repair in refed, caloric restricted mice." Mech. Ageing Dev., (1996). 89:21-43.

Reiser KM: "Nonenzymatic glycation of collagen in aging and diabetes." Proceedings of the Society for Experimental Biology and Medicine (1998) May;218(1):23-37.

Reiser, K., C. McGee, R. Rucker: "Effects of aging and caloric restriction on extracellular matrix biosynthesis in a model of injury and repair in rats." The Journals of Gerontology, (1995). 50A:B40-B47.

Reiter RJ: "Functional diversity of the pineal hormone melatonin: its role as an antioxidant." Experimental and Clinical Endocrinology & Diabetes (1996) 104:10-16.

Reiter RJ: "Oxidative processes and antioxidative defense mechanisms." The FASEB Journal (1995) 9:526-533.

Reiter RJ: "Oxygen radical detoxification processes during aging: the functional importance of melatonin." Aging (Milano) (1995) Oct;7(5) 340-351.

Reiter RJ: "The pineal gland and melatonin in relation to aging: a summary of the theories and of the data." Exp. Gerontol. (1995) 30:199-212.

Reiter RJ, Carneiro RC, Oh CS: "Melatonin in relation to cellular antioxidative defense mechanisms." Horm Metab Res (1997) Aug;29(8) 363-372.

Reiter RJ, Guerrero JM, Escames G, Pappolla MA, Acuna-Castroviejo D: "Prophylactic actions of melatonin in oxidative neurotoxicity." Ann N Y Acad Sci (1997) Oct 15;825 70-78.

Reiter RJ, Pablos MI, Agapito TT, Guerrero JM: "Melatonin in the context of the free radical theory of aging." Ann. N. Y. Acad. Sci. (1996) 786:362-378.

Resnick B: "Dermatologic problems in the elderly." Lippincotts Primary Care Practice (1997) Mar;1(1):14-30.

Resnick NM: "Geriatric incontinence." Urologic Clinics of North America (1996) Feb;23(1):55-74.

Rice MM, Graves AB, McCurry SM, Larson EB: "Estrogen replacement therapy and cognitive function in postmenopausal women without dementia." Am J Med. (1997) Sep 22;103(3A) 26S-35S.

Rich MW: "Epidemiology, pathophysiology, and etiology of congestive heart failure in older adults." Journal of the American Geriatrics Society (1997) Aug;45(8):968-974.

Richardson JS, Zhou Y, Kumar U: "Free radicals in the neurotoxic actions of beta-amyloid." Ann N Y Acad Sci. (1996) Jan 17;777 362-7.

Richter C: "Reactive oxygen and nitrogen species regulate mitochondrial Ca2+ homeostasis and respiration." Biosci Rep (1997) Feb;17(1):53-66.

Robert L: "Aging of the vascular wall and atherogenesis: role of the elastin-laminin receptor." Atherosclerosis (1996) Jun;123(1-2):169-179.

Robert L, Jacob MP, Fulop T: "Elastin in blood vessels." Ciba Foundation Symposium (1995) 192:286-299.

Robinson G: "Cross-cultural perspectives on menopause." Journal of Nervous and Mental Disease (1996) Aug;184(8):453-458.

Roecker EB, Kemnitz JM, Erschler WB, Weindruch R: "Reduced immune responses in rhesus monkeys subjected to dietary restriction." J. Gerontol. Biol. Sci. Med. Sci. (1996) 51(4):B276- B279.

Roos MR, Rice CL, Vandervoort AA: "Age-related changes in motor unit function." Muscle and Nerve (1997) Jun;20(6):679-690.

Rose RC, Richer SP, Bode AM: "Ocular oxidants and antioxidant protection." Proceedings of the Society for Experimental Biology and Medicine (1998) Apr;217(4):397-407.

Rosen CJ, Conover C: "Growth hormone/insulin-like growth factor-I axis in aging: a summary of a National Institutes of Aging-Sponsored Symposium." Journal of Clinical Endocrinology and Metabolism (1997) Dec;82(12):3919-3922.

Rosen CJ : "Growth hormone, IGF-I, and the elderly. Clues to potential therapeutic interventions." Endocrine. (1997) Aug;7(1) 39-40.

Rosenberg IH : "Nutrition and senescence." Nutr Rev. (1997) Jan;55(1 Pt 2) S69-73.

Roses AD: "Apolipoprotein E, a gene with complex biological interactions in the aging brain." Neurobiology of Disease (1997) 4(3-4):170-185.

Ross PD : "Clinical consequences of vertebral fractures." Am J Med. (1997) Aug 18;103(2A) 30S-42S.

Roth GS, Ingram DK, Lana MA: "Slowing ageing by caloric restriction." Nature Medicine (1995) 1(5):414-415.

Rowe JW, Kahn RL: "Successful aging." Gerontologist (1997) Aug;37(4):433-440.

Rubin CD: "Southwestern Internal Medicine Conference: prevention of hip fractures in the elderly." American Journal of the Medical Sciences (1995) Aug;310(2):77-85.

Rubin H: "Cell aging in vivo and in vitro." Mechanisms of Ageing and Development (1997) Oct;98(1):1-35.

Rush CB, Entman SS: "Pelvic organ prolapse and stress urinary incontinence." Medical Clinics of North America (1995) Nov;79(6):1473-1479.

Sasaki H, Sekizawa K, Yanai M, Arai H, Yamaya M, Ohrui T: "New strategies for aspiration pneumonia." Internal Medicine (1997) Dec;36(12):851-855.

Schaefer EJ, Lichtenstein AH, Lamon-Fava S, McNamara JR, Ordovas JM: "Lipoproteins, nutrition, aging, and atherosclerosis." American Journal of Clinical Nutrition (1995) Mar;61(3 Suppl):726S-740S.

Scharbo-Dehaan M : "Hormone replacement therapy." Nurse Pract. (1996) Dec;21(12 Pt 2) 1-13.

Scharffetter-Kochanek K, Wlaschek M, Brenneisen P, Schauen M, Blaudschun R, Wenk J: "UV-induced reactive oxygen species in photocarcinogenesis and photoaging." Biological Chemistry (1997) Nov;378(11):1247-1257.

Schmucker DL, Heyworth MF, Owen RL, Daniels CK: "Impact of aging on gastrointestinal mucosal immunity." Digestive Diseases and Sciences (1996) Jun;41(6):1183-1193.

Schultz AB: "Muscle function and mobility biomechanics in the elderly: an overview of some recent research." J Gerontol A Biol Sci Med Sci (1995) Nov;50 Spec No:60-3.

Schwartz RS: "Sarcopenia and physical performance in old age: introduction." Muscle and Nerve. Supplement (1997) 5:S10-S12.

Seeman TE, McEwen BS, Singer BH, Albert MS, Rowe JW: "Increase in urinary cortisol excretion and memory declines: MacArthur studies of successful aging." Journal of Clinical Endocrinology and Metabolism (1997) Aug;82(8):2458-2465.

Seidman MD, Jacobson GP: "Update on tinnitus." Otolaryngologic Clinics of North America (1996) Jun;29(3):455-465.

Seiler KS, Spirduso WW, Martin JC: "Gender differences in rowing performance and power with aging." Medicine and Science in Sports and Exercise (1998) Jan;30(1):121-127.

Sell DR, Lane MA, Johnson WA, Masoro EJ, Mock OB, Reiser KM, Fogarty JF, Cutler RG, Ingram DK, Roth GS, Monnier VM: "Longevity and the genetic determination of collagen glycoxidation kinetics in mammalian senescence." Proc. Natl. Acad. Sci. USA (1996) 93:485-409.

Shadden BB: "Discourse behaviors in older adults." Seminars in Speech and Language (1997) May;18(2):143-156.

Shah GN, Mooradian AD: "Age-related changes in the blood-brain barrier." Experimental Gerontology (1997) Jul;32(4-5):501-519.

Shepard RJ, Shek PN: "Impact of physical activity and sport on the immune system." Reviews on Environmental Health (1996) Jul;11(3):133-147.

Shephard RJ, Shek PN: "Cancer, immune function, and physical activity." Canadian Journal of Applied Physiology (1995) Mar;20(1):1-25.

Sherratt EJ, Thomas AW, Alcolado JC: "Mitochondrial DNA defects: a widening clinical spectrum of disorders." Clin Sci (Colch) (1997) Mar;92(3):225-235.

Shiflett S, Cooke CE: "Osteoporosis: a focus on treatment." Maryland Medical Journal (1997) Jul;46(6):303-307.

Shinkai S, Konishi M, Shephard RJ: "Aging, exercise, training, and the immune system." Exercise Immunology Review (1997) 3:68-95.

Shmookler Reis RJ, Ebert RH 2nd: "Genetics of aging: current animal models." Experimental Gerontology (1996) Jan;31(1-2):69-81.

Sial S, Coggan AR, Carroll R, Goodwin J, Klein S: "Fat and carbohydrate metabolism during exercise in elderly and young subjects." American Journal of Physiology (1996) Dec;271(6 Pt 1):E983-E989.

Simpkins JW, Green PS, Gridley KE, Singh M, de Fiebre NC, Rajakumar G : "Role of estrogen replacement therapy in memory enhancement and the prevention of neuronal loss associated with Alzheimer's disease." Am J Med. (1997) Sep 22;103(3A) 19S-25S.

Singh I, Marshall MC Jr: "Diabetes mellitus in the elderly." Endocrinology and Metabolism Clinics of North America (1995) Jun;24(2):255-272.

Smith DW: "Evolution of longevity in mammals." Mechanisms of Ageing and Development (1995) Jun 30;81(1):51-60.

Smith JB, Fenske NA: "Cutaneous manifestations and consequences of smoking." Journal of the American Academy of Dermatology (1996) May;34(5 Pt 1):717-732.

Smith MA: "Alzheimer disease." International Review of Neurobiology (1998) 42:1-54.

Smith SC: "Aging, physiology, and vision." Nurse Practitioner Forum (1998) Mar;9(1):19-22.

Smulyan H, Safar ME: "Systolic blood pressure revisited." Journal of American College of Cardiology (1997) Jun;29(7):1407-1413.

Soffa VM: "Alternatives to hormone replacement for menopause. Part 1: Implications for the elderly." Alternative Therapies in Health and Medicine (1996) Mar;2(2):34-39.

Solomon DH, Gurwitz JH: "Toxicity of nonsteroidal anti-inflammatory drugs in the elderly: is advanced age a risk factor?." American Journal of Medicine (1997) Feb;102(2):208-215.

Stabler SP : "Vitamin B12 deficiency in older people: improving diagnosis and preventing disability." J Am Geriatr Soc. (1998) Oct;46(10) 1317-9.

Staron RS: "Human skeletal muscle fiber types: delineation, development, and distribution." Canadian Journal of Applied Physiology (1997) Aug;22(4):307-327.

Steenvoorden DP, van Henegouwen GM: "The use of endogenous antioxidants to improve photoprotection." Journal of Photochemistry and Photobiology. B, Biology (1997) Nov;41(1-2):1-10.

Stevens J, Cai J, Pamuk ER, Williamson DF, Thun MJ, Wood JL: "The effect of age on the association between body-mass index and mortality." New England Journal of Medicine (1998) Jan 1;338(1):1-7.

Strigini L, Ryan T: "Wound healing in elderly human skin." Clinics in Dermatology (1996) Mar;14(2):197-206.

Sulcova J, Hill M, Hampl R, Starka L: "Age and sex related differences in serum levels of unconjugated dehydroepiandrosterone and its sulphate in normal subjects." Journal of Endocrinology (1997) Jul;154(1):57-62.

Sullivan MP, Yalla S: "Detrusor contractility and compliance characteristics in adult male patients with obstructive and nonobstructive voiding dysfunction." Journal of Urology (1996) Jun;155(6):1995-2000.

Sullivan R: "Contributions to senescence: non-enzymatic glycosylation of proteins." Arch Physiol Biochem (1996) Dec;104(7):797-806.

Sunderkotter C, Kalden H, Luger TA: "Aging and the skin immune system." Archives of Dermatology (1997) Oct;133(10):1256-1262.

Susic D: "Hypertension, aging, and atherosclerosis: the endothelial interface." Medical Clinics of North America (1997) Sep;81(5):1231-1240.

Swaab DF: "Ageing of the human hypothalamus." American Journal of Cardiology (1995) Nov 2;76(13):2D-7D.

Swynghedauw B, Besse S, Assayag P, Carre F, Chevalier B, Charlemagne D, Delcayre C, Hardouin S, Heymes C, Moalic JM: "Molecular and cellular biology of the senescent hypertrophied and failing heart." American Journal of Cardiology (1995) Nov 2;76(13):2D-7D.

Tailleux A, Fruchart JC: "HDL heterogeneity and atherosclerosis." Critical Reviews in Clinical Laboratory Science (1996) Jun;33(3):163-201.

Tamayo-Orozco J, Arzac-Palumbo P, Peon-Vidales H, Mota-Bolfeta R, Fuentes F : "Vertebral fractures associated with osteoporosis patient management." Am J Med. (1997) Aug 18;103(2A) 44S-48S.

Tanaka H, Seals DR: "Age and gender interactions in physiological functional capacity: insight from swimming performance." Journal of Applied Physiology (1997) Mar;82(3):846-851.

Taunton JE, Martin AD, Rhodes EC, Wolski LA, Donelly M, Elliot J: "Exercise for the older woman: choosing the right prescription." British Journal of Sports Medicine (1997) Mar;31(1):5-10.

Tenover JS: "Androgen administration to aging men." Endocrinology and Metabolism Clinics of North America (1994);Dec;23(4):877-892.

Teri L, McCurry SM, Logsdon RG: "Memory, thinking, and aging. What we know about what we know." Western Journal of Medicine (1997) Oct;167(4):269-275.

Terman A, Brunk UT: "Lipofuscin: mechanisms of formation and increase with age." APMIS (1998) Feb;106(2):265-276.

Termrungruanglert W, Kudelka AP, Edwards CL, Delclos L, Verschraegen CF, Kavanagh JJ: "Gynecologic cancer in the elderly." Clinics in Geriatric Medicine (1997) May;13(2):363-379.

Thorpe SR, Baynes JW: "Role of the Maillard reaction in diabetes mellitus and diseases of aging." Drugs and Aging (1996) Aug;9(2):69-77.

Thurman JE, Mooradian AD: "Vitamin supplementation therapy in the elderly." Drugs and Aging (1997) Dec;11(6):433-449.

Timiras PS: "Education, homeostasis, and longevity." Experimental Gerontology (1995) May;30(3-4):189-198.

Tohno S, Tohno Y, Minami T, Ichii M, Okazaki Y, Utsumi M, Nishiwaki F, Yamada M: "Difference of mineral contents in human intervertebral disks and its age-related change." Biological Trace Element Research (1996) May;52(2):117-124.

Tohno Y, Utsumi M, Tohno S, Minami T, Okazaki Y, Moriwake Y, Nishiwaki F, Yamada M, Fujii T, Takakura Y: "Age-dependent changes of mineral contents in men and women's calcanei." Biological Trace Element Research (1997) Oct;60(1-2):81-90.

Touitou Y, Bogdan A, Haus E, Touitou C: "Modifications of circadian and circannual rhythms with aging." Experimental Gerontology (1997) Jul;32(4-5):603-614.

Touitou Y: "Effects of ageing on endocrine and neuroendocrine rhythms in humans." Hormone Research (1995) 43(1-3):12-19.

Tuite DJ, Renstrom PA, O'Brien M: "The aging tendon." Scandinavian Journal of Medicine and Science in Sports (1997) Apr;7(2):72-77.

van Venrooij GE, Boon TA: "Extensive urodynamic investigation: interaction among diuresis, detrusor instability, urethral relaxation, incontinence and complaints in women with a history of urge incontinence." Journal of Urology (1994);Nov;152(5 Pt 1):1535-1538.

Van de Kerkhof PC, Van Bergen B, Spruijt K, Kuiper JP: "Age-related changes in wound healing." Clinical and Experimental Dermatology (1994);Sep;19(5):369-374.

Van Scott EJ, Ditre CM, Yu RJ: "Alpha-hydroxy acids in the treatment of signs of photoaging." Clinics in Dermatology (1996) Mar;14(2):217-226.

Vaziri H, Benchimol S: "From telomere loss to p53 induction and activation of a DNA-damage pathway at senescence: the telomere loss/DNA damage model of cell aging." Experimental Gerontology (1996) Jan;31(1-2):295-301.

Veldhuis JD, Iranmanesh A, Weltman A: "Elements in the pathophysiology of diminished growth hormone (GH) secretion in aging humans." Endocrine (1997) Aug;7(1):41-48.

Vermeulen A, Kaufman JM: "Ageing of the hypothalamo-pituitary-testicular axis in men." Hormone Research (1995) 43(1-3):25-28.

Vestal RE: "Aging and pharmacology." Cancer (1997) Oct 1;80(7):1302-1310.

Vilar-Rojas C, Guzman-Grenfell AM, Hicks JJ: "Participation of oxygen-free radicals in the oxido-reduction of proteins."

Arch Med Res (1996) 27(1):1-6.

Villeponteau B: "The heterochromatin loss model of aging." Experimental Gerontology (1997) Jul;32(4-5):383-394.

Vita AJ, Terry RB, Hubert HB, Fries JF: "Aging, health risks, and cumulative disability." New England Journal of Medicine (1998) Apr 9;338(15):1035-1041.

Vitez M: "A look through the eyes of age." The Philadelphia Inquirer Friday, November 20: A1 (1998).

Vrensen GF: "Aging of the human eye lens—a morphological point of view." Comparative Biochemistry and Physiology. Part A, Physiology (1995) Aug;111(4):519-532.

Wagg AS, Lieu PK, Ding YY, Malone-Lee JG: "A urodynamic analysis of age associated changes in urethral function in women with lower urinary tract symptoms." Journal of Urology (1996) Dec;156(6):1984-1988.

Wallace RB: "Cognitive change, medical illness, and crash risk among older drivers: an epidemiological consideration." Alzheimer Disease and Associated Disorders (1997) Jun;11 Suppl 1:31-37.

Wang E: "Regulation of apoptosis resistance and ontogeny of age-dependent diseases." Experimental Gerontology (1997) Jul;32(4-5):471-484.

Ward JA: "Should antioxidant vitamins be routinely recommended for older people?." Drugs and Aging (1998) Mar;12(3):169-175.

Warner HR: "Aging and regulation of apoptosis." Curr Top Cell Regul (1997) 35:107-121.

Wei YH, "Oxidative stress and mitochondrial DNA mutations in human aging." Proc Soc Exp Biol Med. (1998) Jan;217(1):53-63.

Weindruch R: "Caloric restriction and aging." Scientific American (1996) Jan 46-52.

Weindruch R: "The retardation of aging by caloric restriction: studies in rodents and primates." Toxicologic Pathology (1996) Nov;24(6):742-745.

Weindruch R, Marriott BM, Conway J, Knapka JJ, Lane MA, Cutler RG, Roth GS, Ingram DK: "Measures of body size and growth in rhesus and squirrel monkeys subjected to long-term dietary restriction." Am. J. Primatol. (1995). 35:207-228.

Weindruch R, Sohal RS: "Seminars in medicine of the Beth Israel Deaconess Medical Center. Caloric intake and aging." New England Journal of Medicine (1997) Oct 2;337(14):986-994.

Weinstock MA : "Death from skin cancer among the elderly epidemiological patterns." Arch Dermatol (1997) Oct;133(10) 1207-9.

Weng NP, Palmer LD, Levine BL, Lane HC, June CH, Hodes RJ: "Tales of tails: regulation of telomere length and telomerase activity during lymphocyte development, differentiation, activation, and aging." Immunological Reviews (1997) Dec;160:43-54.

Wheaton K, Atadja P, Riabowol K: "Regulation of transcription factor activity during cellular aging." Biochem Cell Biol (1996) 74(4):523-534.

Wick G, Grubeck-Loebenstein B: "Primary and secondary alterations of immune reactivity in the elderly: impact of dietary factors and disease." Immunological Reviews (1997) Dec;160:171-184.

Wilkinson CW, Peskind ER, Raskind MA: "Decreased hypo-

thalamic-pituitary-adrenal axis sensitivity to cortisol feedback inhibition in human aging." Neuroendocrinology (1997) Jan;65(1):79-90.

Wisniewski HM, Silverman W: "Diagnostic criteria for the neuropathological assessment of Alzheimer's disease: current status and major issues." Neurobiology of Aging (1997) Jul;18(4 Suppl):S43-S50.

Witelson SF: "Sex differences in neuroanatomical changes with aging." New England Journal of Medicine (1991) Jul 18;325(3):211-212.

Wohlert AB: "Tactile perception of spatial stimuli on the lip surface by young and older adults." Journal of Speech and Hear Research (1996) Dec;39(6):1191-1198.

Wolf OT, Neumann O, Hellhammer DH, Geiben AC, Strasburger CJ, Dressendorfer RA, Pirke KM, Kirschbaum C: "Effects of a two-week physiological dehydroepiandrosterone substitution on cognitive performance and well-being in healthy elderly women and men." Journal of Clinical Endocrinology and Metabolism (1997) Jul;82(7):2363-2367.

Wolfson L, Judge J, Whipple R, King M: "Strength is a major factor in balance, gait, and the occurrence of falls." J Gerontol A Biol Sci Med Sci (1995) Nov;50 Spec No:64-7.

Wood RJ, Suter PM, Russell RM: "Mineral requirements of elderly people." American Journal of Clinical Nutrition (1995) Sep;62(3):493-505.

Woodhead AD: "Aging, the fishy side: an appreciation of Alex Comfort's studies." Experimental Gerontology (1998) Jan;33(1-2):39-51.

Woodruff-Pak DS: "Classical conditioning." International Review of Neurobiology (1997) 41:341-366.

Wyatt HJ, Fisher RF: "A simple view of age-related changes in the shape of the lens of the human eye." Eye (1995) 9(Pt 6):772-775.

Yaar M, Gilchrest BA: "Human melanocytes as a model system for studies of Alzheimer disease." Archives of Dermatology (1997) Oct;133(10):1287-1291.

Yancik R: "Cancer burden in the aged: an epidemiologic and demographic overview." Cancer (1997) Oct 1;80(7):1273-1283.

Yates AA: "Process and development of dietary reference intakes: basis, need, and application of recommended dietary allowances." Nutrition Reviews (1998) Apr;56(4 Pt 2):S5-S9.

Yin D: "Studies on age pigments evolving into a new theory of biological aging." Gerontology (1995) 41 Suppl 2:159-172.

Ying W: "Deleterious network hypothesis of aging." Medical Hypotheses (1997) Feb;48(2):143-148.

Young A: "Ageing and physiological functions." Philosophical Transactions of the Royal Society of London. Series B: Biologic(1997) Dec 29;352(1363):1837-1843.

Zauderer B: "Age-related changes in renal function." Critical Care Nursing Quarterly (1996) Aug;19(2):34-40.

Zec RF: "The neuropsychology of aging." Experimental Gerontology (1995) May;30(3-4):431-442.

Zheng B, Han S, Takahashi Y, Kelsoe G: "Immunosenescence and germinal center reaction." Immunological Reviews (1997) Dec;160:63-77.

Zs -Nagy I: "The membrane hypothesis of aging: its relevance to recent progress in genetic research." Journal of Molecular Medicine (1997) Oct;75(10):703-714.

BIBLIOGRAPHY
First Edition

Abrams, William B., and Robert Berkow: *The Merck Manual of Geriatrics*, Merck Sharp and Dome Research Laboratories, Rahway, N.J., 1990.

Adams, Inta: "Plasticity of the Synaptic Contact Zone Following Loss of Synapses in the Cerebral Cortex of Aging Humans," *Brain Research*, Vol. 424, pp. 343-351, 1987.

Aging America: Trends and Projections, U.S. Department of Health and Human Services, Washington, D.C., 1991 edition.

Aging America: Trends and Projections, Serial No. 101-J, U.S. Government Printing Office, Washington, D.C., 1990.

Aging America: Trends and Projections, U.S. Department of Health and Human Services, Washington, D.C., 1987-88 edition.

Allgeier, Elizabeth R., and Albert R. Allgeier: *Sexual Interactions*, 3rd edition, D.C. Heath, Lexington, Mass., 1991.

Andres, Reubin, Edwin Bierman, and William R. Hazzard, (eds.): *Principles of Geriatric Medicine*, McGraw-Hill Book Co., New York. 1985.

Andrew, Warren: *The Anatomy of Aging in Man and Animals*, Grune and Stratton, New York 1971.

Arenberg, David, and Elizabeth A. Robertson-Tchabo: "Adult Age Differences in Memory and Linguiistic Integration Revisited," *Experimental Aging Research*, Vol. 11, pp. 187-191, 1985.

Arking, Robert: *Biology of Aging: Observations and Principles*, Prentice Hall, Englewood Cliffs, N. J., 1991.

Ashford, J. Wesson, Paul Kolm, Jerry A. Colliver, Cathy Bekian, and Lee-Nah Hsu: "Alzheimer Patient Evaluation and the Mini-Mental State: Item Characteristic Curve Analysis," *Journal of Gerontology*, Vol. 44., pp. P139-146, 1989.

Avioli, Louis V., (ed.): *The Osteoporotic Syndrome: Detection, Prevention, and Treatment*, 2nd edition, Grune and Stratton, Inc., Harcourt Brace Jovanovich, Publishers, New York, 1987.

Baker, Andrea C., Li-Wen Ko and John P. Blass: "Systemic Manifistations of Alzheimer's Disease," *Age*, Vol. 11, pp. 60-65, 1988.

Barclay, Laurie: "Differential Diagnosis of Dementing Diseases," *Age*, Vol. 11, No. 1, pp. 19-22, 1988.

Barzilai, Nir, Jochanan Stessman, Pinchas Cohen, Gil Morali, David Barzilai, and Eddy Karnieli: "Glucoregulatory Hormone Influence on Hepatic Glucose Production in the Elderly," *Age*, Vol. 12, pp. 13-17, 1989.

Baylor, Ann M., and Waneen W. Spirduso: "Systematic Aerobic Exercise and Components of Reaction Time in Older Women," *Journal of Gerontology*, Vol. 43, pp. P121-126, 1988.

Behnke, John A., Caleb E. Finch, and Gairdner B. Moment, (eds.): *The Biology of Aging*, Plenum Press, New York, 1978.

Berkow, Robert, (ed.): *The Merck Manual of Diagnosis and Therapy*, 14th edition, Merck Sharp and Dohme Research Laboratories, Rahway, N.J., 1982.

Berliner, Harriet: "Aging Skin, Part Two," *American Journal of Nursing*, pp. 1259-1261, November 1986.

Berliner, Harriet: "Aging Skin," *American Journal of Nursing*, pp. 1138-1142, October 1986.

Bigl, V., T. Arendt, S. Fischer, S. Fischer, M. Werner, and A. Arendt: "The Cholinergic System in Aging," *Gerontology*, Vol. 33, pp. 172-180, 1987.

Bittles, A. H., and K. J. Collins: *The Biology of Human Ageing*,

Cambridge University Press, Cambridge, U.K., 1986.

Blumenthal, James A., *et al.*: "Cardiovascular and Behavioral Effects of Aerobic Exercise Training in Healthy Older Men and Women," *Journal of Gerontology*, Vol. 44, no. 5, pp. M147-157, 1989.

Bogardus, Clifton, and Stephen Lillioja: "Where All the Glucose Doesn't Go in Non-Insulin-Dependent Diabetes Mellitus," *New England Journal of Medicine*, Vol. 322, pp. 262-263, 1990.

Borson, Soo, Robert F. Barnes, Richard C. Veith, Jeffrey B. Halter, and Murray A. Raskind: "Impaired Sympathetic Nervous System Response to Cognitive Effort in Early Alzheimer's Disease," *Journal of Gerontology*, Vol. 44, pp. M8-12, 1989.

Bouissou, Hubert, Marie-Therese Pieraggi, Monique Julian, and Theresa Savit: "The Elastic Tissue of the Skin: A Comparison of Spontaneous and Actinic (Solar) Aging," *International Journal of Dermatology*, Vol. 27(Jun.) pp. 327-335, 1988.

Brandt, Lawrence J.: *Gastrointestinal Disorders of the Elderly*, Raven Press, New York, 1984.

Brill, S., T. Kukulansky, E. Tal, L. Abel, Y. Polgin, C. Dassa, and A. Globerson: "Individual Changes in T Lymphocyte Parameters of Old Human Subjects," *Mechanisms of Ageing and Development*, Vol. 40, pp. 71-79, Elsevier Scientific Publishers Ireland Ltd., 1987.

Brookbank, John W.: *The Biology of Aging*, Harper & Row, New York, 1990.

Burke, Jean R., Gary Kamen, and David M. Koceja: "Long-Latency Enhancement of Quadriceps Excitability From Stimultation of Skin Afferents in Young and Old Adults," *Journal of Gerontology*, Vol. 44, pp. M158-163, 1989.

Burns, Elizabeth M., and Kathleen Coen Buckwalter: "Pathophysiology and Etiology of Alzheimer's Disease" pp. 11-29 in *The Nursing Clinics of North America*, Vol. 23, No. 1, W. B. Saunders Co., Harcourt Brace Jovanovich, Inc., Philadelphia, March 1988.

Busse, Ewald W., and George L. Maddox: *The Duke Longitudinal Studies of Normal Aging: 1955-1980*, Springer, New York, N. Y., 1985.

Campbell, A. John, Michael K. Borrie, and Geroge F. Spears: "Risk Factors for Falls in a Community-Based Prospective Study of People 70 Years and Older," *Journal of Gerontology*, Vol. 44(4), pp. M112-117, 1989.

Carola, Robert, John P. Harley, and Charles R. Noback: *Human Anatomy and Physiology*, second edition, McGraw-Hill, New York, 1992.
Carola, Robert, John P. Harley, and Charles R. Noback: *Human Anatomy and Physiology*, McGraw-Hill, New York, 1990.

Charness, Neil: *Aging and Human Performance*,Wiley, New York, 1985.

Chauhan, J., Z. J. Hawrysh, M. Gee, E. A. Donald, and T. K. Basu: "Age-Related Olfactory and Taste Changes and Interrelationships Between Taste and Nutrition," *Journal of the American Dietetic Association*, Vol. 87, pp. 1543-1550, 1987.

Chauhan, J., Z. J. Hawrysh, M. Gee, E. A. Donald, and T. K. Basu: "Age-Related Olfactory and Taste Changes and Interrelationships Between Taste and Nutrition," *Journal of the American Dietetic Association*, Vol. 87, pp. 1543-1550.

Chernoff, Ronni, and David A. Lipschitz: *Health Promotion and Disease Prevention in the Elderly*, Aging Series, Vol. 35, Raven Press, New York, 1988.

Clark, Matt, Mariana Gosnell, Deborah Witherspoon, Janet Huck, Mary Hager, Darby Junkin, Patricia King, Amy Wallace,

and Tracey L. Robinson: "A Slow Death of the Mind," *Newsweek*, Vol. 104, pp. 156-162, 1984.

Clarkson, Priscilla M., and Mary E. Dedrick: "Exercise-Induced Muscle Damage, Repair, and Adaptation in Old and Young Subjects," *Journal of Gerontology*, Vol. 43, pp. M91-96, 1988.

Coleman, Paul D., and Dorothy G. Flood: "Neuron Numbers and Dendritic Extent in Normal Aging and Alzheimer's Disease," *Neurobiology of Aging*, Vol. 8, pp. 521-545, 1987.
Corberand, J., P. Laharrague, and G. Fillola: "Blood Cell Parameters Do Not Change during Physiological Human Ageing," *Gerontology*, Vol. 33, pp. 72-76, 1987.

Costa, Paul T., and Robert R. McCrae: "The Case for Personality Stability," pp. 418-431 in Maddox, George L., and E. W. Busse, (eds.), *Aging: The Universal Human Experience*, Springer, New York, 1987.

Cowley, Geoffrey, and Mary Hager: "Bad News for the Geritol Set," *Newsweek*, Vol. 120, p. 69, Sept. 21, 1992.

Cowley, Geoffrey: "Progress on Parkinson's," *Newsweek*, Vol. 122, p. 68, Dec. 7, 1992.

Cross, Richard J.: "What Doctors and Others Need to Know: Six Rules on Human Sexuality and Aging". *Siecus Report*, Vol. 17, No. 3, pp. 14-16, 1989.

Crowe, M.J., M. L. Forsling, B. J. Rolls, P. A. Phillips, J. G. G. Ledingham, and R. F. Smith: "Altered Water Excretion in Healthy Elderly Men," *Age and Ageing*, Vol. 16, pp. 285-293, 1987.

Danon, D., N. W. Shock, and M. Marois: *Aging; A Challenge to Science and Society*, Oxford University Press, New York, 1981.

Darr, Kevin C., David R. Bassett, Barbara J. Morgan, and D. Paul Thomas: "Effects of Age and Training Status on Heart Rate Recovery After Peak Exercise," *American Journal of Physiology*, Vol. 254(2) pp. H340-H343, 1988.

Davis, Bernard B., and W. Gibson Wood, (eds.): *Homeostatic Functions and Aging*, Aging Series, Vol. 30, Raven Press, New York, 1985.

De Paoli, P., S. Battistin, and G. F. Santini: "Age-Related Changes in Human Lymphocyte Subsets: Progressive Reduction of the CD4 CD45R (Suppressor Inducer) Population," *Clinical Immunology and Immunopathology*, Vol. 48, pp. 290-296, 1988.

Downs, Robert W., personal communication at Williamsburg, Virginia, Nov. 1989.

Ehrman, Jan Stuart: "Use of Biofeedback to Treat Incontinence," *Jounal of the American Geriatrics Society*, Vol 31, pp. 182-184, 1983.

Elliott, James L.: "Swallowing Disorders in the Elderly: A Guide to Diagnosis and Treatment," *Geriatrics*, Vol. 43(1) pp. 95-113, 1988.

Engel, Bernard T.: "Incontinence," pp. 346-347, in Maddox, G. L., (ed.), *The Encyclopedia of Aging*, Springer, New York, 1987.

Escher, Jeffrey E., and Steven Gambert,"Metabolic Bone Disease," *Age*, Vol. 10, pp. 62-69, 1987.

Euans, David W.: "Renal Function in the Elderly," *American Family Physician*, Vol. 38, pp. 147-150, 1988.

Evered, David, and Julie Whelan, (eds.): *Research and the Aging Population*, Ciba Foundation, Wiley, Chichester, England, 1988.

Evered, David, and Julie Whelan, (eds.): *Research and the Ag-

ing Population, Ciba Foundation, Wiley, Chichester, England, 1988.

Facchini, A., Erminia Mariani, Adriana Rita Mariani, S. Papa, M. Vitale, and F. A. Manzoli: "Increased Number of Circulating Leu 11+ (CD 16) Large Granular Lymphocytes and Decreased NK Activity During Human Aging," *Clinical and Experimental Immunology*, Vol. 68, pp. 340-347, 1987.

Falcao, R. P., S. J. Ismael, and E. A. Donadi: "Age-Associated Changes in T Lymphocyte Subsets," *Diagnostic and Clinical Immunology*, Vol. 5, pp. 205-208, 1987.

Fenske, Neil A., and Christine B. Conard: "Aging Skin," *American Family Physician*, Vol. 37(2), pp. 219-30, 1988.

Ferguson, D. B., (ed.): *The Aging Mouth*, Karger, New York, 1987.

Ferrini, Armeda F., and Rebecca L. Ferrini: *Health in the Later Years*, Brown, Dubuque, Iowa, 1989.

Fiatarone, Maria A., John E. Morley, Eda T. Bloom, Donna Benton, George F. Solomon, and Takashi Makinodan: "The Effect of Exercise on Natural Killer Cell Activity in Young and Old Subjects," *Journal of Gerontology*, Vol. 44, M37-45, 1989.

Finch, Caleb E., and Edward L. Schneider, (eds.): *Handbook of the Biology of Aging*, 2nd edition. Van Nostrand Reinhold, New York, 1985.

Finch, Caleb E., and Leonard Hayflick, (eds.): *Handbook of the Biology of Aging*, 1st edition, Van Nostrand Reinhold, New York, 1977.

Findlay, Steven, Doug Podolsky, and Joanne Silberner: "Iron and Your Heart," *U.S. News and World Report*, Vol. 113, pp. 63-68, Sept. 21, 1992.

Fleg, Jerome L.: "Alterations in Cardiovascular Structure and Function with Advancing Age," *The American Journal of Cardiology*, Vol. 57, pp. 33C-44C, 1986.

Fleg, Jerome L.: "The Aging Heart," pp. 253-273, in Wenger, N. K., C. D. Furberg, and E. Pitt, (eds.), *Coronary Heart Disease in the Elderly*, Elsevier, New York, 1986.

Florini, James R., Minireview: Limitations of Interpretation of Age-Related Changes in Hormone Levels: Illustration by Effects of Thyroid Hormones on Cardiac and Skeletal Muscle," *Journal of Gerontology*, Vol. 44, pp. B107-109, 1989.

Foster, Vicky L., Gwenne J. E. Hume, William C. Byrnes, Arthur L. Dickinson, and Steven J. Chatfield: "Endurance Training For Elderly Women: Moderate vs Low Intensity," *Journal of Gerontology*, Vol. 44, pp. M184-178, 1989.

Fox, Kathleen M., Jordan D. Tobin, and Chris C. Plato: "Longitudinal Study of Bone Loss in the Second Metacarpal," in *Calcified Tissue International*, Springer-Verlag New York Inc., New York, Vol. 39, pp. 218-225, 1986.

Freed, Curt R., *et al.*, "Survival of Implanted Fetal Dopamine Cells and Neurologic Improvement 12 to 46 Months after Transplantation for Parkinson's Disease," *New England Journal of Medicine*, Vol. 327, pp. 1549-1555, 1992.

Ganguly, Rama, and Myron R. Szewczuk: "Age and Immunity to Respiratory Tract Infections," *Age*, Vol. 12, pp. 25-35, 1989.

Gerstenblith, Gary, Dale G. Renlund, and Edward G. Lakatta: "Cardiovascular Response to Exercise in Younger and Older Men," *Federation Proceedings*, Vol. 46, pp. 1834-1839, 1987.

Gietzen, Dorothy W., Theodore A. Goodman, Philip G. Weiler, Kay Graf, David R. Fregreau, Joseph R. Magliozzi, Allen R. Doran, and Richard J. Maddock: "Beta Receptor Density in

Human Lymphocyte Membranes: Changes with Aging?," *Journal of Gerontology*, Vol. 46, pp. B130-134, 1991.

Gilchrest, Barbara A.: *Skin and Aging Processes*, CRC Press, Inc. Boca Raton, Fla., 1984.

Gittings, Neil S., and James L. Fozard: "Age Related Changes in Visual Acuity," *Experimental Gerontology*, Vol. 21, pp. 423-433, 1986.

Goldman, Ralph, and Morris Rockstein, (eds.): *The Physiology and Pathology of Human Aging*, Academic Press, New York, 1975.

Goldstein, Iris B., and Dave Shapiro: "Cardiovascular Response During Postural Change in the Elderly," *Journal of Gerontology*, Vol. 45, no. 1, pp. M20-25, 1990.

Gorman, Kevin M., and Joel D. Posner: "Benefits of Exercise in Old Age," *Common Clinical Challenges in Geriatrics*, Vol. 4, no. 1, pp. 181-192, 1988.

Greene, Henry A., and David J. Madden: "Adult Age Differences in Visual Acuity, Stereopsis, and Contrast Sensitivity," *American Journal of Optometry and Physiological Optics*, Vol. 64, pp. 749-753, 1987.

Grimby, G.: "Physical Activity and Effects of Muscle Training in the Elderly," *Annals of Clinical Research*, Vol. 20, pp. 62-66, 1988.

Gunnersen, Debra, and Boyd Haley, "Detection of Glutamine Synthetase in the Cerebrospinal Fluid of Alzheimer Diseased Patients: A Potential Diagnostic Biochemical Marker", *Proceedings of the National Academy of Science*, Vol 89, pp. 11949-11953, 1992.

Hallfrisch, Judith, Denis Muller, Donald Drinkwater, Jordan Tobin, and Reubin Andres: "Continuing Diet Trends in Men: The Baltimore Longitudinal Study of Aging (1961-1987)," *Journal of Gerontology*, Vol. 45, pp. M186-191, 1990.

Hamberg, James M., John E. Yerg, II, and Douglas R. Seals: "Pulmonary Function in Young and Older Athletes and Untrained Men," *Journal of Applied Physiology*, Vol. 65, pp. 101-105, 1988.

Hayslip, Bert, and Kevin J. Kennelly: "Cognitive and Noncognitive Factors Affecting Learning among Older Adults," pp. 73-98 in Lumsden, Barry, D., (ed.), *The Older Adult as Learner*, Hemisphere, New York. 1988.

Hazzard, William R., Reubin Andres, Edwin L. Bierman, and John P. Blass: *Principles of Geriatric Medicine and Gerontology*, 2nd edition, McGraw-Hill, New York, 1990.

Huang, Yon-Peng, Laurent Gauthey, Martine Michel, Myriam Loreto, Michel Paccaud, Jean-Claude Pechere, and Jean-Pierre Michel: "The Relationship Between Influenza Vaccine-Induced Specific Antibody Responses and Vaccine-Induced Nonspecific Autoantibody Responses in Healthy Older Women," *Journal of Gerontology*, Vol 47, pp. M50-55, 1992.

Hyman, L. "Epidemiology of Eye Disease in the Elderly," *Eye*, Vol. 1, pp. 330-341, 1987.

Inglin, Barbara, and Marjorie Woollacott: "Age-Related Changes in Anticipatory Postural Adjustments Associated With Arm Movements," *Jounal of Gerontology*, Vol. 43, pp. M105-113, 1988.

Jankovic, B. D., P. Korolija, K. Isalovic, L. J. Popeskovic, M. C. Pesic, J. Horvat, D. Jeremic, and V. Vajs: "Immunorestorative Effects in Elderly Humans of Lipid and Protein Fractions from the Calf Thymus: A Double Blind Study," *Annals of the New York Academy of Science*, Vol 521, pp. 247-259, 1988.

Joachim, Catharine L., and Dennis J. Selkoe: "Minireview:

Amyloid Protein in Alzheimer's Disease," *Journal of Gerontology*, Vol. 44, pp. B77-82, 1989.

Johnston, Priscilla W., (ed.): *Perspectives on Aging: Exploding the Myths*, Ballinger, Cambridge, Mass., 1981.

Joseph, James A., (ed.): *Central Determinants of Age-Related Declines in Motor Function*, Vol. 15 of *Annals of New York Academy of Sciences*, New York Academy of Sciences, New York, 1988.

Kallman, Douglas A., Chris C. Plato, and Jordan D. Tobin: "The Role of Muscle Loss in the Age-Related Decline of Grip Strength: Cross-Sectional and Longitudinal Perspectives," *Journal of Gerontology*, Vol. 45, pp. M82-88, 1990.

Kart, Carl S., Eileen S. Metress, and James F. Metress: *Aging and Health: Biological and Social Perspectives*, Addison-Wesley Publishing Company, Menlo Park, Calif., 1978.

Katzman, Robert: "Alzheimer's Disease," *New England Journal of Medicine*, Vol. 314, pp. 964-973, 1986.

Kawasaki, Takeshi, Shigetake Sasayama, Shin-Ichi Yagi, Tetsuya Asakawa, and Tadakazu Hirai: "Non-Invasive Assessment of the Age Related Changes in Stiffness of Major Branches of the Human Arteries," *Cardiovascular Research*, Vol. 21, pp. 678-687, 1987.

Kenney, Richard A.: *Physiology of Aging*, Year Book Medical Publishers, Inc., Chicago, 1982.

Kent, Barbara, and Robert N. Butler: *Human Aging Research: Concepts and Techniques, in Aging Series*, Volume 34, Raven Press, New York, 1988.

Kirwan, J. P., W. M. Kohrt, M. A. Staten, D. M. Wojta, and J. O. Holloszy: "Factors Associated with Basal Insulin Levels in Older Individuals," *Gerontologist*, Vol. 30, p. 38A, 1990.

Kline, Donald W.: "Ageing and the Spatiotemporal Discrimination Performance of the Visual System," *Eye*, Vol. 1, pp. 323-329, 1987.

Kochersberger, Gary, Connie Bales, Bruce Lobaugh, and Kenneth W. Lyles: "Calcium Supplementation Lowers Serum Parathyroid Hormone Levels in Elderly Subjects," *Journal of Gerontology*, Vol. 45, pp. M159-162, 1990.

Koerner, Martha E., and Glenda R. Dickinson: "Adult Arthritis," *American Journal of Nursing*, pp. 253-266, Feb. 1983.

Korht, W. M., M. A. Staten, J. P. Kirwan, D. M. Wojta, and J. O. Holloszy: "Insulin Resistance of Aging is Related to Body Composition," *Gerontologist*, Vol. 30, p. 38A, 1990.

Krishnaraj, Rajabather, and Gerald Blanford: "Age-Associated Alterations in Human Natural Killer Cells: 2. Increased Frequency of Selective NK Subsets," *Cellular Immunology*, Vol 114, pp. 137-148, 1988.

Krishnaraj, Rajabather, and Gerald Blanford: "Age-Associated Alterations in Human Natural Killer Cells: 1. Increased Activity as per Conventional and Kinetic Analysis," *Clinical Immunology and Immunopathology*, Vol. 45, pp. 268-285, 1987.

Lakatta, Edward G., Jere H. Mitchell, Ariela Pomerance, and George G. Rowe: "Characteristics of Specific Cardiovascular Disorders in the Elderly," *Journal of the American College of Cardiology*, Vol. 10, pp. 42A-47A, 1987.

Lakatta, Edward G.: "Cardiovascular Reserve and Aging," in Horan, M. J., G. M. Steinberg, J. B. Dunbar and E.C. Hadley, (eds.), *NIH Blood Pressure Regulation and Aging*, Biomedical Information, New York, pp. 51-78, 1986.

Lakatta, Edward G.: "The Aging Heart: Myths and Realities," pp. 179-193, in Elias, J. W., and P. H. Marshall, (eds.), *Cardio-*

vascular Disease and Behavior, Hemisphere, Washington, D.C., 1987.

Lakatta, Edward G.: "The Aging Heart: Myths and Realities," pp. 179-193, in Elias, J. W., and P. H. Marshall, (eds.), *Cardiovascular Disease and Behavior*, Hemisphere, Washington, 1987.

Lakatta, Edward G.: "Why Cardiovascular Function May Decline with Age," *Geriatrics*, Vol. 42, pp. 84-95, 1987.

Lamb, Marion: *Biology of Aging,* Wiley, New York, 1977.

Lesnoff-Caravaglia, Gari, (ed.): *Realistic Expectations for Long Life*, Human Sciences Press, Inc., New York, 1987.

Lexell, Jan, Charles C. Taylor, and Michael Sjostrom: "What is the Cause of the Ageing Atrophy?," *Journal of the Neurological Sciences*, Vol. 84, pp. 275-294, 1988.

Lindsay, Robert: "Managing Osteoporosis: Current Trends, Future Possibilities," *Geriatrics*, Vol. 42, No. 3, pp. 35-40, March 1987.

Lonergan, Edmund T.: "Aging and the Kidney: Adjusting Treatment to Physiologic Change," *Geriatrics*, Vol. 34, pp. 27-33, 1988.

Lyles, Kenneth W., personal communication at Williamsburg, Nov. 1989.

Maddox, G. L., (ed.): *The Encyclopedia of Aging*, Springer, New York, 1987.

Maddox, G. L., (ed.): *The Encyclopedia of Aging*, Springer, New York, 1987.

Maddox, George L., and E.W. Busse: *Aging: The Universal Human Experience*, Springer, New York, 1987.

Mader, Scott L., and Pamela B. Davis: "Effect of Age on Acuts Regulation of Beta- Andrenergic responses in Mononuclear Leukocytes," *Journal of Greontology*, Vol. 44, pp. M168-173, 1989.

Mainster, Martin A.: "Light and Macular Degeneration: A Biophysical and Clinical Perspective," *Eye*, Vol. 1, pp. 304-310, 1987.

Maranto, Gina: "Aging - Can We Slow the Inevitable?," *Discover*, pp. 17-25, December 1984.

Marieb, Elaine N.: *Human Anatomy and Physiology*, 2nd edition, The Benjamin/Cummings, Redwood City, Calif., 1992.

Marieb, Elaine N.: *Human Anatomy and Physiology*, The Benjamin/Cummings, Redwood City, Calif., 1989.

Marshall, John: "The Ageing Retina: Physiology or Pathology," *Eye*, Vol. 1, pp. 282-295, 1987.

Masciee, A. A. M., L. M. Geuskens, W. M. M. Dreissen, J. B. M. J. Jansen, and C. B. H. W. Lamers: "Effect of Aging on Plasma Cholecystokinin Secretion and Gall Bladder Emptying," *Age*, Vol. 11, pp. 136-140, 1988.

Masters, William H., Virginia E. Johnson, and Robert C. Kolodny: *Human Sexuality*, 4th edition, Harper Collins Publishers, New York, 1992.

Masters, Willian H., and Virginia E. Johnson: *Human Sexual Response*, Little, Brown and Company, Boston, 1966.

McCance, Kathryn L., and Sue E. Huether: *Pathophysiology: The Biological Basis for Disease in Adults* and *Children*, The C. V. Mosby, St. Louis, 1990.

Mellerio, J.: "Yellowing of the Human Lens: Nuclear and Cortical Contributions," *Vision Research*, Vol. 27, pp. 1581-1587, 1987.

Merimee, Thomas J.: "Diabetic Retinopathy: A Synthesis of Perspectives," *New England Journal of Medicine*, Vol. 322, pp. 978-984, 1990.

Miller, Myron: "Fluid and Electrolyte Balance in the Elderly," *Geriatrics*, Vol. 42, pp. 65-76, 1987.

Minaker, Kenneth L., Graydon S. Meneilly, James B. Young, Lewis Landsberg, Jeff S. Stoff, Gary L. Robertson, and John W. Rowe: "Blood Pressure, Pulse, and Neurohumoral Response to Nitroprusside-Induced Hypotension in Normotensive Aging Men," *Journal of Gerontology*, Vol. 46, pp. M151-154, 1991.

Mitchell, David R., and Kenneth W. Lyles: "Glucocorticoid-Induced Osteoporosis: Mechanisms for Bone Loss; Evaluation of Strategies for Prevention," *Journal of Gerontology*, Vol. 45, pp. M153-158, 1990.

Montamat, Stephen C., and Albert O. Davies: "Physiological Response to Isoproterenol and Coupling of Bete-Adrenergic Receptors in Young and Elderly Human Subjects," *Journal of Gerontology*, Vol. 44, no. 3, pp. M100-105, 1989.

Montamat, Stephen C., and Albert O. Davies: "Physiological Response to Isoproterenol and Coupling of Bete-Andrenergic Receptors in Young and Elderly Human Subjects," *Jounal of Genontology*, Vol. 44, no. 3, pp. M100-105, 1989.

Morrow, Linda A., Stephen G. Rosen, and Jeffrey B. Halter: "Beta-Adrenergic Regulation of Insulin Secretion: Evidence of Tissue Heterogeneity of Beta-Adrenergic Responsiveness in the Elderly," *Journal of Gerontology*, Vol. 46, pp. M108-113, 1991.

Moskowitz, Roland W., and Marie R. Haug: *Arthritis and the Elderly*, Springer, New York, 1986.

Muniain, M. A., C. Rodrigues, F. Pozueto, A. Romero, R. Matta, J. Perez-Venegas, D. Rodriguez, and M. Garrido: "Chemotaxis in the Elderly: Modifications After the Intake of Ascorbic Acid and Levamisole," *Age*, Vol 11, pp. 88-92, 1988.

Nagel, J. E., and W. H. Adler: "Effects of Aging on Immune Function and Reserve Capacity," pp. 195-207, in Burger, E. J., R. G. Tardiff, J. A. and Bellanti, (eds.), *Advances in Modern Environmental Technology, Vol. XIII- Environmental Chemical Exposures and Immune System Integrity*, Princeton Scientific Publishing Co., Inc., Princeton, N.J., 1987.

Nagel, James E., and Jacques J. Proust: "Age-Related Changes in Humoral Immunity, Complement, and Polymorphonuclear Leukocyte Function," *Review of Biological Research in Aging*, Vol. 3, pp. 147-159, 1987.

Nagel, James E., Kyungeun Han, Patricia J. Coon, William H. Adler, and Bradley S. Bender: "Age Differences in Phagocytosis by Polymorphonuclear Leukocytes Measured by Flow Cytometry," *Journal of Leukocyte Biology*, Vol. 39, pp. 399-407, 1986.

National Research Council, *Recommended Dietary Allowances*, 10th edition, National Academy Press, Washington, D.C., 1989.

Newton, J. P., R. Yemm, and M. J. N. McDonagh: "Study of Age Changes in the Motor Units of the First Dorsal Interosseous Muscle in Man" *Gerontology*, Vol. 34, pp. 115-119. 1988.

Nieman, David C., Diane E. Butterworth, and Catherine N. Nieman: *Nutrition*, Brown, Dubuque, Iowa, 1990.

"Now Its Iron," *Time*, Vol 140, p. 18, Sept. 21, 1992.

Panton, Lynn Bishop, James E. Graves, Michael L. Pollock, James M. Hagberg, and William Chen: "Effect of Aerobic and Resistance Training on Fractionated Reaction Time and Speed of Movement," *Journal of Gerontology*, Vol. 45, pp. M26-31, 1990.

Plato, C. C.: "The Effects of Aging on Bioanthropological Variables: Changes in Bone Mineral Density with Increasing Age," *Collegium Antropologium*, Vol. 11, pp. 59-71, 1987.

Poes, Ollie, (ed.) *Annual Editions: Human Sexuality; 91-92*, The Dushkin Publishing Group, Inc., Guilford, Conn. 1991.

Poggi, Paola, Carla Marchetti, and Roberto Scelsi: "Automatic Morphometric Analysis of Skeletal Muscle Fibers in the Aging Man," *The Anatomical Record*, Vol. 217, pp. 30-34, 1987.

Powers, Douglas C., James E. Nagel, John Hoh, and William H. Adler: "Immune Function in the Elderly," in, *Postgraduate Medicine Immune Function*, Vol. 81, pp. 355-359, 1987.

Pressman, Mark R., and June M. Fry: "What Is Normal Sleep in the Elderly?," *Common Clinical Challenges in Geriatrics*, Vol. 4, pp. 71-80, 1988.

Price, Sylvia, A., and Lorraine M. Wilson: *Pathophysiology: Clinical Concepts of Disease Processes*, 4th edition, Mosby-Year Book, Inc., St. Louis, 1992.

Price, Sylvia, A., and Lorraine M. Wilson: *Pathophysiology: Clinical Concepts of Disease Processes*, 3rd edition, McGraw-Hill Book Company, New York, 1986.

Proust, Jacques J.: "Signal Transduction, Lymphokine Production, Receptor Expression and the Immunology of Aging," *Review of Biological Research in Aging*, Vol. 3, pp. 136-146, 1987.

Proust, Jacques J.: "Signal Transduction, Lymphokine Production, Receptor Expression and the Immunology of Aging," *Review of Biological Research in Aging*, Vol. 3, pp. 136-146, 1987.

Rennie, John: "The Body Against Itself," *Scientific American*, Vol. 261, pp. 106-115, Dec. 1990.

Richey, Monica L., Hobart K. Richey, and Neil A. Fenske: "Aging-Related Skin Changes: Development and Clinical Meaning," *Geriatrics*, Vol. 43(Apr), pp. 49-64, 1988.

Robbins, Stanley L., Marcia Angell, and Vinay Kumar: *Basic Pathology*, W. B. Saunders Company, Philadelphia, 1981.

Robert, C., C. Lesty, and A. M. Robert: "Ageing of the Skin: Study of Elastic Fiber Network Modifications by Computerized Image Analysis," *Gerontology*, Vol. 34, pp. 291-296, 1988.

Rockstein, Morris, and Marvin Sussman: *Biology of Aging*, Wadsworth Publishing Company, Belmont, Calif., 1979.

Rolandi, E., R. Franceschini, V. Messina, A. Cataldi, M. Salvemini, and T. Barreca: "Somatostatin in the Elderly: Diurnal Plasma Profile and Secretory Response to Meal Stimulation," *Gerontology*, Vol. 33, pp. 296-301, 1987.

Rosenthal, Julian: "Aging and the Cardiovascular System," *Gerontology*, Vol. 33: suppl. 1, pp. 3-8, 1987.

Rosenthal, Mark, J., and Georges M. Argoud: "Absence of the Dawn Glucose Rise in Nondiabetic Men Compared by Age," *Journal of Gerontology*, Vol. 44, pp. M57-61, 1989.

Rowe, John W.: "Clinical Consequences of Age-Related Impairments in Vascular Compliance," *American Journal of Cardiology*, Vol. 60, pp. 68G-71G, 1987.

Rudman, Daniel, Axel G. Feller, Hoskote S. Nagraj, Gregory A. Gergans, Pardee Y. Lalitha, Allen F. Goldberg, Robert A. Schlenker, Lester Cohn, Inge W. Rudman, and Dale E. Mattson: "Effects of Human Growth Hormone in Men Over 60 Years Old," *New England Journal of Medicine*, Vol. 323, pp. 1-6, 1990.

Rytel, Michael W.: "Effect of Age on Viral Infections: Possible Role of Interferon," *Journal of the American Geriatrics Society*, Vol. 35, pp. 1092-1099, 1987.

Sawin, Clark T., Harold E. Carlson, Andrew Geller, William P. Castelli, and Pamela Bacharach: "Serum Prolactin and Aging: Basal Values and Changes With Estrogen Use and Hypothyroidism," *Journal of Gerontology*, Vol. 44, pp. M131-135, 1989.

Schaadt, Ole, and Hans Bohr: "Different Trends of Age-Related Diminution of Bone Mineral Content in the Lumbar Spine, Femoral Neck, and Femoral Shaft in Women," *Calcified Tissue International*, Vol. 42, pp. 71-76, 1988.

Schein, Jeff, and Bill Dougal: *Boning Up on Osteoporosis: A Guide to Prevention and Treatment*, National Osteoporosis Foundation, Washington, D.C., 1989.

Schellenberg, Gerard D., *et al.*: "Genetic Linkage Evidence for a Familial Alzheimer's Disease Locus on Chromosome 14," *Science*, Vol. 258, pp. 668-671, Oct. 23, 1992.

Schneider, Edward L., and John W. Rowe, (eds.): *Handbook of the Biology of Aging*, third edition, Academic Press, Inc., San Diego, 1990.

Scott, Robert B., personal communication at Williamsburg, Nov. 1989.

Sebag, J.: "Ageing of the Vitreous," *Eye*, Vol. 1, pp. 254-262, 1987.

Shephard, Roy J., and William Montelpare: "Geriatric Benefits of Exercise as an Adult," *Journal of Gerontology*, Vol. 43, pp. M86-90, 1988.

Shephard, Roy J.: "Physical Training for the Elderly," *Clinics in Sports Medicine*, Vol. 5, No. 3, 1986

Shephard, Roy J.: *Physical Activity and Aging*, Yearbook Medical Publishers, Inc., Chicago, 1978.

Shimokata, Hiroshi, Jordan D. Tobin, Denis C. Miller, Dariush Elahi, Patricia J. Coon, and Reubin Andres, 'Studies in the Distribution of Body Fat: I. Effects of Age, Sex, and Obesity," *Journal of Gerontology*, Vol 44(2), pp. M66-73, 1989.

Ship, Jonathan A., Luren L. Patton, and Carolyn A. Tylenda: "An Assessment of Salivary Function in Healthy Premenopausal and Postmenopausal Females," *Journal of Gerontology*, Vol. 46, pp. M11-15, 1990.

Shock, Nathan W., Richard C. Greulich, Reubin Andres, David Arenberg, Paul T. Costa, Edward G. Lakatta, and Jordan D. Tobin: *Normal Human Aging: The Baltimore Longitudinal Study of Aging*, NIH Publication No. 84-2450, U.S. Department of Health and Human Services, Washington, D.C., 1984.

Smith, Everett L., and Robert C. Serfass, (eds.): *Exercise and Aging: The Scientific Basis*, Enslow Publishers, Hillside, N. J., 1980.

Sorock, Gary S., Trudy L. Bush, Anne L. Golden, Linda P. Fried, Brenda Breuer, and William E. Hale: "Physical Activity and Fracture Risk in a Free-Living Eldery Cohort," *Journal of Gerontology*, Vol. 43, pp. M134-139, 1988.

Spencer, Dennis D., *et al.*: "Unilateral Transplantation of Human Fetal Mesencephalic Tissue into the Caudate Nucleus of Patients with Parkinson's Disease," *New England Journal of Medicine*, Vol. 327, pp. 1541-1548, 1992.

Steinhagen-Thiessen, Elisabeth: "Influences of Age and Training on Bone and Muscle Tissue in Humans and Mice," pp. 133-142, in Bergener, M., Ermini M., Stahelin and H. B., *Dimensions in Aging*, Academic Press, N.Y., 1986.

Stelmach, George E., Jim Phillips, Richard P. DiFabio, and Normand Teasdale: "Age, Functional Postural Reflexes, and Voluntary Sway," *Journal of Gerontology*, Vol. 44 pp. B100-106, 1989.

Sun, Fuchuan, Lawrance Stark, An Nguyen, James Wong, Vasudevan Lakshminarayanan, and Elizabeth Mueller: "Changes in Accommodation with Age: Static and Dynamic," *American Journal of Optometry and Physiological Optics*, Vol. 64, pp. 492-498, 1988.

Supii, Sari: *Aging: The Methuselah Syndrome*, videotape, WGBH Educational Foundation, Boston, 1982.

Tamai, Toshitaka, Tsuguhiko Nakai, Hirotada Takai, Ryuichi Fujiwara, Susumu Miyabo, Mitsuru Higuchi, and Shuhei Kobayashi: "The Effects of Physical Exercise on Plasma Lipoprotein and Apolipoprotein Metabolism in Elderly Men," *Journal of Gerontology*, Vol 43, no. 4, pp. M75-79, 1988.

Tenover, Joyce S., Alvin M. Matsumoto, Donald K. Clifton, and William J. Bremner: "Age-Related Alterations in the Circadian Rhythms of Pulsatile Luteininzing Hormone and Testosterone in Healthy Men," *Journal of Gerontology*, Vol. 43, pp. M163-169, 1988.

Terry, Robert D., (ed.): *Aging and the Brain*, Aging Series, Vol.32, Raven Press, New York, 1988.

Tideiskaar, Reid: "Falls in the Elderly: A Literature Review," *Age*, Vol. 11, no. 3, pp. 112-114, 1988.

Timiras, Paola S., (ed.): *Physiological Basis of Geriatrics*, MacMillan, New York, 1988.

Tonino, Richard P., and Patricia A. Driscoll: "Reliability of Maximal and Submaximal *Parameters of Treadmill Testing for the Measurement of Physical Training in Older Persons*," *Journal of Gerontology*, Vol. 43, pp. M101-104, 1988.

Twomey, L. T., and J. R. Taylor: "Age Changes in Lumbar Vertebrae and Intervertebral Discs," *Clinical Orthopaedics and Related Research*, No. 224, pp. 97-107. 1987.

Tylenda, C. A., J. A. Ship, P. C. Fox, and B. J. Baum. "Evaluation of Submandibular Salivary Flow Rate in Different Age Groups," *Journal of Dental Research*, Vol. 67(9), pp. 1225-1228, 1988.

"Unmasking a Stealthy Cancer," *Time*, Vol. 137, p. 45, May 6, 1991.

U'ren, Richard C.: "Testing Older Patients' Mental Status: Practical Office-Based Approach," *Geriatrics*, Vol. 42, pp. 49-56, 1987.

Van De Graff, Kent. M., and Stuart Ira Fox: *Concepts of Human Anatomy and Physiology*, second edition, Brown, Dubuque, Iowa, 1989.

Vance, Mary Lee: "Growth Hormone for the Elderly," *New England Journal of Medicine*, Vol. 323, pp. 52-55, 1990.

Vericel, E., M. Croset, P. Sedivy, Ph. Courpron, M. Dechavanne, and M. Lagarde: "Platelets and Aging; I-Aggregation, Arachidonate Metabolism and Antioxidant Status," *Thrombosis Research*, Vol. 49, pp. 331-342, 1988.

Vlassara, Helen: "Peripheral Neuropathy and Aging," *Age*, Vol. 11, No.2., pp. 74-78, 1988.

Wantz, Molly, and John E. Gay: *The Aging Process: A Health Perspective*, Winthrop, Cambridge, Mass., 1981.

Wardlaw, Gordon M., and Paul M. Insel: *Perspectives in Nutrition*, Times Mirror / Mosby College Publishing, St. Louis, 1990.

Watson, Ronald R., (ed.): *CRC Handbook of Nutrition in the Aged*, CRC Press, Inc., Boca Raton, Fla., 1985.

Watson, Ronald R., (ed.): *CRC Handbook of Nutrition in the Aged*, CRC Press, Inc., Boca Raton, Fla., 1985.

Wegener, M., G. Borsch, J. Schaffstein, I. Luth, R. Rickels, and D. Ricken: "Effect of Ageing on the Gastro-Intestinal Transit of a Lactulose-Supplemented Mixed Solid-Liquid Meal in Humans," *Digestion*, Vol. 30, pp. 40-46, 1988.

Whitbourne, Susan Krauss: The Aging Body: Physiological Changes and Psychological Consequences, Springer-Verlag, New York, 1985.

Whitehead, William E., Burgio, Kathryn L., and Engel, Bernard T.: "Biofeedback Treatment of Fecal Incontinence in Geriatric Patients," *Journal of the American Geriatric Society*, Vol. 33, pp. 320-324, 1985.

Widner, Hakan, *et al.*: "Bilateral Fetal Mesencephalic Grafting in Two Patients with Parkinsonism Induced by 1-methyl-4-phenyl-1,2,3,6-tetrahydropyridine (MPTP), *New England Journal of Medicine*, Vol. 327, pp. 1556-1563, 1992.

Wurtman, Richard J.: "Alzheimer's Disease," *Scientific American*, Vol. 252, pp. 62-66, 1985.

Young, Gloria, Robert Marcus, Jerome R. Minkoff, Lance Y. Kim, and Gino V. Segre: "Age-related Rise in Parathyroid Hormone in Man: The Use of Intact and Midmolecule Antisera to Distinguish Hormone Secretion from Retention," *Journal of Bone and Mineral Research*, Vol. 2, pp. 367-374, 1987.

Young, Roscoe C., Jr., Denise L. Borden, and Raylinda E. Rachal: "Aging of the Lung: Pulmonary Disease in the Elderly," *Age*, Vol. 10, pp. 138-145, 1987.

INDEX

Page numbers in **boldface** indicate illustrations or tables.

Abnormal changes 10-11
    (See also specific abnormalities and diseases)
Absorption:
    digestive system and 208
    in large intestine 218
    in small intestine 217-218
    stomach and 215
Abuse 10
    age changes from 11
Accessory reproductive structures 265
Accident(s):
    aging and 16
    in wear and tear theory 47
Accommodation 154
    age changes and 155
Accumulation of late-acting error theory 43
Acetylcholine 125 146
    level of Alzheimer's disease and 143
Acid/base balance 76, 91, 93
    in blood 103
    circulatory system and 70-71
    maintaining of urinary system and 256
    monitoring of 104
Action potential 117 175
    age changes and 127
Acute gastritis 216
Acute pancreatitis 232
Adequate Intakes (AIs) 238
Adenosine diphosphate (ADP) 26
Adenosine triphosphate (ATP) 26, 33, 177, 240
    mitochondria and 34
    physical activity and 183
Adipose tissue:
    age changes in 159
    of eye 159
    of skin **51**
Adrenal cortex 299
Adrenal gland 55, 291, **292,** 299
    malfunction of 141
Adrenal medulla 129 147
Advanced glycation end-product (AGE) 32
Aerobic capacity (Vo2max) 182
    alterations in by increased exercise 185
    high level of physical activity
        throughout life and 182-183
    lowered consequences of 182-183
    physical activity and 183
Afterimage 162
Age categories:
    built-in bias among 12
    of cross-sectional study 11
Age changes 3
    accommodation and 155
    in adipose tissue 159
    age-related changes versus 10-11
    in airways 103
    in aldosterone 301
    in alveoli **104**
    in antibodies 324
    in antidiuretic hormone 296
    in appendix 221
    in aqueous humor 153
    in arteries 82-84
    in autonomic motor functioning 128-129
    in B cells 323-324
    beginning of 14
    beneficial 4
    in binocular vision 160, 162
    in blood 92
    of body parts 14-16
    in bones 195-197
    of oral region 212-213
    in brain 131-132
    in brain functions 9, 132-135
    in breasts 282

in bulbourethral glands................................271
in calcitonin..298
in capillaries...88
in central nervous system.....................130-132
changes in hair and...............................54-55
in changes in speed.....................................171
in cholecystokinin......................................299
in choroid...157
in ciliary body..151
in collecting ducts of kidneys....................259
in compliance...104
in cones...157
in conjunctiva..150
in conscious sensation..............................130
in control systems................................104-105
in cornea...150-151
coronary artery disease and.........................78
in defecation..221
in defense mechanisms..............................104
in depth perception....................................160
in dermal fibers and cells.......................57-58
in dermal vessels....................................58-59
detrimental..14-15
in diffusion..108
in digestive system..............................209-210
in energy balance.......................................242
in energy use...242
in esophagus..214
in epidermal-dermal boundary....................61
in external ear...165
in external muscles of the eye...............158-159
in eyelids...159
in female genitalia.....................................282
in female reproductive system..............280-282
in female sex hormones........................304-305
in female sexual activity............................284
in gallbladder...231
in glucagon..307
in glucocorticoids.......................................300
in gravity sensitivity to.............................171
in growth hormone...............................295-296
in hair..54-55
in hearing..167
in heart..75
exercise and..75-76
in immune response.............................323-325
in immune system................................322-325
within individual.....................................12-13
influence of on ventilation.........................103
in inner ear..167
in insulin...307
in iris..151
in joints...202-204
of oral region.......................................212-214
in keratin..53
in keratinocytes...53
in kidneys..258-260
knowing timing and nature of.....................11
in lacrimal gland..160
in Langerhans cells................................53-54
in large intestine..221
in lens of eye.......................................152-153
in liver..229
in localization of sound.............................167
in lymphatics..90
in macrophages....................................322-323
in male reproductive system.................270-271
in male sex hormones..........................302-303
in male sexual activity..........................273-275
in melanocytes..53
in melatonin...297
in memory..132-134
in middle ear...166
in mineralocorticoids.................................301
in motor units..179
in muscle cells......................................177-178
in muscles of oral region......................210-212
in muscle mass...180

in muscle system performance..............181-183
other changes versus............................174-175
performance...181-183
in nails...55-56
in nervous system...7
in optimizing vision..................................162
in oral mucosa...210
in ovaries...281
in oviducts...281
in pancreas...232
in parathormone...................................298-299
in penis..271
in perfusion..107
in personality...135
in plasma..92
in platelets..92
in pressure changes...................................104
in prostate..271
in pupil..151
rate of...14
in red blood cells..92
in reflexes..129-130
in reproductive ducts.................................271
in retina...157
in rotation sensitivity to............................171
in salivary glands.......................................212
in sclera...158
in sebaceous glands......................................60
in seminal vesicles.....................................271
in sensory functioning..........................125-127
in sensory neurons.......................................60
in skin receptors....................................125-126
in sleep..136
in small intestine..................................217-218
in smell sense of.......................................126
social aging and...9
in somatic motor functioning.....................127
in sound production...................................114
in spinal chord...131
in spleen..90
in stomach..215
in strength...127, 180
in subcutaneous layer...........................62-63
in suspensory ligaments............................152
in sweat glands..59
in taste sense of...126
in T cells..322-324
in teeth...210-211
in testes..270-271
in thinking...134-135
in thymosin..299
in thymus...323
in thyrocalcitonin......................................298
in thyroid hormones.............................297-298
true...11
in tubules of kidneys.................................259
in urethra...262
in urinary bladder.....................................261
in urination..262
"usual"...10
in uterus..281
in vagina..282
in veins..88
in vision...160-162
in visual acuity.....................................161-162
in vitamin D production.........................61-62
in vitreous humor......................................153
in voluntary movements............................130
in white blood cells.....................................92
(See also Aging)
Age pigment...45
Age-related macular degeneration (AMD)......163
Age spot...53
Aging..9, 14-15
biological..............................4-5, 5, 7-9
description of.................................3, 4-5
intelligence and.................................12
interactions among types of..................10

methods for studying 11-14
nonhuman studies of 13-14
reasons for studying 1-3, 11
"successful" 10
theories of biological 41-48
types of 4-5, 9-11
variability in 16-17
  (See also Age changes)
AIDS 290
  dementia and 141
Air flow **99**
  during inspiration **95**
  during passive or forced expiration **97**
  speed of 101-103
Air pollution 10-11, 90, 109
  aging and 16
  life expectancy and 20
  panlobar emphysema and 111
Airways **100**
  contributions by during expiration 98-99, 99-101, 103
  lower 93, **94**, **102**
  structure of 103
  upper 93, **94**
  during ventilation 98
Alcohol consumption:
  dementia and 141
  excessive 90
  avoidance of 197
  life expectancy and 20
  high 78
  nutrition and 251
Aldosterone 300-301
Alimentary canal 207
Allergic response 320
  of skin 53-54
Alveolus(i) 93, **94**, 101, **102**
  age changes in 107-108
Alzheimer's disease 11, 141-145
  amyloid in 45-46
  amyloid precursor protein (APP) in 44
  apolipoprotein E and 143   amyloid in 45-46
  beta amyloid and 143
  causes of 142
  chromosome 1 and 145
  chromosome 14 and 145
  chromosome 19 and 145
  chromosome 21 and 145
  diagnosis of 143
  early onset 141
  effects of 142-143
  familial (FAD) 141, 142
  genetic of 144
  late onset 141
  neurofibrillar tangles and 143
  presenilins in 144
  senile plaques in 143
  senile (SDAT) 141
  sporadic 143
  τ-protein in 143
  treatments for 145
  types 141
Amadori product 32
Amino acid(s) 28, 36
  essential 248
Amyloid 45, 132
Amyloid precursor protein (APP) in 144
Anal canal 218, **219**
Androgen-binding protein (ABP) 268, 302
Anemia 215
  dementia and 141
Aneurysm 85
Angina 77
Animals:
  maximum longevity in selective breeding and 18
  studies of biological aging in 13-14
Antacid 197
  with magnesium 222
Antagonistic pleiotrophy theory 43

Anterior pituitary gland **292**, 296
Antibiotic(s):
  diarrhea and 222
Antibody 28, 58, 91, 312, 313, 320
Antidiuretic hormone (ADH) 296
Antigen(s) 313, **319**
  ingestion of by macrophages 318
  processing and presentation of 316
Antigen-specific receptor **314**, **316**, **318**, **320**
  age changes in 323
  formation of 313-315
Antioxidant 32, 44
  vitamins as 250
Anus 90, **208**, 218, **219**, **267**, **276**
Anxiety 135, 186
Aorta **69**, **71**, **72**, **74**, 83, **86**, 106, **139**, **254**
Apolipoprotein E (APOE) 143
Apoptosis 38
Appearance of person 54
  biological aging and 4
Appendix **208**, 218-221, 225, **227**
  age changes in 221
Aqueous humor **150**, 151, 153, 155
  age changes in 153
Areola **280**
Arterial stiffening 78
Arteriole **87**
Arteriosclerosis 11, 84
Artery(ies) 68, **70**, 81, 138
  age changes in 82-84
  hardening of 11
  inner layer of 80
  of integumentary system **51**
  middle layer of: in large arteries 81-83
  in smaller arteries 82-84
  number of 84
  obstructions in 76-77
  outer layer of 82
  stiffening of 129
  structure of **81**
  typical elastic 81
  (See also specific name of artery)
Arthritis 11, 47, 130, 185, 204-206, 300
  (See also Rheumatoid arthritis; Osteoarthritis)
Ascites 230
Astigmatism 151
Atherosclerosis 11, 63, **76**, 77, 83-87, 167, 187
  age-related macular degeneration and 163
  development of 85
  effects of 76-77, 85
  elastin and 85
  elastin peptides and 85
  endothelial dysfunction and 85
  foam cells in 85
  free radical and 85
  glycation and 86
  heat shock protein and 86
  homocysteine and 78
  insulin-like growth factor binding protein
    (IGFBP) and 86
  insulin-like growth factor (IGF) and 86
  iron levels and 80
  lipid peroxides and 85
  mechanisms 85
  Parkinson's disease and 145
  periodontal disease 80
  physical activity and 183
  prevention of 86-87, 137
  strokes and 138-139
  tinnitus and 170
Atherosclerotic plaque 76
Atom 22-23
ATP **27**, **34**
Atrium **69**, 71-72
Atrophic gastritis 215-216
Auerbach's plexus 128, 214
Autoantibody 324
Autoimmune reaction 206, 319

Autoimmune theory 47
Autonomic motor neuron 121
Axillary hair graying of 55
Axon 117-119

B lymphocyte (B cell) 315, **318**
   activation and activities of 320
   age changes in 324
   development of **314**
Baby boomers 3, **5**
Bacteria 111-112
   integumentary system as barrier against 49
   lymph node inactivation of 70
Balance 130
   maintenance of 7
Baldness 55
Baltimore Longitudinal Study of Aging (BLSA) 13
Bartholin's gland **276**, 279
Basal cell papilloma 66
Basal cell carcinoma 66
Basal ganglia **124, 131**
Basal metabolic rate (BMR) 241
Basement membrane 58
   alveolar 106, 107
   capillary 87
Basilar membrane 167
Basophil 91
Beta-amyloid 143
Bedsore 63, **65**, 197
Benign neoplasm 66
Benign prostatic hypertrophy (BPH) 285-287
Bifocal lenses 155
Bile 209, 227
   flow of liver and 226-227
Bile canaliculi 226
Bile duct 226
Bilirubin 227
   unconjugated 229
Binocular vision 160
   age changes in 160, 162
Biological age 4, 5, 9, 11
   average values of for individuals 17-18
   concept of 17-18
   cross-sectional method in study of 11-12
   determination of 20
   longitudinal method in study of 12-13
   studies of in animals 13-14
   what happens during 13-20
   (See also Age changes)
Biological aging theories 41-48
Biorhythm 136
Birth-cohort effect 12
   of longitudinal studies 13
   (See also Baltimore Longitudinal Study of Aging)
Birth control pills combination
   of smoking and intake of 78
Birthrates 2-3, **4**, 326
Births **4**
Bladder urinary 260-261
   age changes in 261
   in female **276**
   in male **267**
   structure of 260-261, 241
Blindness 163-164
Blood 68, 90-92, 192
   age changes in 92
   coughing or spitting up of 110
   iron levels in coronary artery disease and 80
   with oxygen in muscle cell **176**
   plasma in 92
   platelets in 90-92
   red blood cells in 90-92
   white blood cells in 91-92
Blood clot 197
   during atherosclerosis **76**
   formation of 91
   (See also Clot; Thrombus)
Blood flow 70
   coronary 74
   liver and 225-226
   in muscles 179-180
   pathway of in circulatory system **69, 70**
   renal (RBF) 258-259
Blood glucose 78, 307-308
   diabetic retinopathy and 164
   levels of 307-308
Blood pressure **72**, 83, 184
   age changes and 83
   exercise and 187
   high 78
   regulation of urinary system and 256-267
Blood vessel(s) 68
   of dermis 58
      age changes in 58-59
   in Haversian canal **194**
   of kidneys **256**, 257
      age changes in 258
   muscle and 175
   new formation of 56-57
   of skin 51
   supply of brain by 139
   wall of during atherosclerosis 76
Body
   hierarchy of 22-24
   structural basis of 6
Body chemicals 25-28
Body fat percent of 246
Body mass index 243
Body parts:
   aging of 14-16
   biological ages for 18
Body temperature 5, **8**, 84
   age changes and 58-59
   dermal vessels and 58
   fat tissue and 62-63
   regulation of 30-31, 64
   (See also Temperature)
Body weight:
   desirable tables of 243
   energy and 245
Bone(s) 189, 192-194
   age changes in 195-198
   cells of 192-193
   compact **193**, 194
   components of 192-194
   cortical 194
   disease of 198-202
   effects of menopause on 196
   matrix of 195-198
   minerals in 195
   physical activity and 183, 184
   minimizing loss of matrix in 197-198
   of oral region 212
      abnormal changes in 213-214
      age changes in 212
   proteins in 195
   quantity of 195
   spongy 195
   strength of 201
   structure of **193**, 195-196
   trabecular 195, 198
   types of 194-195
   variability in loss of 196-197
   (See also Osteoporosis)
Bone marrow 192, **314**
Bone matrix 192, 195
   building and maintenance of 197
   mineral storage and 193-194
   minimizing decreases in 197-198
Bowel movement 220
Bowman's capsule **255**, 257
Brain 115, **116**
   age changes in 131-132
   blood flow into 137-138, **139**
   dimensions of 131-132
   functions of aging of 132-134

number of neurons in 131
structure of **131**
vision and 148-149
white matter of 120
Breast 265, **266**
age changes in 282
structure of **280**
Breast cancer 288-289
Breathing:
depth of 103
difficulty in 80
work of 96
Bronchiole 101, **102**, 104-105
Bronchitis chronic 110-111
Bronchus(i) **94**, 101, **102**
Bruch's membrane **156**, 157, 163
Bulbourethral gland ... **261**, 266, **267**, 269
age changes in 271
Bullous pemphigoid 326

C. elegans 40
Caffeine avoidance of 197
Calcitonin 298
Calcium 62, 192, 194, 218
consumption of 197
deposit of 83, 85
vitamin D and 50
Calcium theory 46
Calment, Jeanne 18
Caloric restriction 251-252, 327
benefits from 252
effects from 252
exercise and 252
in humans 252
in animals 252
maximum longevity and 252
mean longevity and 252
Calorie 237
Cancer 53, 66-67, 224-225
incidence of 38
metastatic 230
(See also specific types of cancer)
Capillary(ies) 68, 87-88
age changes in 88
of circulatory system 69
of skin 51
structure of capillary exchange and 87
Capillary exchange 87
Carbohydrate(s) 25, **26**, 28, 209, 240
deficiencies of 245
digestible 244
excesses of 245-246
indigestible 245
in intercellular materials 40
manufacturing of 33
recommended dietary intake of 245
uses of 245
Carbon dioxide (CO_2) 91, 96, 101, 103, 107
in gas exchange 93-95
level of 120
monitoring of 104-105
obtaining energy in muscle cell and 176
sleep apnea and 113
snoring and 113
Cardiac adaptability 74, 75
age changes in heart and 75
Cardiac cycle in heart 71-73
Cardiac efficiency 75, 184
Cardiac muscle 73, 173
Cardiac output (CO) 73, 74-76, 184
Cardiac reserve capacity 75
Carotid artery 138, **139**
Careers 327
Cartilage 40, 202, 204
Cataract 152
formation of 163
glycation and 163
free radicals and 163

risk factors for 163
Cavity(ies) (in tooth) 212
CD4 cells 317
CD8 cells 317
Cell(s) 5, 24, 33-40
age changes in:
of dermis 56-57
DNA duplication in 36-39
increase in spacing between 53
structure of **34**
studies of biological aging in 14
(See specific types of cells)
Cell Cycle **37**
Cell division 36-39
Cell membrane 24, 28, 33, **34**, **35**, 44
of muscle cell 175
of nerve cell 127, 132
Cell-mediated response 317
Cellulose 245
Cementum 210, **211**
Central nervous system (CNS) 115, 129
age changes in 130-132
hearing impairments and 169
infections of 140, 143
tinnitus and 170
physical activity and 183
Centrilobar emphysema (CLE) 111
Cerebellar cortex 123-125, **131**
Cerebellum 124,125, **112**
ear impulses and 171
Cerebral cortex 123, **131**, 132
specialized regions of **123**, **124**
Cerebral hemisphere 123, **131**
Cerebrovascular accidents (CVAs) 138
Cerumen 165
Cervical cancer 289
Cervix **276**, 278
Chemical(s)
body 25-29
integumentary system as barrier against 49
skin contact with 53
stratum corneum and 51
Chemical activity constant of cell 24
Chemical reaction **23**, 33
body temperature and 50
energy of 22, 25
free radicals and 44-45
mitochondria and 34
Childhood:
decreased death rates during 2-3
developmental changes occurring before 3
Choking risk of 105
Cholecystokinin (CCK) 231, 299
Cholesterol 28, **30**, 78, 82, 85, 167, 246
high levels of in blood 11
Choroid **150**, 155, **156**, 157
age changes in 157
Chromosome structure **38**
Chronic bronchitis 90, 110-111, 185
Chronic obstructive pulmonary disease (COPD) 185
Chronic pancreatitis 232-233
Chronological age 9
graying of scalp hair and 55
Ciliary body **150**, 151
age changes in 151
Cilium(a) **34**, 97
in respiratory system 104
Circadian rhythm 136, 295
age changes in 137
Circulatory system 68
acid/base balance in 70-71
defense in 70
exercise and 183, 184, 187
main functions for homeostasis in 68-71
pathway of blood flow in 69, 70
pathway of lymph flow 70
physical activity and 83
temperature control 70

transportation in 68-70
   (See also Artery; Atherosclerosis; Blood; Capillary;
   Heart; Lymph; Lymph vessel; Spleen; Vein)
Cirrhosis 90, 229-230
   varicose veins from 89
Clear cell acathoma 66
Clinker theory 45, 46
Clitoris 274, 279
   with erectile tissue **276**
Clonal selection **314**, 315
Clot 73, 76, 85, 88, 92, 192
Cochlea 167
   structure of 168
Collagen fiber(s) 40, 56-57, 62, 82-85
   in arteries and veins 81
   during atherosclerosis 76
   glucose binding of 46
   of skin 51
   in vessel wall 91
Collecting ducts 257-258
   age changes in 259
Colon **208**, 218-220, **227**
Color(s):
   cones of eyes and 155
   distinguishing of 160-161
Colorectal cancer 224-225
Comedo 64
Communication 95, 113-114, 135
Compact bone 194
Compensatory powers 15-16
Complement system 321
Completed stroke 140
Compliance:
   in respiratory system 104
   during ventilation 97
Cone 155
   age changes in 157
Congestive heart failure 80, 89
Conjuctiva 149-150
Constipation 127, 221-222
   causes of diarrhea and 222
   chronic hemorrhoids and 90
   Parkinson's disease and 146
Contact lenses 151, 153, 162, 163
Convergence 158
Cornea **149**, **150**, 155, 159
   age changes in 150-151
Coronary artery(ies) 72, 73, **74**
   diseases of 63, 76-80
   risk factors for 77-80
   functions of 76
Coronary blood flow 74-75
Cortical bone 194-196
Cosmetic aging 9, 63
Cough reflex 99, 103
Cowper's gland 269, 271
Creatinine 92, 253
Crohn's disease 326
Cross-link(s) 46
   between fibers 57
Cross-linkage theories 46
Cross-sectional method in study of biological aging 11-12
Cross-sequential study 13
Crush fracture 198, **199**
Cytoplasm 33, **34**
Cytoskeleton 35, 36
Cytotoxic T cell (cT cell) 316-318
   activation and activities of **319**

D. melanogaster 40
Dead space 98
Death gene theory 43
Death hormone theory 46
Death rate 3, 19, 326
Decubitus ulcer 65
Defecation 220-221
   age changes in 221
Defense mechanisms of body:

circulatory system 70
   immune system as a 312
   nonspecific 313
   in respiratory system 104
   during ventilation 98
Dehydration 127
Delayed-hypersensitivity T cell (dT cell) **316**, 317-318
Delirium 141
Dementia 140-147
   Alzheimer's type (SDAT) 141-145
   causes 141
   incidence 140
   multi-infarct 141
   Parkinson's disease and 146
   strokes and 84
   types 141
   with Lewy bodies 147
Dendrite 117-119
Dendritic spines 132
Deoxyribonucleic acid (DNA) 26, 36
   alteration of 52
   damage of by free radicals **44-45**
   duplication of 36-37
   light damage of 50
   during mitosis 37
   ultraviolet light and 52
Depression 135, 140, 169, 200
   cosmetic aging and 9
Depth perception 160
   age changes in 160, 162
Dermal papilla **51**, **52**, 61
Dermis **51**, 52, 56-61
   blood vessels of 58
      age changes in 58-59
   boundary between epidermis and 61-62
   cells of 56, 57
      age changes in 57, 58
   effects of excess light on 52
   fibers of 56
      age changes in 57-58
   sebaceous glands of 59-60
   sensory neurons of 60
      age changes in 60
   sweat glands of 59
      age changes in 59
Detrusor muscle 260
Developmental biology 3
Developmental change 2-3
DHEA 301
   age changes in 301
   effects of 301
   hazards from 301
   secretion of 301
   sources of 301
   supplementation of 301
Diabetes mellitus (DM) 63, 130-131, 163, 186, 306
   causes of 309
   complications of 309-310
   coronary artery disease and 78
   definition of 308
   excess glucose and 309-310
   incidence of 308-309
   ketoacidosis and 310
   main effects of 309-310
   prevention of 310
   secondary glaucoma and 164
   toenail infections and 56
   treatments for 310-311
   types of 308
Diabetic retinopathy (DR) 164-165
Diaphragm 93-96
Diarrhea 222-223
   causes of 222
Diastolic pressure **72**, 82
Dietary Reference Intakes (DRIs) 238
   Adequate Intakes (AIs) and 238
   basis for 238
   Estimated Average Requirements (EARs) and 238

Tolerable Upper Intake Levels (ULs) and 238
uses of 238
Dietary restriction 251
effects from 251
Diet 25, 234
aging and 16
based on chemical composition 236-238
evaluation and adjustment of 239
food selection 236
nutrition and 234-235
proper 236
for younger and older adults 238-239
recommended dietary allowances and 236, 238
relationships between nutrition and 234-235
sources of energy in 240
Differential mortality 12
Diffusion 87
in respiratory system 95, 107-108
age changes in 108
Digestion 128, 129
chemistry of 209
Digestive system 207, **208**
absorption and 208
age changes in 209-210
conversion of food to usable form and 207
eliminating toxins and wastes and 209
main functions for homeostasis in 207-209
manufacturing materials and 208
oral region of 210-214
storing and converting excess nutrients and 209
supplying nutrients and 207
(See also Esophagus; Large intestine;
Small intestine; Stomach)
Diglyceride 28, 246
Dihydrotestosterone (DHT) 266, 302-303
Disaccharide 26, 244
Disease(s):
age changes from 11
age-related 10-11
of bone 198-202
chronic: aging and 16
mean longevity and 19
of eyes 162-165
of heart 76-80
of immune system 325-326
of joints 204-206
in kidneys 260
of nervous system 137-147
nutrition and 251
prevention of: life expectancy and 20
use of strategies in 10
of reproductive systems 285-290
of respiratory system 109-113
short-term aging and 16
of veins 88-90
Diurnal rhythm 295
Diverticulitis 224
Diverticulosis 224
Dizziness 59, 74, 128, 172, 187
DNA **27**
duplication of 36
Dopamine (DOPA) 146
Double vision 158
Down's syndrome 36, 40
Driving motor vehicles 188
Drugs dementia 141
Drying of skin 63
Ductus deferens 266-269
Duodenal peptic ulcer 216
Duodenum **185**, 206, 299

Ear(s) 148, 165
external 165
age changes in 165
gradual increase in size of 63
inner 167, **168**, **170**
age changes in 167
middle 165-166

age changes in 166
structure of **166**
Ear canal 165, **166**
obstruction of tinnitus and 170
Earwax 165
Economic aging 9
Edema 229
Education 135
aging and 16
life expectancy and 19-20
Ejaculation 272
Ejaculatory duct 267-269
opening of **270**
Elastic fiber:
of lower airways **102**
skin 51
Elastic recoil 96-97
Elasticity 41
Elastin fiber(s) 41, 56-58, 62, 64
in arteries and veins 81
during atherosclerosis 76
of lower airways 102
Elastin peptides 85
Elderly people 1-4
as percent of population **2**
heterogeneity among 16-17
malnutrition among 239-240
planning for 17
rapid increase in number of 1-4
Electrocardiogram (ECG or EKG) 73
Electron 19
Electron transport 34
Embolus 89
Emotion(s) 125, 136
diarrhea and 222
Emotional stress 76
Emphysema 90, 109, 111, 185
centrilobar 111
overall effects of 111
panlobar 111
Employment:
evaluating eligibility for for elderly 17
life expectancy and 20
Endocardium 73
Endocrine glands 46, 291
Endocrine system 291, **292**
comparison of with nervous system 293
effects of starting or increasing exercise on 184, 186
insulin-like growth factor and 295
main functions for homeostasis in 291-293
respiratory system and 103
Endometrial cancer 289
Endometrium **276**, 278
Endoplasmic reticulum 33, **34**, 132
Endothelium 80
during atherosclerosis 76
Endurance 181
high level of physical activity throughout life and 183
Energy 33
body weight and 240-244
of chemical reaction 24
dietary sources of 240
kilocalories and 241
in muscle cells 175-176
obtaining of from molecules 240-241
storage of 240
use of 240-242
age-related changes in 242
Energy balance 241
age-related changes in 242
Environment:
external integumentary system and 49
free radicals and 44
life expectancy and 20
somatic mutation theory and 44
Enzyme(s) 28, 36, 44-45
use of in digestion 209
Epidermis 50-53

age changes in 53-54
boundary between dermis and 61-62
components of 50-53
replacement 53
structure of **52**
Epididymis 266-268
Epiglottis 99, **100**, 212
Erectile tissue **267,** 269, **270**
of clitoris **276**
of penis **261**
Erection 128
of penis 270
Error catastrophe theory 44
Erythropoietin 257
Esophagus **86, 99, 100,** 207, **208,** 214, **216**
abnormal changes in 214-215
age changes in 214
aneurysm pressing on **86**
Essential amino acids 248
Essential fatty acid 247
Estimated safe and adequate daily
dietary intakes (ESADDIs) 238
Estrogen 28, 196-198, 275, 302, 303-304
decline in 79
longevity and 19
Estrogen replacement therapy (ERT) 264, 305-306
alternatives to 306
benefits of 305
risks of 305-306
Evolutionary theory 41
Excitement phase:
of female sexual activity 282-283
age changes in 284
of male sexual activity 272, 273
age changes in 274
Exercise 79, 88, 136, 184-188
aging and 16
aging heart and 75-76
inadequate 10, 11
vigorous level of functioning of organs under 17
Exhaustion avoidance 140
Expiration 95, 96
Expiratory reserve volume (ERV) 96
Eye(s) 148
deterrents to clear vision and 149
diseases of 162-165
external muscle of 158
age changes in 158-159
free radicals in 157
focusing light in **152**
image formation and 148-149
in its orbit **158**
reactive oxygen species and 157
structure of **150**
vision and 148-149
(See also specific components)
Eye muscle diseases of 162-163
Eyeglasses 151, 153, 155, 161, 163
cosmetic aging and 9
Eyelid 159

Facial hair of aging women 55
Fainting 74, 128, 187
Falling 187
hip fractures and 200
reduction of 197
Fallopian tube 277
Familial Alzheimer's disease (FAD) 141
Family history
coronary artery disease and 78
Farsightedness 149
Fasting plasma glucose (FPG) 307
Fat 28, 78, 82, 167, 192, 246
of eye **158,** 159
age changes in 159
in Food Guide Pyramid 237
high level of physical activity throughout life and ... 183
percent body 243

under skin 5
thinning of insulating layer of 7
Fatty acid 28, 246-247
Faulty DNA repair theory 44
Fecal incontinence 142, 221, 223-224
Feces 220
Female reproductive system **260,** 275
age changes in 280-282
breasts of 280
genitalia of 279
menstrual cycles of 278
ovarian cycle of 275, **277**
ovaries of 275
oviducts of 277
structure of **276**
uterus of 277-278
vagina of 278-279
Female sexual activity 282-284
age changes in 284
excitement phase of 282-283
orgasmic phase of 284
plateau phase of 284
resolution phase of 284
Femur **190,** 196
Fiber(s) 40-41, 245
collagen 40
of dermis **51,** 56
age changes in 57
elastin 41
Fiber (dietary) 197, 245-246
Fibrin 92
role of in blood clot 91
Fibroblast 56, 57
Field of view 162
Fight-or-flight response 129
Filtration 88, 257
Financial status economic aging and 9
Fingernail keratin in 55
Flasher 153
Foam cells 85
Floater 153
Follicle 275, 303
Follicle-stimulating hormone (FSH) 302
Food:
choices of 237
conversion of to usable form 207
selection of diet based on 236
during swallowing reflex **100**
Food Guide Pyramid **237**
Fovea centralis **150,** 155, 156
Fracture 112
crush 198-199
decreased risk of 183
of hip 198-200
orthostatic hypotension and 128
treatment after 202
treatment of 197
vertebral osteoporosis and 198-200
Free radical (*FR) 29, 34
chain reaction with 31
chemical reactions with 31
effects of 31
formation of 29
importance in aging 29
in eyes 157
in skin 64
initiation 31
propagation 31
reinitiating 31
sources of 29
termination 31
types of 29
ultraviolet light (UV) and 31, 157
Free radical theory 44
Fungal infections of nails 56
Fungal pneumonia 112
Fungus(i) 112
Future 326

Gag reflex 99, 103
Gallbladder 207, **208**, 226-227, 230-231
  abnormal changes in 231
  age changes in 231
Gallstone 231
Gas exchange 93-94, 108-109
  altered, effects from 108-109
Gastric peptic ulcer 216
Gastric refluxing 214
Gastric ulcer, perforated 194
Gastritis 215-216
Gastrointestinal (GI) tract 207
Gender, longevity and 19
Gene(s) 19, 35-36, 40
  aging and 16
  Alzheimer's disease and 142-144
  control of cell by 35-36
  control of life span by 18
  correcting of errors in 18
  development of coronary atherosclerosis and 78
  in genetic theories 43-44
  graying of hair and 55
  life expectancy and 20
  maximum longevity and 18
  for pattern baldness 55
Genetic heterogeneity among people 14
Genetic (aging) theories 43-44
Genitalia 265
  female 279
    age changes in 282
Gerontology 327
Glare 151, 161, 162
Glaucoma 163-164
Glomerular filtration rate (GFR) 259
Glomerulus **256**, 257
Glottis 98, **100**
Glucagon 232, 306
  abnormal changes in 307-308
  age changes in 307
  blood glucose homeostasis and 307
  control of secretion of 306
  effects of 232
  secretion of 306
  sources of 306
Glucocorticoid(s) 299-300
Glucose **25**
  accumulation of, age changes and 46
  blood levels of 307-308
  diabetes mellitus and 163, 309-310
  energy in muscle cell and **176**
  excess 309-310
Glucose tolerance 184, 186, 307
Glycation theory 46
Glycation 32
Glycogen 25, 177, 183, 184
Glycolic acid 64
Glycolysis 240
Glycoprotein 28
Glycosuria 309
Glycoxydation 31, 32
Goal:
  efficient ways for accomplishment of 15-16
  setting of, for exercise 186-187
Gonorrhea 290
Grave's disease 326
Gravity 165
  detection of 170-171
    age changes in 171
Gray matter 120
  of spinal cord 122
Graying of hair 55
Growth hormone (GH) 184, 295-296

Hair 54-55
  age changes in 54-55
  in armpit 54-55
  biological functions of 54

  cosmetic aging and 9
  decline in color intensity of 55
  in ear **166**
  white 55
Hair follicle **51**, 54, 55
Hard palate 98, **99**
Haversian canal, ... **193**, **194**
Hayflick limit 37
Head 165
  blood flow into **69**, **71**
  respiratory passages in **99**
  sensing rotation of 165, **171**
    age changes in 171
Healing of skin 50, 57-589
Health care 135
  life expectancy and 20
  providing of, for elderly 16
  timely 48
Hearing 123, 165-169
  disorders in 169-170
  external ear and 165
  inner ear and 167
  loss of 10, 169
  middle ear and 166
Hearing aid 169-170
  cosmetic aging and 9
Heart 48, **69**, **70**, 71-75
  advancing age and 78
  blood flow of 73-74, 76-77
  cardiac adaptability and 75
  cardiac cycle of 71-73
  chambers of 71-72
  congestive heart failure and 80
  coronary artery disease and 76-80
  dementia and 141
  diseases of 76-80, 137
  endocardium of 73
  epicardium of 73-74
  failure of 111
  in cardiac adaptability 74-76
  in exercise 75-76
  in resting conditions 75
  internal structure of **71**
  layers of 73
  lipofuscin in 45
  myocardium of 73
  reductions in pumping capacity of 8
  risk factors for 77-80
  valves of 73
  valvular heart disease and 80
Heart attack(s) 10, 48, 74, 77, 112, 113, 187
  atherosclerosis and 84
  disabilities from 11
  likeliness of 75
  physical activity and 183
  in women 19
Heart rate (HR) 72-73, 75
  maximum 75
  resting 75
Heart valves, disease of 80
Heartbeat 72
  irregular 184
Heat
  effects on skin 65
Heat production, muscle system and 174
Heat shock protein 86
Helper T cell (hT cell) **316**, 317, 320
  activation and activities of 318
  age changes in 323
  ratios of 323
  signaling substances from 317
  types of 317
Hemoglobin 91
  glucose binding of 46
Hemorrhagic stroke 138-139
Hemorrhoid 90, 224
  internal versus external **90**
Heparin 57

Hepatic portal system 226, **227**
Heredity (See Gene)
Hernia, sliding hiatal 214-215
Herpes 290, 325
Heterochromatin 36
Heterochromatin loss theory 43
Heterogeneity among elderly 16
Hiatal hernia, sliding 214-215
High blood pressure 11, 80, 81, 127, 163
    coronary artery disease and 78
    hypertensive hemorrhagic strokes and 138
    tinnitus and 170
High-density lipoproteins (HDLs) 78, 183, 184, 247
    physical activity and 183
Hip fracture 198, 200
    osteoporosis and 198-200
Hip joint **190**
Hippocampus 132, 300
Histamine 56, 91
Homocysteine 78
Homeostasis 5, 7, 25
    cells and 24
    circulatory system and 68-71
    deviations from 7
    digestive system and 207-209
    endocrine system and 291-293
    immune system and 312
    integumentary system and 49-50
    keratinocytes and 51
    maintenance of 14-15, 20
    muscle system and 173-174
    nervous system and 115-117
    nutritional 234-235
    reproductive system and 265-266
    reserve capacity of body and 10-11
    respiratory system and 93-95
    skeletal system and 189-192
    urinary system and 253-257
Hormone(s) 28, 46-47, 291, 294
    bone cells and 193
    longevity and 19
    theories about 46-47
    transport of 68
    (See also specific name of hormone)
Hot flash 304
Housing:
    designing of, for elderly 16
    life expectancy and 20
Human leukocyte-associated
    (HLA) receptor 314-316, **319**, 320
Human papillomavirus 290
Humoral response 317
Humpbacked posture 199
Hydrochloric acid (HCl) 215
Hydrocortisone 300
Hydrogen bond **23**
Hydrogen peroxide 29
Hydrogenated fat 246
Hyperglycemia 307
Hypertensive hemorrhagic stroke 138
Hypoglycemia 307
Hypothermia 50, 84, 128
Hysterectomy 290

IL-6 .... 317, 323
    osteoclasts and 193
Image formation 148-149
    conjunctiva and 149
        age changes in 150
    cornea and 150
        age changes in 150-151
Immediate-hypersensitivity reaction 318
Immune deficiency theory 47
Immune response(s) 53, 54, 312, 315-322
    abnormal 325-326
    acquired active immunity and 322
    age changes in 323-324
    consequences of 324-325

antibodies and 321
B-cell activation and 320
cytotoxic T-cell (cT-cell) activities and 317
delayed-hypersensitivity T-cell (dT-cell)
    activities and 317-318
helper T-cell (hT-cell) activities and 317
against infection 10
primary 321
    age changes in 324
secondary 321
    age changes in 324
specificity of 313
suppressor T-cell (sT-cell) activities and 318-320
T-cell specialization and 317

Immune deficiency theory 47
Immune dysregulation theory 47
Immune system 90, 312
    abnormal and disease conditions from 325-326
    antigen-specific receptor formation and 313-315
    B cells and 315, 320-321, 323
    clonal selection and 315
    development of 313-315
    HLA receptor formation and 313-314
    Langerhans cells and 53-54, 313, 322-323
    macrophages and 38, 313, 322-323
    main functions for homeostasis in 312
    structures of **313**
    T cells and 313-315, 317-324
    theories about 47
    thymus and **313**, 315
    unique characteristics of 312-313
Immune theory 47
Immunoglobulin 320
Impaired glucose tolerance (IGT) 307-308
Impotence 286-287
    atherosclerosis and 84
    sildenafil and 287
    Viagra and 287
Income life expectancy and 20
Incontinence:
    fecal 127, 142, 223-224
    urinary 127, 142, 263-264
Indigestion 136
Individual:
    rate of aging of 16
    status of 20
Infection 11, 92, 192
    ability to fight off 9
    bedsores and 65
    in elderly 54
    fungal of nails 56
    immune responses against 10
    integumentary system and 49
    risk of 53
    vaginal 288
Inflammation 56, 91, 192
    of joints 204-205
Inhibin 268, 302
Injury avoidance of osteoporosis and 201
Inspiration 95-96
Inspiratory reserve volume (IRV) 96
Insulin 184-186
    abnormal changes in 307-308
    age changes in 307
    blood glucose homeostasis and 307-308
    control of secretion of 306
    effects of 306
    restoration of levels of 307
    secretion of 232
    sources of 306
Insulin-dependent diabetes mellitus (IDDM) 186, 308-309
Insulin/growth factor imbalance theory 46
Insulin-like growth factor (IGF) 86
Insulin-like growth factor binding protein (IGFBP) 86
Insulin resistance 307
Integumentary system 49-67
    abnormal changes of 63-67

bedsores and 65
components of **51**
cosmetic changes in 63
effects of sunlight on 64-65
main functions for homeostasis in 49-50
major functions of 49
neoplasms and 66-67
providing information and 50
serving as a barrier and 49-50
temperature regulation and 50
vitamin D production and 50
Intellectual ability increases in 4
Intelligence:
    aging and 12-13
    life expectancy and 20
    measurement of in cross-sectional study 12
Intercellular material 40-41
Interleukin-1 (IL-1) 317
Interleukin-2 (IL-2) 317, 318, **320**
Internal urethral sphincter 262
Interstitial cell 266, **268**, 302
    between seminiferous tubules **268**
Interstitial cell-stimulating hormone (ICSH) 302
Interstitial fluid 253
Interstitial lamella **194**
Intervertebral disk **191**, **199**, 202-203
Intervertebral joint **190**
    with osteoarthritis **199**
Intestines 200
    calcium absorption and 50
    (See also Large intestine; Small intestine)
Intraocular pressure (IOP) 164
Intrinsic factor 215
Ion **23**
Iris 128, 149-151
    age changes in 151
Iron 217
    high blood levels of coronary artery disease and 80
Irregular heartbeat 136
Ischemia 137-138
Ischemic stroke 138
Islets of Langerhans 232, 306
Itching of skin 63

Jaundice 229
Jeanne Calment 18
Jogging 186-187
Joint(s) 186, 189
    age changes in 202-204
    diseases of 204-206
    freely movable 203-206
        structure of **203**
    functions of 202
    fused **206**
    immovable 202
    inflammation of 204
    intervertebral **190**
    of oral region 212
    abnormal changes in 213-214
    slightly movable **199**, 202-203
    stiffness from repeated traumatic injury 11
    structure and effects of rheumatoid arthritis on **206**
    symphysis 202
Juvenile-onset diabetes mellitus 308

Kegel exercise 264
Keratin 50-53, 60
    changes in 63
    in fingernails and toenails 55
    hair and 54
Keratinocyte 50-53, 61
    chronic exposure to sunlight and 64-65
    malignant skin neoplasms and 66-67
    size shape and internal structure of 53
Ketoacidosis 310
Ketosis-prone diabetes mellitus 308
Kidney(s) 200, 253-260
    abnormal changes in 260

age changes in 258-260
blood flow of 77, 258-259
blood vessels of **255**, 257-259
collecting ducts of 257-259
diseases of 80, 260
    atherosclerosis and 84
functional capacity of 10
functions of 258
glomerular filtration rate and 259
nephrons of 258
reduction in function of 3, 140
renin and 259
structure of **255**
tubules of 257-259
vitamin D activation and 259
waste elimination by 73
Kilocalorie (kcal) 237, 241
Knauss, Sarah 18
Knee-jerk reflex 121
Kupffer's cell 226, 229
    in sinusoid **226**

Labium(a) majora 275, **276**, **279**
Labium(a) minora 275, **276**, **279**
Lacrimal apparatus 159
Lacrimal fluid 159-160
    diseases of 162
Lacrimal gland 159-160
    age changes in 160
Lactase 217
Lactic acid 76, 241
Lactose 217
Lactose intolerance 217
Langerhans cell 52-54, 64, 66, 312, 313
    age changes in 54, 322
Language 113
    age changes in 113
Large intestine 207, 208, 218
    abnormal changes in 221-225
    absorption in 218-220
    age changes in 221
    appendix and 221, 225
    cancer of 224-225
    constipation and 221-222
    defecation and 220-221
    diarrhea and 222-223
    diverticulosis and 224
    diverticulitis and 224
    fecal incontinence and 223-224
    hemorrhoids and 224
    movements in 220
    regions of **219**
    secretion in 218-220
Larynx **94**, 99, 13-14
Laser treatment with 64, 163, 165
Laxative(s) 221
    diarrhea and 222
Left ventricle 72, 71, 74, **106**
    of heart 69
Leiomyoma 290
Lens 151-153, 155
    during accommodation 154
    age changes in 152-153
    focusing light and 151-152
Leukocyte 91
Levodopa 146
Leydig's cell 266, 301-302
Life 2-3, 18
    quality of 20-21
Life expectancy 2-4, 18-20
    of population 19
    (See also Longevity)
Ligament 40, 46, 204
    attachment of 195
    of freely movable joint 204
Light 150
    in cross-linkage theories 46
    excess melanin and 52-53

focusing of 149, **152**
integumentary system as barrier against 50
quality of in vision 161
Light intensity required age changes in 160-161
Lipid 28, 33, 209, 240, 246
cholesterol and 246
deficiencies of 247
excesses of 247
fatty acids and 246
free radicals and 44
manufacturing of 33
monoglycerides and 28, 246
recommended dietary intakes of 246-247
tri- di- and monoglycerides and 28, 246
uses of 28, 246
Lipid hydroperoxide 29
Lipid peroxide (LP) 29
Lipofuscin 45
Lipoprotein 28, 78, 247
Liver 207, **208**, 225-230
abnormal changes in 229
age changes in 229
bile flow of 226
blood flow of 225-226
cancer and 230
capillary plexus in **227**
cirrhosis and 229-230
functions of 227-229
malfunction of 141
Liver lobule structure of **226**
Liver spot 53
Localization of sound 167
age changes in 167
Locus coeruleus 131
Loneliness social aging and 9
Longevity:
maximum (XL) 18
nutrition and 251-252
mean (ML) 18-20
factors affecting 19-20
in United States 19-20
Longitudinal method in study of biological aging 12-13
Loop of Henle **255-256**
Low-density lipoproteins (LDLs) 247
high blood coronary artery disease and 78
physical activity and 183
Lower airways 93, **102**
Lung(s) **69**, 93-94, 107, 189, 190
aging 16, 104-105
blood flow into **70**, 77
circulation of **70**
lymphatic capillaries in **70**
perfusion of 106-107
stiffening of 3
waste elimination by 73
Lung cancer 109-110
Lung volume 96-98
Luteinizing hormone (LH) 302
Lymph 68, 87, 312
formation of capillary exchange and 87
pathway for flow of in circulatory system **70**
spleen and 90
Lymph capillary 68, **70**, 87
Lymph node(s) 47, 68, 70, 90, 312, **314**
Lymph vessel 50, 68, 70, **102**, 312
Lymphocyte(s) 47, 92, 312, 313
production of **314**
Lymphokine 317-319

Macrophage 56, 90, 312, 313, **316**
age changes in 322-323
development of **314**
Macula **150**, 155, **156**, 161, 163, **170**
Magnesium:
antacids with 222
ketoacids and 310
Male reproductive system **266, 267**
bulbourethral glands of 269

ducts of 268
ductus deferens of 268
ejaculatory duct of 268
epididymis of 268
glands of 269
interstitial cells of 266
penis of 269-270
prostate gland of 269
pubic hair of 54, 270
seminal vesicles of 269
seminiferous tubules of 266
sperm production in 266, 268
structure of **267**
testes of 266
urethra of 268
vessels of 266
Male sex hormones 55
Male sexual activity 271-272
age changes in 273-275
excitement phase of 272-274
orgasmic phase of 272-274
plateau phase of 272-274
refractory period of 272-275
resolution phase of 272-274
Malignant melanoma 66-67
Malignant neoplasm 66
of skin 66-67
Maillard reaction 32
Malnutrition:
consequences of 235-236
dementia and 140
among elderly 239-240
problems from 235
reduction and prevention of 239-240
use of supplements and 239-240
Mammary gland **266**
(See also Breast)
Mammogram 289
Manipulating maximum longevity 18
Manual dexterity reduction in 60
Marriage life expectancy and 20
Mass peristalsis 220
Mast cell **52**, 56, 58
Maturity-onset diabetes mellitus 308
(See also Diabetes mellitus)
Maximum heart rate 75
Maximum longevity (XL) 18
effects from changes of 327, 328
manipulating 18
nutrition and 251-252
Mean longevity (ML) 18-20
effects from changes of 327, 328
in United States 2-4, 16-17
reasons for changes in 2, 327
Medication:
dementia and 141
nutrition and 251
tinnitus and 170
Meissner's corpuscle 125-126
Melanin 51-53, 66-67
Melanocyte(s) 51-53, 64
aging and 53
clumps of 53
in hair follicle 55
Melanoma malignant 67
Melatonin 297
age changes in 297
as antioxidant 297
control of 297
effects of 297
free radicals and 297
supplementation with 297
Memory 116-117, 125, 132-134
age changes in 132-133
eposodic 133
explicit 133
of immune system 294, 324
implicit 133

incidental 133
long-term 132
loss of 9
procedural 133
short-term 132, 184
Memory B cell (mB cell) 320
Memory T cell (mT cell) 321
Men:
aging in 55
coronary artery disease and 79
hearing loss in 169
sex hormones in 301-303
Menopause 55, 196, 275, 280-281, 304-305
age changes in sex hormones after 304
age changes in sex hormones before 303-304
coronary artery disease and 79
effects of on bone 196
osteoporosis after 198
Menstrual (uterine) cycle 278
Methionine 78
Metastatic cancer 109, 230
Microbe(s) 53
integumentary system as barrier against 49
stratum corneum and 51
Micturition 262
Middle ear 165-166, **168**
age changes in 166
infection of 170
Mineral(s) 25, 40
of bone 193-194
characteristics of 250
deficiencies of 250
excesses of 250
maintaining concentrations
of urinary system and 254-255
sources of 250
storage of skeletal system and 192
toxic 192
Mineralocorticoid(s) 300-301
Minute volume 96-98, 107-108
Misuse:
age changes from 11
of parts of body 10
Mitochondrial theory 45
Mitochondrial DNA theory 45
Mitochondrion(a) 33-34, 117
DNA in 35
in muscle cell **176**, 177
Mitosis 37-39
Mole(s) 66
dark decrease in number of 53
Molecule(s) 22-32
of cytoplasm 33
effect of excess light on 52
obtaining energy from 240-241
physical properties of 33
repairing of 18
Monocyte 92, 313
production of **314**
Monoglyceride 28, 246
Monosaccharide 25, 244
Monounsaturated fatty acid 28, 246
Mons pubis **276**, 279
Motor end plate **118**, 175, **176**
Motor neuron(s) 121
axon of **176**, 179
cell body of **179**
of dermis 60
of skin **51**
speed of impulse in physical activity and 183
upper 124
Motor unit 178-179
changes in 179
Mouth 207
(See also Oral mucosa; Oral region)
Movement(s):
of cell 24
in large intestine 220-221

muscle system and 173
skeletal system and 191
in small intestine 217
speed of muscle system performance and 181
physical activity and 183
in stomach 215
Mucociliary escalator 99
Mucopolysaccharide 28, 40, 58
Mucus:
of nasal cavities 98
in respiratory system 103
Multi-infarct dementia 141
Multiple sclerosis 325
Muscle cell 175
age changes in 178
free radicals from 177
Type I 176-177
Type IIA 177
Type IIB 177
types 176-177
Muscle system **8**, 175, 179, 183, 210
blood flow in 179-180
cardiac 173
cells of **8**, 24
age changes in 177-178
contraction of 176-177
energy and 175-176
number of 177-178
changes in **179**
structure of **174**, 175
thickness of 177
physical activity and 183
components of 175
coordination of 140
cramping of 59
of digestive system 212
diseases of 162-163
effects of starting or increasing exercise on 184-186
external of eye 158
age changes in 158-159
functioning of 175
heat production and 174
main functions for homeostasis in 173-174
mass of 84, 183
changes in 180-181
motor units of 178-179
movement and 173
myoglobin in 175-176
of oral region abnormal changes in 214
oxygen and 175-176
performance of 181-183
reductions in power of 10
repair of 178
respiratory 104
skeletal 173, **174**, 191
reflex pathways and 121-122
smooth (visceral) 173, 215
soreness of 183
strength of 183
changes in 180-181
structure of 175
support and 173-174
thickness of 183
three types of 173
weakness of 140
Muscle-nerve interaction 178-179
Myasthenia gravis 326
Myelin **118**, 120, 129
Myocardial infarction (MI) 77, 137
Myocardium 73
Myoglobin **176**
Myometrium **276**, 278

Nails 54-56
age changes in 55-56
function of 55
fungal infections of 56
Nasal cavity **94**, 98, **159**

Nasopharynx 98, **99**
National Institute on Aging 13
Natural killer cell (NK cell) 312, 317
Nearsightedness 149
Negative energy imbalance 241
Negative feedback 54
　integumentary system and 50
　in nervous system 115
　reduction in functioning of 7
　steps in 5-7
　for thermoregulation **8**
Neoplasm:
　of integumentary system 66-67
　(See also Cancer)
Nephron **255**, 257-258
　filtration reabsorption and secretion by **256**, 257
Nerve(s) 54, 115, **116**, 179
　of dermis 60
　of integumentary system **51**
Nerve cell (See Neuron)
Nerve impulse 82, 115
Nerve-muscle interaction 178-179
Nervous system 115, **116**
　age changes in 7
　communicating and 115
　comparison of with endocrine system 293
　conscious sensation and 123
　coordinating and 116
　decline in sensory function of 10
　diseases of 139-147
　effects of starting or increasing exercise on 184-185
　high level of physical activity and 183
　higher-level functions in 125
　main functions for homeostasis in 115-117
　monitoring and 115
　monitoring of body temperature by 5, **8**
　organization of 120-121
　pathways of 121-125
　reflex pathways of 121-122
　reflexes and 121-122
　remembering and 116-117
　respiratory system and 101-103
　stimulating and 116
　thinking and 117
　voluntary movements and 123-125
Network theory 41, 47
Neurofibril **118**, 132
Neurofibrillar tangle 132
Neuroglandular junction 119
Neuroglia 120
Neuromuscular junction 119
Neuromuscular transmission age changes and 127
Neuron(s) 115
　age changes in 127-129
　autonomic motor 128-129
　components of 117
　conduction in 117
　first-order **123**
　motor **51**, 60, 121, 124-125, 178
　olfactory 126
　operations of 117-119
　processing of retina **156**, 157
　reception in 117
　second-order **123**
　sensory **51**, **52**, 54, 60, 120, 125-127
　structure of **118**
　transmission in 119
Neurotransmitter 119-121
　Alzheimer's disease and 141-145
　sympathetic 129
Neutrophil 91
Nitric oxide (*NO) 29
Noise 167
*NO 81
　in arteries 81, 83
*O2 82
　in arteries 82
Nonenzymatic glycosylation 309

Non-insulin-dependent diabetes mellitus
　(NIDDM) 186, 308
　cause of 309
　(See also Diabetes mellitus)
Non-ketosis-prone diabetes mellitus 308
　(See also Diabetes mellitus)
Norepinephrine 15, 75, 82, 83, 103, 105, 129, 136
　production of 131
Normal people average values of biological age for 17
Nose gradual increase in size of 63
Nucleic acid 26-28, 209
Nucleotide 26, 33
　sequence of 26
Nucleus of Meynert 131
Nucleus pulposus 202
Nurturing life expectancy and 20
Nutrient(s) 61
　excess storage and conversion of 209
　manufacturing of 208
　supplying of digestive system and 207
　transportation of 68
Nutrient density 242
Nutrition 20
　alcohol and 251
　disease and 251
　good 48
　maximum longevity and 251-252
　medications and 251
　need for homeostasis of 234-236
　poor 11
　cosmetic aging and 9
　relationships between diet and 234-235
Nutritional programs developing of for elderly 16
Nutritional status evaluation of 239

Obesity 80, 243
　consequences of 243
　coronary artery disease and 79
　correction 244
　definition of 243
　prevention of 223
Occupation aging and and 16
Older people (See Elderly people)
Olfactory neuron 126
Opacity 152
Open-angle glaucoma 164
Optic disk 150, 155, **156**
Optic nerve **149**, **150**, 155, **156**, **158**
Oral cavity **94**, 98, 207, **208**, 210
Oral glucose tolerance test (OGTT) 307, 308
Oral mucosa 210
　abnormal changes in 212
　age changes in 210
Oral region 210-214
　bones of 212
　　abnormal changes in 212-214
　age changes in 212
　joints of 212
　　abnormal changes in 213-214
Organ(s) **24**, 25
　levels of functioning of under resting conditions 17
Organ of Corti 166-168
Orgasmic phase:
　of female sexual activity 284
　　age changes in 284
　of male sexual activity 272
　　age changes in 274
Orthostatic hypotension 128
Osmotic pressure:
　regulation of urinary system and 253-254
Osteoarthritis 10
　effects of 205
　intervertebral joints with **199**
　treatment for 205
Osteon **193**, 194, 195-196
Osteoporosis 48, 183, 174-202
　causes of 200
　diagnosis of 200

effects of 198-200
hip fractures and 200
incidence of 198
intrinsic risk factors of 200-201
modifiable risk factors of 200
postmenopausal 198
recommendations for reducing risk of 197
senile 198
treatments for 201-202
type I or II 198
vertebral fractures and 198-199
Otolith 170
Ovarian cancer 289
Ovarian cycle 275, **277**
Ovarian follicle 276
development of 277
Ovary 265, 266, 275, 276, 292
age changes in 281
structure of 277
surgical removal of 79
Overflow incontinence 263
Overweight person 243
consequences of 243
correction of 243-244
definition of 243
prevention of 243-244
Oviduct 275, 276, 277
age changes in 281
Ovulation 275
Oxidative phosphorylation 34
Oxygen (O₂) 22, 29-31, 72, 74, 76, 91, 96, 101-103, 107
in blood 106
in gas exchange 93-95
mitochondria 33-34
monitoring of 104-105
in muscle cell 175-176
red blood cell production and 192
regulating levels of urinary system and 257
sleep apnea and 113
snoring and 113
transportation of 68
Oxygen debt 176
Oxyhemoglobin 107

Pacemaker 73
Pacinian corpuscle 125
Pain sensation of 60
Pancreas 207, **208, 227, 228,** 231-232, **292**
abnormal changes in 232-233
age changes in 232
cancer of 233
diarrhea and 222
pancreatitis and 232-233
Pancreatitis 231-233
Panlobar emphysema (PLE) 111
Pannus 206
calcified **206**
Pap smear 290
Paralysis 239
spinal cord and 189-190
Parasympathetic motor neuron 121
Parathormone 298-299
Parathyroid gland **292**
Parkinson's disease 145-147
causes of 145
dementia and 141
diagnosis of 141
effects of 145-146
nervous system changes in 146
primary 145
secondary 145
treatments for 146-147
Partially hydrogenated fat 246
Penis 128, 266, 267, 269
age changes in 271
erectile tissue of **261**
erection of 270
structure of 270

Pentosidine 32
Peptic ulcer 216, 218
in stomach and duodenum **216**
Percent body fat 186, 243
Perfusion 68, 94, 106-107
age changes in in respiratory system 107
Period effect 12
of longitudinal study 13
Periodontal disease 79-80, 210
Periodontal membrane **211**
Peripheral nervous system (PNS) 115, 120-121
Peristalsis 212
Peroxynitrite (ONOO⁻) 29
Pernicious anemia 215
Personality 124, 1335
age changes in 135
coronary artery disease and 79
life expectancy and 20
Perspiration 59, 192
pH 70, 71
homeostasis of urinary system and 256
Phagocytosis 93, 98, 99, 207, **208,** 210
Pharynx 74, 79, 81, 184, 185, 187
Phospholipid 28, 33
Phosphorus 26-28, 192, 194
Photoaging of skin 64
beta carotene and 64
glycolic acid 64
selenium and 64
treatment 64
tretinoin 65
vitamin C and 64
vitamin E and 64
Physical activity 183-184, 197
high level of throughout life 183
muscle mass and 183
overview of 183-184
specific effects of 183-184
.Vo2max and 182-183
Physical fitness program planning
of for elderly 16, 187-188
Physical inactivity coronary artery disease and 79
Physiological theory 41
Pigmented epithelium 155, 157, 163
Pineal gland 136, 292, 297
Pituitary gland 46, 291, **292**
Plasma 90-92
buffers in 71
Plasma cell 320
Platelet 91-92, 192
Plaque (circulatory system) 76, 85
formation of 76
Plaque (nervous system) 132
Plateau phase:
of female sexual activity 283
age changes in 284
of male sexual activity 272
age changes in 274
Pneumonia 109, 111-112, 197
dust and vapors and 112
microbial 111-112
Pneumothorax 111
Polydipsia 309
Polymorphonuclear leukocyte (PMN) 91-92
Polyphagia 309
Polysaccharide 26, 244
Polyunsaturated fatty acid 28, 246
Polyuria 309
Population:
life expectancy of 19
trends of 1-4, 327-328
Positive energy imbalance 241
Postmenopausal osteoporosis 198
Postmenopausal woman 281, 280-281
Potassium (ion) 17
ketoacids and 310
Prednisone 300
Premenopausal woman 280-281

Presbycusis 169
  compensatory techniques for 169-170
  effects of 169
  prevention of 169
Presbyopia 152
Presenilin (PS) 144
Prevention:
  importance of 10-11
  (See also specific abnormalities and diseases)
Primary lung cancer 109
Progeroid syndrome 37, 40
Progesterone 28, 275, 303
  decline in 79
  longevity and 19
Program theory 41
Prolapse uterine 281
Proprioceptor 121, 172
Prostate cancer 287-288
Prostate gland **261**, 262, 266-270
  age changes in 271
  cofactor and 29
  surgery of 264
Protein(s) 25-33, 209, 240, 247-248
  body temperature and 50
  of bone 195
  damage of by free radicals 44-45
  deficiencies of 248
  in error catastrophe theory 44
  excess of 197, 248
  in intercellular materials 40-41
  light damage of 50, 52
  manufacturing of 33
  packaging of 33
  structure 29
  recommended dietary intakes of 248
  uses of 28, 248
Protein-carbohydrate malnutrition (PCM) 245
Psychological aging 9
Psychological aspects of person's life
  integumentary system and 49
Pubic hair 54-55, 266
  of male 270
Pulmonary abscess 89
Pulmonary artery 72, **74**, 106
Pulmonary capillary 106
Pulmonary circulation 49, 87
Pulmonary congestion 55
Pulmonary edema 55, 60
Pulmonary embolism 68, 90, 93
Pulmonary fibrosis 93
Pulmonary vein 49, 51, 52, 82, 86, 87
Pulmonary vessel 73
Pupil 109, 126, 127
  age changes in 127
Pyloric sphincter 92, 194

Quality of life 18

Race mean longevity and 19
Radiation:
  harmful 10
  life expectancy and 19
  somatic mutation theory and 44
  in wear and tear theory 47
Rapid eye movement (REM) 136
Rate of living theory 44
Reaction time 183
  muscle system performance and 181
Reactive oxygen species (ROS) 29, 34
  in arteries 83
Receptor (molecule) 28, 33, 119
  antigen-specific 314-316, 318-320
  human leukocyte-associated (HLA) 314, 315, 318-320
Receptor for advanced glycation end-product (RAGE) .. 32
Recommended daily allowances 238
Recommended dietary allowances (RDAs) 236-237
  of carbohydrates 245
  of lipids 246-247

  of proteins 248
  of water 249
Reactive oxygen species (ROS) 29, 34
Recovery time 183
Rectum 90, 208, 218, **219**, **267**, **276**
Red blood cells (RBCs) 91, 92, 106, 192
  buffers in 71
Red bone marrow 199, 312
Reflex(es) 10, 121-123
  acquired 121
  age changes in 129-130
  deep tendon 129
  of eyes 159
  pathways of 121-123
  stretch 129-130
  withdrawal 123
Refraction 150, 152
Refractory period of male sexual activity 272, 273
  age changes in 274
Regional enteritis 326
Remembering nervous system and 116, 117
Renal blood flow (RBF) 258-259
Renin 259, 300
Replicative senescence 37
Reproductive hormone theory 46
Reproductive system 265
  homeostasis and 265-266
  main functions of 128, 265-266
  (See also Female reproductive system; Female sexual
   activity; Male reproductive system; Male sexual
   activity)
Reserve capacity 15
  of body 10
  aging and 14
Residual volume (RV) 97
Resolution phase:
  of female sexual activity 284
    age changes in 284
  of male sexual activity 272
    age changes in 274
Respiratory passages in head and neck **99**
Respiratory rate 96, 98, 101-102
Respiratory system 93, **94**
  acid/base balance and 71
  age changes of 14, 103-105
  control systems in 98, 101-105
  errors in 113
  diffusion in 107-108
  diseases of 109-112
  effects from altered gas exchange in 108-109
  effects of starting or increasing exercise on 184-186
  failure of 111
  gas exchange and 93-95
  main functions for homeostasis in 93-95
  perfusion in 106-107
  requirements for ventilation in 98-103
  sound production and 95, 113-114
  structures associated with **94**
  ventilation in 95-98
Respiratory volume changes in 105
Resting conditions:
  age changes in heart and 75
  heart rate and 75
Retina 149, 150, 155, 159
  age changes in 157
  blood vessels of 150
  cones in 155
  damage of 187
  layers and regions of 155
  rods in 156
  sensory layer of 156
  structure of 156
Reversible ischemic neurological deficits (RINDs) 140
Rheumatic fever 80
Rheumatic heart disease 326
Rheumatoid arthritis (RA) 10, 205-206
  effects of 206
  on joint structure **206**

treatments for 206
Ribonucleic acid (RNA) 26, 35
in error catastrophe theory 44
light damage of 50
ultraviolet light and 52
Right ventricle of heart 69, **71**, 72, **74**, **106**
Risk factor(s) 11
for coronary artery disease 77-80
osteoporosis and 200-201
for strokes 138-139
Rod 156-157
age changes in 157
Rotation:
detection of semicircular canal and 171
of head 165, 171
age changes in 171
Running 88, 186

Saliva 210, 211-212
secretion of 121
Salivary duct 210
Salivary glands 115, 207, 208, 210, 211-212
abnormal changes in 213
age changes in 212
autonomic control of 121
Salt 126
Sarah Knauss 18
Sarcopenia 180
Saturated fat 28, 246
Saturated fatty acid 30
Schwann cell 118, 120
Sclera 150, 156, 157-158
age changes in 60
Scrotum 266, 267
Sebaceous (oil) gland 51, 54, 59-60
age changes in 60
sunlight and 64
Sebum 59-60, 63
Secondary diabetes mellitus 308
Self-antigen 315
Self-esteem lowered 9, 200
Self-recognition 312
Semen 269
Semicircular canal **166**, **170**, 171
detection of rotation and **171**
Seminal vesicle 266, **267**, 269
age changes in 271
Seminiferous tubule 266-268
structure of 268
Senescence 4
Senile dementia of Alzheimer's type (SDAT) 141
causes of 142-144
diagnosis of 143
effects of 142-143
treatments for 145
Senile macular degeneration 163
Senile osteoporosis 198
Senile plaque (SP) 143
Senses 125-127
(See also Sensory function; Smell sense of; Taste sense of)
Sensory function:
age changes in 125-127, 130
conscious 123
decline in 10, 140
Sensory neuron 51, 52, 54, 84, 120
of dermis 60
age changes in 60
Sensory portion of peripheral nervous system 121
Sensory retina 155
Sex chromosome longevity and 19
Sex hormone(s) 28
age-related decreases in levels of 55
apocrine gland activity and 59
in men 55, 301-303
in women 303-306
estrogen replacement therapy and 305-306
Sex hormone-binding globulin (SHBG) 302
Sexual activity 266

aging and 16
enjoyment of 284-285
frequency of 284-285
(See also Female sexual activity; Male sexual activity)
Sexually transmitted diseases (STDs) 290
Shingles 325
Short-term memory 132, 184
greatest decline in 133-134
Sigmoid colon **208**, 218, **219**
Sildenafil 287
effects of 287
male impotence and 287
Skeletal muscle (fiber) 118, 173, 174
cross section of **6**
(See also Muscle system)
Skeletal system **190**
aging respiratory system and 104
blood cell production and 192
components of 192
effects of starting or increasing exercise on 184-186
high level of physical activity and 183
main functions for homeostasis in 189-192
mineral storage and 192
movement and 191
protection from trauma and 190-191
support and 189-190
Skeleton aging 197
Skill muscle system performance and 181
Skin 312
bedsores and 65
coloration of in elderly 53
drying of 63
effects from heat 65
effects of sunlight on 63-65
epidermis of 50-53
age changes in 53-54
excessively dry 63
healing of 57
integumentary system and 49
neoplasms and 66-67
physical wear and 50
receptors in 125-126
scaliness of 53
sunlight and 50, 63-65
wrinkling of 9, 63
excess light and 51-52
Skin cancer 52-53, 64-67
in elderly 53-54
infections in 54
precursors of 53
Skin neoplasm 66-67
Sleep 135-136, 184
age changes in 136
regulation of 131
tinnitus and 170
Sleep apnea 109, 113, 136
Sliding hiatal hernia 214
Slightly movable joint 199, 202-203
age changes in 202-203
Small intestine 207, 208, 219, 227
abnormal changes in 218
absorption in 216-218
age changes in 217-218
lactase secretion and 217
movements in 217
peptic ulcer and 218
secretion in 217
Smell sense of 123, 126
Smoking 11, 109, 112-113
avoidance of 197
chronic cough and 90
coronary artery disease and 78
development of coronary atherosclerosis and 77
effects from 112
free radicals and 112
life expectancy and 20
taste sensations and 126
Smooth muscle 84, 173, 215

cell of 85
Snellen eye chart 161
Social relationships:
    integumentary system and 49
    positive 20
Society influence of elderly on 2-4, 326-328
Sodium (ion) 25, 117
    ketoacids and 310
Soft palate 98, **100**
Somatic motor neuron 121, **124**, 178
Somatic mutation theory 44
Somatic (voluntary) motor pathway 124
Somatic mutation theory 44
Somatomedin C 295
Somatopause 295
Sorbitol 164, 309
    cataracts and 163
Sound localization of 167
Sound production 93, 95, 113-114
    age changes in 114
Sound vibrations pathway of **168**
Squamous papilloma 66
Squamous cell carcinoma 66
Specificity of immune response 313
Speech 113
Speed:
    age changes and 171
    detection of changes in 170-171
    of movement 183
    muscle system performance and 181
Sperm 265
    mature **268**
    production of 266, 268
Spermatogenesis 266
Spermiogenesis 268
Sphincter internal and external 90, 220, 262
Spinal cord 60, 115, **116**, **124**, **131**, **179**, 189-191, 198, **199**
    age changes in 131
    skeletal support for 191
Spinal nerve 199
Spleen **227**, 312, 314
    lymphatics and 90
Spongy bone 193-195
Spouse death of 10
Squamous papilloma 66
Squamous cell carcinoma 66
Stamina 183
    muscle system performance and 181-182
Starch 24
Steroid(s) 28, **30**, 33
Stimulation:
    active response to of cell 24
    nervous system and 116
Stimulus(i) 115
    detection of 170-171
Stochastic theory 41
Stomach **100**, 207, **208**, 215, **227**
    absorption and 215
    acute gastritis and 215
    age changes in 215
    atrophic gastritis and 216
    movements in 215
    peptic ulcer and 216
    secretion in 215
Stratum corneum 50-53
Strength 183
    effects of muscle mass on 180
    of nails 55
Strenuous activity 7, 9
Stress 11, 19, 186
    coronary artery disease and 79
    emotional 79
Stress incontinence 263
Stricture:
    of esophagus 214
Stroke(s) 11, 48, 84, 111, 113, 137-138, 183, 187
    causes of 138-139
    completed 140

dementia and 141
    disabilities from 11
    hemorrhagic 138-139
    hypertensive hemorrhagic 138-139
    ischemic 138
    signs of 139-140
    symptoms of 139-140
    treatments for 140
    types of 138-14
Stroke volume (SV) 72, 76
Structural protein 36
Subcutaneous layer 62
    age changes in 62-63
    fat tissue of 62
    of integumentary system 49, **51**
    loose connective tissue of 62
Substrate control 294
"Successful" aging 10
Sugar(s) 24, 25, 244
    in nucleic acids 26
Sunburn 52, 54, 63-65
Sunlight 11
    effects of on skin 63-65
    exposure to cosmetic aging and 9
    hair and 54
    Langerhans cells and 54
    UV light in 52
    vitamin D production and 61
Superoxide radical ($^*O_2^-$) 29
Superior vena cava **69**, **71**, **74**
Supplements ....
    DHEA 301
    growth hormone 296
    melatonin 297
    nitric oxide and 287
    side effects from 287
    use of malnutrition and 239-240
Support:
    muscle system and 173-174
    skeletal system and 189-190
Suppressor T cell (sT cell) 315-317
Surfactant 101
    layer of in lungs 106
Survival biological aging and 4, 7
Survival curves 326, 328
Suspensory ligaments 150-152
    during accommodation 154
    age changes in 152-153
Suture joint **190**
Swallowing (reflex) 99, 100, 103, 128
Sweat 59, 63
Sweat gland(s) 54, 59, 84, 115
    age changes in 59
    apocrine 59, 165
    autonomic control of 121
    beneficial age changes of 4
    duct of **51**
    eccrine **51**, 59
Sympathetic motor neuron 121
Sympathetic nervous system 103
Sympathetic neurotransmitter 129
Symphysis joint 202
Synapse 119
    aging and 132
Synovial cavity 203-204, **206**
Syphilis 290
System in hierarchy of body structure 24-25
Systemic circulation **69**
Systemic vessel 106-107
Systolic pressure 72, 81, 82, 85

T lymphocyte (T cell) 313, 315, **316**
    age changes in 322-324
    development of **314**
    specialized formation of **316**
Tan development of 53
Target of hormone 294
Target sensitivity insulin and 307

Taste sense of 126
Tau-protein 143
Tear(s) 162
Telomerase 36
Telomere 35, 38
Telomere theory 43
Temperature 82
  proteins and 28
  regulation of: circulatory system and 70
  integumentary system and 50
  negative feedback for **8**
  by water molecules 25
  (See also Body temperature)
Temporomandibular joint (TMJ) **190**, 212
Testes 265-267, **292**
  age changes in 270-271
  structure of 268
Testosterone 28, 55, 200, 266
Theories of aging 41
  accumulation of late-acting error 43
  antagonistic pleiotropy 43
  autoimmune 47
  biological aging of 41-48
  calcium 46
  characteristics 41
  clinker 45
  error catastrophe 44
  evolutionary 41
  faulty DNA repair 44
  free radical 44
  glucocorticoid 46
  glycation 46
  heterochromatin loss 43
  immune 47
  immune deficiency 47
  immune dysregulation 47
  importance of 48
  insulin 46
  insulin/growth factor imbalance 46
  mitochondrial 45
  mitochondrial DNA 45
  network 41, 47
  physiological 41
  programmed 41
  rate of living 44
  reproductive hormone 46
  somatic mutation 44
  stochastic 41
  telomere 43
  wear and tear 47
Thinking 105, 134
  age changes in 134-135
  crystallized intelligence 134
  fluid intelligence 134
  nervous system and 117
  slowed 9
Thirst 78
Thoracic cavity 93, **94**
Thromboembolus 89
Thrombus 89
  formation of 138
Thymus 292, 299, 312-315
  age changes in 322-323
Thyrocalcitonin 298
Thyroid gland 291, **292**
  malfunction of 141
Thyroid hormone(s) 297-298
  excess of 298
  Grave's disease and 298
  inadequate production of 298
Thyroxine (T$_4$) 297-298
Tidal volume (TV) 96
  increase in 103
Time-lag study 12
Tine test 318
Tinnitus 170
Toenails:
  aging and disease changes in 55-56

keratin in 55
Tolerable Upper Intake Levels (ULs) 238
Tongue **99, 100, 208,** 212
Tooth (teeth) **208,** 210-211
  abnormal changes in 212-213
  age changes in 212, 213
  structure of 211
Topically applied medication 58
Total lung capacity (TLC) 98
Touch perception hair and 54-55
Toxin(s) 33
  elimination of digestive system and 209
  life expectancy and 20
  lymph node inactivation of 70
  removal of urinary system and 253
  somatic mutation theory and 44
  from spoiled food 222
  tinnitus and 170
  in wear and tear theory 47
Trabecula(ae) 195
Trabecular bone 194-195
  postmenopausal osteoporosis and 198
Trachea **94,** 99-100
Transient ischemic attacks (TIAs) 140
Transurethral resection of prostate (TURP) 286
Trauma 130
  integumentary system as barrier against 50
  protection from skeletal system and 190-191
Treatment:
  for benign prostatic hypertrophy 286
  for breast cancer 289
  for cervical cancer 290
  for diabetes mellitus 310-311
  of esophageal malfunctioning 214-215
  after fracture 202
  of fractures 197
  for impotence 287
  individualized for elderly 16
  for osteoarthritis 205
  for osteoporosis 201-202
  for prostate cancer 287-288
  for rheumatoid arthritis 206
  for urinary incontinence 263-264
  (See also other specific abnormalities and diseases)
Tremors of hands arms and legs 146
Tretinoin 65
Trifocal lenses 155
Triglyceride(s) 28, 30, 78, 240, 246
Triiodothyronine (T$_3$) 297-298
Tuberculosis 112, 325
Tubules of kidneys 257
  age changes in 259
Tumor 109
  secondary glaucoma and 164
Type I or II diabetes mellitus 308
Type I or II osteoporosis 198

Ulcer decubitus 65
  (See also specific types of ulcers)
Ulcerative colitis 222, 326
Ultraviolet (UV) light 61
  melanin and 52
Underweight 244
  consequences of 244
  correction of 244
  prevention of 244
Unsaturated fat 28, 246
Unsaturated fatty acid **30**
Upper airways 93
Urea 92, 228, 241
Ureter 253, **254, 255,** 260, **267, 276**
Urethra 253, **254,** 261-262, 266, **267, 270, 276**
  age changes in 262
  of male **261,** 268
  structure of **261**
Urge incontinence 263
Urinary bladder 253, **254, 255,** 260-261, **270**
  age changes in 261

in female **276**
in male **267**
structure of 260-261
Urinary incontinence 127, 142, 262, 263-264
complications of 263
contributing factors of 263
mixed 263
prevention of 263-264
treatments for 263-264
types of 263
Urinary system 253, **254**
acid/base balance and 71
activating vitamin D and 257
main functions for homeostasis in 253-257
maintaining acid/base balance and 256
maintaining mineral concentrations and 254-256
regulating blood pressure and 256-257
regulating osmotic pressure and 253-254
regulating oxygen levels and 235-257
removing wastes and toxins and 253
(See also Kidney)
Urination 262
age changes 263
Urine 192, **256**, 257
excessive formation and elimination of 78
production of 129
voiding of 136
Urogenital diaphragm **270**
U.S. Recommended Daily Allowance
(U.S. RDA) 238
Uterine cycle 278
Uterine fibroid 290
Uterine prolapse 281
Uterine tube **266**, 277
Uterus **266**, 275, **276**, 277-278
age changes in 281
UV light 64
eyes and 157

Vaccine 322
Vagina **266**, 275, **276**, 278-279
age changes in 282
Vaginal infection 288
Valve(s):
of heart 71-73
of veins **81**, 88
Valvular heart disease 80
Varicose vein 89-90
formation of 80
Vas deferens **266**, 268
Vein(s) 68-70, **87**, 88
age changes in 88
diseases of 88-89
hemorrhoids and 90
of integumentary system **51**
normal **90**
structure of **81**
Ventilation:
age changes and 103-105
control of 101-103
errors involving 113
expiration in 96-98
inspiration in 95-96
rate of 96-98
requirements for 98-101, 103
Ventricle 72
Ventricular contraction or relaxation **72**
Vertebra(ae) **94**, **190**, 198-199
body of **191**
fractures of 198
old 199
trabecular bone and 96
Vertebral column 190
Vertebral fractures osteoporosis and 198-200
Vertigo 172
Very low density lipoproteins (VLDLs) 247
Viagra 287

effects of 287
male impotence and 287
nitric oxide and 287
side effects from 287
Viral infection 222
Viral pneumonia 112
Virus(es) 53, 112
integumentary system as barrier against 49
Vision 123, 148-149
age changes in 160-162
altered 140
binocular 160
age changes in 162
clear deterrents of 149
optimizing 162
pathways used for **149**
Visual acuity 155, 161-162
age changes in 161-162
Vital capacity (VC) 96
Vitamin(s):
antioxidants 250
characteristics of 249
deficiencies of 250
excesses of 250
fat-soluble 249
homocysteine and 250
integumentary system and 49
sources of 249-250
Vitamin B6 79
Vitamin B12 .... 79, 215-217
Vitamin D 50, 57, 58, 197, 200
activation of 259
urinary system and 235
production of 49, 61
age changes in 61-62
integumentary system and 50
water-soluble 249
Vitamin E free radicals and 44
Vitreous humor 150, 153, 155, **156**, 159
age changes in 153
damage of 187
.Vo2max (See Aerobic capacity)
Vocal cords **99**, 113-114
Vocabulary 113, 135
age changes in 113, 135
Voiding 262
Voluntary motor pathway **124**
Voluntary movement:
age changes in 130
reduced control of 137

Walking 117, 130, 185
brisk 186
difficult or painful 56
Wart 67
Waste(s) 61
elimination of: digestive system and 209
urinary system and 253
Water 22-23, 73, 248-249
of body 25
in cytoplasm 33
deficiencies of 249
excesses of 249
fatty acids and 28
integumentary system as barrier against 49
keratin and 53
in mucopolysaccharides 58
in plasma 90
recommended dietary intakes of 249
uses of 248-249
in digestion 209
Water molecule 25
Water-soluble vitamin 249
Wear and tear theory 47
Web esophageal 214
Weight gain 78
Weight loss 78

Werner's syndrome 36, 38, 40
White blood cells (WBCs) 56, 58, 70, 91-92, 192, 313
White matter of brain 120
White population mean longevity of 19
Withdrawal reflex 123
Women:
  aging in.. 55
  hearing loss in 169
  loss of scalp hair in... 55
  sex hormones in 303-306

Work of breathing 96
World population.. 328, 329
Wound healing speed of 53, 78
Wrinkling of skin 63
  excess light and 52

Yellow bone marrow 192, **193**

Zinc ketoacids and.. 310